Frontiers in Probability and the Statistical Sciences

For further volumes:
http://www.springer.com/series/11957

Somnath Datta • Dan Nettleton
Editors

Statistical Analysis of Next Generation Sequencing Data

Springer

Editors
Somnath Datta
Department of Bioinformatics
and Biostatistics
University of Louisville
Louisville, KY, USA

Dan Nettleton
Department of Statistics
Iowa State University
Ames, IA, USA

ISBN 978-3-319-07211-1 ISBN 978-3-319-07212-8 (eBook)
DOI 10.1007/978-3-319-07212-8
Springer Cham Heidelberg New York Dordrecht London

Library of Congress Control Number: 2014942692

Printed on acid-free paper

Springer is part of Springer Science+Business Media (www.springer.com)

To our families:
Susmita and Anisha
Karen, Sam, Kate, and Owen

Preface

The last 10 to 15 years have seen a great surge of statistical research motivated by data from high-throughput genomic assays. Microarray technology, in particular, has driven many statistical advances in high-dimensional data analysis. Novel and older concepts of error rate control for testing multiple hypotheses, various adaptations of empirical Bayes methods, penalized and shrinkage regression techniques, etc., have all found new glory. It is not unjust to say that statisticians have done their fair share of work in developing the field of array-based bioinformatics and biomedical research.

In the last five or so years, next generation sequencing (NGS) technology has been changing the face of biomedical research, replacing the old (micro)-array technology in many ways. Overall, NGS offers a more accurate and cost effective means of studying a variety of genomic signals with a wide range of applications. With any new high-throughput technology come new data analytic challenges. Statisticians are in a unique position to make a difference in this exciting new interdisciplinary area by offering valid methods for studying signals in noisy data and the means to compare signals across multiple experimental conditions. Statistical methods for this relatively new data type have been sufficiently developed to warrant compilation of this book, which in turn may generate further interest in NGS technology amongst statisticians and lead to additional advances in the field.

The idea of editing a volume on statistical methods for analyzing NGS data first came to us about two years ago. We discussed the possibility during a conference on NGS data analysis that took place at Iowa State University. Subsequently, we approached several prominent researchers with extensive experience in the area. We were fortunate to have received overwhelming support and commitment from many individuals and their research teams. As a result, we now have this exciting volume consisting of twenty chapters written by statistical experts with first-hand knowledge in the field of NGS data analysis.

The first chapter of the book provides an introduction and an overview of NGS technologies, statistical challenges, and data analysis techniques. The next six chapters discuss design issues and inferential techniques for analyzing gene expression data as measured by next generation sequencing of RNA (RNA-seq).

Mapping of expression QTL is discussed next, followed by normalization of RNA-seq data. Statistical clustering and classification methods specially geared toward RNA-seq data are covered in Chaps. 10 and 11. Chapters 12 and 13 present different aspects of isoform detection using RNA-seq data. Another important NGS data type, CHIP-seq, is covered in the next two chapters. Other specialized applications of NGS technology—such as genotype calling, metagenomic analysis, detection of copy number changes and other structural variations, analysis for paired samples, and analysis of rare variants—are discussed in the last five chapters of this book.

This volume has been written primarily for statisticians who are interested in conducting methodological research in this area. No prior knowledge of NGS or genomics is assumed. Most of the required concepts from genomics and biochemistry have been explained, and references have been provided for a deeper understanding of such concepts. Scientists and practitioners dealing with NGS data will also find this book useful. Powerful software tools for NGS data analysis are illustrated in several chapters. Also, many chapters from this book could be used in a one to two semester graduate-level course in statistical bioinformatics.

We wish to thank the outstanding researchers who provided chapters for this book. We appreciate their hard work and their willingness to make the revisions we requested. Reading their work has enhanced our knowledge of the field, and we hope many other readers will benefit from the contributions of the authors.

Louisville, KY, USA Somnath Datta
Ames, IA, USA Dan Nettleton

Acknowledgements

We thank the series editors for accepting our book proposal and all the contributors for their hard work making this book a reality. We also thank the entire Springer team for their help and encouragement, especially, Marc, Jon, Hannah, and Susan.

About the Editors

Somnath Datta is professor and vice chair of Bioinformatics and Biostatistics at the University of Louisville. He is fellow of the American Statistical Association, fellow of the Institute of Mathematical Statistics, and elected member of the International Statistical Institute. He has contributed to numerous research areas in statistics, biostatistics, and bioinformatics.

Dan Nettleton is professor and Laurence H. Baker endowed chair of Biological Statistics in the Department of Statistics at Iowa State University. He is fellow of the American Statistical Association and has published research on a variety of topics in statistics, biology, and bioinformatics.

About the Editors

Contents

Chapter 1
Statistical Analyses of Next Generation Sequencing Data: An Overview

Riten Mitra, Ryan Gill, Susmita Datta, and Somnath Datta

Abstract Next generation sequencing (NGS) is a significant technological advance in biomedical sciences. The sequencing platforms have advanced rapidly to the point that several genomes can now be sequenced simultaneously in a single instrument run in under two weeks. Its applications range from detecting transcription factor binding sites and quantifying gene expression to discovering methylation patterns and comparing genomes. We discuss and review some of the major NGS platforms that are currently in use. Some of these platforms like Illumina represent the fastest evolving genomic technologies in terms of cost, throughput and speed. However, despite overcoming the limitations of first generation platforms and microarray based studies, the generated data is not free of noise. The sources of noise are diverse and complex depending on the generating platform and sequencing chemistry. For example, errors can creep in from any intermediate sequencing steps like ligand adaption, fragmentation, Polymerase Chain Reaction (PCR) amplification and nucleotide removal. In methods like Chromatin Immunoprecipitation Sequencing (ChIP-Seq), non-specific binding is a major source of noise. All of this raises novel statistical and computational challenges, e.g., in basecalling and differential profiling. In this chapter, we point out the critical challenges that arise in NGS data analysis and provide an objective overview of the existing literature. As we shall see, NGS is not only transforming genomics but driving new methodological development in several branches of quantitative science.

R. Mitra • S. Datta • S. Datta (✉)
Department of Bioinformatics and Biostatistics, School of Public Health and Information Science, University of Louisville, 485 E. Gray St., Louisville, KY 40205, USA
e-mail: r0mitr01@louisville.edu; susmita.datta@louisville.edu; somnath.datta@louisville.edu

R. Gill
Department of Mathematics, University of Louisville, Louisville, KY 40292, USA
e-mail: rsgill01@louisville.edu

S. Datta and D. Nettleton (eds.), *Statistical Analysis of Next Generation Sequencing Data*, Frontiers in Probability and the Statistical Sciences, DOI 10.1007/978-3-319-07212-8_1,

1.1 Introduction

1.1.1 DNA: The Unit of Sequencing

Deoxyribonucleic acid, more commonly known as DNA, is a complex doubly stranded molecule present in the nucleus of all living cells. The two strands are held together by weak thermodynamic forces. They form the familiar helical structure, with the double strands winding around a helix like the railing of a spiral staircase. DNA is often referred to as the building block of life.

Structurally, each strand of the DNA is a polymer. The polymer is a chain of linked monomer units called nucleotides. A nucleotide consists of a five-carbon sugar (deoxyribose), a nitrogen containing base attached to the sugar, and a phosphate group. There are four different types of nucleotides found in DNA, differing only in the nitrogenous base. These four nucleotides are assigned single letter abbreviations to represent the four bases. A stands for adenine, G is for guanine C, refers to cytosine and T denotes thymine. The terms nucleotides and bases are interchangeably used in the sequencing literature.

The deoxyribose sugar of the DNA backbone has five carbons and three oxygens. The carbons are numbered $1'$, $2'$ and so on. The hydroxyl groups of the sugars are joined at the $5'$- and $3'$-carbon ends to the phosphate groups by ester links, also known as phosphodiester bonds. Thus, the DNA backbone can be thought of as an alternating sugar-phosphate sequence. Each strand in the backbone is associated with a direction (known as polarity) from top to bottom, determined by the ending and starting carbons. We commonly refer to them as $5'$ or $3'$ ends. The two polynucleotide chains or strands run in opposite directions within the double helix.

The bases of the individual nucleotides are on the inside of the helix, stacked on top of each other, resembling the steps of the spiral staircase. The bases on both strands are paired by hydrogen bonds. A forms two hydrogen bonds with T on the opposite strand, and G forms three hydrogen bonds with C on the opposite strand. Due to such complementarity, the sequence of bases on one strand uniquely determine the entire set of base pairs. The length of a DNA fragment is conventionally measured by the number of base pairs it has—in kBp or mBp (Kilo/Mega base pairs).

What we refer to as a DNA primary sequence is essentially a collection of these bases. They are the repository of all biological information. It is well known that the order in which the bases occur in a DNA determines the information to make proteins. The protein synthesis from DNA is in fact a two-step procedure. The first step is "transcription" by which information is read from sequence of bases to make amino acids and RNA. The next step is "translation" by which these RNA form proteins. These proteins, in turn, regulate all biological processes, ranging from survival to reproduction and regulation of other proteins. In summary, the bases code for proteins, and the proteins are the chief executors of all important cellular functions. However, not all bases in DNA code for proteins. The special coding subunits are called genes. This two step module of protein synthesis is often referred to as the "central dogma of molecular biology". The central dogma is the cornerstone of modern genetics.

Thus, in order to grasp the genetic code, one must know the order in which the bases occur. Sequencing is precisely the process of determining the bases of a given DNA sequence.

1.1.2 Sequencing Technologies: An Overview

The established role of DNA as a primary coding unit of life has made genome sequencing highly relevant for almost any kind of biological research. Sequence information can now be used to identify, diagnose and potentially develop treatments for several genetic diseases. With high rates of growth in pharmacology and medicine, fast and easy sequencing methods has become the need of the hour. Next generation sequencing [64] is currently meeting this demand at a tremendous pace. In fact, it has revolutionized the fields of computational biology, evolution and medicine. This has indirectly triggered a competition among several companies, each trying to come up with faster and cheaper sequencing platforms. In terms of usage, today's platforms can be classified into two broad categories: high-end and bench-top. High end platforms (e.g Illumina-HiSeq) typically have bulky instruments and offer high setup costs, high throughput, and longer reads. Hence they are more suitable for large sequencing centers or core facilities. Bench-top instruments (Ion PGM, MiSeq) are less costly and more appropriate for microbial applications. Irrespective of their sizes, most of these platforms share a common three-step module of library preparation, template amplification and sequencing chemistry. We outline these steps below and present Fig. 1.1 as a pictorial illustration of the entire process.

Library preparation begins with the extraction and purification of genomic DNA. The extracted DNA is then broken into several overlapping fragments. The size of the fragments are selected to provide comprehensive and uniform coverage of the target genome. While older fragmenting protocols used mechanistic methods like nebulization and ultrasonication, newly developed enzymatic methods cleave the DNA chemically using fragmentase enzymes. Mechanically generated fragments undergo an extra step of repair and end-polishing. Enzymatic methods, on the other hand, typically require less input DNA and offer faster sample processing. The cleavage is done in a time-dependent manner, allowing the user to obtain fragments of the desired length.

The amplification step clones the DNA molecules in the library and prepares them for sequencing. This step is required since the size of sample DNA extracted from experiments is typically too small for raw detection. The polymerase chain reaction (PCR) is a biochemical technology designed for such amplification. It acts on a single or a few copies of a specific region of the target DNA and generates thousands to millions of copies of a particular DNA sequence. Most PCR methods

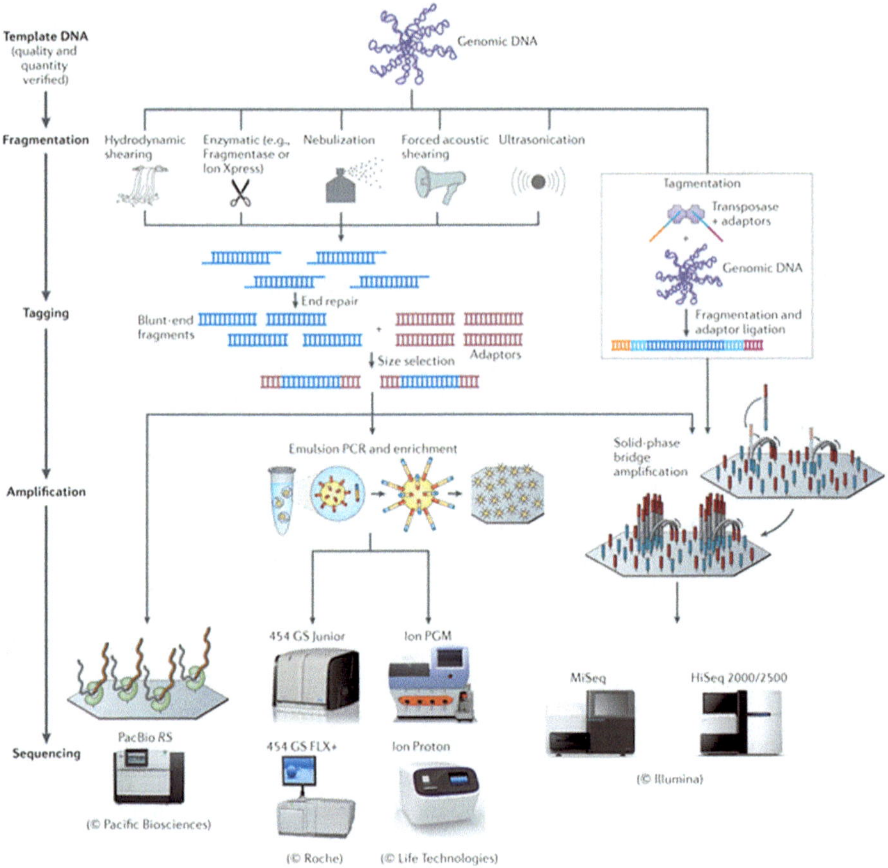

Fig. 1.1 A general workflow of the next generation sequencing technology. The last step is the sequencing process, which vary across different platforms. This picture is reproduced from Loman et al. [41]

typically amplify DNA fragments of between 0.1 and 10 kilo base pairs (kb), although some techniques allow for amplification of fragments up to 40 kb in size. A PCR requires the following key components. First, it must employ an available DNA template that contains the DNA region (target) to be amplified. A DNA polymerase enzyme is required to catalyze the reaction. The next step requires adding of primers that are complementary to the 3′ ends of the DNA targets. Primers are basically nucleic acid strands that serve as a starting point for DNA synthesis. They are required for DNA replication because the polymerase enzymes can add new nucleotides only to an existing strand of DNA. Finally, certain types of nucleotides containing triphosphates (Deoxynucleoside triphosphates or dNTPs) act as substrates for the reaction. A chain reaction ensues leading to the synthesis of replicate DNA strands.

Tagging of the amplified DNA fragments follows. In this step, platform-specific adaptors are ligated to the ends of fragments. These adaptors act as primer-binding sites for the subsequent template amplification and sequencing. *Tagmentation* is a new alternative technique that fragments DNA and incorporates sequence tags simultaneously with the help of transposase enzymes. Nextera (only available for the Illumina platform) is the only system currently providing tagmentation facilities. Paired-end tagging, which allows for tagging at both ends, has become a fairly common method. For details of these tagging technologies, we refer the readers to [23].

Sequencing chemistry is the final step in the next generation sequencing (NGS) protocol. The base pairs from ends of the fragments are read in this step. Sequencing both ends of DNA fragments has become common and is now a part of many standard platforms. This is referred to as "paired-end sequencing". In single-end sequencing protocols, however, only one end is sequenced. The end sequences would later be mapped to locations in the genome through mapping algorithms. However, in single-end sequencing, some reads may not be mapped uniquely. This leads to decreased efficiency and increased since these ambiguous sequences would normally be discarded. Paired end sequencing addresses this problem. By mapping one half of the pair uniquely to a single location in the genome, it determines the location of the other ambiguous half.

As stated earlier, the details of the sequencing chemistries vary across platforms. We shall discuss them in detail while describing the individual platforms.

1.1.3 Downstream Applications

"Reads" are the immediate products of sequencing and the final products of all next generation sequencing platforms. A read is a contiguous stretch of bases sequenced from one or both ends of the DNA fragment. Recording the base sequence of a read is known as "base-calling". A natural next step is to reconstruct the original sequence from these reads. For this, we have to rely on computational algorithms to merge the sequenced DNA fragments and recover the original genome. This process is called genome assembly. This is a necessary challenging; there is no alternative to processing reads since it is impossible to sequence the whole DNA in one go. Genome assembly can be done in two ways. The first way is by mapping or aligning to a "reference genome". The reference genome is an existing sequence against which reads are aligned. At the end of alignment, each read would correspond to a particular position in the reference genome. This leads to a sequence that is similar but not necessarily identical to the reference sequence. The second method is "de-novo" assembly. Under this method, novel sequences are constructed without the aid of a reference genome. They usually involve complex computational approaches like construction of de Bruijn graphs using k-mers (for short reads) or using an overlap-layout-consensus approach (for longer reads). De novo methods are free of the errors associated with alignment tools or with the reference genome itself. It can discover novel individual variations that could be suppressed while aligning to a reference. On the other hand, reference genome mapping is relatively faster and far less resource-intensive.

Apart from "discovering" the sequence per se, sequencing has several important and immediate applications. For example, one might be interested in knowing how certain regions of DNA are functionally different from the rest. The ability of sequencing to discover such special functional regions has profoundly impacted many scientific disciplines like medicine, pharmacology and evolution. For example, one would be interested in genomic regions that are more susceptible to binding by certain drugs or proteins; or detecting DNA segments unbounded by nucleosomes. For this, one needs to couple the NGS technology with state-of-the-art tools for extraction of the relevant genomic portions. This has led to the development of an array of technical platforms. The ChIP-Seq method is a good example of this kind. It is primarily used to analyze protein interactions with DNA. As the name suggests, it combines chromatin immunoprecipitation (ChIP) with massively parallel DNA sequencing to identify the binding sites of DNA-associated proteins. The protein of interest (POI) is cross-linked with the parts of DNA. The entire DNA is fragmented by sonication and the cross linked portions form a DNA protein complex. These fragments are then filtered and purified by removing the bound proteins (by heating). The fragments are then fed into the sequencing step. Again, the outputted reads form the basis of downstream analysis. Any such analysis would act upon a normalized version of the number of reads mapped to particular region. The normalized counts would represent the susceptibility of protein binding in that particular region.

Finally, the capabilities of NGS are not just restricted to genome sequencing. The same techniques can be applied to sequence the bases of any other nucleic acids like RNA. From the central dogma, we know that RNA synthesizes proteins after transcription. The expression of genes is quantified by the amount of mRNA transcripts produced by them. To measure this amount (also known as "transcript abundance"), one would require to sequence the RNA fragments and map them to the genome. Here the number of mapped reads would represent the intensity of expression. The application of NGS to sequence RNA transcripts is called RNA-seq. It has emerged as a powerful and appealing alternative to microarrays for measuring gene expression.

It is no surprise that the new technological developments discussed above is also fundamentally modifying older statistical methods for analyzing genomic data. One reason for this is the marked change in the nature and distribution of the observations For example, in gene expression analysis, statisticians would now have to deal with the read counts instead of log-hybridization ratios traditionally used for microarrays.

Thus depending on the requirement, a typical post-sequencing workflow would consist of steps like base-calling, alignment techniques, de novo genome assembly and estimating transcript abundance. Each of these steps require sophisticated mathematical and statistical techniques and are separate topics in themselves. We shall provide a short overview of these specific areas in Sects. 1.2–1.4. An earlier overview of the NGS and technology and data analysis was provided in [14]. Before describing these methods, we shall discuss some individual NGS platforms and their sequencing techniques in greater detail.

1.2 Examples of NGS Platforms

1.2.1 SOLiD by Applied Biosystem

This technology uses a special PCR called emulsion polymerase chain reaction (em-PCR) to amplify copies of templates of DNA molecules. Initially, the templates are arranged into microscopic magnetic chips called beads. The amplified templates are sequenced in parallel.

For sequencing, SOLiD uses a unique ligation based strategy. Under this method, each data point represents two adjacent bases, and each base is sequenced twice. A sequencing round begins with the addition of a universal primer. All fragments attached to the same magnetic bead will have this universal primer attached. Thus the starting sequence of every fragment is both known and identical. A cycle begins by addition of a mixture of fluorescent probes. The first two bases of some of these probes (starting at the $3'$ end) turn out to be complementary to the nucleotides about to be sequenced. These probes get ligated to the primer while the remaining unbound ones are washed out. The fluorescent signal from the bound probe is then measured. Finally the primer and probes are all reset for the next round. Now the $5'$-end of the new universal primer will match to the base just preceding the earlier base. The entire sequencing step consists of five rounds and each round spans about 5–7 cycles.

The most useful feature of this technique is the double interrogation strategy. This causes a single nucleotide polymorphism (SNP) to result in a two-color change. However a measurement error would result in a single color change. Thus only adjacent color differences can represent a true SNP. This leads to a better discrimination between true polymorphisms and erroneous sequences. Also, SOLiD beads are typically small and very densely packed (100 million beads per sequencing run). Consequently, this platform can produce approximately 20 Gb of short-read sequence data (25–50 bases) per run. Hence it is more preferable for re-sequencing than de novo assembly.

However, decoding the raw data from SOLiD can be problematic. Since SOLiD encodes by two bases, the concept of 'color space' becomes necessary. A color space is an alternative representation of the base sequence. The conventional nucleotide code using the letters A C G and T is replaced by colors. Previously, in Sanger sequencing, each color represented a single nucleotide and was automatically translated. In the SOLID system, each color represents four potential two base combinations. Direct translation of color reads into base reads is not recommended as sequencing errors might be carried forward resulting in a frameshift of the base calls. To reduce such errors, it is deemed best to convert the base reference sequence itself into color-space. Reconversion into nucleotide base space is done after the sequence is aligned to a reference genome encoded in color space. This conversion from the color space to sequence and to a mapped reference genome can be challenging. Overall, the error rate of this platform is significantly higher than for traditional Sanger sequencing.

1.2.2 Illumina/Solexa Genome Analyzer

The Illumina sequencers (including the Genome Analyzers I/II/IIe/IIx and the new Miseq and HiSeq) represent one of the most widely used platforms for NGS experiments. This platform was first introduced by Solexa in 2006 and later on re-branded as Illumina Genome Analyzer (GA).

This technology does not depend on em-PCR to amplify the template DNA strands like SOLiD. Instead, DNA molecules are first attached to primers on a hollow glass slide called the flow cell. Adaptor-ligated template molecules now flow into this cell. DNA polymerase enzyme starts replication of the template at the $3'$-end of the primer, and copies the opposite strand. This newly synthesized strand serves as templates for further isothermal amplification. Finally we get clusters of DNA molecules on the flow cell.

Illumina follows a sequencing strategy called "sequencing by synthesis". Briefly, a single-stranded DNA fragment in the cluster is copied with the use of enzymes making the fragment double stranded. Starting at one end of the DNA fragment, the enzyme sequentially adds a single nucleotide that is the match of the nucleotide on the single strand. The synthesis proceeds by adding a mixture of dNTPs attached with four terminator nucleotides (A, C, G, and T) to the amplified fragments. Each dNTP is fluorescently labeled with a different color corresponding to its constituent base. The four bases then compete for binding sites on the template DNA to be sequenced. Each DNA strand within a cluster incorporates one of the nucleotides. This nucleotide is the same for all strands within a single cycle. The non-incorporated molecules are washed away and a laser is applied to chemically remove the terminators and the fluorophores. A detecting device then records the fluorescent color corresponding to the sequenced base. The process is repeated for several cycles until the entire DNA molecule is sequenced. Since the technique relies on generating reverse complimentary copies of the template it is also referred to as "reversible dye-based termination".

Between 2008 and 2010 there were several technical updates to the Genome Analyzer (GA) platform in mechanics, chemistry and software. In 2009 GAIIx series replaced GA-I instruments offering outputs of 85 GB/run. In early 2010, Illumina launched HiSeq 2000—a high end instrument that handles 600 GB per run. HiSeq can currently sequence 1.6 billion 100-base paired-end reads in a 10.8 day run and handle 120 million clusters per lane. MiSeq, a bench top sequencer was launched in 2011 and shares most technologies with HiSeq. It uses a smaller flow cell and hence a reduced imaging time and dramatically reduced run times.

The template amplification is carried out directly on the instrument which makes it popular among bench top instruments. HiSeq 2000 is now equipped with HiSeq control software (HCS), a real-time analyzer software (RTA-to perform basecalling directly on instrument) and BaseSpace. The latter is Illumina's own genomics cloud computing environment for secure analysis, archiving and sharing of NGS data. An app-store was recently added to BaseSpace to minimize transfer time to the cloud through compression and serialization technologies. This feature is directly integrated with all MiSeq and HiSeq instruments, enabling automated software installation and upgrading. Among high-end platforms, HiSeq 2000 has currently the lowest sequencing cost at 0.02 dollars per million bases.

1.2.3 Ion Semiconductor Sequencing

This technology was released in early 2010. The sequencing is conducted on a set of ion semiconductor chips, each containing an array of microwells. These microwells are equipped with one single-stranded template DNA and one DNA polymerase molecule. An ion sensitive layer is placed beneath each well. The sequencing starts by flooding the micro wells with a single species of deoxyribonucleotide triphosphate (dNTPs). Complementary dNTPs get sequentially affixed into the strand complementary to the template. The resulting process of DNA polymerization releases a hydrogen ion which triggers a corresponding sensor. The electric pulses from the sensors are then transmitted to a computer and gets directly translated into a DNA sequence without any intermediate steps. The final steps of signal processing and DNA assembly are carried out in an embedded software. The ion based approached is innovative and has completely removed the use of modified nucleotides or optical instruments for this platform. This has led to a rapid sequencing speed and reduced the upfront and operating costs. However two major limitations exist. First, the technology cannot handle repeats of the same nucleotide (e.g. GGGGG) which are present on the template strand. In such cases, multiple nucleotides are included leading to release of more hydrogen ions per cycle. This causes a greater pH change and a proportionally greater electronic signal. Hence, high repeat numbers cannot be effectively distinguished from repeats of a similar but different number. Another major disadvantage is the short size of their reads (400 base pairs per run). Their throughput is also currently lower than that of other sequencing technologies, although the developers hope to change this by increasing the density of the chip. Ion semiconductor sequencing may also be referred to as ion torrent sequencing, pH-mediated sequencing, silicon sequencing, or semiconductor sequencing in NGS literature. Despite its shortcomings, this technology is well suited to whole genome sequencing in bacteria.

1.2.4 Single Molecule Real-Time Sequencing

Single-molecule real-time (SMRT) is a "third-generation" sequencing method developed by Pacific Biosciences [44]. A SMRT chip consists of millions of nano-photonic confinement structures called zero-mode waveguides (ZMWs). A ZMW is an extremely tiny hole fabricated in a 100 nm metal film deposited on a glass substrate. An active DNA polymerase and a stranded DNA template is fixed to its bottom. Light enters through the ZMW hole and creates a visualization chamber small enough to record a single nucleotide. The sequencing now proceeds by incorporating dyed bases to the template strand by the action of DNA polymerase. A fluorescence pulse is produced by the polymerase retaining the nucleotide with its dyed fluorophore. This reaction is observed in real time. SMRT provides information for both signal strengths and signal differences across time. This feature could be used to study structural properties of sequences and is highly relevant for epigenetic studies. The final sequencing results are generated through consensus analysis, i.e., by averaging the sequence information from multiple reads for each reference position. Murray et al. [50] used SMRT sequencing to generate six full bacteria methylomes.

SMRT is also being employed in several resequencing projects. Resequencing of candidate genes is essential to detect mutations linked with various congenital diseases. Smith et al. [65] used SMRT to assess the presence of activating internal tandem duplication mutations in FLT3—a therapeutic target in acute myeloid leukemia. In August 2012, scientists from the Broad Institute published an evaluation of SMRT sequencing in the context of SNP calling [1]. Compared to second generation platforms, SMRT has several advantages. It provides significantly longer reads (average read length is 1,300 bp) and has a low error profile. The absence of any PCR step reduces any bias or artifacts due to amplification. Although the throughput is lower than any second-generation sequencer, it is more appropriate for clinical laboratories, especially for microbiology research.

1.3 Statistical Tools for Using Sequence Reads

Downstream analysis of NGS data consists of a sequence of steps. Depending on the application, these usually include a combination of quality monitoring, base-calling, alignment to a reference genome, de novo genome assembly and estimating transcript abundance. Each of these steps require sophisticated mathematical and statistical techniques and are separate topics in themselves. In this chapter, we shall provide a short overview of the important developments in these specific areas.

1.3.1 Data Quality and Reproducibility

While many earlier papers examined and proved the reliability and reproducibility of the next generation sequencing platforms, some studies found systematic difficulties with the reads obtained from next generation sequencing platforms.

Marioni et al. [42] showed that Illumina sequencing can be a good alternative to microarray technology for studying mRNA expression levels. However, they were concerned that the technical variance associated with Illumina would affect the final inference of differentially expressed genes. To check this, they proposed a statistical testing procedure to determine the lane to lane variability in sequencing results. The authors concluded that next generation sequencing data from the Illumina platform are highly reproducible with very few systematic differences between technical replicates. They also found that the results are better than microarray technology, especially for genes with high expression. High correlation between gene counts provided strong evidence for the reliability of the replicates.

In order to test the "lane" effect, they modeled the number of reads mapped to gene j for lane k and sample i as a Poisson random variable with mean $\mu_{ik} = c_{ik}\lambda_{ijk}$, where c_{ik} is the total rate at which sample i produced reads at lane k and λ_{ijk} is the rate of mapping of gene j in the kth lane for sample i relative to other genes. The terms λ_{ijk} sum to 1 across all genes for each i and k. In order to test the null hypothesis that the lane effect corresponding to λ_{ijk} for lane k remains constant across all the L lanes, they computed a goodness-of-fit statistic across L lanes. A qq plot of the quantiles showed evidence of a "lane effect" for only a small percentage of genes (0.5 %). The authors also proved that using the Poisson model for read counts from this technology could identify a significantly higher number of genes to be differentially expressed compared to the corresponding microarray data using the same False Discovery Rate (FDR) cutoff. They used quantitative PCR (qPCR) to examine discrepancies in the two platforms. Results of the qPCR study agreed more with Illumina study than with the microarrays.

Additionally, there are multiple studies which showed good correlation between microarray and RNA-seq results [2,80]. Fu et al. [22] compared the relative accuracy of RNA-seq and microarrays with protein expression data from the adult human cerebellum using 2D-LC MS/MS. They found that the RNA-seq data provided more accurate estimation of absolute transcript levels. Wall et al. [71] compared next generation sequencing with traditional capillary-based sequencing by a simulation model. They concluded that next generation sequencing offers better coverage over capillary-based sequencing. However, they suggested combining sequencing technologies such as FLX and Solexa to obtain optimal performance at a modest cost.

In contrast with the above mentioned results, some studies such as [6] showed that two-channel microarrays were more sensitive in identifying genes with low expression than RNA-seq data. Williams et al. [75] also suggested that microarrays showed better correlation with the synthetic miRNA data compared to RNA-seq data.

Dohm et al. [16] considered two Solexa read data sets and observed that error rates were greater at the end of reads (0.3 % at the beginning compared with 3.8 % at the end) and that incorrect basecalls are often preceded by base G. Also, base substitution errors with A to C were significantly higher (ten times more frequent) than the C to G substitution. Similar scenarios were observed by [8]. They considered data from the control lane of an Illumina ChIP-Seq experiment and reported the A to T miscall to be the most common error in their calibration study.

These conflicting experimental outcomes emphasize the importance of determining NGS performance measures based on some standardized objective benchmarks. The measures include accuracy, detection limits, reproducibility, dynamic range, and several other diagnostics. A comprehensive effort in this direction was initiated by the external RNA Control Consortium (ERCC). A set of external RNA standards for microarray, qPCR and sequencing experiments was developed [5, 15, 57]. Jiang et al. [27] suggested that RNA-seq data generated from GA-II in several modENCODE and ENCODE experiments should be used as ERCC RNA-seq benchmarks. They employed such data as spike-in controls in their study to determine the sensitivity and biases.

Oshlack et al. [53] considered three sequencing data sets from Illumina and SOLiD platforms. For each data set, they demonstrated an important result: if gene expression was measured by aggregated tag counts (directly obtained from RNA-seq) for each gene, then the length of the transcript would highly correlate with the ability to call differentially expression. Bullard et al. [9] studied the effects of different systematic sources of variability in measuring the differential expression of genes using the platforms such as mRNA-seq data from Illumina sequencing, microarray and quantitative real time PCR assay data. All these measurements were based on the Microarray quality control project (MAQC). Moreover, they emphasized that using an auto-calibration instead of Illumina's standard way of reserving one flowcell lane for the control can improve the mapping quality of the reads resulting in a more efficient and cost-effective experimental design. Additionally, they suggested different normalization strategies to overcome the systematic biases.

Trimar et al. [68] discussed the quality control issue for next generation sequencing data in the context of DNA methylation. Next generation sequencing allowing unbiased methylome profiling of a large number of patient cohorts can be very useful for biomarker discovery. They showed that post sequencing quality control matrices [11] can help in excluding poor quality samples from the analysis. This would lead to reduced noise and greater accuracy in identifying differentially methylated regions. They also suggested direct removal of some data due to validity concerns. The validity would usually depend on the enrichment of methylated fragments before sequencing. Another critical recommendation was to check the statistical reproducibility of the data by replicating sequencing lanes. They advised that this replicability should be computationally tested before going forward with any downstream analysis. They also formalized the enrichment of certain regions in the genome known as "CpG islands".

The CpG islands are segments characterized by high CpG dinucleotide content compared to the rest of the genome. It has been observed that CpG islands of promoters are unmethylated when the genes are expressed. This observation has led to the hypothesis that methylation of CpG sites in the promoter of a gene may inhibit gene expression. Trimar et al. [68] defined the enrichment of the CpG parameter by comparing the frequency of CpGs in the sequenced sample with respect to the reference genome. Statistical reproducibility was then determined by the high Pearson correlation coefficient of the enrichment score at two random partitions of the sequenced sample. A benchmark of breath and strength of methylation signal (confidence in methylation calls) was created using 5×CpG coverage. The latter represents the fraction of the total number of CpG loci that correspond to five or more reads in the sample.

Accumulation of B-tails at the read ends is another major problem that persists even in the state-of-the-art technologies HiSEQ and GA-II. B-tails are low quality $3'$ ends (marked with a B-tail in the quality string). These were observed at distinct locations, and in several cases only on one strand. Illumina recommends excluding this portion of the read in further analysis. Extreme cases of reads entirely composed of Bs are also common. From a recent study [47], the fraction of bases lying within B-tails was found to be 13.8 % for HiSeq data and 25.8 % for GAIIx data. The B-tail errors could also be sequence specific and are often attributed to an artifact of sequencing called "phasing".

Errors like these and several others may bias conclusions drawn from NGS data. Though these artifacts are embedded into the specific technologies, a detailed understanding of their sources could help us counteract them through statistical methods. We discuss such approaches in the next section.

1.3.2 Basecalling

Improved basecalling techniques are required to account for some of the errors described above. For a long time, *Phred* was the conventional and only computer-based program for base calling.

For the Illumina platform, images obtained from a charged coupled device record fluorescence intensities in each of the four nucleotide channels for each cycle of the sequencing-by-synthesis procedure. In a typical intensity matrix the rows represent the channels and the columns contain the cycles. Illumina's standard basecaller Bustard converts these intensities into concentrations by multiplying the observed intensities by the inverse of an estimated crosstalk matrix to adjust for the correlation between the four channels. As the cycles progress, loss of fragment copies leads to reduced intensities and consequently reduced concentrations. Thus, Bustard rescales the concentrations in each cycle by a factor proportional to the reciprocal of the average concentration for the cycle so that all cycles have the same average concentration. Finally, Bustard uses a Markov chain to model the probabilities of the events that one base is correctly synthesized during a cycle, that no new base is synthesized (known as phasing or lagging), and that two bases are synthesized (known as pre-phasing or leading), and it adjusts the rescaled concentrations based on estimates of the transition probabilities in the Markov chain. These adjusted concentrations are then used to make basecalls and assign quality scores for the bases. More details about Bustard can be found in [8, 29, 33, 38, 74].

Recently several algorithms have been proposed to improve the standard Illumina basecalls. Alta-Cyclic [19] used a support vector machine (SVM) to classify each nucleotide as one of the four possible bases. This basecaller first trains the SVM using a known reference genome in one of the flow cell lanes. The aim here is to find optimal phasing deconvolution parameters as well as internal parameters related to dynamic estimates for the crosstalk. Then the optimized SVM is applied to classify the bases in the other lanes and obtain quality scores. On the other hand, Rolexa, a basecaller for Solexa [59], used a probabilistic model. The algorithm corrects for positional bias, phasing, rephrasing and crosstalk. Then it estimates conditional probabilities of each base given a quadruple of intensities. The quadruple is modeled as a mixture of four multivariate normal random variables. As a byproduct, it also computes entropies for each basecall and uses them to identify and remove ambiguous basecalls.

Swift [74] and BING [35] provided alternative image processing and basecalling pipelines. Swift includes several modifications to Illumina's image processing procedures including the way it handles crosstalk and removal of the background noise in the image. Among other modifications, BING includes the option to perform pixel-based basecalling as to opposed cluster-based basecalling. Bravo et al. [8] proposed a probabilistic model for the log intensity which includes read and base-cycle effects and latent indicator variables for each possible base in a given read and cycle. They obtained posterior probabilities for each base as well as estimates of the other effects using the Expectation-Maximization algorithm. BayesCall [29] assumed a full Bayesian model for the nucleotide bases, concentration of active templates, and observed fluorescence intensities. A notable feature is the incorporation of cycle dependent parameters. The model parameters are estimated using MCEM [73], ECM [45], and simulated annealing [34]. The final quality scores are based on the maximum a posteriori estimates.

Recently, for GA-II, a cycle dependent basecaller named Ibis (Improved base identification system) was developed. It used multiclass-SVM and based its information about phasing at a given cycle on intensities on the preceding and subsequent cycles. In a recent comparison with other basecallers considered by [38], Ibis clearly emerged as the fastest and the most accurate algorithm and maintained its superiority consistently across different genomes. An updated version called freeIbis [58] has been recently published. AYB (All Your Base), yet another basecaller, was recently proposed by [43]. This used an iteratively reweighted least squares algorithm to fit a cluster-specific multivariate statistical regression model for the intensity matrix.

For the SOLiD platform, data points collected in color space represent two adjacent bases. Rsolid [76] proposed a quantile normalization procedure to improve the basecaller provided by the manufacturer at that time. SOLiD Exact Call

Chemistry—the 5500 Series SOLiD [43] system uses convolutional coding theory with error-correcting codes and resolves ambiguous read positions. Recently, [70] proposed a supervised learning method using a multi-class support vector machine for basecalling data from the SOLID 5500/5500 XL platform.

By assigning quality scores to the calls, one can assess the performances of the various basecallers. *Phred* is the conventional and the mostly widely used quality scoring program. The phred algorithm analyzes the sequences in a four-phase procedure to estimate several parameters about peak shape and peak resolution. It then uses these parameters to look up a corresponding quality score in some known lookup tables. The accuracy of Phred quality scores have been verified for a wide variety of sequencing platforms. Quality scores range from 4 to about 60, with higher values corresponding to higher quality. The quality score Q is logarithmically linked to error probability of basecall, P, by the formula $Q = -\log_{10} P$.

1.3.3 Alignment and Assembly Tools

After the bases in sequence reads have been called, mapping tools for assembling the reads are needed, and several algorithms have been proposed to align the reads to a reference genome. MAQ [40] uses a Eland-like hashing technique to align the reads. Then it uses a Bayesian statistical model to produce phred-scaled quality scores for the read alignments. This score is based on posterior probabilities and is ten times the common logarithm of the probability of incorrect assignment. MAQ efficiently combines the mapping quality information with the quality scores and utilizes mate-pair information for paired-end read alignment in diploid samples. Bowtie [36] has recently emerged as a widely used alignment tool. It employs a Burrows-Wheeler transform from string matching theory with an efficient backward search. This allows high-quality alignments with double indexing to prevent excessive backtracking. Bowtie's procedure is greedy and hence sub-optimal. But it includes options which allow users to trade efficiency for accuracy. BWA [39] is yet another algorithm which uses the computational advantages of the Burrows-Wheeler transform while allowing for inexact matching and gapped alignment.

For RNA seq data, a major challenge lies in identifying novel splice junctions. The sampled RNA also has several attributes such as single nucleotide polymorphisms (SNPs) and indels (insertions or deletions of certain bases). Tophat [67] was the first software designed to discover such junctions ab initio. The algorithm works in two steps. First, it maps all reads to the reference genome using Bowtie. All reads that do not map to the genome are set aside as "initially unmapped" (IUM) reads. It then assembles the mapped reads using MAQ [40] to construct an initial consensus. Sequences flanking potential donor/acceptor splice sites within neighboring regions are joined to form potential splice junctions. The IUM reads are then indexed and aligned to these splice junction sequences. A new version of the original software named Tophat2 was released in early 2013. TopHat2 has the additional capacity to align variable length reads and allow variable-length indels with respect to the reference genome.

1.3.4 Post-Alignment Analysis

Processing of NGS data is incomplete without statistical procedures connecting the mapped reads to relevant biological hypotheses. This requires developing rigorous tools for normalization, peak calling, enrichment testing etc. These tools would apply to a wide range of applications, e.g. detecting activated histone marks, exploring DNA methylation patterns, estimating RNA transcript abundance and so on.

Post-alignment transcriptome analysis encompasses a broad range of research; see, e.g., [51] in yeast; [13] and [42] in human; [49] in mouse; [69] in butterfly, and so on. Transcriptome analysis typically involves solving two subproblems—(1) aligning reads to the transcriptome and (2) estimation of the abundance of unique transcripts.

The analysis begins by normalizing the mapped read counts. RPKM (reads per kilobase per million mapped reads) was originally introduced by [49] for this purpose. Based on transcript lengths and the sequencing depths, this RPKM measure can compare the expression measures across different genes and samples. Since mapped reads are frequently shared by multiple isoforms, using the normalized reads to estimate transcript abundance is another big challenge. Older count based abundance models avoid this issue by assuming all transcripts have a single isoform and reads are uniquely mappable to transcripts. One ad hoc approach is to allocate fractions of multi-mapped reads to target transcript isoforms equally. Another approach is to allocate fractions of the reads in proportion to the coverage of uniquely mapped reads divided by the length of the transcript isoforms ("rescue" method) [49].

More realistic multi-read models [26, 78] addressed the case for multiple isoforms. Jiang et al. [26] was one of the first model-based approaches in this direction. The major assumption was that the number of reads coming from an exon of a certain length is Poisson where the mean is a normalized function of the exon length. The primary estimable parameters were the relative abundances of different transcripts. Concave optimization was used to maximize the likelihood. With the advent of paired-end sequencing, such models were slowly replaced by more complex ones which included length distribution of fragments along with transcript lengths. Recall that in paired-end tagging, reads correspond to both ends of the sequenced fragments. Thus, when the reads in a pair map upstream and downstream of an alternatively spliced exon, the inclusion and exclusion isoforms will typically imply different intervening insert lengths. Such evidence is utilized to sharpen the inference on abundance estimation. The first insert length model was published in [67] which essentially extended the approach of Jiang and Wong to paired-end reads. Their algorithm was made available through a comprehensive software called *Cufflinks* in early 2011. For a review of general insert length models we refer the readers to [30,52,61]. Insert length models implicitly assume that reads are filtered based on length independent of their sequence. Conditional on insert length, the sampling of transcripts is modeled as uniform. The read pairs are then probabilistically assigned to isoforms that are consistent with both individual reads. These assignments are weighted by the relative probability of observing the given insert length.

Katz et al. [30] introduced a Bayesian RNA-seq algorithm called MISO and explicitly showed how the distribution of the insert lengths affects our knowledge of splicing events. They based their length distribution on the implied length of read pairs which map to large constitutive intronless regions such as 3′ UTRs. Mezilini et al. [46] identified the importance of solving the problems of alignment and estimation simultaneously through an iterative algorithm. They pointed out that accurate estimation of isoform abundance is extremely difficult if not all isoforms are known since the read pairs from unknown isoforms can affect the estimation of known ones. IsoInfer/IsoLasso [21] is a similar algorithm that solves the two subproblems by computing a large set of possible isoforms and then using lasso [66] to select a best subset.

A number of broad data analytic tools to handle different types of NGS data have been developed over last five or so years. As for example, the software package F-seq [7] produces continuous signals along a chromosome by kernel smoothing the high-throughput sequencing read counts. The user can visualize the resulting signal directly in the UCSC Genome Browser. A number of specific features such as transcription factor binding sites (ChIP-Seq) or regions of open chromatin (DNase-seq) can be identified this way. Often a statistical distribution such as the gamma distribution is fit to the maximum F-seq signal in an enriched region to determine if a peak is significant; see, e.g., [72]. Other tools besides mapping [28] and alignment [36] include MACS (Model-based Analysis of ChIP-Seq)—a more robust predictor of binding sites; a tool for enrichment analysis [20]; tools for ChIP-Seq data [60,79]; tools for protein-DNA interactions [63]; and a tool for DNA copy number variation detection [77]. Pepke et al. [56] provided a useful review of methods for RNA-seq and ChIP-Seq data; also see [25]. More recent reviews of RNA-seq data analytic tools are provided in [12,24]. Methods related to splicing based on RNA-seq data are reviewed in [3]. Computational and statistical tools for ChIP-Seq data are reviewed in [32]. A number of post-alignment data analytic tools have been incorporated into R and Bioconductor packages. These are discussed in the next section.

1.3.5 R and Bioconductor Packages

We briefly describe a selected number of R and Bioconductor packages useful for developing a basic NGS data analysis pipeline. More statistically oriented packages for specific and advanced analyses are introduced in other chapters of this book.

The ShortRead package [48] is quite popular in working with raw short read outputs of standard NGS platforms. It can perform quality assessment, data manipulation and provide high-level data summary useful for subsequent statistical analysis. It can also work with a number of standard alignment programs such as Bowtie [36], MAQ [40] and ELAND (Illumina's proprietary alignment program). The ShortRead package can be used in conjunction with [54, 55]. The last two packages are useful for efficient manipulation of big strings. Lawrence et al. [55] has tools for representation, manipulation, and analysis tools of large sequences and subsequences; it can also attach additional information to such subsequences. The rtracklayer package [37] is useful for interfacing R with genome browsers.

The biomaRt package [17] provides an interface to BioMart databases (e.g. Ensembl, COSMIC, Wormbase, Gramene etc). Using this tool, users can gain direct access to a diverse set of data in a simple and uniform manner. They are able to mine these databases and obtain gene annotations. An infrastructure package BSgenome [54] is available for accessing, analyzing, creating, or modifying Biostrings-based genome data packages. The rGADEM package is a de novo motif discovery tool for large-scale genomic sequence data. It depends on the seqLogo package for plotting sequence logos for DNA sequence alignments. Visualization of NGS data by means of Hilbert curves is possible using the HilbertVis package [4]. Plotting tools for NGS and other genomic data are provided in GenomeGraphs [18].

ChIP-Seq is an important technology for detecting transcription factor binding sites and epigenetic marks. It uses NGS platforms as a part of its workflow. Amongst various packages for analyzing ChIP-seq experiments are the chipseq [31], ChIPpeakAnno [81], BayesPeak [10] and the Rcade package (http://www.bioconductor.org/packages/release/bioc/vignettes/Rcade/). Tools for design and analysis of ChIP-Seq experiments are provided in chipseq which corrects for background signals and improves tag alignment. The ChIPpeakAnno package has tools for batch annotation of the identified peaks from ChIP-Seq experiments. As the name suggests, BayesPeak is a peak detection software based on Bayesian methods. Rcade provides an integrated analysis of ChIP-Seq together with differential gene expression summary.

Several statistical methods and related R packages for differential gene expression analysis based on RNA-seq data have been developed over the years. The packages DESeq and EdgeR are popular choice amongst users of RNA-seq. BaySeq is a Bioconductor package that identifies differential expression using next-generation sequencing data via empirical Bayesian methods. A number of these packages and the corresponding statistical methods are described in greater details in Chapter 2 of this book.

1.4 Conclusion

Decoding the DNA sequence is fundamental to all branches of biological research. Therefore it is essential to have a widely available technology for accurate and complete sequencing. Early DNA sequencing capability relied on some version of Sanger sequencing [62]. Despite its widespread use in scientific laboratories in the 1990s, this "first generation" technology suffered from inherent limitations in throughput, speed and cost. To remedy these deficiencies, next generation sequencing arrived in 2005 and quickly rose to prominence within the span of few years. It represents a fundamental departure from the previous generation, primarily in its ability to sequence millions of reads in parallel. For today's genome researchers, a wide variety of platforms is available to choose from. We reviewed these various platforms and discussed their relative merits with respect to cost, speed and suitability for certain kinds of biological experiments. Competition between manufacturers is increasingly leading to lower costs and improved speed. It has been estimated that sequencing capability is now accelerating at a rate faster than Moore's law.

To keep pace with the ever growing technology considerable demands are being placed on the IT infrastructure, storage capabilities and data tracking. Cloud service is now getting used for efficient information exchange and project management. More importantly, sophisticated computational and statistical software are in demand for analyzing the gigabytes of generated data. A critical challenge is that the measurement of gene expression or TF binding affinity is now based on direct counts of reads rather than on hybridization to probes. This has reduced the noise caused by cross-hybridization and the bias caused by the variation in probe binding efficiency. However, the read data is accompanied by its own unique features and problems.

Overall, NGS presents some novel methodological issues for statisticians. Mapping or constructing de-novo genome assembly demands fast computational algorithms for alignment. Analyzing mapped reads is the next major challenge since the count data is hard to fit by conventional Gaussian distributions. Basecalling and differential profiling increasingly requires sophisticated hierarchical models. One critical concern is the curse of dimensionality. For example, in RNA-seq experiments we often have very few replicate samples, where conventional methods for quantifying uncertainty fail. Bayesian hierarchical models partially mitigate this problem by borrowing strength across different genes and/or experimental units. NGS is significantly impacting network analysis as well . Deciphering bimolecular pathways through proteomic or genomic networks is one of the holy grails of modern biology. Pathway inference has huge implications for a wide range of applications including targeted drug therapy, personal genomics and pharmacokinetics. However, traditional graphical models fail to account for non-Gaussian count distributions and miss the correlation between the true biological signals. Again, small n large p issues emerge as a major challenge in such network inference. Bayesian models utilizing novel graphical priors become highly relevant in this context. Some of these issues are discussed at length in later chapters.

Apart from the primary goal of mere sequencing, NGS technologies have a broad range of applications. With the rapid decline in sequencing cost, personal genomics is soon to become a reality. The platforms are now maturing to the point where NGS is being considered by many laboratories for routine diagnostic use. NGS platforms have now made genomic analysis possible for any organism, allowing comparison across individuals and ecotypes. Recent applications of deep sequencing in microbial genomes highlight the striking impact it has on the fields of evolutionary biology and metagenomics. Thus NGS has an enormous potential to transform current genomic research and enhance our fundamental understanding of biological processes. It is equally critical both for biologists and quantitative scientists to acquire a detailed comprehension of this technology and tap into this powerful resource.

References

[1] Abecasis, G., Altshuler, D., Auton, A., Brooks, L., Durbin, R., Gibbs, R.A., Hurles, M.E., McVean, G.A., Bentley, D., Chakravarti, A., et al.: A map of human genome variation from population-scale sequencing. Nature **467**(7319), 1061–1073 (2010)

[2] Agarwal, A., Koppstein, D., Rozowsky, J., Sboner, A., Habegger, L., Hillier, L.W., Sasidharan, R., Reinke, V., Waterston, R.H., Gerstein, M.: Comparison and calibration of transcriptome data from rna-seq and tiling arrays. BMC Genom. **11**(1), 383 (2010)

[3] Alamancos, G.P., Agirre, E., Eyras, E.: Methods to study splicing from high-throughput rna sequencing data. Meth. Mol. Biol., **1126**, 357–397 (2014)

[4] Anders, S.: Visualization of genomic data with the hilbert curve. Bioinformatics **25**(10), 1231–1235 (2009)

[5] Baker, S.C., Bauer, S.R., Beyer, R.P., Brenton, J.D., Bromley, B., Burrill, J., Causton, H., Conley, M.P., Elespuru, R., Fero, M., et al.: The external rna controls consortium: a progress report. Nat. Meth. **2**(10), 731–734 (2005)

[6] Bloom, J.S., Khan, Z., Kruglyak, L., Singh, M., Caudy, A.A.: Measuring differential gene expression by short read sequencing: quantitative comparison to 2-channel gene expression microarrays. BMC Genom. **10**(1), 221 (2009)

[7] Boyle, A.P., Guinney, J., Crawford, G.E., Furey, T.S.: F-seq: a feature density estimator for high-throughput sequence tags. Bioinformatics **24**(21), 2537–2538 (2008). doi:10.1093/bioinformatics/btn480

[8] Bravo, H.C., Irizarry, R.A.: Model-based quality assessment and base-calling for second-generation sequencing data. Biometrics **66**(3), 665–674 (2010)

[9] Bullard, J.H., Purdom, E., Hansen, K.D., Dudoit, S.: Evaluation of statistical methods for normalization and differential expression in mrna-seq experiments. BMC Bioinform. **11**, 94 (2010). doi:10.1186/1471-2105-11-94

[10] Cairns, J., Spyrou, C., Stark, R., Smith, M.L., Lynch, A.G., Tavare, S.: Bayespeak: an r package for analysing chip-seq data. Bioinformatics **27**(5), 713–714 (2011)

[11] Chavez, L., Jozefczuk, J., Grimm, C., Dietrich, J., Timmermann, B., Lehrach, H., Herwig, R., Adjaye, J.: Computational analysis of genome-wide dna methylation during the differentiation of human embryonic stem cells along the endodermal lineage. Genome Res. **20**(10), 1441–1450 (2010)

[12] Chen, G., Wang, C., Shi, T.: Overview of available methods for diverse rna-seq data analyses. Sci. China Life Sci. **54**(12), 1121–1128 (2011)

[13] Cloonan, N., Grimmond, S.M.: Transcriptome content and dynamics at single-nucleotide resolution. Genome Biol. **9**(9), 234 (2008). doi:10.1186/gb-2008-9-9-234

[14] Datta, S., Datta, S., Kim, S., Chakraborty, S., Gill, R.S.: Statistical analyses of next generation sequence data: a partial overview. J. Proteonomics Bioinform. **3**(6), 183 (2010)

[15] Devonshire, A., Elaswarapu, R., Foy, C.: Evaluation of external rna controls for the standardisation of gene expression biomarker measurements. BMC Genom. **11**(1), 662 (2010)

[16] Dohm, J.C., Lottaz, C., Borodina, T., Himmelbauer, H.: Substantial biases in ultra-short read data sets from high-throughput dna sequencing. Nucleic Acids Res. **36**(16), e105 (2008). doi:10.1093/nar/gkn425

[17] Durinck, S., Moreau, Y., Kasprzyk, A., Davis, S., De Moor, B., Brazma, A., Huber, W.: Biomart and bioconductor: a powerful link between biological databases and microarray data analysis. Bioinformatics **21**(16), 3439–3440 (2005)

[18] Durinck, S., Bullard, J., Spellman, P.T., Dudoit, S.: Genomegraphs: integrated genomic data visualization with r. BMC Bioinform. **10**(1), 2 (2009)

[19] Erlich, Y., Mitra, P.P., delaBastide, M., McCombie, W.R., Hannon, G.J.: Alta-cyclic: a self-optimizing base caller for next-generation sequencing. Nat. Meth. **5**(8), 679–682 (2008). doi:10.1038/nmeth.1230

[20] Fejes, A.P., Robertson, G., Bilenky, M., Varhol, R., Bainbridge, M., Jones, S.J.M.: Findpeaks 3.1: a tool for identifying areas of enrichment from massively parallel short-read sequencing technology. Bioinformatics **24**(15), 1729–1730 (2008). doi:10.1093/bioinformatics/btn305

[21] Feng, J., Li, W., Jiang, T.: Inference of isoforms from short sequence reads. J. Comput. Biol. **18**(3), 305–321 (2011). doi:10.1089/cmb.2010.0243

[22] Fu, X., Fu, N., Guo, S., Yan, Z., Xu, Y., Hu, H., Menzel, C., Chen, W., Li, Y., Zeng, R., et al.: Estimating accuracy of rna-seq and microarrays with proteomics. BMC Genom. **10**(1), 161 (2009)

[23] Fullwood, M.J., Wei, C.L., Liu, E.T., Ruan, Y.: Next-generation dna sequencing of paired-end tags (pet) for transcriptome and genome analyses. Genome Res. **19**(4), 521–532 (2009)

[24] Garber, M., Grabherr, M.G., Guttman, M., Trapnell, C.: Computational methods for transcriptome annotation and quantification using rna-seq. Nat. Meth. **8**(6), 469–477 (2011)

[25] Ghosh, D., Qin, Z.S.: Statistical issues in the analysis of chip-seq and rna-seq data. Genes **1**(2), 317–334 (2010)

[26] Jiang, H., Wong, W.H.: Statistical inferences for isoform expression in rna-seq. Bioinformatics **25**(8), 1026–1032 (2009). doi:10.1093/bioinformatics/btp113

[27] Jiang, L., Schlesinger, F., Davis, C.A., Zhang, Y., Li, R., Salit, M., Gingeras, T.R., Oliver, B.: Synthetic spike-in standards for rna-seq experiments. Genome Res. **21**(9), 1543–1551 (2011)

[28] Johnson, T.: Bayesian method for gene detection and mapping, using a case and control design and dna pooling. Biostatistics **8**(3), 546–565 (2007). doi:10.1093/biostatistics/kxl028

[29] Kao, W.C., Stevens, K., Song, Y.S.: Bayescall: a model-based base-calling algorithm for high-throughput short-read sequencing. Genome Res. **19**(10), 1884–1895 (2009). doi:10.1101/gr.095299.109

[30] Katz, Y., Wang, E.T., Airoldi, E.M., Burge, C.B.: Analysis and design of rna sequencing experiments for identifying isoform regulation. Nat. Meth. **7**(12), 1009–1015 (2010). doi:10.1038/nmeth.1528

[31] Kharchenko, P.V., Tolstorukov, M.Y., Park, P.J.: Design and analysis of chip-seq experiments for dna-binding proteins. Nat. Biotech. **26**(12), 1351–1359 (2008)

[32] Kim, H., Kim, J., Selby, H., Gao, D., Tong, T., Phang, T.L., Tan, A.C., et al.: A short survey of computational analysis methods in analysing chip-seq data. Hum. Genom. **5**(2), 117–123 (2011)

[33] Kircher, M., Stenzel, U., Kelso, J., et al.: Improved base calling for the illumina genome analyzer using machine learning strategies. Genome Biol. **10**(8), R83 (2009)

[34] Kirkpatrick, S.: Optimization by simulated annealing: Quantitative studies. J. Stat. Phys. **34**(5–6), 975–986 (1984)

[35] Kriseman, J., Busick, C., Szelinger, S., Dinu, V.: Bing: biomedical informatics pipeline for next generation sequencing. J. Biomed. Informat. **43**(3), 428–434 (2010)

[36] Langmead, B.: Aligning short sequencing reads with Bowtie. Curr. Protoc. Bioinform. **32**, 11–17 (2010)

[37] Lawrence, M., Gentleman, R., Carey, V.: rtracklayer: an r package for interfacing with genome browsers. Bioinformatics **25**(14), 1841–1842 (2009)

[38] Ledergerber, C., Dessimoz, C.: Base-calling for next-generation sequencing platforms. Briefings Bioinform. **12**(5), 489–497 (2011)

[39] Li, H., Durbin, R.: Fast and accurate short read alignment with burrows-wheeler transform. Bioinformatics **25**(14), 1754–1760 (2009). doi:10.1093/bioinformatics/btp324

[40] Li, H., Ruan, J., Durbin, R.: Mapping short dna sequencing reads and calling variants using mapping quality scores. Genome Res. **18**(11), 1851–1858 (2008). doi:10.1101/gr.078212.108

[41] Loman, N.J., Constantinidou, C., Chan, J.Z., Halachev, M., Sergeant, M., Penn, C.W., Robinson, E.R., Pallen, M.J.: High-throughput bacterial genome sequencing: an embarrassment of choice, a world of opportunity. Nat. Rev. Microbiol. **10**(9), 599–606 (2012)

[42] Marioni, J.C., Mason, C.E., Mane, S.M., Stephens, M., Gilad, Y.: Rna-seq: an assessment of technical reproducibility and comparison with gene expression arrays. Genome Res. **18**(9), 1509–1517 (2008). doi:10.1101/gr.079558.108

[43] Massingham, T., Goldman, N.: All your base: a fast and accurate probabilistic approach to base calling. Genome Biol. **13**, R13 (2012)

[44] McCarthy, A.: Third generation dna sequencing: pacific biosciences' single molecule real time technology. Chem. Biol. **17**(7), 675–676 (2010). doi:10.1016/j.chembiol.2010.07.004

[45] Meng, X.L., Rubin, D.B.: Maximum likelihood estimation via the ecm algorithm: a general framework. Biometrika **80**(2), 267–278 (1993)

[46] Mezlini, A.M., Smith, E.J., Fiume, M., Buske, O., Savich, G.L., Shah, S., Aparicio, S., Chiang, D.Y., Goldenberg, A., Brudno, M.: ireckon: simultaneous isoform discovery and abundance estimation from rna-seq data. Genome Res. **23**(3), 519–529 (2013)

[47] Minoche, A.E., Dohm, J.C., Himmelbauer, H.: Evaluation of genomic high-throughput sequencing data generated on illumina hiseq and genome analyzer systems. Genome Biol. **12**(11), R112 (2011). doi:10.1186/gb-2011-12-11-r112

[48] Morgan, M., Anders, S., Lawrence, M., Aboyoun, P., Pagès, H., Gentleman, R.: Shortread: a bioconductor package for input, quality assessment and exploration of high-throughput sequence data. Bioinformatics **25**(19), 2607–2608 (2009)

[49] Mortazavi, A., Williams, B.A., McCue, K., Schaeffer, L., Wold, B.: Mapping and quantifying mammalian transcriptomes by rna-seq. Nat. Meth. **5**(7), 621–628 (2008). doi:10.1038/nmeth.1226

[50] Murray, I.A., Clark, T.A., Morgan, R.D., Boitano, M., Anton, B.P., Luong, K., Fomenkov, A., Turner, S.W., Korlach, J., Roberts, R.J.: The methylomes of six bacteria. Nucleic Acids Res. **40**(22), 11,450–11,462 (2012)

[51] Nagalakshmi, U., Wang, Z., Waern, K., Shou, C., Raha, D., Gerstein, M., Snyder, M.: The transcriptional landscape of the yeast genome defined by rna sequencing. Science **320**(5881), 1344–1349 (2008). doi:10.1126/science.1158441

[52] Nicolae, M., Mangul, S., Măndoiu, I.I., Zelikovsky, A.: Estimation of alternative splicing isoform frequencies from rna-seq data. Algorithms Mol. Biol. **6**(1), 9 (2011). doi:10.1186/1748-7188-6-9

[53] Oshlack, A., Wakefield, M.J.: Transcript length bias in rna-seq data confounds systems biology. Biol. Direct. **4**, 14 (2009). doi:10.1186/1745-6150-4-14

[54] Pages, H.: Bsgenome: infrastructure for biostrings-based genome data packages. R Package Version 1.32.0 (2014)

[55] Lawrence, M., Huber, W., Pagès, H., Aboyoun, P., Carlson, M., Gentleman, R., Morgan, M., Carey, V.: Software for computing and annotating genomic ranges. PLoS Comput. Biol., **9**, (2013)

[56] Pepke, S., Wold, B., Mortazavi, A.: Computation for chip-seq and rna-seq studies. Nat. Meth. **6**(11 Suppl), S22–S32 (2009). doi:10.1038/nmeth.1371
[57] Reid, L.H.: Proposed methods for testing and selecting the ercc external rna controls. BMC Genom. **6**(1), 1–18 (2005)
[58] Renaud, G., Kircher, M., Stenzel, U., Kelso, J.: freeibis: an efficient basecaller with calibrated quality scores for illumina sequencers. Bioinformatics **29**(9), 1208–1209 (2013)
[59] Rougemont, J., Amzallag, A., Iseli, C., Farinelli, L., Xenarios, I., Naef, F.: Probabilistic base calling of solexa sequencing data. BMC Bioinform. **9**, 431 (2008). doi:10.1186/1471-2105-9-431
[60] Rozowsky, J., Euskirchen, G., Auerbach, R.K., Zhang, Z.D., Gibson, T., Bjornson, R., Carriero, N., Snyder, M., Gerstein, M.B.: Peakseq enables systematic scoring of chip-seq experiments relative to controls. Nat. Biotech. **27**(1), 66–75 (2009). doi:10.1038/nbt.1518
[61] Salzman, J., Jiang, H., Wong, W.H.: Statistical modeling of rna-seq data. Stat. Sci. **26**(1), 62–83 (2011)
[62] Sanger, F., Nicklen, S., Coulson, A.R.: Dna sequencing with chain-terminating inhibitors. Proc. Natl. Acad. Sci. **74**(12), 5463–5467 (1977)
[63] Sharon, E., Lubliner, S., Segal, E.: A feature-based approach to modeling protein-dna interactions. PLoS Comput. Biol. **4**(8), e1000,154 (2008). doi:10.1371/journal.pcbi.1000154
[64] Shendure, J., Ji, H.: Next-generation dna sequencing. Nat. Biotech. **26**(10), 1135–1145 (2008). doi:10.1038/nbt1486
[65] Smith, C.L., Migliaccio, I., Chaubal, V., Wu, M.F., Pace, M.C., Hartmaier, R., Jiang, S., Edwards, D.P., Gutiérrez, M.C., Hilsenbeck, S.G., Oesterreich, S.: Elevated nuclear expression of the smrt corepressor in breast cancer is associated with earlier tumor recurrence. Breast Cancer Res. Treat. **136**(1), 253–265 (2012). doi:10.1007/s10549-012-2262-7
[66] Tibshirani, R.: Regression shrinkage and selection via the lasso. J. Roy. Stat. Soc. Ser. B (Methodological) **58**, 267–288 (1996)
[67] Trapnell, C., Pachter, L., Salzberg, S.L.: Tophat: discovering splice junctions with rna-seq. Bioinformatics **25**(9), 1105–1111 (2009). doi:10.1093/bioinformatics/btp120
[68] Trimarchi, M.P., Murphy, M., Frankhouser, D., Rodriguez, B.A., Curfman, J., Marcucci, G., Yan, P., Bundschuh, R.: Enrichment-based dna methylation analysis using next-generation sequencing: sample exclusion, estimating changes in global methylation, and the contribution of replicate lanes. BMC Genom. **13**(Suppl 8), S6 (2012)
[69] Vera, J.C., Wheat, C.W., Fescemyer, H.W., Frilander, M.J., Crawford, D.L., Hanski, I., Marden, J.H.: Rapid transcriptome characterization for a nonmodel organism using 454 pyrosequencing. Mol. Ecol. **17**(7), 1636–1647 (2008). doi:10.1111/j.1365-294X.2008.03666.x
[70] Viswanath, S., Yang, C.: Color call improvement in next generation sequencing using multiclass support vector machines. BMC Bioinform. **13**(Suppl 18), A3 (2012)
[71] Wall, P.K., Leebens-Mack, J., Chanderbali, A.S., Barakat, A., Wolcott, E., Liang, H., Landherr, L., Tomsho, L.P., Hu, Y., Carlson, J.E., Ma, H., Schuster, S.C., Soltis, D.E., Soltis, P.S., Altman, N., dePamphilis, C.W.: Comparison of next generation sequencing technologies for transcriptome characterization. BMC Genom. **10**, 347 (2009). doi:10.1186/1471-2164-10-347
[72] Wang, D., Rendon, A., Wernisch, L.: Transcription factor and chromatin features predict genes associated with eqtls. Nucleic Acids Res. **41**(3), 1450–1463 (2013)
[73] Wei, G.C., Tanner, M.A.: A monte carlo implementation of the em algorithm and the poor man's data augmentation algorithms. J. Am. Stat. Assoc. **85**(411), 699–704 (1990)
[74] Whiteford, N., Skelly, T., Curtis, C., Ritchie, M.E., Löhr, A., Zaranek, A.W., Abnizova, I., Brown, C.: Swift: primary data analysis for the illumina solexa sequencing platform. Bioinformatics **25**(17), 2194–2199 (2009). doi:10.1093/bioinformatics/btp383
[75] Willenbrock, H., Salomon, J., Søkilde, R., Barken, K.B., Hansen, T.N., Nielsen, F.C., Møller, S., Litman, T.: Quantitative mirna expression analysis: comparing microarrays with next-generation sequencing. RNA **15**(11), 2028–2034 (2009)

[76] Wu, H., Irizarry, R.A., Bravo, H.C.: Intensity normalization improves color calling in solid sequencing. Nat. Meth. **7**(5), 336–337 (2010)

[77] Xie, C., Tammi, M.T.: Cnv-seq, a new method to detect copy number variation using high-throughput sequencing. BMC Bioinform. **10**, 80 (2009). doi:10.1186/1471-2105-10-80

[78] Xing, Y., Yu, T., Wu, Y.N., Roy, M., Kim, J., Lee, C.: An expectation-maximization algorithm for probabilistic reconstructions of full-length isoforms from splice graphs. Nucleic Acids Res. **34**(10), 3150–3160 (2006)

[79] Zhang, Z.D., Rozowsky, J., Snyder, M., Chang, J., Gerstein, M.: Modeling chip sequencing in silico with applications. PLoS Comput. Biol. **4**(8), e1000,158 (2008). doi:10.1371/journal.pcbi.1000158

[80] Zhang, Y., Malone, J.H., Powell, S.K., Periwal, V., Spana, E., MacAlpine, D.M., Oliver, B.: Expression in aneuploid drosophila s2 cells. PLoS Biol. **8**(2), e1000,320 (2010)

[81] Zhu, L., Gazin, C., Lawson, N., Pagès, H., Lin, S., Lapointe, D., Green, M.: Chippeakanno: a bioconductor package to annotate chip-seq and chip-chip data. BMC Bioinform. **11**(1), 237 (2010)

Chapter 2
Using RNA-seq Data to Detect Differentially Expressed Genes

Douglas J. Lorenz, Ryan S. Gill, Ritendranath Mitra, and Susmita Datta

Abstract RNA-sequencing (RNA-seq) technology has become a major choice in detecting differentially expressed genes across different biological conditions. Although microarray technology is used for the same purpose, statistical methods available for identifying differential expression for microarray data are generally not readily applicable to the analysis of RNA-seq data, as RNA-seq data comprise discrete counts of reads mapped to particular genes. In this chapter, we review statistical methods uniquely developed for detecting differential expression among different populations of RNA-seq data as well as techniques designed originally for the analysis of microarray data that have been modified for the analysis of RNA-seq data. We include a very brief description of the normalization of RNA-seq data and then elaborate on parametric and nonparametric testing procedures, as well as empirical and fully Bayesian methods. We include a brief review of software available for the analysis of differential expression and summarize the results of a recent comprehensive simulation study comparing existing methods.

2.1 Introduction: RNA-seq Data

RNA-seq is a next generation sequencing (NGS) procedure of the entire transcriptome by which one can measure the expression of several features such as gene expression, allelic expression, and intragenic expression. The number of reads mapped to a given gene or transcript is considered to be the estimate

D.J. Lorenz • R. Mitra • S. Datta (✉)
Department of Bioinformatics and Biostatistics, School of Public Health and Information Science, University of Louisville, 485 E. Gray St., Louisville, KY 40205, USA
e-mail: susmita.datta@louisville.edu

R.S. Gill
Department of Mathematics, University of Louisville, Louisville, KY 40292, USA

S. Datta and D. Nettleton (eds.), *Statistical Analysis of Next Generation Sequencing Data*, Frontiers in Probability and the Statistical Sciences, DOI 10.1007/978-3-319-07212-8_2,

Table 2.1 Table of read counts from a hypothetical RNA-seq experiment

	Population 1			Population 2	
Gene	Sample 1	Sample 2	Sample 3	Sample 1	Sample 2
1	22	26	15	66	44
2	4	1	20	1	4
3	75	113	281	116	97
⋮	⋮	⋮	⋮	⋮	⋮
10,000	0	9	0	1	2
Total	824,015	782,345	1,345,387	693,428	923,450

In this example, there are $K = 2$ populations, $J_1 = 3$ samples in the first population, $J_2 = 2$ samples in the second population, and $G = 10,000$ genes. The final row lists the cumulative read counts for each sample, frequently referred to as the library size for a sample

of the expression level of that feature using this technology [24]. Microarray technology has been the method of choice to measure gene expression since the nineteen-nineties. However, RNA-seq is generally acknowledged to be a better platform for transcription profiling for several reasons [8, 22, 25, 26, 28, 43, 50]. RNA-seq is believed to have a wider range of signal detection. The resolution of microarray expression measures cannot go beyond the probe level. In contrast, the majority of the reads from NGS technology map to the reference genome with single base resolution and consequently RNA-seq can be evaluated at single-base resolution. Moreover, in microarray technology one needs to have knowledge of the target sequences to construct the probe sets. Hence, RNA-seq is more suitable for the discovery of novel transcripts.

The end-product of a RNA-seq experiment is a sequence of read counts, typically represented as a matrix with rows representing genes and columns representing samples from one or more populations, as in Table 2.1. When RNA-seq data are generated from two or more populations, interest often is in the detection of differentially expressed genes among the populations, i.e., genes for which read count distributions differ among populations. Methods for detecting differential expression in microarray data are well-established but generally not applicable to RNA-seq data, as the data from a RNA-seq experiment are discrete counts rather than continuous measures of expression levels.

A challenge in the detection of differential expression for RNA-seq data results from the way in which reads are mapped to features such as genes, transcripts or exons. One of the issues is that the expression quantification from short reads using RNA-seq data depends on the length of the features; longer features usually produce more reads. Normalization by dividing by the length of the transcript [25] alleviates this problems somewhat but not completely [54]. The expression value used by Mortazavi et al. [25] is referred to as Reads per Kilobase per Million reads (RPKM). Differential RNA-seq analysis using an empirical Bayes procedure by the *limma* method [38] uses log-counts per million (log-cpm), analogous to the log-intensity values in microarray studies.

Differential expression analysis may also be affected by the sequence depth of the NGS data generation. Sequence depth can be calculated as $N \times L/G$, where N is the number of reads, L is the average read length and G is the length of the original genome. For example, $N = 8$ reads for a genome with $G = 2,000$ base pairs at average length $L = 500$ will have a sequence depth of 2 or $2\times$ redundancy. This also is equivalent to the percentage of genome covered by reads and the average number of times a base is read. Higher coverage can improve the power to identify differential expression using RNA-seq data. However, read counts are subject to technical variation in which the overall read count for a sample, referred to as the library size, can substantially vary among repeated NGS experiments on the same sample. In order to accommodate this source of variability, log-cpm values need to be adjusted by accounting for mean-variance trends typically observed in RNA-seq data, particularly among genes with lower counts. Zero counts are augmented by a small positive value to avoid taking the logarithm of zero, ensuring non-missing log-cpm and reducing the variability at lower count values.

An additional challenge is that some of genes may exhibit very large read counts while the rest of the reads are distributed among the remaining genes. Hence, even if library sizes are identical between samples, some genes may mask the expression of others which may be moderately equivalently expressed. Thus, the expression signals of genes or transcripts in RNA-seq data not only depend on sequence depth, but also are dependent on the expression levels of other transcripts. Because of this and the technical variation of NGS experiments noted above, raw read counts from different populations are not necessarily directly comparable in an analysis of differential expression without adjustment for technical variation. In other words, simply viewing the count for a given gene and sample as proportional to the sample's total read count is problematic because a few genes may have extremely large counts that artificially inflate a sample's total read count. Alternative complex normalization schemes for RNA-seq data have been proposed by Bullard et al. [6], Anders and Huber [1], and Robinson and Oshlack [32]. In these methods, there are additional sample specific normalizations combined with library sizes. There are other methods of normalization as well. A thorough evaluation of many normalization methods for RNA-seq data is provided in Dillies et al. [11]. Trimmed mean of M-values normalization (TMM) [32] and the normalization scheme provided by Anders and Huber [1] are among the easiest to use and provide a decent solution to the normalization problem of RNA-seq data. However, even these methods assume that very few genes are differentially expressed between different populations and those are equivalently spread between the up- and down-regulated genes. Other types of normalization strategies deal with the GC content of the reads. Normalization for this specific reason transforms RNA-seq data in such a way that it no longer remains count data and should be dealt with differently in terms of further analysis for finding differentially expressed genes. *Cufflinks/Cuffdiff* [48] provides a normalization scheme in their integrated differential analysis algorithm. For a more thorough discussion of normalization methods, we encourage the reader to consult the chapter on normalization in this volume.

The focus of this chapter is to provide a comprehensive review of the methods related to the analysis of differential expression for RNA-seq data. In recent years, a number of reviews of RNA-seq data analysis methods have been published, and they all are effective in communicating the current status of the analysis of RNA-seq data [2,40,53]. Much further work will be devoted to developing statistical methods for the detection of differentially expressed genes for RNA-seq data. In this chapter, we review statistical methods for detecting differential expression in RNA-seq data, including the application of techniques for analyzing microarray data to RNA-seq data, parametric and nonparametric tests, and empirical and fully Bayesian methods. We summarize the results of several simulation studies, including a recently published thorough examination of several of these methods. We briefly describe some existing open source R and Bioconductor software for testing differential expression for RNA-seq data. We conclude the chapter with a discussion section.

2.2 Statistical Methods for Testing Differential Expression

For consistency of notation in what follows, we have established a single unifying notation for the RNA-seq read counts. As a result, the notation we use here is frequently different from the source works. We consider read counts for G genes measured in K populations. Let Y_{ijg} denote the number of RNA-seq reads mapped to gene g in replicate j of population i, where $1 \leq i \leq K$, $1 \leq j \leq J_i$, and $1 \leq g \leq G$. We will generally refer to "genes" as that which are being tested for differential expression, with the understanding that other features (transcripts, exome expression, etc.) may be tested as well. While the developments below will focus on detection of differential expression between two populations, several of the methods have natural extensions permitting the comparison of more than two populations.

2.2.1 *Simple Approaches*

An early treatment [6] of differential expression for RNA-seq data examined the performance of Fisher's exact test and test statistics derived from generalized linear models used to derive and normalize expression counts. We temporarily extend our notation and let Y_{ijgk} denote the read count for gene g along lane k in sample j of population i. A Poisson generalized linear model for Y_{ijgk} is

$$\log\left(E[Y_{ijgk} \mid d_{ijk}]\right) = \log\left(d_{ijk}\right) + \lambda_{ijg} + \theta_{ijgk}, \tag{2.1}$$

relating the logarithm of the expected read count for gene g in lane k as a linear function of the gene g rate in sample j of population i (λ_{ijg}), an offset term adjusting for variation in lane depths (d_{ijk}), and other unspecified technical effects that vary by gene, lane, and sample (θ_{ijgk}). Tests of differential expression are derived from this model through a likelihood ratio test (LRT) or t-tests of the maximum likelihood estimates (MLE) of the expression parameters λ_{ijg}. The performance of these tests as well as Fisher's exact test in detecting differentially expressed genes was evaluated on a gold standard data set [7]. Two variants of the GLM-derived t-tests—one using the variance of the MLE of λ_{ijg} and one using variance calculated via the delta method—exhibited reduced detection rates. Fisher's exact test and the LRT performed equivalently and exhibited uniformly greater true positive rates (TPR) than the t-tests. The authors noted that screening genes based on read counts improved the performance of both the t-test and LRT. When genes with read counts lower than 20 were filtered out, detection rates for the LRT and t-test greatly improved and were roughly equivalent. The filtering threshold, however, was arbitrarily selected and tested only on the single gold standard data set.

A recently developed R software package, *DEGseq* [51], also employs Fisher's exact test as well as the two versions of the likelihood ratio test noted by Bullard et al. [6]. Additionally, *DEGseq* introduces two tests based on the thresholding of plots of log fold change as a function of mean log expression level (MA plots) commonly used in microarray data, one for analyses based on single samples in each population and one for analyses based on technical replicates. These MA plot-based tests are based upon binomial assumptions for the read counts and a normal approximation of the conditional distribution of the log count ratio between populations (M) and average of log counts (A) between populations.

Another simple two-sample test can be constructed by assuming a Poisson distribution for the read counts. To this end, suppose that the $Y_{ijg} \sim POI(c_{ij}\lambda_{ig})$, where λ_{ig} represents the relative rate parameter for gene g in population i and c_{ij} is a replicate-specific constant. The constant c_{ij} is included to account for variation in read intensity among biological replicates, which can artificially inflate overall library sizes for replicates with high intensity. The within-population and overall read counts are defined as $Y_{i\cdot g} = \sum_j Y_{ijg}$ and $Y_{\cdot\cdot g} = \sum_{i,j} Y_{ijg}$, which follow $POI(\sum_j \lambda_{ig} c_{ij})$ and $POI(\sum_{i,j} \lambda_{ig} c_{ij})$ distributions, respectively, under the Poisson assumption for the individual read counts. The null hypothesis for testing differential expression for gene g is that of equal relative rates of expression, which takes the form $H_{0,g} : \lambda_{1g} = \lambda_{2g}$. Under the null, the conditional distribution of the read count for gene g in population i ($Y_{i\cdot g}$) given the total read count for gene g ($Y_{\cdot\cdot g}$) is binomial with size $Y_{\cdot\cdot g}$ and success probability $\pi_0 = \sum_j c_{ij} / \sum_{i,j} c_{ij}$, which is common to all G genes. The test of $H_{0,g}$ is then any binomial test (e.g. asymptotic, exact, Clopper-Pearson) of $Y_{i\cdot g}$ successes in $Y_{\cdot\cdot g}$ trials against null probability π_0. Adjustment of p-values from the G tests to control the false discovery rate (FDR) can be achieved via the Benjamini–Hochberg [4] correction, or any other suitable method.

Fisher's exact test, GLM-based tests, the MA plot tests of *DEGSeq*, and the conditional binomial test have received little attention, in large part due to the practical infeasibility of assumptions about the marginal or conditional distributions of the read counts. In particular, the Poisson assumption for read count distributions and the binomial assumption for conditional read count distributions have proven infeasible for real data. The variation in replicate samples is typically far greater than that modeled by the Poisson distribution even after adjustment for read intensity. Other tests for differential expression have focused on extensions of the Poisson model for read counts or alternative discrete probability distributions.

2.2.2 Tests Based on Extensions of the Poisson Distribution

Srivastava and Chen [41] proposed a test of differential expression based upon the generalized Poisson distribution. In terms of RNA-seq data, the generalized Poisson model is

$$P(Y_{ijg} = y) = \lambda_{ig}\,(\lambda_{ig} + \theta_{ig}y)^{y-1}\,e^{-\lambda_{ig}-\theta_{ig}y}/y!, \qquad (2.2)$$

where λ_{ig} is the read intensity parameter for gene g in population i and θ_{ig} is a parameter referred to by the authors as the average bias caused by the sample preparation and sequencing process. The authors note that the bias parameter θ_{ig} serves as a shrinkage factor relative to the Poisson distribution, as $E[Y_{ijg}] = \lambda_{ig}(1-\theta_{ig})^{-1}$ and $Var[Y_{ijg}] = \lambda_{ig}(1-\theta_{ig})^{-3}$. To construct a likelihood ratio test based on the generalized Poisson (GP) model, the intensity and sequencing-bias parameters $(\lambda_{ig}, \theta_{ig})$ are first estimated freely. The intensity parameters are then estimated under the restriction $\lambda_{2g} = w\lambda_{1g}$, where w represents a normalization constant accounting for different sequencing depths between populations. In practice, this normalization constant w is chosen as the ratio of the total amount of sequenced RNA in the two populations. This in turn is estimated in each population as a weighted sum over all genes of the unrestricted MLE of the λ_{ig}, with weights defined by gene lengths. The LRT statistic calculated from the restricted ($\lambda_{i2} = w\lambda_{i1}$) and unrestricted likelihoods approximately follows the χ_1^2 distribution. Using a standard data set [37], the GP test was shown to be more sensitive than the Poisson LRT as well as LRT derived from generalized linear models under Poisson, negative binomial, and quasi-Poisson distributions. The generalized Poisson distribution does permit negative intensities λ_{ig} which are not interpretable in a practical sense. The authors note that the GP test fails when data produce a negative estimate of λ_{ig} as likelihoods become zero and maximum likelihood estimation fails, a notable drawback to the applicability of the GP test.

Auer and Doerge [3] introduced the two-stage Poisson model (TSPM), in which gene counts are first screened for overdispersion and different test statistics are calculated for genes determined to be overdispersed/not overdispersed. In the first stage of TSPM, genes are filtered so that those with small cumulative counts over replicates and populations are not given further consideration. The authors arbitrarily select 10 as the cutoff, but do note that the cutoff can be varied based on the number of replicates and overall read intensity. After filtering, a random effects Poisson model is fitted to the gene counts assuming no overdispersion, and an adjusted score statistic is calculated to test the null hypothesis of no overdispersion per gene, $H_{0g} : \phi_g = 1$, where ϕ_g is the overdispersion parameter for gene g, $1 \leq g \leq G$. The quantiles of the adjusted score statistic are compared to theoretical quantiles from the χ_1^2 distribution. Genes for which the adjusted score statistic is greater than the upper bound of the Working-Hotelling simultaneous confidence band for the theoretical χ_1^2 quantiles are classified as overdispersed. All other genes are classified as not overdispersed. In the second stage of the TSPM, genes classified as overdispersed are tested using a likelihood ratio test derived from fitting overdispersed quasi-likelihood models under the null and alternative hypotheses of no differential expression and differential expression, respectively. Genes classified as not overdispersed in stage 1 are tested using a standard likelihood ratio test from a Poisson model. The authors recommend that corrections for FDR control be applied separately within the sets of genes found to be overdispersed and not overdispersed as a power-saving strategy, diverging from common implementation of methods for FDR control. In a simulation study, the authors show that the TSPM exhibited improved power over a negative binomial model and a quasi-likelihood approach in settings where some genes were overdispersed and others not.

Pounds et al. [30] proposed two procedures for identifying differentially expressed genes using both a likelihood ratio test with a Poisson distribution and a quasi-likelihood model which adjusts for overdispersion. Both procedures are based on the adaptive histogram estimator of empirical Bayesian probabilities of no differential expression and of no overdispersion. The Assumption Adequacy Averaging (AAA) procedure uses the law of total probability to estimate the empirical Bayesian probabilities of no differential expression for each gene. These estimates are based on a weighted average of the empirical Bayesian probabilities of no differential expression for the gene using the Poisson and quasi-likelihood models, with weights based on the empirical Bayesian probability of no overdispersion. The Empirical Best Test (EBT) procedure alternatively selects the best test based on the empirical Bayesian probabilities of no overdispersion for each gene. The EBT procedure then applies the adaptive histogram estimator to obtain the empirical Bayesian probabilities based on the set of p-values for the tests for differential expression, using the best test for each individual gene. The authors present simulation studies which evaluate the performance of these two procedures based on various performance metrics and scenarios, and also compare them to the Poisson model, the quasi-likelihood model, TSPM, and negative binomial and Bayesian tests discussed below. The authors also discuss some nice theoretical properties of the two proposed procedures.

2.2.3 Negative Binomial and Quasi-Likelihood Tests

Rather than extend the Poisson distribution [41] or work around overdispersion via screening [3], several authors have proposed differential expression methods based on the negative binomial distribution. The use of the negative binomial distribution was motivated by observation that real RNA-seq data sets typically exhibited greater variability than could be modeled via the Poisson distribution. Robinson and Smyth [33] assume a negative binomial distribution for the read counts for all genes with a common dispersion parameter, (i.e.) $Y_{ijg} \sim NB(\mu_{ijg}, \phi)$, where $\mu_{ijg} = m_{ij}\lambda_{ig}$, m_{ij} the library size for sample j in population i, and λ_{ig} a relative abundance parameter for gene g in population i, which is assumed to be common to the replicate samples within a population. The dispersion parameter ϕ is estimated by maximizing the conditional likelihood given the sum of the counts in each population. This conditional maximization is straightforward when library sizes are assumed to be equal within each population. When this is not the case, a quantile adjustment is applied to the library sizes, adjusting observed counts to the geometric mean of the replicates. These adjusted library sizes are then used in the maximization of the conditional likelihood for the dispersion parameter, a process referred to as quantile adjusted conditional maximum likelihood (qCML) estimation. The null hypothesis for the test of differential expression is the equality of the relative abundance parameters, $H_{0g} : \lambda_{1g} = \lambda_{2g}, g = 1, \ldots, G$. The authors suggest an exact negative binomial test based on the same quantile adjustment used in estimating the dispersion parameter, in which the "pseudosum" of adjusted counts for a given population is conditioned on the pseudosum of counts across populations, and a p-value calculated as the probability of observing counts greater than those observed.

The assumption of a dispersion parameter ϕ common to all genes is frequently biologically implausible. As such, Robinson and Smyth [34] extended their original negative binomial approach and suggested the use of gene specific dispersion parameters ϕ_g, so that the distributional assumption on read counts becomes $Y_{ijg} \sim NB(\mu_{ijg}, \phi_g)$. The authors suggested estimation of the ϕ_g via a weighted likelihood approach, approximating an empirical Bayes procedure. The weighted likelihood for ϕ_g is defined as the weighted sum of the likelihood with gene-specific overdispersion (ϕ_g) and the common likelihood function with common overdispersion (ϕ). The weight parameter α determines the weight assigned to the common likelihood relative to the gene-specific likelihood. In practice, the parameter α is selected based on a Bayesian normal hierarchical model for the gene-specific dispersion parameters ϕ_g. The authors demonstrate that when dispersions do not differ among genes, this approach results in greater values of α, which gives greater weight to the common likelihood in the weighted likelihood equations and thus shrinks the gene-specific ϕ_g to a common value. A simulation study demonstrated that the ability of the exact test [33] to detect differentially expressed genes improved when the empirical Bayes estimation of the gene-specific dispersion parameters was implemented, and was equivalent to the performance of a Wald test from an overdispersed log-linear model when genes were commonly overdispersed. Further, the exact test with empirical Bayes adjustment was better able to control false discovery rates when gene-specific overdispersion was introduced.

Anders and Huber [1] noted that in practice, dispersion often varies with expected read count, and suggested an extended negative binomial model in which the variances of the read count are defined as a nonparametric function of their expectation. Formally, $Y_{ijg} \sim NB(\mu_{ijg}, \phi_\mu)$, where, as in the Robinson–Smyth approach, $\mu_{ijg} = m_{ij}\lambda_{ig}$, and m_{ij} is a library size parameter accounting for the sampling depth in replicate j in population i. The notation ϕ_μ is understood to imply that dispersion varies in an unspecified fashion with the expectation. Under this approach, $Var(Y_{ijg}) = \mu_{ijg}(1 + \phi_\mu \mu_{ijg})$, which departs from the Robinson–Smyth [33] negative binomial approach for which $Var(Y_{ijg}) = \mu_{ijg}(1 + \phi \mu_{ijg})$. As noted above, Robinson and Smyth [34] extended the standard negative binomial approach by estimating gene-specific dispersion parameters via empirical Bayes weighted likelihood estimation, in which gene-specific dispersion parameter estimates were shrunk toward a common dispersion. Anders and Huber [1] employ a gamma-family generalized linear local regression to model the mean-dispersion relationship. The null hypothesis in the test of differential expression, $H_{0g} : \lambda_{1g} = \lambda_{2g}$, is tested via an exact test constructed similarly to the Robinson and Smyth test. The Robinson and Smyth approach adjusts counts by qCML to achieve equal pseudocounts per replication. The equality of the pseudocounts is then used in the construction of exact negative binomial test statistics. In contrast, Anders and Huber approximated the distribution of the sum of negative binomial random variables assuming unequal library sizes. The authors demonstrated their method on four standard data sets, noting that both approaches were effective at controlling false discovery rates, while a Poisson-based χ^2 test failed. The authors note that the overall sensitivities of their test and the common-dispersion version of the Robinson and Smyth test were roughly equivalent. However, the Robinson and Smyth test was less conservative for weakly expressed genes and more conservative for strongly expressed genes, an apparent product of the flexibility of the nonparametric variance estimator in the Anders and Huber test.

Di et al. [9] applied a generalized negative binomial distribution, known as the negative binomial power (NBP) distribution, to test for differential expression. The NBP distribution is a gamma mixture of Poisson distributions; if $Y|Z \sim POI(Z)$ and $Z \sim \Gamma$ with mean μ and variance $\phi\mu^\alpha$, then marginal distribution of Y is NBP. The authors note that by assuming NBP-distributed read counts, $Var(Y_{ijg}) = \mu_{ig}(1 + \phi(\mu_{ig})^{\alpha-1})$. While the dispersion parameter is common to all genes, the mean-variance relationship is given flexibility via the power parameter α. This is in contrast to the Robinson–Smyth and Anders–Huber approaches, in which the dispersion parameters themselves are varied. The NBP tests is constructed as an exact test based on the NBP assumption. The null hypothesis is $\lambda_{1g} = \lambda_{2g}$, where, as in the other negative binomial tests, $\mu_{ijg} = m_{ij}\lambda_{ig}$, and m_{ij} represents the library size for replicate j in population i. Under the assumption of equal library sizes, the authors estimate the relative frequency parameters λ_{ig} as simple averages over replicates weighted by the common library size. The dispersion parameters ϕ and α are estimated via maximum likelihood conditional on the sum of read counts within each population and the estimated λ_{ig}. The exact test is constructed in the fashion of Robinson and Smyth [33], based on the conditional distribution of the read count sums in one population given the read count sum over both populations. To permit varying library sizes, the authors randomly sample read counts to force equal active library sizes, a process they term "thinning". A simulation study was conducted in which read counts were simulated from the Poisson and several variants of the negative binomial distribution, under different assumptions on the functional form of the negative binomial variance. The authors noted that each of the negative binomial tests, including their own, appeared adequate at controlling the false discovery rate under their simulation settings, while the NBP test appeared to be most powerful, particularly under a simulation model in which the log-dispersion parameter was defined as a quadratic function of the log-mean.

Lund et al. [23] noted that while methods based on extensions of the Poisson distribution or the negative binomial distribution provide added flexibility in modeling read count overdispersion, these methods fail to properly account for uncertainty arising from estimating this overdispersion. In general, this results in overly liberal tests of differential expression and skewed p-value distributions when genes are not differentially expressed. The authors suggest modeling read counts via quasi-likelihood (QL) by defining the read count variance to be proportional to a user-defined function—$Var(Y_{ijg}) = \Phi_g V_g(\mu_{ijg})$, where Φ_g is a quasi-dispersion parameter to be estimated from the data, and the variance function $V_g()$ must possess a corresponding quasi-likelihood function satisfying $\partial l(\mu_{ijg}|y_{ijg})/\partial \mu_{ijg} = (y_{ijg} - \mu_{ijg})/V_g(\mu_{ijg})$. Differential expression is tested through a quasi-likelihood ratio test, for which three methods for estimating the QL dispersion parameter Φ_g are discussed. The first is a standard deviance-based estimator. The second is an empirical Bayes estimator, adapted from an approach introduced by Smyth [38], which borrows information across genes in estimating gene specific dispersions by placing a scaled inverse χ^2 prior distribution on the QL dispersion parameter. The third approach accounts for mean-variance relationships in the read counts by fitting a cubic spline of the logarithm of the deviance-based QL dispersion estimator against the log-average counts. A preliminary estimator of the QL dispersion is derived from the spline function, and the aforementioned empirical Bayes approach of Smyth is employed to arrive at the spline-based estimator of the QL dispersion. The authors note that the latter two methods, termed QLShrink and QLSpline, can be characterized as shrinkage estimators—weighted averages of the deviance-based and Bayesian or spline estimators. Lund et al. [23] conducted a simulation study demonstrating the liberal nature of existing Poisson and negative binomial tests, and noted that of the three proposed QL methods, the spline-based method (QLSpline) appeared to perform best.

2.2.4 Other Methods

Parametric approaches to modeling RNA-seq data based on discrete distributions for counts can be adversely affected by model misspecifications and the presence of outliers. A nonparametric approach to the identification of differentially expressed genes in RNA-seq data was proposed by Li and Tibshirani [20]. A modified two-sample Wilcoxon statistic

$$T_g^* = \frac{1}{S}\sum_{s=1}^{S}\left\{\sum_j R_{1jg}(Y'^s) - \frac{J_1(J+1)}{2}\right\} \tag{2.3}$$

based on a multiple Poisson sampling procedure over S iterations is used to examine the differential expression of the gth feature (gene) in two-class data. As in the previous section, Y_{ijg} denotes the RNA-seq count for the gth gene in the jth experimental observation in population i, J_i is the number of observations in the ith population for $i = 1,2$, and we define $J = J_1 + J_2$. The rank statistic $R_{ijg}(Y)$ gives the rank of Y_{ijg} in the set $Y = \{Y_{11g},\ldots,Y_{1J_1g},Y_{21g},\ldots,Y_{2J_2g}\}$. The use of equation (2.3) requires equal sequencing depths, so the authors suggest Poisson sampling of the read counts, replacing original counts Y_{ijg} with random variables Y'_{ijg} resampled from a Poisson distribution with mean $\bar{d}Y_{ijg}/d_{ij}$ for $i = 1,2$ and $j = 1,\ldots,J_i$ where the d_{ij} represent the original sequencing depths for replicate j in population i, and $\bar{d} = \left(\prod_{i,j} d_{ij}\right)^{1/n}$ is the geometric mean of all sequencing depths. This Poisson sampling procedure is repeated S times and the resulting average test statistic is computed to alleviate limitations resulting from the additional randomness introduced by resampling and by tie-breaking procedures for the rank statistic. Since the distribution of the average of the Wilcoxon statistics is complicated, the false discovery rate (FDR) is estimated based on a permutation plug-in estimate. The FDR estimates for this test are more conservative than for parametric alternatives, and were shown to be accurate in simulated data with outliers for which some parametric models greatly underestimated the FDR. In overdispersed data sets with outliers, parametric methods often identified features with a small number of very large count values as differentially expressed, whereas the Li and Tibshirani test tended to identify features where the counts in one class were consistently larger than the counts in the other class.

Tarazona et al. [44] introduced a nonparametric approach designed to be robust against sequencing depth effects. The empirical distributions of fold-change differences $M^g = \log_2(\tilde{Y}_{1\cdot g}/\tilde{Y}_{2\cdot g})$ and absolute expression differences $D^g = |\tilde{Y}_{1\cdot g} - \tilde{Y}_{2\cdot g}|$ are used to estimate the probability that the gth gene is differentially expressed, where the $\tilde{Y}_{i\cdot g}$ represent cumulative read counts normalized to correct for different sequencing depths and adjusted to avoid zero counts. Genes are declared to be differentially expressed if the estimated probability exceeds a specified threshold; 0.8 is used by the authors. The empirical probabilities are computed using technical replicates when available, or through technical replicates simulated from the multinomial distribution when not available. Tarazona et al. [44] examined the effect of sequencing depth on the identification of expressed genes via their nonparametric test, sequencing noise, transcript length, and genes declared to be differentially expressed. A thorough comparison to other novel methods [1, 14, 35] as well as Fisher's Exact Test was made. The authors found that the number of differentially expressed genes as well as the length, fold-change, and expression level of the discovered genes strongly depended on the sequencing depth for the parametric methods, while their nonparametric method was relatively consistent. Further, the authors noted an increase in the number of false positives as the sequencing depth increased for the parametric methods, which was also found by Li and Tibshirani [20], while their nonparametric method was able to control the rate of false discovery.

Recently, a Markov random field approach was proposed by Yang et al. [52]. Consider the set $X = \{x_1, \ldots, x_G\}$ of binary random variables defining indicators x_i which equal 1 if a gene is differentially expressed and equal 0 otherwise. A vector $Y = \{y_1, \ldots, y_G\}$ of observed discretized FDRs are computed for the individual genes using the Anders and Huber [1] test, and the joint probability of X given Y is modeled as proportional to the product $\prod_{(i,j)\in E} \psi_{(i,j)}(x_i, x_j) \prod_{i=1}^{G} \phi_i(x_i)$ where E is the set of vertices with coexpressed gene database (COXPRESdb) correlations $c_{i,j}$ larger than a specified value [27] and $\psi_{(i,j)}(x_i, x_j) = e^{c_{i,j}}$ if $x_i = x_j$ and 1 otherwise. The unary function $\phi_i(x_i)$ are defined to be $P(x_i = 1|y_i)/P(x_i = 0|y_i)$ if $P(x_i = 1|y_i) > P(x_i = 0|y_i)$ and $x_i = 1$, $P(x_i = 0|y_i)/P(x_i = 1|y_i)$ if $P(x_i = 0|y_i) > P(x_i = 1|y_i)$ and $x_i = 0$, and 1 otherwise. It is shown that these clique potential functions of this pairwise Markov random field model are selected so that maximum a posteriori estimation of the differentially expressed genes is reduced to a maximum flow problem discussed in Kolmogorov and Zabih [15]. By including information about the dependence of gene expressions, Yang et al. [52] show through simulation studies and real data examples that this method exhibited improved sensitivity without a loss of precision. Through the inclusion of additional coexpression information, this method additionally helped remove bias against detection of genes with low read counts.

Zhou et al. [55] proposed a beta-binomial model where the probabilities that a single read in each sample is mapped to gene g is a vector $\theta_{g\cdot}$ of beta random variables for which the logits of the expected values are modeled linearly by XB_g. The design matrix X is flexible and can include columns indicating group assignments for experimental conditions as well as any other desired covariates. The vector of regression coefficients B_g corresponds to the effects of the variables in the columns of X for the gth gene. Two approaches are considered—(1) a free model where the likelihood function is directly maximized, and (2) a shrinkage approach with a constrained model where the overdispersion ϕ_g of the beta distribution is modeled as a polynomial function of the mean. The authors additionally suggest an automatic correction for outliers. While other penalized approaches and the constrained model offer some advantages for very small sample sizes, simulation studies and a real data example support direct parametric modeling with the free model.

2.2.5 Bayesian and Empirical Bayes Approaches

A number of fully Bayesian and empirical Bayes methods have been developed for analyzing differential expression. Typically inferences on differential expression span across multiple genes and conditions, each characterized by its own set of parameters. It is frequently natural to express these parameters as a mixture over two latent states. The states may imply the presence or absence of differential effects and hence define the primary objects of inference. In Bayesian approaches, such gene-specific parameters are assigned prior distributions, which are in turn indexed by a common hyper-parameter. The model is then completed by assuming specific sampling models for normalized count data conditional on these parameters. As in frequentist settings, these distributions are chosen to allow for overdispersion, which poses a critical challenge in analyzing RNA-seq data. All Bayesian models typically follow this common hierarchy.

However, empirical Bayes and fully Bayesian methods differ sharply in their approaches to inference and shrinkage. The former estimates the relevant hyper-parameters directly from the data and through this combined estimate, pools information among genes. In contrast, fully Bayesian methods borrow strength by fixing the hyper-parameter at the highest level of the Bayesian hierarchy and sharing the parameters themselves across different levels. For example, one could achieve some shrinkage by simply assuming a common probability for the presence of indicators. More generally, the extent and nature of shrinkage vary with the desired level. Shrinkage is highly relevant in differential expression settings, where we have multiple genes but very few replicates per gene. In the following discussion, we shall review some commonly used empirical Bayes approaches introduced by van de Wiel et al. [49], Leng et al. [19], and Hardcastle and Kelly [14], and conclude by describing a fully-Bayesian method [17].

The sampling model considered by van de Wiel et al. [49] is a zero-inflated negative binomial regression: $Y_{ijg} \sim ZI - NB(\mu_{ijg}, \phi_g, w_{0g})$ and $\mu_{ijg} = h^{-1}(\beta_{g0} + \sum_k \beta_{gk} x_{ijk})$, where g indexes the genes, h is a link function, ϕ_g the negative binomial overdispersion parameter, and w_{0g} a zero-inflation parameter. The zero-inflation parameter is defined to be a probability mixing the negative binomial distribution $NB(\mu_{ijg}, \phi_g)$ with probability $1 - w_{0g}$ and a point mass at zero with probability w_{0g}. The regression coefficients are permitted to have their own normal random effects. The covariates typically correspond to different conditions or populations corresponding to possible differential expression. In assigning priors, van de Wiel et al. examined several different choices. Both flat and mixture priors were considered for β_{gl}, while the prior for $\log(\phi_g)$ was assumed to be a mixture. Each parameter family had its own associated set of hyper-parameters. A conventional method of estimating hyper-parameters in an empirical Bayes framework is by maximizing the marginal likelihood. As an alternative, van de Wiel et al. [49] utilize the fact that the likelihood estimator α approximately satisfies $\pi_\alpha(\cdot) = (1/G)\sum_{g=1}^{G} \pi_\alpha(\cdot|\mathbf{Y}_g)$, where G is the number of genes and $\mathbf{Y}_g$ the vector of read counts for gene g. This approximation can be seen by setting the derivative of the log-marginal likelihood to 0. Since the model includes multiple parameter families (e.g overdispersion, regression coefficients), this generic procedure was extended to an iterative algorithm which conditioned on a given set of parameters at each step. Shrinkage of overdispersion is treated separately. Since the overdispersion and mean are intertwined in NB models, a univariate shrinkage of the former may not work. The authors suggest shrinking the individual ϕ_g through a prior that regresses them against the gene counts. Specifically, they assume $\phi_g = h(c_g) + \varepsilon_g$ where c_g is the log of the gene count and the function h is left unspecified and estimated via LOESS. Initial values required by this iterative algorithm are fixed at the posterior mean estimates of the ϕ_g obtained under a flat prior. Having obtained these estimates, the shrinkage prior was assigned as $\phi_g - \hat{\phi}_g \sim N(0, \sigma^2)$ where σ^2 was also estimated from the iterative procedure. The authors also suggest the importance of the zero-inflation component in this context, describing it as a potential reason for overdispersion. Indeed, including factors accounting for zero inflation was shown to effectively account for the residual trends of ϕ_g in simulation settings. Finally, posterior estimates of the specific contrasts involving the regression coefficients are computed, and then Bayesian and local false discovery rates are applied to these estimates to infer differential expression.

The approach of Hardcastle and Kelly [14] deals directly with the latent indicators of differential expression. In the most general version of this approach, a broad space of models is encompassed, each corresponding to a hypothesis to be tested. For simplicity of exposition, we consider here just two exclusive models: (1) no differential expression and (2) differential expression. Each gene in the data set then has an associated latent indicator identifying whether it is differentially expressed. A key difference with the method of van de Wiel et al. [49] is that Hardcastle and Kelly [14] do not explicitly estimate a hyper-parameter. Instead, their method estimates the entire prior distribution through resampling and quasi-likelihood. The pooling of prior probabilities for the different indicators is done through iterative estimation. The sampling model in this approach is negative binomial, with the probabilities weighted by library sizes. Posterior probabilities are obtained as the final step.

Leng et al. [19] introduced an empirical Bayes method that not only models differential expression among genes but also among isoforms of the same gene. In this setup, let Y_{ijgl} denote the read counts in isoform l of gene g in sample j of population i. This count is assumed to follow a negative binomial distribution, where the parameters of the negative binomial can vary across genes, isoforms, and biological conditions. The prior distribution of the negative binomial mean-variance ratio is assumed to be $Beta(\alpha, \beta^{I_g})$, where I_g denotes a grouping of genes. The hyper-parameter α is shared across all isoforms and genes, while β varies by gene group (I_g). These gene groups can be defined freely to provide flexibility to the approach; for example, genes can be grouped by the number of their isoforms. As in other differential expression approaches, the full model was expressed as mixture over two latent states. In the EB step, the four global hyper-parameters (each pair corresponding to a state) are estimated via the EM algorithm. Conditioned on these estimates, the state-specific posterior probabilities are calculated.

Lee et al. [17] proposed a fully Bayesian hierarchical model that diverged from existing approaches in that the cumulative read count of gene g at each genomic position l in population i is explicitly modeled as $Y_{i \cdot gl} \sim Bin(Y_{\cdot\cdot gl}; p_{gl})$, independently across the positions. The binomial probability p_{gl} is modeled by adding another layer in the hierarchy and assuming $p_{gl} \sim (1 - w_{gl})Beta(\alpha_g; \beta_g) + w_{gl}Beta(.5, .5)$. In this formulation, w_{gl} expresses the outlier effect, while the α_g and β_g are gene specific and centered around a mixture prior. This mixture is over three possible indicators encoding for high, low, and non-differential expression. The parameters corresponding to each indicator are assigned their own Gaussian priors. These priors allow for the usual inter-gene pooling as in previous hierarchical setups. However, this method implements full posterior inference using MCMC methods. Final results are obtained by direct posterior sampling of the latent indicators. This approach offers a number of advantages. First, prior normalization of the mapped read counts is not required. Rather, normalization and differential calling are done simultaneously via the model. Second, this approach effectively downweights outliers at the position level through the w_{gl}. The authors showed that this step played a significant role in increasing the specificity and sensitivity of differential expression calls. Third, the pooling across positions increased the effective sample size per gene per sample. Importantly, this model uses each position in the gene as a data point, thus we have multiple observations per gene in the absence of replicates. This can be relevant for many cost-prohibitive RNA-seq studies where replicates are difficult to obtain.

2.3 Software for Differential Expression in RNA-seq Data

Several of the novel methods for detecting differential expression in RNA-seq data have associated software packages, most of which have been released via the open source R [31] and Bioconductor [12] software environments. Below we provide a brief summary of R and Bioconductor implementations of the different techniques for detecting differential expression in RNA-seq data. We do not discuss other methods such as Fisher's exact test, two sample t-tests, GLM-derived tests, and methods for microarray data analysis applied to RNA-seq data. The package names we use in the discussion below can be used to load the R and Bioconductor libraries for the associated methods, via the commands `library(`*pkgname*`)` for R packages (after local installation) and `biocLite(`*pkgname*`)` for Bioconductor packages.

The general convention for formatting RNA-seq data for use in frequentist analyses is as a $G \times J$ matrix for G genes measured in J samples, with the columns typically arranged so that the first few columns are read counts of replicates from population 1 and the remaining columns read counts of replicates from population 2. Most functions for detecting differential expression accept two arguments at a minimum—the matrix of read counts and a vector defining a population identifier for the columns (e.g. 1 or 2).

The R package *GPseq* [42] implements the generalized Poisson test via the function `estimate_differential_expression`. The interface for this function differs somewhat from other implementations, in that the function accepts an annotated read count matrix as well as exon and gene annotation matrices for unraveling the annotated read count matrix. The *GPseq* package also includes functions to calculate chi-square goodness-of-fit tests for the generalized Poisson distribution and workhorse functions for the generalized Poisson likelihood and likelihood ratios, and functions for permutation tests of the generalized Poisson test statistic. The other Poisson test, based on the two-stage Poisson model [3], is not available through an R library, rather as an R function downloadable from the authors' website (http://www.stat.purdue.edu/~doerge/software/TSPM.R). The function reads in a matrix of read counts and indicators defining populations for the columns of the matrix, and returns adjusted and unadjusted p-values as well as vectors of indicators defining genes found to be overdispersed. The R functions used to implement the procedures introduced by Pounds et al. [30] utilize some of the code for TSPM and are available on the personal website (http://www.stjuderesearch.org/site/depts/biostats/software/ebshtpasced).

DEGseq is a Bioconductor package implementing Fisher's exact test, two likelihood ratio tests, and tests based on MA plots [51], all through the function `DEGseq`. This function also does not follow the convention of accepting matrices of read counts. Rather, `DEGseq` accepts mapping files for samples from two populations as well as arguments specifying characteristics of the RNA-seq data files. Additional arguments specify the differential expression test to be conducted and customize the characteristics of said tests, such as p- and q-value thresholds and thresholds for tests derived from MA plots.

Libraries for the negative binomial tests [1, 9, 33, 34] are available in R and Bioconductor. The Robinson–Smyth test can be found in the Bionconductor package *edgeR* [35]. To obtain the Robinson–Smyth test, users of *edgeR* format a matrix of counts into a package-specific object that is then fed to the function `estimateCommonDisp`, which estimates the common dispersion parameters and outputs a matrix of pseudocounts and pseudo-library sizes. The object created by this function is fed to the function `exactTest` which calculates p-values from the exact negative binomial tests based on the quantile-adjusted counts. Additional functions in the *edgeR* library provide tests based on the assumption of gene-specific dispersion parameters, utility functions for RNA-seq data, workhorse functions for estimation and testing, and additional functions for the analysis of RNA-seq data. We refer the reader to this book's chapter on the *edgeR* package for further details. The test of Anders and Huber [1] is available via the Bioconductor package *DESeq*. To test differential expression using *DESeq*, users must create a package-specific object containing the read count matrix via the function `newCountDataSet`, normalize the counts using the function `estimateSizeFactors`, estimate overdispersion using `estimateDispersions`, and then conduct the negative binomial test using `nbinomTest`. *DESeq* includes additional functions for conducting the negative binomial test directly on count matrices, as well as functions for graphics (e.g. MA plots), variance stabilizing transformations, and negative binomial GLM tests per gene. The R package *NBPSeq* [10] implements the NBP test [9]. The function `nbp.test` accepts a matrix of read counts and vector of indicators for group membership. Normalization of read counts by the random resampling process ("thinning" as termed by the authors) is accomplished internally within `nbp.test`. Additional functions in *NBPSeq* estimate the negative binomial dispersion parameters (`estimate.disp`) and normalization factors (`estimate.norm.factors`), perform exact negative binomial tests (`exact.nb.test`) and GLM-based tests (`nb.glm.test`), and perform utility functions on package-specific objects. Most of these functions are workhorses for `nbp.test`. The quasi-likelihood approach of Lund et al. [23] is implemented in the R package *QuasiSeq*, which requires the *edgeR* library. The function `QL.fit` accepts a matrix of read counts and a list containing design matrices for full and reduced models. Additional options permit customization of the QL model and estimation of dispersion parameters. The list object returned by `QL.fit` can be fed to the function `QL.results`, which produces lists of p-values and q-values.

The nonparametric approach of Li and Tibshirani [20] is implemented by the R package *samr* [46]. The function `SAMseq` is specifically designed for the analysis of count data, whereas `samr` and other functions in the package are designed for microarray data analysis. `SAMseq` permits flexibility in the type of analysis to be conducted via the `resp.type` argument, which can be used to request paired and unpaired two-class comparisons, comparison of three or more classes, analysis of association with a quantitative predictor, and analysis of a survival outcome. Other functions in *samr* can be used to estimate sequencing depths and normalize read counts. Registered academic users can also download a supplementary Addin for Microsoft Excel from the developers web page (http://www-stat.stanford.edu/~tibs/SAM/). The nonparametric test of Tarazona et al. [44] is implemented in the Bioconductor package *NOISeq* [45] using the functions `noiseq` and `noiseqbio`. These functions, which operate on package-specific objects containing the read counts, include options for handling data with technical and biological replicates, as well as data with no replicates. The function outputs a list of differentially expressed genes based on the desired threshold probability. This package also provides several exploratory plots for biotype detection, sequencing depth and expression quantification, and sequencing bias that are useful for detecting potential problems that need to be corrected by normalization procedures and several plots which summarize the differentially expressed genes identified by the algorithm.

The Markov random field approach of Yang et al. [52], termed *MRFSeq*, is implemented as C++ code and distributed from the author's website (http://www.cs.ucr.edu/~yyang027/mrfseq.htm). *MRFSeq* depends upon the coexpressed gene database COXPRESdb [27], available at http://coxpresdb.jp, and *DESeq*, the Bioconductor package for the negative binomial test of Anders and Huber [1]. The beta-binomial test of Zhou et al. [55] is available in the R package *BBSeq*, available only from the author's webpage (http://www.bios.unc.edu/research/genomic_software/BBSeq/). The separate functions `free.estimate` and `constrained.estimate` compute parameter estimates and estimate p-values based on the corresponding likelihood and shrinkage approaches discussed in the previous section. An additional utility function (`outlier.flag`) is included to identify potential outliers among the read counts.

Among the Bayesian methods, the multiple shrinkage priors approach of van de Wiel et al. [49] is implemented in the R package *ShrinkBayes*, available from the primary author's webpage (http://www.few.vu.nl/~mavdwiel/ShrinkBayes.html). The function `ShrinkSeq` is used to fit the multiple shrinkage priors model based on specification of a model formula, the model parameters to be shrunk, whether or not a mixture prior for overdispersion is to be implemented, and the family of distributions used to fit the data (zero-inflated negative binomial being the default). Since *ShrinkSeq* is computationally intensive, parallel computing is implemented, and the user is permitted to specify the number of processors to be used in parallel. Formal documentation of the functions comprising *ShrinkSeq* are unavailable, but thorough examples of code usage are provided in the package documentation. Use of the *ShrinkBayes* package requires the installation of *inla* [36], an R package for Bayesian modeling via integrated nested Laplace approximation.

The Bioconductor package *baySeq* [13] implements the empirical Bayes method of Hardcastle and Kelly [14]. The functions `getPriors` and `getLikelihood` are the two most important functions in this package. The first constructs the empirical priors by bootstrapping, while the second yields posterior probabilities. *baySeq* offers a fair amount of choice in analysis, e.g., in the number of bootstrap samples and in techniques for re-estimating priors. *baySeq* can be run in parallel mode, via the independent R package *snow* [47] for networking workstations. *EBSeq* [18] is the Bioconductor package implementing the method of Leng et al. [19]. The `EBtest` function in this package uses the EM algorithm to obtain posterior probabilities for the detection of two-condition differential expression. The function `EBMultitest` extends this utility for multiple conditions. The underlying model in *EBSeq* is assumed to be negative binomial. *EBSeq* offers the users a range of simulated datasets upon which to test the algorithm. The R package *BMDE* implements the fully Bayesian method of Lee et al. [17], and is available for download at http://health.bsd.uchicago.edu/yji/soft.html. Since *BMDE* uses full posterior inference, it is able to provide the entire set of posterior samples, allowing the users to choose their own posterior summaries. Unlike the empirical Bayes algorithms mentioned above, *BMDE* relies on certain hyper-parameter settings. The users are provided the flexibility to choose them and examine the sensitivity of results based on selections for the hyper-parameters.

2.4 Comparison of Methods for Detecting Differential Expression

In most of the source works for the methods detailed in Sect. 2.2, simulation studies and/or analyses of live RNA-seq data sets were conducted to evaluate the detection capabilities of the proposed methods and to make comparisons to existing methods.

These simulation studies were largely designed to highlight special features of the proposed methods and demonstrate the superiority of these methods under specific conditions. More general comparative simulation studies have been conducted to compare these methods; we discuss three such studies below. We note that these comparative studies generally implemented default settings that were defined in the software packages corresponding to each method, that these default settings can change over time with new package version releases, and that the conclusions reached by each the comparative studies may be version specific.

Bullard et al. [6] compared the performance of Fisher's exact test and three tests derived from the generalized linear model in (2.1)—the likelihood ratio test (LRT), and t-tests based on the GLM-derived variance and the delta method variance. The authors compared RNA-seq data from two biological samples from the MicroArray Quality Control (MAQC) Project [37]. The detection capability of these four tests were compared using the results of analysis of 375 genes by qRT-PCR gold standard for differential expression. The authors found that the LRT and Fisher's exact test performed comparably in detecting differential expression, while the two t-tests were also comparable but exhibited substantially reduced detection rates relative to the LRT and Fisher tests. A notable contribution of this paper was the impact of filtering genes with low read counts on the detection of differential expression. After removing 186 genes with read counts less than 20 and repeating the analysis of the MAQC data, the authors noted that the detection rate of both the LRT and the t-test with GLM-based variance improved greatly and, in particular, the detection rate of the t-test was roughly equivalent to that of the LRT.

Kvam et al. [16] conducted a comparative study of the two-stage Poisson model [3] and three tests based on the negative binomial distribution—*edgeR* [35], *DESeq* [1], and *baySeq* [14]. The authors simulated data under four models—Poisson read counts with half of the genes simulated from an overdispersed Poisson model, following a simulation conducted by Auer and Doerge [3] to evaluate the TSPM, counts generated from the Poisson or negative binomial distribution with mean and dispersion parameters estimated from a known plant data set [21], and counts generated from a data set of human lymphoblastoid cell lines [29] with randomly-induced differential expression. The authors noted that the three negative binomial tests *edgeR*, *DESeq*, and *baySeq* performed similarly under each simulation setting. The performance of the TSPM test was notably affected by the number of replicates simulated, as detection capability was severely reduced for two replicates per population. Further, the TSPM notably underperformed relative to the negative binomial tests when all counts were simulated from the negative binomial distribution. An analysis of a plant data set [21] showed that *edgeR* and *DESeq* largely identified the same genes as differentially expressed, while most of the genes identified by the TSPM were not declared differentially expressed by *edgeR* or *DESeq*.

A recently published study by Soneson and Delorenzi [40] comprehensively examined via simulation the performance of nine tests—*DESeq*, *edgeR*, *NBPSeq*, *TSPM*, *baySeq*, *EBSeq*, *NOISeq*, *SAMSeq*, *ShrinkSeq*—and two tests based on the empirical Bayes linear model *limma* [38, 39] after variance-stabilizing or logarithmic transformation. Using the negative binomial distribution with common dispersion between the two populations as a foundation for simulating RNA-seq data, the authors compared the performance of the 11 tests and noted the impact on performance of mixing in Poisson-simulated counts, adding high-count outlier genes, varying the number of differentially expressed genes, the direction of differential expression (up- or down-regulated), sample size, and altering the dispersion parameter in one of the populations.

Under these multiple simulation scenarios, the authors compared the methods in terms of true positive rates (TPR), ranking of differential expression, type I error control, and false discovery rate control. We paraphrase the general characteristics of each test here, and refer the reader to the source work [40] for more detailed explanations. Among the negative binomial tests, *DESeq* was generally conservative, exhibiting low detection capability but strong FDR control, even in the presence of outliers except for when sample sizes were small (two per population). Both *edgeR* and *NBPSeq* were liberal, particularly when outliers were present. *edgeR* exhibited greater sensitivity than *NBPSeq* in most settings, and became less liberal under large sample sizes while *NBPSeq* was liberal for all sample sizes. Both were poor at controlling the FDR, and *NBPSeq* often ranked truly non-differentially expressed genes as the most differentially expressed. The hallmark characteristic of the *TSPM*, which relies on asymptotic theory for its test of differential expression, was its sample-size dependence. For small samples the *TSPM* was poor at controlling FDR and ranking differentially expressed genes, although performance improved greatly with minimal increases in sample size and outliers were generally non-problematic. The *TSPM* performed poorly in terms of differential expression rankings when all genes were overdispersed, but this was improved when non-overdispersed genes were mixed in.

When differential expression occurred in a uniform direction (e.g. all genes up-regulated in one population), *baySeq* exhibited highly variable performance for each metric (TPR, FDR control, type I error control). This effect was mitigated when differential expression was mixed. *baySeq* was largely conservative with good FDR control, except when sample sizes were low. *EBSeq* provided a liberal test with good sensitivity and poor FDR control, and was particularly resistant to the effect of outliers. Control of the FDR for *NOISeq* was unevaluated due to lack of clarity in how thresholds could be set, but it was noted that *NOISeq* was particularly adept at ranking genes when populations were differentially overdispersed. *SAMSeq* was non-sensitive at low sample sizes, but power rapidly increased with sample size, and *SAMSeq* was particularly resistant to the presence of outliers. *ShrinkSeq* exhibited high sensitivity and poor FDR control at default settings, but featured a user-controlled fold-change thresholding procedure that could conceivably offer stronger FDR control.

limma with transformation exhibited strong control of type I error that was resistant to outliers. Control of FDR was also strong and resistant to outliers, except under settings in which a large proportion of genes were uniformly upregulated in one population and when populations were differentially overdispersed. The *limma* method was relatively conservative, particularly under low sample sizes, where no genes were declared differentially expressed when only two samples per population were available.

In addition to the simulation study, Soneson and Delorenzi [40] analyzed RNA-seq data from two mouse strains [5] to compare methods. *ShrinkSeq* and *SAMSeq* called the most genes as differentially expressed, while *baySeq*, *DESeq*, and *EBSeq* were particularly conservative. Among the negative binomial methods and *TSPM*, all genes called as differentially-expressed by *DESeq* were also called by one or more of the other methods. *NBPSeq*, *TSPM*, and *edgeR* called a substantial number of the same genes, but also called a non-trivial number of distinct genes not called by the other methods. Genes called by *baySeq* were a subset of those called by the log-transformed *limma* method, and the genes called by the variance-stabilized *limma* method contained most genes called by log-transformed *limma*. Genes called by *EBSeq* were effectively a subset of the variance-stabilized *limma* method, although *EBSeq* called a substantial number of unique genes. A resampled analysis of one of the mouse-strains, under which no genes would be expected to be differentially expressed, showed the tendency of *TSPM* to be too liberal, as the average number of genes called differentially expressed by *TSPM* was far greater than the other methods.

2.5 Discussion

The challenge in analyzing RNA-seq data, particularly in the detection of differential expression, has three primary sources. The first is the inherent problem with the technology; the second is the laboratory or experimental errors causing technical variation across samples. However, these sources of error are usually present in any relatively new technology. The third and the most important challenge is that current costs of producing RNA-seq data are prohibitive to the generation of many biological replicates, which poses a problem for statistical data analysis. Very small sample sizes for a typical RNA-seq study prevent the appropriate use of asymptotic statistical inference commonly employed for count data analysis. Frequently, due to these reasons, estimated false discovery rates (FDR) are not less than the selected FDR cut-off. Thus, asymptotic tests are adversely affected by small sample size in the analysis of RNA-seq data.

Small sample sizes (two samples per condition) imposed problems also for the methods that were indeed able to find differentially expressed genes, thereby leading to false discovery rates sometimes widely exceeding the desired threshold implied by the FDR cut-off. For the parametric methods, this may also be due to inaccuracies in the estimation of mean and dispersion parameters. In the previous section, we noted that TSPM stood out as the method being most affected by sample size, potentially due to the use of asymptotic statistics. Currently, RNA-seq experiments are often too expensive to allow extensive replication in scientific experiments. Hence, we strongly suggest that the differentially expressed genes found between small sample studies be interpreted with caution and that the true FDR may be several times higher than the selected FDR threshold. The negative binomial methods [1, 9, 33, 34] tests are based on similar principles and work relatively well. However, due to differences in the estimation of the overdispersion parameters, lists of differentially expressed genes produced by these methods at the same FDR level were different.

In Sect. 2.4, we summarized the results of a detailed comparison of many existing methods and the resulting guidelines to users about the suitability of one method over others for a given data type. We advocate that those testing differential expression in RNA-seq data be cognizant of the characteristics of their data, particularly with regard to the simulation settings evaluated by Soneson and Delorenzi [40]—sample size, direction of regulation, presence of outliers, degree and variability of overdispersion. Awareness of these characteristics will permit a more informed choice of test for differential expression. We also advocate that analysts not rely on a single test of differential expression nor on a single setting for a given test, and rather perform several tests or several settings of a given test based on their suitability for the data set at hand and compare lists of differentially expressed genes. We have also provided brief descriptions of existing software and their respective functionality in analysis of RNA-seq data. We hope that this review will provide a comprehensive description of the current status of the analysis of RNA-seq data.

References

[1] Anders, S., Huber, W.: Differential expression analysis for sequence count data. Genome Biol. **11**, R106 (2010)

[2] Anders, S., McCarthy, D.J., Chen, Y., Okoniewski, M., Smyth, G.K., Huber, W., Robinson, M.D.: Count-based differential expression analysis of RNA sequencing data using R and bioconductor. Nat. Protocol. **8**, 1765–1786 (2013)

[3] Auer, P.L., Doerge, R.W.: A two-stage poisson model for testing RNA-seq data. Stat. Appl. Genet. Mol. Biol. **10(1)**, 26 (2011)

[4] Benjamini, Y., Hochberg, Y.: Controlling the false discovery rate: a practical and powerful approach to multiple testing. J. Roy. Stat. Soc. Ser. B **57**, 289–300 (1995)

[5] Bottomly, D., Walter, N.A., Hunter, J.E., Darakjian, P., Kawane, S., Buck, K.J., Searles, R.P., Mooney, M., McWeeney, S.K., Hitzermann, R.: Evaluating gene expression in C57BL/6J and DBA/2J mouse striatum using RNA-seq and microarrays. PLoS One **6**(3), e17820 (2011)

[6] Bullard, J.H., Purdom, E., Hansen, K.D., Dudoit, S.: Evaluation of statistical methods for normalization and differential expression in mRNA-seq experiments. BMC Bioinform. **11**, 94 (2010)

[7] Canales, R.D., Luo, Y., Willey, J.C., Austermiller, B., Barbacioru, C.C., Boysen, C., Hunkapiller, K., Jensen, R.V., Knight, C.R., Lee, K.Y., et al.: Evaluation of DNA microarray results with quantitative gene expression platforms. Nat. Biotech. **24**(9), 1115–1122 (2006)

[8] Cloonan, N., Forrest, A.R.R., Kolle, G., Gardiner, B.B.A., Faulkner, G.J., Brown, M.K., Taylor, D.F., Steptoe, A.L., Wani, S., Bethel, G., et al.: Stem cell transcriptome profiling via massive-scale mRNA sequencing. Nat. Meth. **5**, 613–619 (2008)

[9] Di, Y., Schafer, D.W., Cumbie, J.S., Chang, J.H.: The NBP negative binomial model for assessing differential gene expression from RNA-seq. Stat. Appl. Genet. Mol. Biol. **10**(1), 24 (2011)
[10] Di, Y., Schafer, D.W, Cumbie, J.S., Chang, J.H. NBPSeq: negative binomial models for RNA-sequencing data. R Package Version 0.1.8. (2012). http://CRAN.R-project.org/package=NBPSeq
[11] Dillies, M.A., Rau, A., Aubert, J., Hennequet-Antier, C., Jeanmougin, M., Servant, N., Keime, C., Marot, G., Castel, D., Estelle, J., et al.: A comprehensive evaluation of normalization methods for Illumina high-throughput RNA sequencing data analysis. Brief Bioinform. (2012). doi:10.1093/bib/bbs046
[12] Gentleman R., Carey V.J., Bates D.M., Bolstad B., Dettling M., Dudoit S., Ellis B., Gautier L., Ge Y., Others: Bioconductor: open software development for computational biology and bioinformatics. Genome Biol. **5**, R80 (2004)
[13] Hardcastle, T.J.: baySeq: empirical Bayesian analysis of patterns of differential expression in count data. R Package Version 1.16.0. (2012)
[14] Hardcastle, T.J., Kelly, K.A.: baySeq: empirical Bayesian methods for identifying differential expression in sequence count data. BMC Bioinform. **11**, 422 (2010)
[15] Kolmogorov, V., Zabih, R.: What energy functions can be minimized via graph cuts? IEEE Trans. Pattern Anal. Mach. Intell. **26**, 147–159 (2004)
[16] Kvam, V.M., Liu, P., Si, Y.: A comparison of statistical methods for detecting differentially expressed genes from RNA-seq data. Am. J. Botany **99**(2), 248–256 (2012)
[17] Lee, J., Ji, Y., Liang, S., Cai, G., Muller, P.: On differential gene expression using RNA-seq data. Cancer Inform. **10**, 205–215 (2011)
[18] Leng, N.: EBSeq: an R package for gene and isoform differential expression analysis of RNA-seq data. R Package Version 1.2.0 (2013)
[19] Leng, N., Dawson, J., Thomson, J., Ruotti, V., Rissman, A., Smits, B., Haag, J., Gould, M., Stewart, R., Kendziorski, C.: EBSeq: an empirical bayes hierarchical model for inference in RNA-seq experiments. Technical Report 226. Department of Biostatistics and Medical Informatics, University of Wisconsin, Madison (2012). http://www.biostat.wisc.edu/Tech-Reports/pdf/tr_226.pdf
[20] Li, J., Tibshirani, R.: Finding consistent patterns: a nonparametric approach for identifying differential expression in RNA-seq data. Stat. Meth. Med. Res. **22**(5), 519–536 (2011)
[21] Li, P., Ponnala, L., Gandotra, N., Wang, L., Si, Y. Tausta, S.L., Kebrom, T.H., et al. The developmental dynamics of the maize leaf transcriptome. Nat. Genet. **42**, 1060–1067 (2010)
[22] Lister, R., O'Malley, R.C., Tonti-Filippini, J., Gregory, B.D., Berry, C.C., Millar, A.H., Ecker, J.R.: Highly integrated single-base resolution maps of the epigenome in Arabidopsis. Cell **133**, 523–536 (2008)
[23] Lund, S.P., Nettleton, D., McCarthy, D.J., Smyth, G.K.: Detecting differential expression in RNA-sequence data using quasi-likelihood with shrunken dispersion estimates. Stat. Appl. Genet. Mol. Biol. **11**(5), Article 8 (2012)
[24] Marioni, J.C., Mason, C.E., Mane, S.M., Stephens, M., Gilad, Y.: RNA-seq: an assessment of technical reproducibility and comparison with gene expression arrays. Genome Res. **18**, 1509–1517 (2008)
[25] Mortazavi, A., Williams, B.A., McCue, K., Schaeffer, L., Wold, B.: Mapping and quantifying mammalian transcriptomes by RNA-seq. Nat. Meth. **5**, 621–628 (2008)
[26] Nagalakshmi, U., Wang, Z., Waern, K., Shou, C., Raha, D., Gerstein, M., Snyder, M.: The transcriptional language of the yeast genome defined by RNA sequencing. Science **320**(5881), 1344–1349 (2008)
[27] Obayashi, T., Kinoshuta, K.: Coxpresdb: a database to compare gene coexpression in seven model animals. Nucleic Acids Res. **39**, D1016–D1022 (2011)
[28] Pan, Q., Shai, O., Lee, L.J., Frey, B.J., Blencowe, B.J.: Deep surveying of alternative splicing complexity in the human transcriptome by high-throughput sequencing. Nat. Genet. **40**, 1413–1415 (2008)

[29] Pickrell, J.K. , Marioni, J.C., Pai, A.A., Degner, J.F., Engelhardt B.E., Nkadori, E., Veyrieras, J.B., et al.: Understanding mechanisms underlying human gene expression variation with RNA sequencing. Nature **464**, 768–772 (2010)

[30] Pounds, S.B., Gao, C.L., Zhang, H.: Empirical Bayesian selection of hypothesis testing procedures for analysis of sequence count expression data. Stat. Appl. Genet. Mol. Biol. **11**(5), Article 7 (2012)

[31] R Core Team: R: A Language and Environment for Statistical Computing. R Foundation for Statistical Computing, Vienna, Austria (2013). http://www.R-project.org/

[32] Robinson, M.D., Oshlack, A.: A scaling normalization method for differential expression analysis of RNA-seq data. Genome Biol. **11**, R25 (2010)

[33] Robinson, M.D., Smyth, G.K.: Moderated statistical tests for assessing differences in tag abundance. Bioinformatics **23**, 2881–2887 (2007)

[34] Robinson, M.D., Smyth, G.K.: Small-sample estimation of negative binomial dispersion, with applications to SAGE data. Biostatistics **9**, 321–332 (2008)

[35] Robinson, M.D., McCarthy, D.J., Smyth, G.K.: edgeR: a bioconductor package for differential expression analysis of digital gene expression data. Bioinformatics **26**, 139–140 (2010)

[36] Rue, H., Martino, S., Chopin, N.: Approximate Bayesian inference for latent Gaussian models using integrated nested Laplace approximations (with discussion). JRSSB **71**(2), 319–392 (2009)

[37] Shi, L., Reid, L.H., Jones, W.D., Shippy, R., Warrington, J.A., Baker, S.C., Collins, P.J., de Longueville, F., Kawasaki, E.S., Lee, K.Y., et al.: The microarray quality control (MAQC) project shows inter- and intraplatform reproducibility of gene expression measurements. Nat. Biotech. **24**, 1151–1161 (2006)

[38] Smyth, G.K.: Linear models and empirical bayes methods for assessing differential expression in microarray experiments. Stat. Appl. Genet. Mol. Biol. **3**, Article 3 (2004)

[39] Smyth, G.K.: Limma: linear models for microarray data. In: Gentleman, R., Carey, V., Dudoit, S., Irizarry, R., Huber, W. (eds.) Bioinformatics and Computational Biology Solutions Using R and Bioconductor, pp. 397–420. Springer, New York (2005)

[40] Soneson, C., Delorenzi, M.: A comparison of methods for differential expression analysis of RNA-seq data. BMC Bioinform. **14**, 91 (2013)

[41] Srivastava, S., Chen, L.: A two-parameter generalized Poisson model to improve the analysis of RNA-seq data. Nucleic Acids Res. **38**(17), e170 (2010)

[42] Srivastava, S., Chen, L.: GPseq: using the generalized Poisson distribution to model sequence read counts from high throughput sequencing experiments. R Package Version 0.5. (2011). http://CRAN.R-project.org/package=GPseq

[43] Sultan, M., Schulz, M.H., Richard, H., Magen, A., Klingenhoff, A., Scherf, M., Seifert, M., Borodina, T., Soldatov, A., Parkhomchuk, D., et al.: A global view of gene activity and alternative splicing by deep sequencing of the human transcriptome. Science **321**, 956–960 (2008)

[44] Tarazona, S., García-Alcalde, F., Dopazo, J., Ferrer, A., Conesa, A.: Differential expression in RNA-seq: a matter of depth. Genome Res. **21**, 2213–2223 (2011)

[45] Tarazona, S., Furio-Tari, P., Ferrer, A., Conesa, A.: NOISeq: Exploratory analysis and differential expression for RNA-seq data. R Package Version 2.2.1 (2012)

[46] Tibshirani, R., Chu, G., Narasimhan, B., Li, J.: samr: SAM: significance analysis of microarrays. R Package Version 2.0. (2011). http://CRAN.R-project.org/package=samr

[47] Tierney, L., Rossini, A.J., Li, N., Sevcikova, H.: snow: simple Network of Workstations. R Package Version 0.3–13 (2013). http://CRAN.R-project.org/package=snow

[48] Trapnell, C., Williams, B.A., Pertea, G., Mortazavi, A., Kwan, G., van Baren, M.J., Salzberg, S.L., Wold, B.J., Pachter, L.: Transcript assembly and quantification by RNA-seq reveals unannotated transcripts and isoform switching during cell differentiation. Nat. Biotech. **28**, 511–515 (2010)

[49] van de Wiel, M.A., Leday, G.G.R., Pardo, L., Rue, H., van der Vaart, A.W., Van Wieringen, W.N.: Bayesian analysis of RNA sequencing data by estimating multiple shrinkage priors. Biostatistics **14**, 113–128 (2012)

[50] Wang, Z., Gerstein, M., Snyder, M.: RNA-seq: a revolutionary tool for transcriptomics. Nat. Rev. Genet. **10**, 57–63 (2009)
[51] Wang, L., Feng, Z., Wang, X., Wang, X., Zhang, X.: DEGseq: an R package for identifying differentially expressed genes from RNA-seq data. Bioinformatics **26**, 136–138 (2010)
[52] Yang, E., Girke, T., Jiang, T.: Differential gene expression analysis using coexpression and RNA-seq data. Bioinformatics **29**(17), 2153–2161 (2013). doi:10.1093/bioinformatics/btt363
[53] Yendrek, Y.R., Ainsworth, A.A., Thimmaruram, J.: The bench scientist's guide to statistical analysis of RNA-seq data. BMC Res. Notes **5**, 506 (2012)
[54] Young, M.D., Wakefield, M.J., Smyth, G.K., Oshlack, A.: Gene ontology analysis for RNA-seq: accounting for selection bias. Genome Biol. **11**, R14 (2010). doi:10.1186/gb-2010-11-2-r14
[55] Zhou, Y., Xia, K., Wright, F.A.: A powerful and flexible approach to the analysis of RNA sequence count data. Bioinformatics **27**(19), 2672–2678 (2011)

Chapter 3
Differential Expression Analysis of Complex RNA-seq Experiments Using edgeR

Yunshun Chen, Aaron T.L. Lun, and Gordon K. Smyth

Abstract This article reviews the statistical theory underlying the edgeR software package for differential expression of RNA-seq data. Negative binomial models are used to capture the quadratic mean-variance relationship that can be observed in RNA-seq data. Conditional likelihood methods are used to avoid bias when estimating the level of variation. Empirical Bayes methods are used to allow gene-specific variation estimates even when the number of replicate samples is very small. Generalized linear models are used to accommodate arbitrarily complex designs. A key feature of the edgeR package is the use of weighted likelihood methods to implement a flexible empirical Bayes approach in the absence of easily tractable sampling distributions. The methodology is implemented in flexible software that is easy to use even for users who are not professional statisticians or bioinformaticians. The software is part of the Bioconductor project.

This article describes some recently implemented features. Loess-style weighting is used to improve the weighted likelihood approach, and an analogy with quasi-likelihood is used to estimate the optimal weight to be given to the empirical Bayes prior. The article includes a fully worked case study with complete code.

3.1 Introduction

With the dramatic drop in sequencing costs provided by the Next Generation sequencing technologies in past few years, RNA-seq has now supplanted microarrays as the technology of choice for genome level expression profiling of RNA samples [17,24,28]. RNA-seq data is typically summarized by counting the number of sequence reads that map to genomic features of interest [9]. In this article we will

Y. Chen • A.T.L. Lun • G.K. Smyth (✉)
Walter and Eliza Hall Institute of Medical Research, 1G Royal Parade, Parkville, VIC 3052, Australia
e-mail: yuchen@wehi.edu.au; alun@wehi.edu.au; smyth@wehi.edu.au

S. Datta and D. Nettleton (eds.), *Statistical Analysis of Next Generation Sequencing Data*, Frontiers in Probability and the Statistical Sciences, DOI 10.1007/978-3-319-07212-8_3,

assume that the aim is to conduct a gene-level analysis, but similar analyses could be done for exons or exon-junctions or other genomic constructs. One very common problem is to use the read counts to identify genes that are differentially expressed between experimental conditions.

This article reviews the statistical theory underlying the edgeR software package [23] for differential expression analysis of RNA-seq data. Rigorous and statistically powerful analysis of RNA-seq data requires careful attention to a number of issues. The read counts are discrete integers that show strong mean-variance relationships. Different genes show different levels of variability, but the number of replicate samples from which variability is estimated can be very small indeed. Meanwhile, experiments may involve complex experimental designs with multiple treatment factors and other experimental variables.

edgeR uses negative binomial based models to capture the quadratic mean-variance relationship that can be observed in RNA-seq data, and to distinguish between biological and technical sources of variation [15]. By technical variation, we mean that associated with the sequencing technology whereas biological variation refers to changes in expression levels between experimental subjects. Information is shared between genes to estimate biological variation reliably even when the number of replicates is very small [22]. Conditional likelihood methods are used to avoid bias when estimating the level of variation [15, 22]. Empirical Bayes methods are used to allow gene-specific variation estimates while borrowing information between genes [15, 21]. A key feature of the edgeR package is the use of weighted likelihood methods to implement a flexible empirical Bayes approach in the absence of easily tractable sampling distributions. Finally, generalized linear models are used to accommodate arbitrarily complex designs, and the conditional likelihood and empirical Bayes procedures are generalized to work in this context [15].

This article also describes some recent additions to the package, not previously described in published form. In particular, loess-style weighting is used to improved the weighted likelihood approach, and an analogy with quasi-likelihood [11] is used to estimate the optimal weight to be given to the empirical Bayes prior. The article includes a fully worked case study.

The edgeR package is part of the Bioconductor project [7]. Some advanced numerical algorithms are used to ensure reliable convergence of the iterative algorithms, and some of the core code has been implemented in C++ for speed and numerical stability. The package can be installed from the Bioconductor website http://www.bioconductor.org.

3.2 The Negative Binomial Model

3.2.1 Summarizing an RNA-seq Experiment with a Count Matrix

In a typical RNA-seq experiment, purified RNA is converted to cDNA and sequenced on one of the high-throughput platforms. Millions of short 'read' sequences ranging from 25 to 300 base pairs in length are generated from one (single-end) or both (paired-end) ends of the cDNA fragments. These sequences must be aligned (or *mapped*) to a reference genome or transcriptome. Summarization is then performed by counting the number of reads mapped to known genomic features such as genes or exons. For simplicity, we will refer to these features as 'genes' although any genomic interval can be used. This results in a table of read counts for tens of thousands of genes across a number of samples. These samples are associated with a variety of treatment conditions that we want to compare.

Table 3.1 shows an example of the matrix of read counts for a very simple RNA-seq experiment. The dataset consists of two groups (wild-type and mutant), each of which contains samples from two mice, i.e., two biological replicates. After sequencing, reads for each sample are mapped to the mouse genome and summarized into gene-level counts. The final RNA-seq expression profile is represented by a table of read counts for tens of thousands of genes in all four mice samples (Table 3.1). The aim of this experiment is to identify differentially expressed genes between wild-type and mutant mice.

In this article, the total number of genes is denoted by G and the total number of samples is denoted by n. Hence, the table of read counts from an RNA-seq experiment is a $G \times n$ matrix of non-negative integers. We refer to the set of read counts for a sample as a *library* and the total number of reads in the library as the

Table 3.1 Table of read counts for a simple RNA-seq experiment with four samples

	Wild-type		Mutant	
	Sample 1	Sample 2	Sample 3	Sample 4
Gene 1	24	31	76	59
Gene 2	0	3	7	2
Gene 3	1,988	1,125	3,052	2,450
Gene 4	5	0	0	1
...	...	...	...	...
Total	22,341,961	20,739,175	15,669,423	23,711,320

Each column corresponds to a sample from a mouse with a wild-type or mutant genotype. Each row corresponds to a gene in the mouse genome. Each entry is set at the number of reads mapped to a particular gene in a particular sample. The sum of counts in each column is the library size for the corresponding sample

library size. For a particular gene g, let y_{gi} denote the read count in the ith sample. The expected value of y_{gi} given the experimental conditions and the sequencing depths is then

$$E(y_{gi}) = \mu_{gi} = \lambda_{gi} \cdot N_i, \tag{3.1}$$

where N_i is the library size and λ_{gi} is the expected proportion of reads mapped to gene g in the ith sample.

In the above example, we have $\lambda_{g1} = \lambda_{g2} = \lambda_g^W$ and $\lambda_{g3} = \lambda_{g4} = \lambda_g^M$ where λ_g^W and λ_g^M are the expected proportion of reads mapped to gene g in the wild-type and the mutant groups, respectively. Then, the aim of the differential expression analysis is to test

$$H_0 : \lambda_g^W = \lambda_g^M \qquad \text{against} \qquad H_1 : \lambda_g^W \neq \lambda_g^M, \tag{3.2}$$

for each gene $g = 1, 2, \ldots, G$.

3.2.2 Distinguishing Technical from Biological Variation

Two levels of variation can be distinguished in any RNA-seq experiment. First, there is the basic variability in the expression level of each gene from one biological sample to another, even when the experimental conditions have not been changed. Second, because expression levels can never be measured perfectly, there is a certain level of technical variation arising from measurement error. RNA-seq provides the possibility of disentangling these two sources of variation. Unlike microarrays, RNA-seq can do this without technical replicates of the same RNA samples, because the level of technical variation from sequencing is of a predictable nature.

Let π_{gi} be the fraction of all cDNA fragments in the ith sample that originate from gene g. This can be viewed as the true unobserved expression level of gene g in individual sample i. Given π_{gi} and the library size N_i, the expected count is $E(y_{gi}|\pi_{gi}) = \pi_{gi}N_i$. The read counts for any given gene are usually considered to follow a Poisson law under repeated sequencing runs of the same RNA sample [14], so it is reasonable to suppose that $\text{var}(y_{gi}|\pi_{gi}) = \pi_{gi}N_i$ also. This represents technical variability associated with the sequencing technology.

Let us further suppose that π_{gi} varies between biological replicates in such a way that the coefficient of variation (CV) remains constant for any given gene. This implies that $E(\pi_{gi}) = \lambda_{gi}$ and $\text{var}(\pi_{gi}) = \phi_g\lambda_{gi}^2$, where ϕ_g is the squared CV and λ_{gi} is the population mean proportion for gene g given the experimental conditions applied to sample i. The unconditional variance of y_{gi} can then be derived as

$$\text{var}(y_{gi}) = E_\pi[\text{var}(y|\pi)] + \text{var}_\pi[E(y|\pi)] = \mu_{gi} + \phi_g\mu_{gi}^2 \tag{3.3}$$

where $\mu_{gi} = \lambda_{gi} N_i$ is the population mean of y_{gi}. Dividing both sides by μ_{gi}^2 gives

$$\mathrm{CV}^2(y_{gi}) = 1/\mu_{gi} + \phi_g \quad (3.4)$$

The first term is the squared CV of y_{gi} given π_{gi} and the second is the squared CV of π_{gi}. In other words,

$$\text{Total CV}^2 = \text{Technical CV}^2 + \text{Biological CV}^2 . \quad (3.5)$$

This partition of CV^2 into technical and biological components was first derived by [15].

We call $\phi_g^{1/2}$ the biological coefficient of variation (BCV). BCV represents the coefficient of variation with which the true abundance of the gene varies between replicate RNA samples. It represents the CV that would remain between biological replicates if sequencing depth could be increased indefinitely. Note that the technical CV decreases as the size of the counts increases whereas the BCV does not. Thus, the BCV is likely to be the dominant source of uncertainty for high-count genes. Reliable estimation of the BCV is therefore crucial for realistic assessment of differential expression in RNA-seq experiments.

3.2.3 *Generalized Linear Models Accommodate Complex Experiments*

Generalized linear models (GLMs) are an extension of classical linear models to non-normally distributed response data [16, 18]. We use GLMs to accommodate complex experimental designs with multiple explanatory factors. GLMs allow the responses to follow any linear exponential family of probability distributions, and each distribution family is characterized by it mean-variance relationship. In our case, the quadratic mean-variance relationship shown above in (3.3) determines the negative binomial distribution family for read counts. We assume therefore that

$$y_{gi} \sim \mathrm{NB}(\mu_{gi}, \phi_g) , \quad (3.6)$$

where μ_{gi} is the mean and ϕ_g is now the negative binomial dispersion parameter. The assumption of negative binomial variation for y_{gi} is equivalent to assuming that the true gene abundances π_{gi} follow a gamma distributional law across replicate RNA samples.

We use a log-linear model to represent the influence of the treatment conditions and the library sizes on the expected count sizes for any gene. Recall that μ_{gi} is the product of the expression proportion λ_{gi} and the library size. We suppose that λ_{gi} can be represented by a log-linear model,

$$\log \lambda_{gi} = x_i^T \beta_g , \quad (3.7)$$

where x_i is a covariate vector indicating the treatment conditions applied to sample i and β_g is a vector of regression coefficients by which the covariate effects are mediated for gene g. It follows that

$$\log \mu_{gi} = x_i^T \beta_g + \log N_i. \tag{3.8}$$

Gathering the covariate vectors x_i into a design matrix X, the vector of linear predictors for gene g is the matrix product $X\beta_g$. The standard GLM method would use Fisher-scoring to estimate the parameter vector β_g. This is usually successful but can fail to converge for some datasets. edgeR enhances the usual Fisher-scoring algorithm with a Levenberg damping modification to ensure that the sequence of iterations converges for all genes and all datasets [15]. The modified algorithm forces a reduction in the residual deviance at each iteration. The sequence of deviances is monotonic and bounded, and so always converges unless floating point inaccuracies intervene first.

In the simple example shown in Sect. 3.2.1, the design matrix might take the form

$$X = \begin{pmatrix} 1 & 0 \\ 1 & 0 \\ 1 & 1 \\ 1 & 1 \end{pmatrix}. \tag{3.9}$$

In that case the first regression coefficient β_{g1} would represent the log-expression proportion in the wild-type group and the second coefficient β_{g2} would represent the log-fold change in expression in the mutant group relative to wild-type. In the notation of Sect. 3.2.1, $\beta_{g1} = \log \lambda_g^W$ and $\beta_{g2} = \log(\lambda_g^M / \lambda_g^W)$. The hypothesis of interest in this example is

$$H_{0g} : \beta_{g2} = 0 \qquad \text{against} \qquad H_{1g} : \beta_{g2} \neq 0, \tag{3.10}$$

and this hypothesis is tested for all genes.

edgeR provides the ability to test whether any contrast of the regresssion coefficients equal to zero. Specifically, one can test the null hypothesis $H_0 : c^T \beta_g = 0$ where c is an arbitrary contrast vector. By default, hypotheses are tested using the usual asymptotic chisquare approximation to the likelihood ratio statistic, although edgeR also offers two more conservative F-test approximations as alternative options.

3.3 Empirical Bayes Dispersion Estimation

3.3.1 Overview

Accurate estimation of the dispersion parameter ϕ in the negative binomial model is vital for fitting GLMs and assessing differential expression. Given that an RNA-seq dataset often has a small number of samples, traditional univariate estimators of ϕ

tend to perform poorly [22]. Maximum likelihood estimators (MLEs) in particular tend to underestimate dispersion parameters because they make no adjustment for the fact that the mean is estimated from the same data [22].

The differential expression analysis of an RNA-seq experiment with a one-way layout was studied by Robinson and Smyth [21, 22] who proposed a quantile-adjusted conditional maximum likelihood method for dispersion estimation. This approach is available in edgeR via the `estimateTagwiseDisp` function, but is restricted to experiments with a one-way layout, i.e., to experiments with only one experimental factor.

In this chapter, we will focus on the general case in which RNA-seq experiments may involve multiple treatment conditions and blocking variables. Dispersion estimation for complex experimental designs was studied by McCarthy et al. [15]. Their method is based on the idea of an adjusted profile likelihood proposed by Cox and Reid [4].

3.3.2 *Cox-Reid Adjusted Profile Likelihood*

For general RNA-seq experiments with multiple factors, negative binomial dispersions are estimated using the Cox-Reid (CR) adjusted profile likelihood method [4, 15]. The CR method is based on the idea of approximate conditional likelihood which reduces to residual maximum likelihood (REML). Briefly, REML removes the effect of nuisance parameters which allows unbiased estimation of the dispersion. This accounts for all systematic sources of variation in the model.

For the purpose of estimating the dispersion, ϕ_g is the parameter of interest whereas the regression coefficients β_g and the means μ_{gi} are nuisance parameters. One condition of the CR method is that the nuisance parameters are assumed to be orthogonal to the parameter of interest, i.e., the Fisher information matrix must be block diagonal [4]. It can be shown that orthogonality between β_g and ϕ_g follows here from the fact that ϕ_g appears only in the variance function and not in the mean of the negative binomial GLMs [26].

The Cox-Reid adjusted profile likelihood (APL) for ϕ_g is the penalized log-likelihood, i.e.,

$$\mathrm{APL}_g(\phi_g) = \ell(\phi_g; y_g, \hat{\beta}_g) - \frac{1}{2}\log\det(\mathscr{I}_g), \tag{3.11}$$

where y_g is the vector of counts for gene g, $\hat{\beta}_g$ is the estimated coefficient vector, ℓ is the log-likelihood function and $\mathscr{I}_g$ is the Fisher information of β_g evaluated at $\hat{\beta}_g$ and ϕ_g.

Note that the $\hat{\beta}_g$ is the MLE of β_g given ϕ_g. Thus, $\hat{\beta}_g$ is also a function of ϕ_g. This means that the log-likelihood ℓ can be considered as a profile likelihood ℓ_p which depends only on ϕ_g, i.e., $\ell(\phi_g; y_g, \hat{\beta}_g) = \ell_p(\phi_g; y_g)$. Similarly, the adjustment term $\mathscr{I}_g$ can be treated as a function of ϕ_g. Maximization of $\mathrm{APL}_g(\phi_g)$ can then be used to obtain an estimate for ϕ_g.

3.3.3 *Weighted Likelihood Empirical Bayes*

The empirical Bayes method is one of the most powerful tools in data analysis. The aim is to estimate the prior distribution from the data and then apply the standard Bayesian approach to obtain posterior estimates. Empirical Bayes estimation has been shown to outperform classical maximum likelihood estimates for high dimensional problems [5, 6, 25].

The cost of RNA-seq experiments often limits RNA-seq studies to only a small number of replicate libraries. This makes it difficult to obtain reliable dispersion estimates. The situation is further complicated by the fact that different genes may have different dispersions. For microarray data, this problem has been solved by applying an empirical Bayes strategy [25] where information is shared across genes or probes to stabilize the gene-wise variance estimates. It is tempting to apply a similar approach to RNA-seq data. Unfortunately, the direct empirical Bayes approach to stabilize the dispersion estimates is not applicable in the case of RNA-seq data since there is no conjugate prior distribution for the negative binomial dispersion ϕ.

One way to approximate the empirical Bayes strategy is to use a weighted likelihood. It can be shown that an empirical Bayes estimator is equivalent to an estimate obtained by maximizing a weighted likelihood function on a set of observations [3, 27]. This result provides an opportunity to implement an approximation of the empirical Bayes method for RNA-seq data.

3.3.3.1 Common Dispersion

The simplest approach of sharing information between genes is to assume that all genes share a same dispersion value ϕ, which is called the *common dispersion* [15, 22]. It can be estimated by maximizing the common APL, which is defined as

$$\text{APL}_C(\phi) = \frac{1}{G}\sum_{g=1}^{G} \text{APL}_g(\phi), \tag{3.12}$$

where G is the total number of genes in the dataset.

The common APL can be considered as a special weighted likelihood in which the likelihoods for each gene have equal weights. Hence, all genes contribute equally to the estimation of this common dispersion. A common dispersion can be estimated in edgeR via the `estimateGLMCommonDisp` function.

3.3.3.2 Trended Dispersion

The common dispersion approach is almost certainly too simple. It is far more likely that some genes have larger or smaller dispersion values than other genes.

It has been found in many RNA-seq datasets that genes with lower expression level tend to have larger dispersions, and vice versa. Hence, it is reasonable to assume that the dispersion values depend on the gene-wise expression levels and can be modelled by a mean-dispersion trend [1]. In edgeR, the dispersion values obtained from the mean-dispersion trend are referred to as the *trended dispersion*, and in principle genes with the same expression level (or the same mean) should have the same trended dispersion.

The trended dispersion can also be estimated by the weighted likelihood approach. Given an RNA-seq dataset, the overall expression level of each gene is calculated as an average across all samples and expressed as an average log count-per-million (logCPM) using the `aveLogCPM` function. This average is computed by a simple GLM, taking into account the common dispersion and the library sizes. Then, all the genes are sorted according to their average logCPM values. For a particular gene g, a locally shared APL denoted $\text{APL}_{S_g}(\phi_g)$ is formed by averaging the APLs of the set of genes, denoted C_g, that are nearest to gene g in average logCPM. By default, the neighbourhood set C_g is chosen to contain at least 25 % of all genes, and the proportion is automatically increased if the total number of genes in the dataset is small. This ensures that each set C_g contains enough genes (and hence sufficient information) to represent the dispersion trend locally.

A graduated weighting approach was used to account for the relevance in expression level between gene g and other genes in the set C_g. The weight for the APL of gene a in C_g, denoted w_a, is determined by the tricube function, i.e.,

$$w_a = (1 - |x_a|^3)^3, \tag{3.13}$$

where $-1 < x_a < 1$ represents the scaled difference in average logCPMs for genes g and a. In other words, the closer the expression levels of genes g and a are, the smaller $|x_a|$ will be, and thus the larger w_a will be. This process can be repeated for all the genes in the set to obtain

$$\text{APL}_{S_g}(\phi_g) = \frac{\sum_{a \in C_g} w_a \cdot \text{APL}_a(\phi_g)}{\sum_{a \in C_g} w_a}, \tag{3.14}$$

as the locally shared APL for gene g. This is equivalent to fixing ϕ to a constant, fitting a loess curve of degree 0 through those $\text{APL}_a(\phi)$ for $a = 1, 2, \ldots, G$, and using the fitted value as the final value of the locally shared APL at ϕ for each gene. The trended dispersion for gene g can then be estimated by maximizing $\text{APL}_{S_g}(\phi_g)$.

3.3.3.3 Gene-Specific Dispersion

The trended dispersion approach would be sufficient if the true dispersions followed the mean-dispersion trend and genes with the same expression level had identical dispersion. This however is rarely true for real datasets and in practice dispersions are gene-specific. An individual dispersion therefore should be estimated for each individual gene, yet we are faced with the problem that the data from a single gene are often insufficient for reliable estimation of this dispersion. We need therefore a method that allows each gene to have its own dispersion estimate while still gaining information from the other genes. This can be achieved by an empirical Bayes approach that combines individual and shared information to obtain stable dispersion estimators. Such an approach has the effect of squeezing the genewise dispersions towards a pooled estimate, resulting in more stable inference when the number of samples is small.

The problem with directly applying the empirical Bayes approach is that there is no conjugate prior for the negative binomial dispersion ϕ_g. Thus, a weighted likelihood method has been proposed to approximate the empirical Bayes strategy for RNA-seq count data [15, 21]. To estimate the gene-specific dispersion, the weighted APL for a particular gene g is constructed as

$$\mathrm{APL}_{W_g}(\phi_g) = \mathrm{APL}_g(\phi_g) + G_0 \cdot \mathrm{APL}_{S_g}(\phi_g), \tag{3.15}$$

where $\mathrm{APL}_g(\phi_g)$ is the gene-wise APL using the information from gene g only, $\mathrm{APL}_{S_g}(\phi_g)$ is the locally shared APL for gene g, and G_0 is the weight assigned to the $\mathrm{APL}_{S_g}(\phi_g)$. The gene-specific dispersion ϕ_g is then estimated by maximizing $\mathrm{APL}_{W_g}(\phi_g)$. This weighted APL approach is described in Fig. 3.1.

In empirical Bayes terms, the locally shared APL, $\mathrm{APL}_{S_g}(\phi_g)$, can be interpreted as the prior distribution for ϕ_g, and the $\mathrm{APL}_g(\phi_g)$ as the likelihood from the direct observed data. This means that the $\mathrm{APL}_{W_g}(\phi_g)$ can be interpreted as the posterior distribution for ϕ_g, which is a compromise between the prior and the observation. In the weighed likelihood approach, the prior distribution for ϕ_g can be thought of as arising from prior observations on a set of G_0 genes. Hence, the prior weight G_0 is referred to as the *prior number of observations*.

The optimal choice for G_0 depends on the variability of the dispersions. Large values are best when the dispersions are a constant for all the genes or they closely follow the mean-dispersion trend. Smaller values are recommended when the dispersions are more variable among different genes. If $G_0 = 0$, no information is borrowed from other genes. This means that the gene-specific dispersion for a particular gene is purely estimated from its gene-wise APL. If G_0 is set to be infinitely large, information from that individual gene will be ignored. This means that the gene-specific dispersion will be fully determined by its locally shared APL such that the result will be the same as the trended dispersion. This information borrowing strategy can be viewed as shrinking individual dispersion estimates towards the dispersion trend (Fig. 3.2) where the value of G_0 represents the amount of shrinkage.

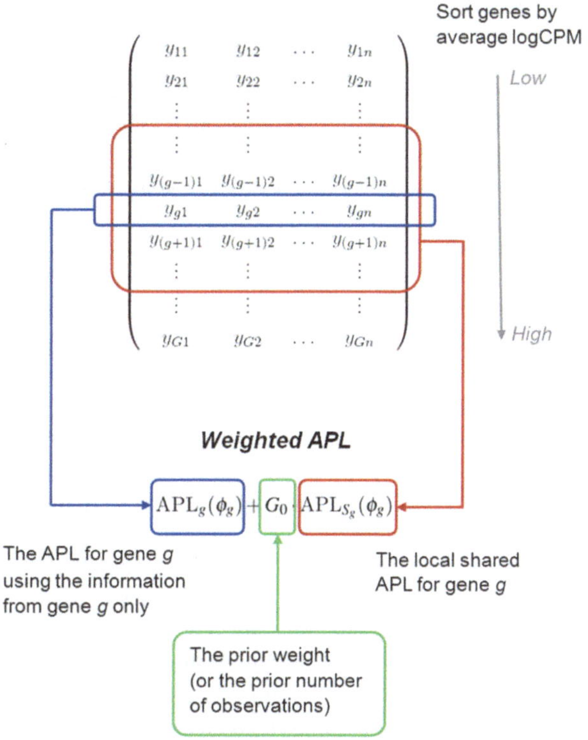

Fig. 3.1 Genes are sorted by their expression level. The gene-specific dispersion for a particular gene g is estimated by maximizing the weighted APL, i.e., the weighted average between the gene-wise APL and the locally shared APL. The weight assigned to the locally shared likelihood is denoted by G_0 which can be interpreted as the prior number of observations

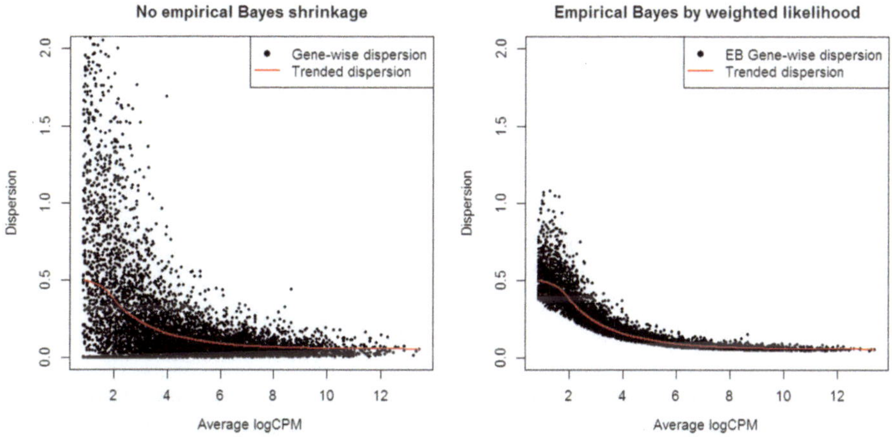

Fig. 3.2 The empirical Bayes shrinkage by weighted likelihood on simulated data. The plot on the *left* shows the dispersion estimates without empirical Bayes shrinkage. For each gene, the gene-wise dispersion estimate is obtained using the information of that gene only. The plot on the *right* shows the gene-wise dispersion estimates after empirical Bayes shrinkage. Gene-wise dispersion estimates are squeezed towards the dispersion trend which represents the use of prior information

3.3.4 Estimating Prior Weight

As mentioned previously, there is no conjugate prior for the genewise dispersion parameters. This means that there is no automatic estimation for the prior number of observations G_0. Thus, an alterative approach must be used. To account for the fact that more samples result in more gene-wise information, we write G_0 as

$$G_0 = \frac{d_0}{d_g}, \tag{3.16}$$

where d_0 is the *prior degrees of freedom* and d_g is the (known) residual degrees of freedom for gene g. The prior degrees of freedom represents the precision of the prior and does not depend on the total number of samples. The prior degrees of freedom can also be viewed as a measure of the consistency of the genewise dispersions. If the dispersions tend to be very gene-specific, then d_0 should be small and the prior will be vague. If the genewise dispersions tend to be consistent, i.e., close to the global trend, then d_0 should be large making the prior very informative. Once we estimate the d_0, we can easily calculate the prior weight G_0 in the weighted likelihood to obtain the best estimator for ϕ_g.

One way to estimate the prior degrees of freedom under the GLM framework is to use a quasi-likelihood in which the uncertainty of the variance can be absorbed into an overdispersion parameter. In GLM theory, the variance function $V(\mu)$ uniquely specifies a probability distribution such as the Poisson or negative binomial distribution. The quasi-likelihood variance function can then be written as

$$\text{var}(y_{gi}) = \sigma_g^2 \cdot V(\mu_{gi}), \tag{3.17}$$

where σ_g^2 is a factor that we will call the *quasi-dispersion parameter*. Note that the quasi-likelihood function is not a log-likelihood corresponding to any actual probability distribution. Instead, it can be used to describe a function that has similar properties to a log-likelihood function.

Following [11], we assume that the prior distribution for σ_g^2 is a scaled inverse χ^2-distribution with degrees of freedom d_0 and scaling factor $s_0^2 d_0$, i.e.,

$$\sigma_g^2 \sim s_0^2 \cdot \frac{d_0}{\chi_{d_0}^2}, \tag{3.18}$$

where s_0^2 can be considered as a prior mean for the quasi-dispersion. Our aim is to estimate d_0, which represents the precision of the prior distribution for σ_g^2.

Write D_g for the residual deviance of the generalized linear model fitted to the read counts for gene g. The mean residual deviance

$$s_g^2 = \frac{1}{d_g} D_g \tag{3.19}$$

is an estimator of σ_g^2. It can be shown [3] using the saddlepoint approximation [8] that the mean deviance s_g^2 follows approximately a χ^2-distribution with degrees of freedom d_g and scaling factor σ_g^2/d_g , i.e.,

$$s_g^2|\sigma_g^2 \sim \sigma_g^2 \cdot \frac{\chi_{d_g}^2}{d_g}. \tag{3.20}$$

To make this approximation more accurate, a special calculation is required for the residual degrees of freedom d_g when some of the fitted values are exactly zero. In particular, we ensure that any experimental condition for which the counts are all zero does not contribute to d_g. This is because such counts will have fitted values exactly zero and will make zero contribution to the residual deviance regardless of the value of the dispersion. This calculation is a refinement on the procedure of Lund et al. [11], and serves to make s_g^2 more nearly unbiased for σ_g^2 in the presence of zero counts.

The values of s_0^2 and d_0 can be estimated from the marginal distribution of s_g^2, which is scaled F-distribution,

$$s_g^2 \sim s_0^2 \cdot F_{d_g,d_0}, \tag{3.21}$$

where F_{d_g,d_0} denotes the F-distribution with degrees of freedom d_g and d_0 [11, 25]. Estimators of s_0^2 and d_0 can then be obtained by the method of moments [25].

In the main edgeR analysis pipeline, the quasi-likelihood is used only to estimate d_0. We assume that it is reasonable to use the same d_0 for empirical Bayes estimation of the negative binomial dispersions ϕ_g as for the quasi-dispersions σ_g^2. This allows us to calculate the prior weight G_0 required for (3.15) from (3.16) using d_g and the quasi-likelihood estimate for d_0.

3.4 Case Study: Transcriptional Program Regulation by IRF4

3.4.1 Experimental Design

We now demonstrate by way of a case study how the statistical theory in Sects. 3.2 and 3.3 is applied in practice to analyze RNA-seq datasets. The case study includes the complete R code used to undertake the analysis. The data are from a study on the transcription factor IRF4 [13]. In the study, it was found that IRF4 regulated the expression of key molecules required for the aerobic glycolysis of effector T cells and was essential for the clonal expansion and maintenance of effector function of antigen-specific CD8+ T cells [13].

One part of this study was to identify the transcriptional program regulated by IRF4 during the TCR affinity-driven population expansion of CD8+ T cells. To investigate this, T cells were harvested from Irf4$^{+/+}$ wild-type or Irf4$^{-/-}$ knock-out mice. The knock-out mice have a mutation which prevents the Irf4 gene from producing a viable protein. T cells were stimulated with high-affinity peptides (N4) or low-affinity peptides (V4). RNA was extracted from the cells and profiled using RNA-seq.

The study can be viewed as a 2×2 factorial experiment with 2–3 replicates for each combination of IRF4 and affinity peptide conditions. There are nine RNA samples in all. As is usual for an edgeR analysis, we start with experimental information about each RNA sample contained in a data frame called `targets`. The data frame was created using a spreadsheet and read into R using `readTargets`. It contains the two experimental factors, Genotype and Treatment, as well as the identifier for each sample on the public ENA repository:

```
> targets
        ENA       Label Genotype Treatment
1 SRR953136 WT.N4.rep1        WT        N4
2 SRR953137 WT.N4.rep2        WT        N4
3 SRR953138 WT.V4.rep1        WT        V4
4 SRR953139 WT.V4.rep2        WT        V4
5 SRR953140 KO.N4.rep1        KO        N4
6 SRR953141 KO.N4.rep2        KO        N4
7 SRR953142 KO.N4.rep3        KO        N4
8 SRR953143 KO.V4.rep1        KO        V4
9 SRR953144 KO.V4.rep2        KO        V4
```

The aim is to detect genes that are differentially expressed (DE) between different conditions.

3.4.2 Mapping Reads to the Mouse Genome

The RNA samples were sequenced on an Illumina HiSeq 2000 at the Australian Genome Research Facility. Paired end sequencing was used, and reads were 100 bases long. This means that the first and last 100 bases of each RNA fragment were sequenced. Fragments were up to about 600 bases long in total.

The raw sequence reads are available either in SRA format from the Gene Expression Omnibus (http://www.ncbi.nlm.nih.gov/geo) as series GSE49929 or in FastQ format from the European Nucleotide Archive (ENA) (http://www.ebi.ac.uk/ena) as series SRP028864. We analyse here gzipped FastQ files downloaded from ENA. There are a total of 11 samples under ENA series SRP028864, the first 9 of which are analyzed here.

We start with a data frame of file names in R:

```
> files
                Forward                Reverse           SAM
1 SRR953136_1.fastq.gz SRR953136_2.fastq.gz SRR953136.sam
2 SRR953137_1.fastq.gz SRR953137_2.fastq.gz SRR953137.sam
```

```
3 SRR953138_1.fastq.gz SRR953138_2.fastq.gz SRR953138.sam
4 SRR953139_1.fastq.gz SRR953139_2.fastq.gz SRR953139.sam
5 SRR953140_1.fastq.gz SRR953140_2.fastq.gz SRR953140.sam
6 SRR953141_1.fastq.gz SRR953141_2.fastq.gz SRR953141.sam
7 SRR953142_1.fastq.gz SRR953142_2.fastq.gz SRR953142.sam
8 SRR953143_1.fastq.gz SRR953143_2.fastq.gz SRR953143.sam
9 SRR953144_1.fastq.gz SRR953144_2.fastq.gz SRR953144.sam
```

Each row corresponds to an RNA sample. The first column gives the name of the file containing the sequences of the forward strand ends of the RNA fragments. The second column gives the name of the file containing the reverse strand reads.

The paired reads were mapped to the mouse genome using the Subread aligner [10]. The aligner uses the reads from both ends of each fragment to locate the fragment on the genome.

```
> library(Rsubread)
> align("mm9", readfile1=files$Forward, readfile2=files$Reverse,
+        "gzFASTQ", output_file=files$SAM, tieBreakQS=TRUE)
```

This code also uses an index ("mm9") of the mouse genome. The index was created from the NCBI37/mm9 (July 2007) build of the mouse genome using the `buildindex` command of the subread package [10]. The mm9 index file can be downloaded from the Subread website http://subread.sourceforge.net.

The number of reads (forward and reverse) varies from 12 million to 19 million for each sample. For this dataset, the proportion of reads successfully mapped to the genome was more than 99 % for all samples. This suggests good quality RNA samples and successful alignment:

```
> propmapped(file$SAM)
        Samples NumTotal  NumMapped PropMapped
1 SRR953136.sam 13164036  13089886       0.994
2 SRR953137.sam 13007946  12932901       0.994
3 SRR953138.sam 12919854  12849910       0.995
4 SRR953139.sam 12334822  12262014       0.994
5 SRR953140.sam 12454324  12370667       0.993
6 SRR953141.sam 18595382  18487656       0.994
7 SRR953142.sam 19119234  19008197       0.994
8 SRR953143.sam 13217130  13125153       0.993
9 SRR953144.sam 13273338  13200580       0.995
```

3.4.3 Fragment Counts for Each Gene

Now we compute a table of genewise counts. This is a two-step process. First the mapped reads are converted into mapped RNA fragments. A pair of forward and reverse reads is considered to represent an RNA fragment whenever they map to compatible nearby locations on the genome. The fragment is then assigned to a gene whenever the fragment overlaps at least one exon of the gene. This computation is done by the `featureCounts` function of the Rsubread package [9]:

```
> fc <- featureCounts(files$SAM, isPairedEnd=TRUE)
```

By default, the function uses RefSeq annotation from the National Center for Biotechnology Information (NCBI) giving the start and end positions of each exon [19]. The output is a matrix of counts, one row for each NCBI Entrez Gene identifier and one column for each RNA sample.

3.4.4 Creating a DGEList Object

The edgeR package stores data in a simple list-based data object called a DGEList. edgeR provides a range of generic functions and methods for such data objects, but they can at the same time be manipulated like ordinary lists in R. The main components of a DGEList object are a matrix of integer counts, a data frame of sample information and an optional data frame of gene annotation.

```
> library(edgeR)
> y <- DGEList(counts=fc$counts, group=targets$Genotype)
> colnames(y) <- targets$Label
```

There are entries for 26,310 genes and 9 samples:

```
> dim(y)
[1] 26301     9
```

Note the application of standard generic functions `colnames` and `dim` which have methods defined for DGEList objects. Many other generic functions in R that are applicable to matrices or data frames also have methods for `DGEList` objects.

The library sizes are automatically computed by `DGEList` as the total number of assigned RNA fragments for each sample. The number of mapped fragments is slightly less than half the total number of mapped reads shown in Sect. 3.4.2, and the number of fragments assigned to genes is about 80 % of that.

```
> y$samples
             group lib.size norm.factors
WT.N4.rep1      WT  5038159            1
WT.N4.rep2      WT  4966457            1
WT.V4.rep1      WT  5026320            1
WT.V4.rep2      WT  4665370            1
KO.N4.rep1      KO  4703442            1
KO.N4.rep2      KO  6975408            1
KO.N4.rep3      KO  7271163            1
KO.V4.rep1      KO  4726829            1
KO.V4.rep2      KO  4995218            1
```

Many edgeR functions will accept an ordinary matrix of counts, but a DGEList object is more convenient because it automatically collates a variety of related information. For example, subsetting the above DGEList object `y` by column would automatically subset both the counts and the sample information at the same time.

3.4.5 Filtering and Normalization

Genes with counts that are all zero or all very low are usually not of interest in a differential expression analysis for two reasons. The biological reason is that a gene must be expressed at some minimal level before it is likely to be translated into a protein or to be biologically important. The statistical reason is that very low counts provide little statistical information to distinguish between the null and alternative hypotheses. In this particular dataset, we consider a gene to be expressed at a reasonable level in a sample if its count-per-million (CPM) value is above 1, which is equivalent to having 5–7 fragments in that sample. A gene is kept in the analysis if it is sufficiently expressed (CPM > 1) in at least two samples:

```
> CPM <- cpm(y)
> keep <-  rowSums(CPM > 1) >= 2
> y <- y[keep, ]
```

The filtering rule doesn't use the experimental design information, yet will keep any gene that is expressed in both the samples for any combinations of genotype and treatment condition.

After filtering, there are 12,347 genes remaining and most of the counts are greater than zero:

```
> dim(y)
[1] 12347     9
> head(y$counts)
          WT.N4.rep1 WT.N4.rep2 WT.V4.rep1 WT.V4.rep2 KO.N4.rep1
27395            305        291        430        499        599
18777            510        527        653        642        404
21399            333        361        445        608        424
108664           194        124        230        281        264
12421            326        355        158        210        193
100504079         15         15          3         10         23
          KO.N4.rep2 KO.N4.rep3 KO.V4.rep1 KO.V4.rep2
27395            702        895        785        671
18777            888        724        585        544
21399            710        806        771        572
108664           398        444        334        340
12421            388        263        175        237
100504079         36         10          5         10
```

It is also useful to compute relative scaling factors for the libraries by

```
> y <- calcNormFactors(y)
> y$samples
           group lib.size norm.factors
WT.N4.rep1    WT  5038159        1.033
WT.N4.rep2    WT  4966457        1.013
WT.V4.rep1    WT  5026320        0.964
WT.V4.rep2    WT  4665370        0.986
KO.N4.rep1    KO  4703442        1.009
KO.N4.rep2    KO  6975408        1.015
```

```
KO.N4.rep3     KO   7271163          1.039
KO.V4.rep1     KO   4726829          0.931
KO.V4.rep2     KO   4995218          1.016
```

The `calcNormFactors` function returns the DGEList data argument back with only the norm.factors changed. The scaling factors here represent compositional differences between the shape of the count distributions for the samples. The normalization factors multiply to unity. Factors below 1 indicate that an excessive number of fragments have been assigned to a small number of very highly expressed genes in that library, meaning that less sequencing depth is available for the remaining genes [20].

3.4.6 Gene Annotation

The summarized counts from Rsubread include Entrez Gene IDs as rownames. The Entrez IDs link to gene-specific information from the NCBI database [12]. To get more details such as gene symbol and chromosome number, we use the annotation file 'Mus_musculus.gene_info' obtained from the NCBI website (ftp://ftp.ncbi.nih.gov/gene/DATA/GENE_INFO/Mammalia).

```
> anno <- read.delim(file="Mus_musculus.gene_info",
+                    header=FALSE, skip=1)
```

We add selected annotation columns to the DGEList object:

```
> m <- match(rownames(y), anno[,2])
> y$genes <- anno[m, c(2,3,7)]
> colnames(y$genes) <- c("GeneID", "Symbol", "Chr")
> head(y$genes)
      GeneID  Symbol Chr
7060   27395  Mrpl15   1
4165   18777  Lypla1   1
5899   21399   Tcea1   1
24191 108664 Atp6v1h   1
625    12421  Rb1cc1   1
```

3.4.7 Data Exploration

A multiple dimensional scaling (MDS) plot can be used to check the dissimilarities among the samples:

```
> plotMDS(y, col=as.numeric(targets$Genotype))
```

`plotMDS` is a generic function defined in the limma package with a method defined for DGEList objects. The distance between each pair of samples is calculated as the *leading fold change*, defined as the root-mean-square of the largest

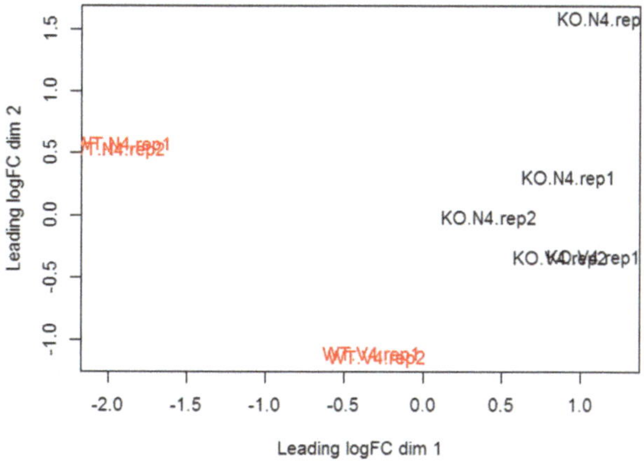

Fig. 3.3 The MDS plot of the IRF4 RNA-seq dataset. Samples are separated by the genotype (IRF4 wild-type and knock-out) in the first dimension. A separation by the affinity peptide level (N4 and V4) is also observed in the second dimension

500 $\log_2$-fold changes between that pair of samples. The MDS plot is shown in Fig. 3.3. Samples are well separated by the genotype condition (i.e., IRF4 wild-type and knock-out) in the first dimension. A separation by the affinity peptide level (N4 and V4) is also observed in the second dimension. All the replicates are close to each other except for the ones in the IRF4 knock-out (KO) with high-affinity peptides (N4).

3.4.8 *The Design Matrix*

We create a design matrix to capture all the experimental information. In this case study, the IRF4 genotype conditions (KO and WT) and the affinity peptide levels (N4 and V4) divide the data into four separate groups. The design matrix can be constructed using the `model.matrix` function as described below.

```
> fac <- paste(targets$Genotype, targets$Treatment, sep=".")
> fac <- factor(fac)
> design <- model.matrix(~0+fac)
> colnames(design) <- levels(fac)
> design
  KO.N4 KO.V4 WT.N4 WT.V4
1     0     0     1     0
2     0     0     1     0
3     0     0     0     1
4     0     0     0     1
5     1     0     0     0
6     1     0     0     0
7     1     0     0     0
8     0     1     0     0
9     0     1     0     0
attr(,"assign")
[1] 1 1 1 1
```

```
attr(,"contrasts")
attr(,"contrasts")$fac
[1] "contr.treatment"
```

We use this simple group-mean parametrization instead of a classic factorial model because it allows contrasts between the groups to be extracted in a simple and transparent way.

3.4.9 Estimating Dispersions

Now we can proceed to dispersion estimation. The `estimateDisp` function implements the weighted likelihood empirical Bayes strategy described earlier in this chapter. It takes the data object and the design matrix as arguments, and inserts the common, trended and genewise (tagwise) dispersions into the data object:

```
> y <- estimateDisp(y, design)
```

The common dispersion of 0.051 is equivalent to a overall BCV of 23 %:

```
> y$common.dispersion
[1] 0.051
```

The gene-specific dispersions vary between 0.024 and 1.1:

```
> summary(y$tagwise.dispersion)
   Min. 1st Qu.  Median    Mean 3rd Qu.    Max.
  0.024   0.034   0.046   0.065   0.073   1.100
```

The estimated prior degrees of freedom for this dataset is 6.9:

```
> y$prior.df
[1] 6.9
```

This can be compared to the residual degrees of freedom d_g, which is equal to 5 for most genes in this dataset. The prior degrees of freedom is slightly greater than the residual degrees of freedom, meaning that slightly more weight is being given to the global trend rather than the individual gene when estimating each genewise dispersion.

The BCV plot, as shown in Fig. 3.4, shows the common, trended and genewise dispersions as a function of average logCPM.

```
> plotBCV(y)
```

Recall that the BCV is the square root of the dispersion. Most of the gene-specific BCVs cluster around the BCV trend, which decreases and then asymptotes to a constant value as the gene expression level increases.

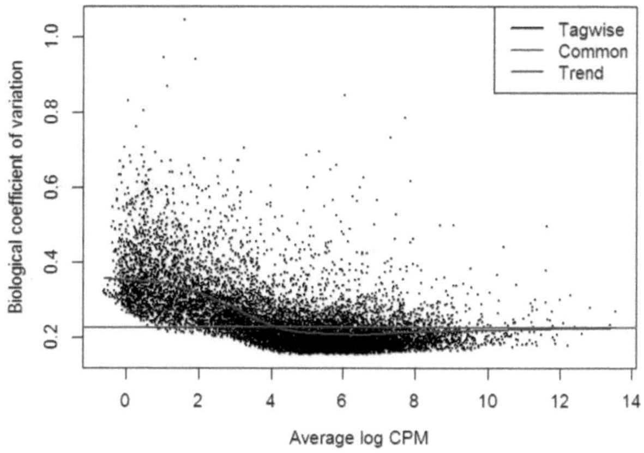

Fig. 3.4 The BCV plot of the IRF4 RNA-seq dataset. Gene-specific BCVs cluster around the BCV trend, which decreases and then asymptotes to a constant value as the gene expression level increases

3.4.10 Detecting Differentially Expressed Genes

In this study, one particular comparison of interest is between IRF4 wild-type (WT) cells stimulated with high-affinity peptide (N4) and WT cells stimulated with low-affinity peptide (V4). To find genes that are DE for this comparison, the first step is to fit genewise negative binomial GLMs using the gene-specific dispersions estimated above:

```
> fit <- glmFit(y, design)
```

Then likelihood ratio statistics are computed for the comparison of interest:

```
> lrt <- glmLRT(fit, contrast=c(0,0,1,-1))
```

Here the contrast argument specifies that the third and fourth groups are to be compared.

The `topTags` function collates results for the most significant genes:

```
> topTags(lrt)
Coefficient:  1*WT.N4 -1*WT.V4
      GeneID  Symbol Chr logFC logCPM    LR   PValue       FDR
1505   13813   Eomes   9 -5.70   7.07 225.7 5.29e-51 6.53e-47
10096  60596 Gucy1a3   3  5.88   4.93 179.9 5.06e-41 3.12e-37
2549   16001   Igf1r   7  3.75   4.88 127.4 1.51e-29 6.20e-26
14239  68404    Nrn1  13  4.16   5.41 109.4 1.30e-25 4.01e-22
30622 236915 Arhgef9   X  5.91   3.38  98.6 3.09e-23 7.62e-20
27600 140795  P2ry14   3 -3.86   4.70  92.9 5.54e-22 9.78e-19
3811   18186    Nrp1   8  3.97   4.99  92.9 5.54e-22 9.78e-19
2157   14945    Gzmk  13 -3.40   3.72  85.0 2.94e-20 4.54e-17
35406 380797    Ighd  12  3.70   3.59  83.4 6.56e-20 9.00e-17
34084 320407   Klri2   6  3.83   3.63  78.2 9.35e-19 1.15e-15
```

Local false discovery rates (FDR) are calculated using the Benjamini-Hochberg (BH) method [2]. By default, `topTags` displays the top 10 genes, but can be asked

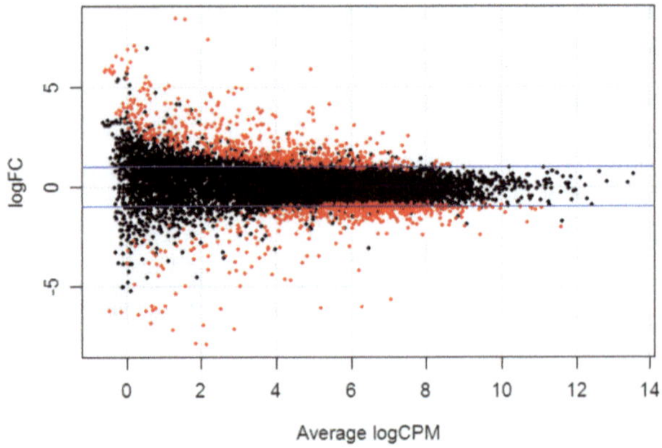

Fig. 3.5 The smear plot of the IRF4 RNA-seq dataset comparing IRF4 wild-type (WT) cells stimulated with high-affinity peptide (N4) and with low-affinity peptide (V4). DE genes are highlighted in *red*. The *blue lines* indicate twofold up or down

to select any number. By ranking all genes, we can see that there are 1,181 genes detected as DE at an FDR cutoff of 1%:

```
> tp <- topTags(lrt, n=Inf)
> sum(tp$table$FDR < 0.01)
[1] 1181
```

A smearplot (a form of MA-plot) can be produced to display the DE results graphically (Fig. 3.5):

```
> DE <- tp$table[tp$table$FDR < 0.01,]$GeneID
> plotSmear(lrt, de.tags=DE, cex = 0.4)
> abline(h=c(-1, 1), col="blue")
```

The axes of the plot correspond to the `logCPM` and `logFC` columns of the results table.

3.4.11 Session Information

The following output shows the R session and package versions used for this case study:

```
> sessionInfo()
R version 3.0.2 (2013-09-25)
Platform: i386-w64-mingw32/i386 (32-bit)

locale:
[1] LC_COLLATE=English_Australia.1252
[2] LC_CTYPE=English_Australia.1252
[3] LC_MONETARY=English_Australia.1252
[4] LC_NUMERIC=C
[5] LC_TIME=English_Australia.1252
```

```
attached base packages:
[1] splines   stats     graphics  grDevices utils     datasets
[7] methods   base

other attached packages:
[1] locfit_1.5-9.1 edgeR_3.4.0    limma_3.18.3   Rsubread_1.12.6

loaded via a namespace (and not attached):
[1] grid_3.0.2      lattice_0.20-24
```

Acknowledgements Thanks to Wei Shi for providing the fragment counts and alignment code for the IRF4 data, and to Davis McCarthy who programmed the original implementation of the loess local likelihood trend described in Sect. 3.3.3.

References

[1] Anders, S., Huber, W.: Differential expression analysis for sequence count data. Genome Biol. **11**(10), R106 (2010). doi:10.1186/gb-2010-11-10-r106
[2] Benjamini, Y., Hochberg, Y.: Controlling the false discovery rate: a practical and powerful approach to multiple testing. J. R. Stat. Soc. Series B **57**, 289–300 (1995)
[3] Chen, Y.: Differential expression analysis of complex RNA-seq experiments. Ph.D. thesis, University of Melbourne (2013)
[4] Cox, D.R., Reid, N.: Parameter orthogonality and approximate conditional inference. J. R. Stat. Soc. Series B **49**, 1–39 (1987)
[5] Efron, B.: Robbins, empirical Bayes and microarrays. Ann. Stat. **31**(2), 366–378 (2003)
[6] Efron, B., Morris, C.: Stein's estimation rule and its competitors: an empirical Bayes approach. J. Am. Stat. Assoc. **68**(341), 117–130 (1973)
[7] Gentleman, R.C., Carey, V.J., Bates, D.M., Bolstad, B., Dettling, M., Dudoit, S., Ellis, B., Gautier, L., Ge, Y., Gentry, J., Hornik, K., Hothorn, T., Huber, W., Iacus, S., Irizarry, R., Leisch, F., Li, C., Maechler, M., Rossini, A.J., Sawitzki, G., Smith, C., Smyth, G.K., Tierney, L., Yang, J.Y., Zhang, J.: Bioconductor: open software development for computational biology and bioinformatics. Genome Biol. **5**(10), R80 (2004)
[8] Jørgensen, B.: The Theory of Dispersion Models. Chapman & Hall, London (1997)
[9] Liao, Y., Smyth, G.K., Shi, W.: featureCounts: an efficient general-purpose read summarization program. Bioinformatics **30**, 923–930 (2014)
[10] Liao, Y., Smyth, G.K., Shi, W.: The Subread aligner: fast, accurate and scalable read mapping by seed-and-vote. Nucleic Acids Res. **41**(10), e108 (2013)
[11] Lund, S., Nettleton, D., McCarthy, D., Smyth, G.: Detecting differential expression in RNA-sequence data using quasi-likelihood with shrunken dispersion estimates. Stat. Appl. Genet. Mol. Biol. **11**(5), Article 8 (2012)
[12] Maglott, D., Ostell, J., Pruitt, K., Tatusova, T.: Entrez Gene: gene-centered information at NCBI. Nucleic Acids Res. **39**, D52–D57 (2011)
[13] Man, K., Miasari, M., Shi, W., Xin, A., Henstridge, D., Preston, S., Pellegrini, M., Belz, G., Smyth, G., Febbraio M Kallies, A.: IRF4 is essential for T cell receptor affinity mediated metabolic programming and clonal expansion of T cells. Nat. Immunol. **14**, 1155–1165 (2013)
[14] Marioni, J.C., Mason, C.E., Mane, S.M., Stephens, M., Gilad, Y.: RNA-seq: an assessment of technical reproducibility and comparison with gene expression arrays. Genome Res. **18**, 1509–1517 (2008). doi:10.1101/gr.079558.108
[15] McCarthy, D.J., Chen, Y., Smyth, G.K.: Differential expression analysis of multifactor RNA-Seq experiments with respect to biological variation. Nucleic Acids Res. **40**(10), 4288–4297 (2012)

[16] McCullagh, P., Nelder, J.A.: Generalized Linear Models, 2nd edn. Chapman & Hall/CRC, Boca Raton (1989)
[17] Mortazavi, A., Williams, B.A., McCue, K., Schaeffer, L., Wold, B.: Mapping and quantifying mammalian transcriptomes by RNA-seq. Nat. Meth. **5**(7), 621–628 (2008)
[18] Nelder, J.A., Wedderburn, R.W.M.: Generalized linear models. J. R. Stat. Soc. Series A (General) **135**(3), 370–384 (1972). http://www.jstor.org/stable/2344614
[19] Pruitt, K., Tatusova, T., Brown, G., Maglott, D.: NCBI reference sequences (RefSeq): current status, new features and genome annotation policy. Nucleic Acids Res. **40**, D130–D135 (2012)
[20] Robinson, M.D., Oshlack, A.: A scaling normalization method for differential expression analysis of RNA-seq data. Genome Biol. **11**(3), R25 (2010). doi:10.1186/gb-2010-11-3-r25
[21] Robinson, M.D., Smyth, G.K.: Moderated statistical tests for assessing differences in tag abundance. Bioinformatics **23**(21), 2881–2887 (2007)
[22] Robinson, M.D., Smyth, G.K.: Small-sample estimation of negative binomial dispersion, with applications to SAGE data. Biostatistics **9**(2), 321–332 (2008)
[23] Robinson, M., McCarthy, D., Smyth, G.: edgeR: a Bioconductor package for differential expression analysis of digital gene expression data. Bioinformatics **26**(1), 139–140 (2010)
[24] Shendure, J., Aiden, E.L.: The expanding scope of DNA sequencing. Nat. Biotechnol. **30**(11), 1084–1094 (2012)
[25] Smyth, G.: Linear models and empirical Bayes methods for assessing differential expression in microarray experiments. Stat. Appl. Genet. Mol. Biol. **3**(1), Article 3 (2004)
[26] Smyth, G., Verbyla, A.: Adjusted likelihood methods for modelling dispersion in generalized linear models. Environmetrics **10**(6), 695–709 (1999)
[27] Wang, X.: Approximating Bayesian inference by weighted likelihood. Can. J. Stat. **34**(2), 279–298 (2006)
[28] Wang, Z., Gerstein, M., Snyder, M.: RNA-Seq: a revolutionary tool for transcriptomics. Nat. Rev. Genet. **10**(1), 57–63 (2009). doi:10.1038/nrg2484

Chapter 4
Analysis of Next Generation Sequencing Data Using Integrated Nested Laplace Approximation (INLA)

Andrea Riebler, Mark D. Robinson, and Mark A. van de Wiel

Abstract Integrated Nested Laplace Approximation (INLA), implemented in the R-package `r-inla`, is a very versatile methodology for the Bayesian analysis of next generation sequencing count data: it can account for zero-inflation, random effects and correlation across genomic features. We demonstrate its use and provide some insights on its approximations of marginal posteriors. In high-dimension settings like these, INLA is in particular attractive in combination with empirical Bayes. We show how to apply this by estimating priors from the output of INLA. We extend this methodology to estimation of joint priors for a limited number of parameters, which effectuates multivariate shrinkage. Joint priors are useful for appropriate inference when two or more parameters are likely to be strongly correlated. Two examples serve as illustrations: (1) joint inference for differential zero-inflation and means between two groups; (2) correlated group effects on mRNA expression. For both simulated and real data we show that multivariate shrinkage may lead to improved marker selection. We end with a discussion on the use of this INLA-based method within the spectrum of other available methods.

A. Riebler
Department of Mathematical Sciences, Norwegian University of Science and Technology, N-7491 Trondheim, Norway
e-mail: andrea.riebler@math.ntnu.no

M.D. Robinson
Institute of Molecular Life Sciences, University of Zurich, Winterthurerstrasse 190, CH-8057 Zurich, Switzerland
e-mail: mark.robinson@imls.uzh.ch

M.A. van de Wiel (✉)
Department of Epidemiology and Biostatistics, VU University Medical Center and Department of Mathematics, VU University, Amsterdam, The Netherlands
e-mail: mark.vdwiel@vumc.nl

S. Datta and D. Nettleton (eds.), *Statistical Analysis of Next Generation Sequencing Data*, Frontiers in Probability and the Statistical Sciences, DOI 10.1007/978-3-319-07212-8_4,

4.1 Introduction

The rapid development of next generation sequencing (NGS) technologies has opened up new possibilities for biological and medical investigation. To account for the discrete nature of the data produced by NGS, new statistical approaches have been proposed. Bayesian approaches are interesting, because they account for uncertainty in hyperparameters, and, more importantly, are able to incorporate prior information from historical data or expert knowledge. However, in high-dimensional settings full Bayesian inference is often practically unfeasible when the posterior distribution is analytically unavailable. Sampling approaches, such as Markov chain Monte Carlo techniques, would be required but are computationally intractable. Furthermore, application-specific implementations and convergence checks would be needed, which might be one reason for the rare use of Bayesian approaches. In contrast, empirical Bayes statistical methods are often computationally lighter and may be used to borrow strength across data.

Rue et al. [10] proposed a deterministic approach based on integrated nested Laplace approximations (INLA) for full Bayesian inference in latent Gaussian models. A wide range of count models, such as binomial, Poisson, negative binomial and their zero-inflated extensions, are possible within INLA. Furthermore, it allows flexible study designs where fixed effects can be combined with different random effects. This allows the incorporation of potential dependence structures along the genome. Uncertainty regarding hyperparameters is thereby integrated in a full Bayesian way. Since no sampling is involved, INLA is reasonably fast. Van de Wiel et al. [15] used INLA successfully as a central component for analyzing RNA sequencing (RNA-seq) count data.

The goal of this chapter is to introduce INLA and sketch its potential for analyzing NGS data, in particular for complex inferential settings, such as inference for multiple correlated parameters. In Sect. 4.2, we give a short introduction to latent Gaussian models and INLA, while Sect. 4.3 points out potential advantages of INLA, but at the same time describes limitations. We shall introduce the R-package `ShrinkBayes` [15], which extends the functionality of INLA. In particular, a combination of INLA and empirical Bayes for multiparameter shrinkage is implemented. Section 4.4 extends the work of van de Wiel et al. [15] on RNA-seq to the use of multivariate shrinkage priors. The use of such priors and corresponding posterior distributions is proposed to accommodate dependence between parameters, which allows for simultaneous estimation and inference. In Sect. 4.5, results based on two simulated scenarios are presented, which show that while multivariate shrinkage provides only small gain in terms of ranking, it is clearly beneficial for feature selection at a given false discovery rate. In Sect. 4.6, we apply the methodology to lymphoblastoid cell lines of the International HapMap project [6]. We conclude with a discussion and an outline of the potential of INLA for other NGS data analyses.

4.2 INLA: A Deterministic Framework for Bayesian Inference in Latent Gaussian Models

4.2.1 Theoretical Background

Data arising from NGS experiments are represented as counts. In RNA-seq, for example, the number of tags mapping to certain genomic features is counted [2]. Consider a structured additive regression model for RNA-seq counts $\mathbf{Y} = (Y_{11},\ldots,Y_{mn})^\top$, Here, the count Y_{ij}, with $i = 1,\ldots,m$ denoting variables (genomic features) and $j = 1,\ldots,n$ samples, is assumed to follow a distribution $F_{\mu_{ij},\boldsymbol{\gamma}_i}$ in an exponential family. The mean μ_{ij} is thereby linked to the linear predictor η_{ij} using a link function $g(\cdot)$, so that $\mu_{ij} = g^{-1}(\eta_{ij})$. The distribution function $F_{\mu_{ij},\boldsymbol{\gamma}_i}$ can be controlled by further hyperparameters $\boldsymbol{\gamma}_i$, including, for example, overdispersion, zero-inflation, variance or correlation parameters. The linear predictor η_{ij} is then additively described by

$$\eta_{ij} = \beta_{i0} + \sum_{h=1}^{H} \beta_{ih} x_{jh} + \sum_{l=1}^{L} f^l(u_{il}), \tag{4.1}$$

where x_{jh} is the value of the hth covariate for sample j and f^l is an unknown function of covariate l, which takes value u_{il} for variable i. It can be used to model random effects and (genomic) dependencies between different features.

A latent Gaussian model is obtained by assigning a Gaussian prior distribution to the latent field $\boldsymbol{v} = \{\{f^l(\cdot)\}, \beta_{i0}, \{\beta_{ih}\}\}$. This Gaussian prior depends on the inverse covariance matrix, the precision matrix, $\mathbf{Q}(\boldsymbol{\tau})$ with hyperparameters $\boldsymbol{\tau}$. Finally, prior distributions are assigned to the hyperparameters $\boldsymbol{\theta}$, which is the set containing all elements of $\boldsymbol{\gamma} = (\boldsymbol{\gamma}_i)_{i=1}^m$ and $\boldsymbol{\tau}$.

When analyzing RNA-seq data with a model of type (4.1), we are usually interested in the differential behavior between samples. Hence, one focus may be on the posterior estimates of the regression coefficients, and contrasts between those for different samples. This means we are interested in single components of $\boldsymbol{v}$, or linear combinations of those. To assess the model at hand, posterior marginal distributions for single hyperparameters θ_k, such as zero-inflation or overdispersion, should be inspected. If random effects are included, such as random walks of first or second order, which model the dependence between neighboring genomic features, we would like to analyze the fitted pattern along the genome. All these statistics are based on the posterior distribution of the Gaussian latent field $\boldsymbol{v}$ and the hyperparameters $\boldsymbol{\theta}$, which is given as:

$$\mathrm{p}(\boldsymbol{v}, \boldsymbol{\theta} \mid \mathbf{Y}) \propto \mathrm{p}(\mathbf{Y} \mid \boldsymbol{v}, \boldsymbol{\theta})\, \mathrm{p}(\boldsymbol{v} \mid \boldsymbol{\theta})\, \mathrm{p}(\boldsymbol{\theta}). \tag{4.2}$$

The posterior marginal distributions of single components v_ℓ of $\boldsymbol{v}$ or θ_k of $\boldsymbol{\theta}$ can then be derived from (4.2) as

$$p(v_\ell \,|\, \mathbf{Y}) = \int_{\boldsymbol{\theta}} \int_{\boldsymbol{v}_{-\ell}} p(\boldsymbol{v}, \boldsymbol{\theta} \,|\, \mathbf{Y}) d\boldsymbol{v}_{-\ell} d\boldsymbol{\theta}, \tag{4.3}$$

$$p(\theta_k \,|\, \mathbf{Y}) = \int_{\boldsymbol{v}} \int_{\boldsymbol{\theta}_{-k}} p(\boldsymbol{v}, \boldsymbol{\theta} \,|\, \mathbf{Y})\, d\boldsymbol{\theta}_{-k} d\boldsymbol{v} \tag{4.4}$$

where $\boldsymbol{v}_{-\ell}$ and $\boldsymbol{\theta}_{-k}$ denote the vectors $\boldsymbol{v}$ and $\boldsymbol{\theta}$ without the ℓth and kth element, respectively. The computation of high dimension integrals is challenging, when the integral cannot be solved analytically. Hence, sampling-based approaches have been commonly used [3]. However, these approaches are time-consuming and in particular in high-dimensional settings, as we consider here, computationally intractable.

With INLA, Rue et al. [10] proposed an efficient deterministic computational method based on accurate approximations. A central component in their approach is the Laplace approximation, as proposed by Tierney and Kadane [13]. The Laplace approximation is often used to approximate integrals appearing in the estimation of moments, but can also be used to approximate marginal posterior densities.

INLA uses the fact that (4.3) can be written as

$$p(v_\ell \,|\, \mathbf{Y}) = \int_{\boldsymbol{\theta}} p(v_\ell \,|\, \boldsymbol{\theta}, \mathbf{Y})\, p(\boldsymbol{\theta} \,|\, \mathbf{Y}) d\boldsymbol{\theta},$$

and approximates this term by the finite sum

$$\tilde{p}(v_\ell \,|\, \mathbf{Y}) = \sum_t \tilde{p}(v_\ell \,|\, \boldsymbol{\theta}_t, \mathbf{Y}) \tilde{p}(\boldsymbol{\theta}_t \,|\, \mathbf{Y}) \Delta_t.$$

Here, $\tilde{p}(v_\ell \,|\, \boldsymbol{\theta}, \mathbf{Y})$ and $\tilde{p}(\boldsymbol{\theta} \,|\, \mathbf{Y})$ denote approximations of $p(v_\ell \,|\, \boldsymbol{\theta}, \mathbf{Y})$ and $p(\boldsymbol{\theta} \,|\, \mathbf{Y})$, respectively, and the sum is computed over suitable support points θ_t with appropriate weights Δ_t. Since

$$p(\boldsymbol{v}, \boldsymbol{\theta}, \mathbf{Y}) = p(\boldsymbol{v} \,|\, \boldsymbol{\theta}, \mathbf{Y})\, p(\boldsymbol{\theta} \mid \mathbf{Y})\, p(\mathbf{Y})$$

it follows that

$$p(\boldsymbol{\theta} \mid \mathbf{Y}) \propto \frac{p(\boldsymbol{v}, \boldsymbol{\theta}, \mathbf{Y})}{p(\boldsymbol{v} \,|\, \boldsymbol{\theta}, \mathbf{Y})}$$

for all $\boldsymbol{v}$. INLA uses a Laplace approximation to approximate $\boldsymbol{\theta} \mid \mathbf{Y}$ as

$$\tilde{p}(\boldsymbol{\theta} \mid \mathbf{Y}) \propto \frac{p(\boldsymbol{v}, \boldsymbol{\theta}, \mathbf{Y})}{\tilde{p}_G(\boldsymbol{v} \,|\, \boldsymbol{\theta}, \mathbf{Y})} \Big|_{\boldsymbol{v}=\boldsymbol{v}^\star(\boldsymbol{\theta})} . \tag{4.5}$$

Here, $\tilde{p}_G(.)$ represents a Gaussian approximation of the full conditional distribution of $\boldsymbol{v}$, and $\boldsymbol{v}^\star(\boldsymbol{\theta})$ denotes the mode of $p(\boldsymbol{v} \,|\, \boldsymbol{\theta}, \mathbf{Y})$ as determined by an iterative optimization algorithm. As shown in Rue and Held [9], a full conditional distribution can be well approximated by a Gaussian distribution by matching the mode and curvature at the mode, so that (4.5) is a sensible choice. For $\mathrm{p}(v_\ell \mid \boldsymbol{\theta}, \mathbf{Y})$ three different approximation models are available. They differ not only in accuracy, but also in computational efficiency; for details we refer to [10]. As default, a simplified Laplace approximation is used. Posterior marginals for $\mathrm{p}(\theta_k \mid \mathbf{Y})$ can be obtained similarly from $\tilde{\mathrm{p}}(\boldsymbol{\theta} \mid \mathbf{Y})$. For a discussion of the accuracy of INLA we refer to [10]. In practice, the approximation error of INLA was found to be small compared to the Monte Carlo error, see for example [5, 11].

4.2.2 *The* R*-package* `r-inla`

The R-package `r-inla` allows the user to combine different types of likelihood functions with different regression models. Similar to model formulations based on `lm()` or `glm()` in R, the linear predictor $\boldsymbol{\eta}$ of a model is defined using a *formula*. That is, we use a formula notation:

```
formula = y ~ a + b + a:b
      + f(c, model="iid", hyper=list(...), ...) + ...
```

to specify the model of interest. Here, `y` is the response of length $n \cdot m_1$, where $m_1 \leq m$ denotes the number of features for which a joint fit is desired (for example, $m_1 = 1$ refers to the univariate case). Moreover, `a`, `b` denote fixed effects and their interaction terms are included using "`:`". Gaussian random effects are included using the `f()` function. Here, we define an independent and identically distributed (`iid`) random effect for each observation, where `c` represents an index $1, \ldots, n \cdot m_1$. More complex models such as random walks, user-defined Gaussian models and even multivariate random effects are supported. Prior distributions for the hyperparameters of a random effect are defined within the corresponding `f()` function using the argument `hyper=list(...)`. Once the linear predictor is specified, the function `inla()` is called with the formula object as first argument:

```
result = inla(formula, data=data.frame(y,a,b,c, ...),
   family="zeroinflatednbinomial1",
   control.family=list(hyper=list(...), ...),
   control.fixed=list(correlation.matrix=TRUE, ...),
   control.compute=list(config=TRUE, ...),
   num.threads=4, ...).
```

Using the second argument `family`, we define the likelihood family. Settings for potential hyperparameters $\boldsymbol{\gamma}_i$ can be specified in the `control.family()` argument. Additional specifications for the fixed effects are listed in `control.fixed`. Here, we specify that the correlation matrix between the fixed effects should be returned. This is needed, for example, if we would like to look at the joint posterior marginal distributions of several fixed effects for one sample. Using the argument `control.compute`, different options regarding the output generation are available. By setting `config=TRUE`, INLA saves all internal approximations. The `r-inla` package is an interface to the core `INLA` program which is implemented in `C` and parallelized using OpenMP. By using the argument `num.threads` we can specify the number of threads, which the `r-inla` program will use, but it is usually more efficient to parallelize computations over genomic features.

After the `inla(.)`-call finishes, all output is saved in the `result` object. Marginal posteriors of fixed effects are saved in `result$marginals.fixed` and those of random effects in `result$marginals.random`. Summary information including posterior mean, standard deviation and posterior quantiles is found in `result$summary.fixed` and `result$summary.random`, respectively.

Linear combinations between different latent components can be specified within the `inla()`-call and become part of the latent field $\boldsymbol{v}$. Non-linear combinations or multivariate posterior marginals are not directly available. However, the function

```
inla.posterior.sample(n=1000, result)
```

may be used to generate 1,000 samples, say, from the approximated joint posterior distribution of the fitted model, as stored in `result`. These samples can then be further processed to derive quantities of interest. For further details, we refer the interested reader to www.r-inla.org.

4.3 Combining INLA with Empirical Bayes

Flexibility, computational efficiency (relative to other Bayesian methods) and accuracy make `INLA` a very suitable method for analysis of RNA-seq and other count-based sequencing data. In particular the versatility in terms of count models is attractive: besides the often used negative binomial (Poisson-gamma) likelihood model, it provides alternatives like Poisson-Gaussian, beta-binomial and zero-inflated versions thereof. `INLA`, however, is not designed for high-dimensional data, and hence lacks some functionality for this purpose. The `R`-package `ShrinkBayes` [15] aims at providing such functionality.

In the following sections, all models are assumed to be univariate in the feature space $i = 1, \ldots, m$, which allows the use of some notational shortcuts: $\boldsymbol{y}$ denotes any random count vector of length n, whereas $\mathbf{Y}_i = (Y_{i1}, \ldots, Y_{in})$ specifically refers to the ith data row. In addition, we do not use a feature index i for any parameter, or sets of those including $\boldsymbol{\theta}$ and $\boldsymbol{v}$. Then, `ShrinkBayes` extends `INLA` by:

- Implementation of Bayesian multi-parameter shrinkage by empirical Bayes-type estimation of parameters of the prior distributions. The estimation method relies only on marginal posterior distributions: for any parameter $\alpha = \theta_k$ or $\alpha = v_\ell$ its prior $\mathrm{p}(\alpha)$ can be approximated by the empirical point-wise mean (over the data for feature i, $\mathbf{Y}_i$) of posteriors:

$$p(\alpha) = \int p(\alpha|\mathbf{y})P(\mathbf{y})d\mu(\mathbf{y}) \approx \frac{1}{m}\sum_{i=1}^{m} p(\alpha|\mathbf{Y}_i),$$

where $\mu()$ is a suitable measure for the data generating process. Iteration is then required, because the latter posteriors depend on the prior as well (hence, in the iteration on the current estimate of the prior).

- Extension of the class of admissible prior distributions for one central parameter of interest towards nonparametric or mixture priors, which may contain a point mass. For a smooth new prior p_{new} it uses a simple re-scaling argument to recompute posterior distributions from the `INLA`-results:

$$P_{\text{new}}(\alpha|\mathbf{y}) \propto P_{\text{inla}}(\alpha|\mathbf{y}) \cdot \frac{p_{\text{new}}(\alpha)}{p_{\text{inla}}(\alpha)},$$

where the integral required to compute the proportionality constant is conveniently computed by means of the function `inla.expectation`. For mixture priors it combines the `INLA`-results from the fits under the separate mixture components and the corresponding marginal likelihoods.

- Providing a Bayesian estimate of FDR [BFDR;16] allowing easy interpretation of significant results. Note that BFDR is actually an estimate of the False Discovery Proportion (FDP), rather than FDR, but we use the conventional terminology here. Basically, $\text{BFDR}(t) = E[\text{lfdr}|\text{lfdr} < t]$, where local FDR, lfdr, equals the posterior probability of the parameter to be included in the null-domain $\mathcal{H}_0$: $\text{lfdr} = P(\alpha \in \mathcal{H}_0|\mathbf{y})$.

Multi-parameter shrinkage can be particularly beneficial for relatively small studies: shrinkage of a nuisance parameter diminishes its influence and consumption of degrees of freedom when it is unimportant across features, shrinkage of dispersion-related parameters leads to more reproducible results, and shrinkage of the parameter of interest leads to better inference in terms of FDR estimation.

The versatility of `INLA` allows the extension of `ShrinkBayes` in various directions, such as data integration, gene set testing and multi-parameter inference using multivariate priors. Below we detail the latter. A disadvantage of `INLA` (relative to MCMC) is that, by default, it only approximates marginal posterior distributions. The illustration below, however, also shows that summaries and samples of the posterior can be computed with the most recent version of `INLA`.

4.4 Extension: Bayesian Multivariate Shrinkage

In many practical problems, joint inference on multiple parameters is desirable. In a frequentist GLM-setting, likelihood-ratio tests may serve this purpose, although these generally do not account for potential relations between the parameters and are not straightforward in application when the parameters are a mix of regression parameters and hyper-parameters (e.g. zero-inflation or dispersion). Below we illustrate how to extend INLA for the purpose of multivariate shrinkage and inference. Two practical problems motivate this work. First, in an RNA-seq setting, we wish to jointly infer the difference in mean and difference in zero-inflation. Excess of zeros is a phenomenon observed by many for RNA-seq data, but often not modeled explicitly, partly because it complicates inference. INLA nicely allows for differential zero-inflation by use of strata. Combined inference for differential means and differential zero-inflation may lead to more power than inference for differential means only [14], and may render more meaningful parameter estimates.

Second, we wish to improve inference when correlation in the parameters is likely to occur. For example, suppose a fixed set of microRNAs may regulate genes of a given pathway in a coordinated manner or two pathological tumor staging covariates are available, and one would prefer not to a priori select one of the two.

We will consider three types of models. Initially, we do not distinguish between regression parameters and hyper-parameters, and simply denote all parameters for which joint inference is desired by $\boldsymbol{\alpha} \subset \bigcup_k \{\theta_k\} \cup \bigcup_\ell \{v_\ell\}$. For example, let β_1 and β_2 denote the means of group 1 and group 2, respectively, and ω_1 and ω_2 their group-specific zero-inflation parameters. Then $\boldsymbol{\alpha} = \{\beta, \omega\}$, with $\beta = \beta_2 - \beta_1$ and $\omega = \omega_2 - \omega_1$ includes the differences in mean and zero-inflation between the two groups. The three models share the same core model $\mathscr{M}$ and differ only by the prior on $\boldsymbol{\alpha}$. The core model $\mathscr{M}$ may depend on the context, and hence is not specified here. Example models which contain a zero-inflated negative binomial likelihood, a log-linear regression for the mean and appropriate priors are given in [15]. We do emphasize that in our setting $\mathscr{M}$ should contain an intercept; hence, group-related parameters are interpreted as deviations from the intercept. Below we simply identify the three models of interest by the prior on $\boldsymbol{\alpha}$; null model $\mathscr{M}_0$ with a point mass on $\mathbf{0}$, and alternative models $\mathscr{M}_1^U$ and $\mathscr{M}_1^J$ with univariate (U) priors on the R single components of $\boldsymbol{\alpha}$, or a joint (J) prior for $\boldsymbol{\alpha}$:

$$\mathscr{M}_0 : \pi(\boldsymbol{\alpha}) = \delta(\mathbf{0})$$

$$\mathscr{M}_1^U : \pi(\boldsymbol{\alpha}) = \pi^U(\boldsymbol{\alpha}) = \prod_{r=1}^{R} \pi_r(\alpha_r)$$

$$\mathscr{M}_1^J : \pi(\boldsymbol{\alpha}) = \pi^J(\boldsymbol{\alpha}),$$

where δ is the delta-Dirac function to denote a point mass. Ultimately, we are interested in obtaining posteriors under the mixture model (allowing some abuse of notation):

$$\mathscr{M}_{\text{mixt}}^J = q_0 \mathscr{M}_0 + (1 - q_0)\mathscr{M}_1^J. \tag{4.6}$$

For this, the marginal likelihoods under models $\mathscr{M}_0$ and $\mathscr{M}_1^J$ are required. The auxiliary model $\mathscr{M}_1^U$ is needed for obtaining the desired results from INLA, which only allows univariate priors as input. Below we show how to re-scale a marginal likelihood obtained from `INLA` under univariate priors to obtain a marginal likelihood under the joint prior. The marginal likelihoods under model $\mathscr{M}_1^J$ and $\mathscr{M}_1^U$ are denoted by $\mathrm{p}^J(\boldsymbol{y})$ and $\mathrm{p}^U(\boldsymbol{y})$, respectively. Then, we have:

$$\begin{aligned}\mathrm{p}^J(\boldsymbol{y}) &= \int \mathrm{p}^J(\boldsymbol{y}|\boldsymbol{\alpha})\pi^J(\boldsymbol{\alpha})d\boldsymbol{\alpha} = \int \mathrm{p}^U(\boldsymbol{y}|\boldsymbol{\alpha})\pi^J(\boldsymbol{\alpha})d\boldsymbol{\alpha} = \mathrm{p}^U(\boldsymbol{y}) \int \frac{\mathrm{p}^U(\boldsymbol{\alpha}|\boldsymbol{y})\pi^J(\boldsymbol{\alpha})}{\prod_{r=1}^R \pi_r(\alpha_r)} d\boldsymbol{\alpha} \\ &= \mathrm{p}^U(\boldsymbol{y}) E_{\mathrm{p}^U(\boldsymbol{\alpha}|\boldsymbol{y})}\left[\frac{\pi^J(\boldsymbol{\alpha})}{\prod_{r=1}^R \pi_r(\alpha_r)}\right]. \end{aligned} \tag{4.7}$$

Likewise, it is easy to show that

$$\mathrm{p}^J(\boldsymbol{\alpha}|\boldsymbol{y}) \propto \mathrm{p}^U(\boldsymbol{\alpha}|\boldsymbol{y}) \frac{\pi^J(\boldsymbol{\alpha})}{\prod_{r=1}^R \pi_r(\alpha_r)}. \tag{4.8}$$

Both (4.7) and (4.8) require numerical integration, which is explained below. Also, we need to estimate the unknown prior parameters in (4.6): q_0 and the parameters of $\pi^J(\boldsymbol{\alpha})$. After transforming all components of $\boldsymbol{\alpha}$ that are not yet on a Gaussian scale to a Gaussian scale, e.g. by a logistic-transformation on the zero-inflation parameters, it seems reasonable to use $\pi^J(\boldsymbol{\alpha}) = N(\mathbf{0}, \boldsymbol{\Sigma})$. Note that in `INLA` the Gaussian scale is default for all parameters in $\boldsymbol{v}$. Likewise, on the internal scale, the hyper-parameters for zero-inflation and over-dispersion are endowed with Gaussian priors in `INLA`. However, for other hyper-parameters, e.g. precisions with log-Gamma priors, a different multivariate distribution may be required (e.g. constructed by use of a copula). The methodology below applies to any multivariate parametric form.

Iterative estimating of q_0 and $\boldsymbol{\Sigma}$ is performed analogously as in [15]: sample from the empirical (point-wise) mean of current posteriors under mixture model $\mathscr{M}_{\mathrm{mixt}}^J$: $\mathrm{p}_{\mathrm{mixt}}^J(\boldsymbol{\alpha}|\mathbf{Y}_i)$, where index $i = 1, \ldots, m$ represents a feature and $\boldsymbol{\alpha}$ denotes the corresponding instance of $\boldsymbol{\alpha}$; then, $\hat{q}_0$ is simply the fraction of features for which $\boldsymbol{\alpha} = \mathbf{0}$ and $\hat{\Sigma}$ is the usual covariance estimate pooled across all features for which $\boldsymbol{\alpha} \neq \mathbf{0}$. Note that given the current estimates of $q_0, \mathrm{p}^J(\boldsymbol{y}), \mathrm{p}^J(\boldsymbol{\alpha}|\boldsymbol{y})$ and the marginal likelihood under null-model $\mathscr{M}_0$, $\mathrm{p}_0(\boldsymbol{y})$ (available from `INLA`), it is straightforward to compute $\mathrm{p}_{\mathrm{mixt}}^J(\boldsymbol{\alpha}|\boldsymbol{y})$ using Bayes' rule:

$$\mathrm{p}_{\mathrm{mixt}}^J(\boldsymbol{\alpha}|\boldsymbol{y}) = \begin{cases} q_0\, \mathrm{p}_0(\boldsymbol{y})/\mathrm{p}(\boldsymbol{y}), & \text{for } \boldsymbol{\alpha} = \mathbf{0}; \\ (1-q_0)\, \mathrm{p}^J(\boldsymbol{y})\, \mathrm{p}^J(\boldsymbol{\alpha}|\boldsymbol{y})/\mathrm{p}(\boldsymbol{y}), & \text{for } \boldsymbol{\alpha} \neq \mathbf{0}, \end{cases} \tag{4.9}$$

with $\mathrm{p}(\boldsymbol{y}) = q_0\, \mathrm{p}_0(\boldsymbol{y}) + (1-q_0)\, \mathrm{p}^J(\boldsymbol{y})$.

The integral in (4.7) is approximated by a Monte Carlo sum, so we use the fact that the integral can be written as an expectation. We generate n samples from the *joint* posteriors $p^U(\boldsymbol{\alpha}|\boldsymbol{y})$ using the function `inla.posterior.sample`, then compute $w_s = \pi^J(\boldsymbol{\alpha}^s)/\prod_{r=1}^R \pi_r(\alpha_r^s)$ for each sample $\boldsymbol{\alpha}^s$ and compute $\sum_{s=1}^S w_s/S$. The function `inla.posterior.sample` is very convenient here, in particular when $\boldsymbol{\alpha}$ consists of both regression and hyper-parameters. Use of this function requires to set `config=TRUE` in the `control.compute` argument of the `inla` function. The sampling by `inla.posterior.sample` is, however, somewhat slow when many samples are required for many features $i = 1, \ldots, m$. When $\boldsymbol{\alpha}$ consists of fixed effect parameters only, there is a very fast and often accurate alternative: as indicated before, `INLA` computes posterior correlations between the components of $\boldsymbol{\alpha}$ when setting `correlation.matrix=TRUE` in the `control.fixed` argument of the `inla` function. These correlations are used to approximate the joint posterior by a multivariate Gaussian, which is then used for extremely fast sampling. The algorithm to perform inference under mixture model $\mathscr{M}^J_{\text{mixt}}$ is given below.

Algorithm

1. Fit models $\mathscr{M}_0$ and $\mathscr{M}^U_1$ with `INLA`.
2. Initiate q_0, e.g. $q_0 = 0.9$ and $\pi^J(\boldsymbol{\alpha}) = \prod_{r=1}^R \pi_r(\alpha_r)$.
3. Start iterative estimation of q_0 and $\pi^J(\boldsymbol{\alpha})$ until convergence

   a. Compute $\mathrm{p}^J_{\text{mixt}}(\boldsymbol{\alpha}|\mathbf{Y}_i)$ for $i \in \mathscr{I}$ using (4.7)–(4.9). Here, $\mathscr{I}$ denotes a random subset of the entire set of feature indices $\{1, \ldots, m\}$ with $|\mathscr{I}| < m$, see Remark 3(a) below.
   b. Sample from the empirical mixture of the posteriors $\mathrm{p}^J_{\text{mixt}}(\boldsymbol{\alpha}|\mathbf{Y}_i), i \in \mathscr{I}$.
   c. Re-estimate $q_0, \pi^J(\boldsymbol{\alpha})$ and recompute $\mathrm{p}^J(\boldsymbol{\alpha}|\mathbf{Y}_i)$.

4. Compute $\mathrm{p}^J_{\text{mixt}}(\boldsymbol{\alpha}|\mathbf{Y}_i)$ for all $i = 1, \ldots, m$.
5. Compute local false discovery rate (lfdr): $\text{lfdr} = \mathrm{p}^J_{\text{mixt}}(\mathbf{0}|\mathbf{y})$ and, for any threshold t, the Bayesian False Discovery Rate: $\text{BFDR}(t) = E[\text{lfdr}|\text{lfdr} < t]$.

Remarks.

1. For numerical stability in the computation of (4.7) we prefer the univariate priors to be not very vague, but still cover a reasonably large range.
2. Hence, the initial joint prior is just the product prior of Gaussians.
3. Convergence is assessed by successive evaluation of the total log-marginal likelihood, which equals $\sum_{i=1}^{|\mathscr{I}|} \log \mathrm{p}(\mathbf{Y}_i)$.

   a. The random set $\mathscr{I}$ should be large enough for the empirical Bayes estimation of the priors to work, but may be smaller than m for computational convenience. We use $|\mathscr{I}| = \min\{m, 10^4\}$.
   b. Generate, e.g., five samples for each $i \in \mathscr{I}$.
   c. Use (4.8) for computing $\mathrm{p}^J(\boldsymbol{\alpha}|\mathbf{Y}_i)$.

4. This may be time-consuming when $\boldsymbol{\alpha}$ contains regression and hyper-parameters (requiring use of `inla.posterior.sample`). One may consider to first apply an initial liberal significance screening using univariate priors.
5. Hence, for threshold t, BFDR is simply computed by averaging all lfdrs smaller than t.

4.5 Multivariate Shrinkage Improves Feature Selection at Given FDR

We conducted a simulation study for two settings: (I) four correlated regression parameters β_ℓ and (II) correlated differential mean group parameter β_g and differential zero-inflation parameter ω_g (on the internal logit-scale). Setting I is a 4-group comparison with four samples per group for 5,000 features, 1,000 of which are differential. The four differential parameters β_ℓ are drawn from a $N(\mathbf{0},\boldsymbol{\Sigma})$ with $\Sigma_{\ell,\ell} = 1$ and $\Sigma_{k,\ell} = \rho$ for $k \neq \ell = 1,\ldots,4$, with $\rho = 0.4, 0.8$. We assume $\ell = 1,\ldots,4$ represent 4 levels of a factor. Any data point corresponding to level ℓ is drawn from a negative binomial distribution with log-mean equal to $\beta_\ell + 5$ and equal overdispersion varying over features according to a log-normal distribution with $\mu = \sigma = 0.5$ (roughly based on values observed in real data). Setting II is a 2-group comparison with ten samples per group for 1,000 features, 200 of which are differential. Here, β_g and ω_g are drawn from a bivariate $N(\mathbf{0},\boldsymbol{\Sigma}')$ with $\Sigma'_{g,g} = 1$ and $\Sigma'_{g,h} = \rho'$ for $g \neq h = 1,2$, with $\rho = -0.4, -0.8$. The negative correlation reflects the realistic situation of an increase in group mean to go hand-in-hand with a decrease in zero-inflation. For both groups each data point is drawn from a zero-inflated negative binomial distribution with zero-inflation equal to $\exp(\omega_g)/(1+\exp(\omega_g))$, log-mean equal to $\beta_g + 2$ and equal overdispersion varying over features according to a log-normal distribution with $\mu = \sigma = 0.5$.

The iterative procedures described above seem fairly adequate in estimating the parameters of the multivariate priors. The estimate of the proportion of non-differential features, $\hat{q}_0$, ranges from 0.79 to 0.88 across the four simulations, whereas for $|\rho| = 0.4, 0.8$ $\hat{\rho}_{\mathrm{I}} = 0.32, 0.51$ and $\hat{\rho}_{\mathrm{II}} = -0.43, -0.61$ for settings I and II, respectively. The somewhat conservative estimate of $\hat{q}_0$ (true $q_0 = 0.8$) is probably caused by features with small values of the differential parameters for which the null-model is preferred. For the cases $|\rho| = 0.8$ the estimate of ρ is dampened by the noise in the data.

In setting I, we compared the results from multivariate shrinkage with those of univariate shrinkage, which only shrinks the variances of β_ℓ using the same procedures. We also compare with two popular p-value based methods: `edgeR` v3.2.3, [8] and `DESeq` v1.12.0, [1], using their default settings. Both methods provide likelihood ratio tests in the negative binomial setting for K-sample inference. The ROC-curves in Fig. 4.1 show that the shrinkage-based methods render somewhat better feature rankings than the p-value based methods, in particular when the correlation is strong. Quality of ranking is improved only marginally when using multivariate rather than univariate shrinkage. However, when considering the actual number of selected features at target BFDR $= 0.1$, multivariate shrinkage seems beneficial, see Table 4.1.

In setting II, we compared the results from multivariate shrinkage with those of univariate shrinkage and those of a simple alternative model with *group-independent* zero-inflation and *differential* group mean. Again, the ROC-curves in Fig. 4.2 show only small qualitative differences in terms of rankings, but when considering the

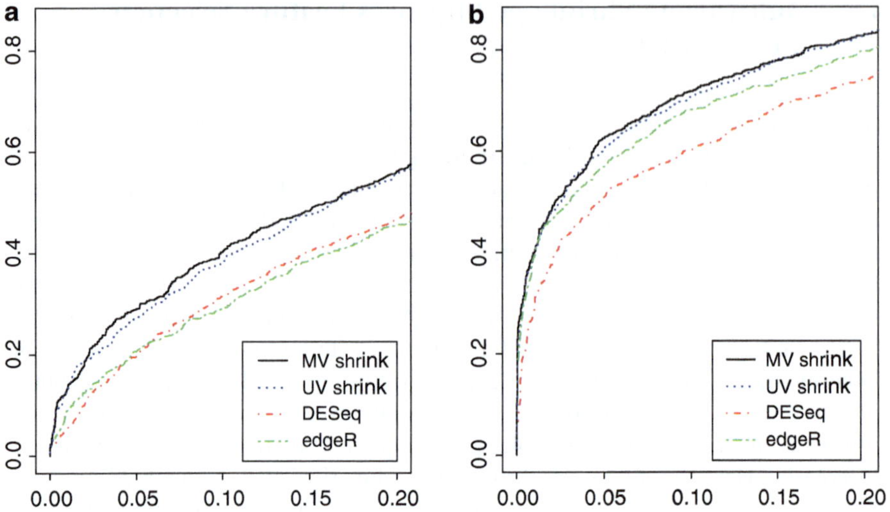

Fig. 4.1 ROC-curves (x-axis: FPR vs y-axis: TPR) for simulation setting I, correlated β's, for $\rho = 0.4$ (**a**) and $\rho = 0.8$ (**b**)

Table 4.1 Selection results for simulation setting I

	# significant		True FDP	
	MV	UV	MV	UV
$\rho = 0.4$	339	293	0.053	0.044
$\rho = 0.8$	38	32	0.105	0.125

 # significant number of significant features at target BFDR$= 0.1$, *UV* univariate Gaussian prior approach, *MV* multivariate Gaussian prior approach, *True FDP* true false discovery proportion of the significant features

actual number of selected features at target BFDR $= 0.1$ in Table 4.2, differential zero-inflation plus multivariate shrinkage is somewhat superior when the correlation is strong ($\rho = -0.8$; most detections and comparable BFDRs). For weaker correlation ($\rho = -0.4$), the simple alternative model (ZI+UV), is competitive: its corresponding true BFDR is closer to the target, but it also detects fewer features.

In both settings, the small gain in terms of ranking is not unexpected: the prior (which is the same for all features) is unlikely to have a strong impact on the ranking and in setting II the group mean parameter of the alternative model can also partly pick up differences in zero-inflation. For actual selection, however, the prior is known to be very important in a multiple testing setting [12], which explains the larger differences in the selection results.

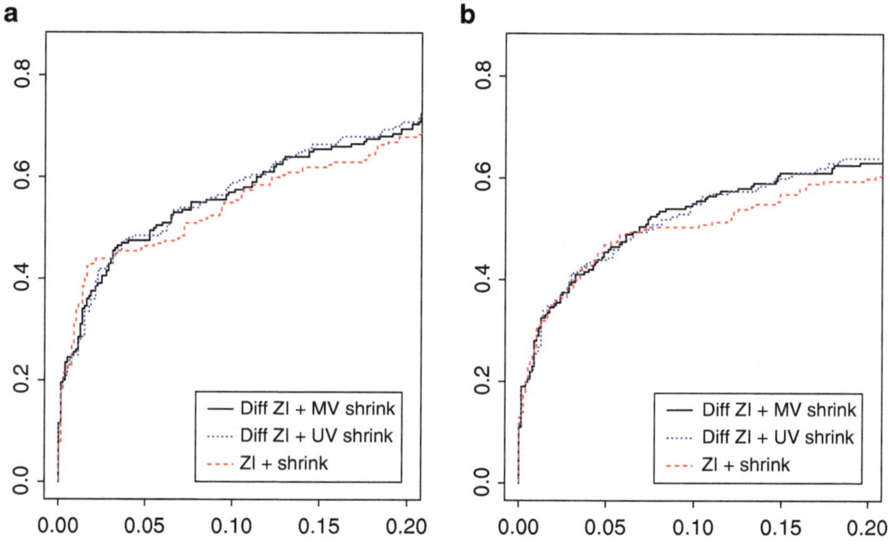

Fig. 4.2 ROC-curves (x-axis: FPR vs y-axis: TPR) for simulation setting II, differential zero-inflation correlated with differential β, for $\rho = -0.4$ (**a**) and $\rho = -0.8$ (**b**)

Table 4.2 Selection results for simulation setting II

	# significant			True FDP		
	diff ZI+MV	diff ZI+UV	ZI+UV	diff ZI+MV	diff ZI+UV	ZI+UV
$\rho = -0.4$	65	61	58	0.138	0.148	0.103
$\rho = -0.8$	68	62	61	0.132	0.145	0.131

 # significant number of significant features at target BFDR $= 0.1$, *UV* univariate Gaussian prior approach, *MV* multivariate Gaussian prior approach, *diff ZI* model with differential zero-inflations, *ZI* model with group-independent zero-inflation, *True FDP* true false discovery proportion of the significant features

4.6 RNA-seq Analysis of Lymphoblastoid Cell Lines

As a proof of principle, we analyzed RNA-seq data from lymphoblastoid cell lines generated as part of the International HapMap project from 69 unrelated Nigerian individuals [6]. We contrasted 29 males to 40 females. Given this contrast, we focus on chromosomes X and Y: one expects the majority of differences here; hence, a full genome analysis could largely diminish power in a multiple testing setting, in particular because X and Y are relatively small chromosomes. As a reference, we do add chromosome 1 to compare the findings for chromosomes X and Y to that of chromosome 1. The entire count matrix was normalized using edgeR's `calcNormFactors` function, with "`method=TMM`" [7]. This function produces after multiplication by the relative library size a normalization factor for each individual. Raw counts were divided by this factor and rounded to integers again. Modest filtering on the minimum number of non-zeros was applied, which should be at least 10. The zero-inflated negative binomial likelihood [15] is used for the analysis. Since we aim to assess the effect of allowing group-dependent zero-inflation instead of group-independent zero-inflation, we focus on features with at least a modest number of zeros, minimally 10 %. This renders 1,811 features, of which 66 % are located on chromosome 1, 28 % on chromosome X and the remaining 6 % on chromosome Y.

At BFDR ≤ 0.1, the multivariate shrinkage model $\mathscr{M}_1^J$ with differential zero-inflation (diffZI+MV) identifies 18 significant features, of which only 1 is located on chromosome 1, 4 are located on chromosome X and 13 are located on chromosome Y. The difference in zero-inflation proportions can be large: its estimate exceeds 0.3 for 50 % of those features, which supports the explicit modeling of differential zero-inflation. The estimated bivariate prior for differential zero-inflation ($\omega = \omega_2 - \omega_1$; on the logit-scale) and differential group effect ($\beta = \beta_2 - \beta_1$) indeed contains a quite strong negative correlation: $\rho = -0.50$. This indicates that multivariate shrinkage is relevant here. For all these 18 features, the estimate of ω is in line with that of β in the sense that either ω is very small (and β is not) or they are in opposite directions. This facilitates interpretation. Note that the multivariate prior with negative correlation can in fact help to avoid 'detecting' features with contradictive estimates of ω and β.

We compared the results from diffZI+MV with those from the more standard model: group-independent zero-inflation with univariate shrinkage for the differential group effect (ZI+UV). Of course, the two models are rather competitive in terms of power, in particular because the differential group effect in the standard model may partly pick up a difference in zero-inflation as well. We define a quasi-False Positive Rate, quasiFPR, as the ratio of the number of detected features located on chromosome 1 and #P: the total number of detections. We are aware that, for RNA, differences between males and females may exist on the autosomes; yet quasiFPR quantifies to what extent a method succeeds in prioritizing the sex chromosomes over the autosomes from the data only. Figure 4.3a shows quasiFPR for $10 \leq$#P≤ 40, which covers the range of number of detections for any $0.05 \leq$ BFDR ≤ 0.2. Note that, locally, an increase in number of selected positives, reduces the quasiFPR when only the denominator of this quantity increases, which explains the zigzag pattern. We observe that diffZI+MV is better able to prioritize the sex chromosomes features than ZI+UV. When including chromosomes other than 1, we observed similar performances: e.g. quasiFPR is 30–40 % lower for diffZI+MV than for ZI+UV when including chromosomes 2 and 3.

Since outliers are known to be a potential problem, in particular for features in the low range spectrum, we briefly study the effect of creating one single outlier for 25 features in the lymphoblastoid data set; the outlier was created by multiplying the largest count for the specific feature by 10. Given the sparse signal in the data set ($\hat{q}_0 = 0.982$) (and the majority of features with an outlier lies on chromosome 1) it seems reasonable to assume that any detection of these 25 is false. Figure 4.3b shows the FPR for $10 \leq$#P≤ 40, hence the same range as before. We observe that diffZI+MV is more robust against outliers than ZI+UV. The intuitive reason for this

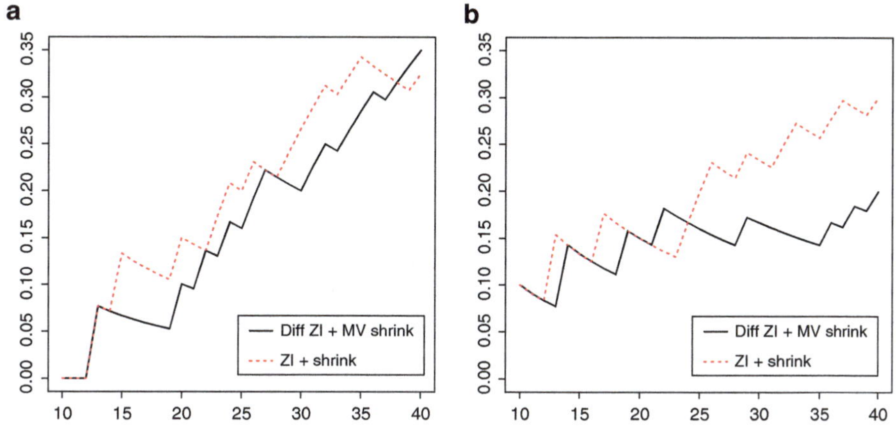

Fig. 4.3 Number of detections (x-axis) versus quasiFPR (**a**) and FPR (**b**) in the chromosome 1+X+Y data set, and the same data set with 25 artificial outliers, respectively

is that the diffZI+MV model recognizes that the $\beta_2 - \beta_1$ difference is not supported by the zero-inflation difference, $\omega_2 - \omega_1$: the apparent increase in β_2 (if the outlier is in the second group) is countered by an increase in ω_2 to better fit the zeros, which does not comply well with the a priori negative correlation, and leads to a decrease in marginal likelihood.

4.7 Discussion

We demonstrated the potential of `INLA` for the analysis of sequencing data. Here, we focused mostly on flexibility in terms of prior-specification by extending `INLA` to allow multivariate priors. In addition, `INLA` provides a large amount of flexibility regarding specification of the observation model, the regression model, and the definition of hyperparameter settings. In the following, we will shortly summarize these three levels.

For count data, several likelihood functions such as Poisson, binomial, negative binomial and their zero-inflated variations are available. Furthermore, it is possible to combine several likelihood functions, so that different groups of observations can have different likelihoods [4]. For example, one subset may follow a Poisson likelihood and the rest a Gaussian likelihood. This is an interesting feature for the coupling of two (or more) observed processes which are assumed to share the same covariates or latent structure. A potential application might be data integration where different high-throughput sequencing experiments are used to learn about a biological phenomenon.

The linear predictor can be additively composed by fixed effects and different types of random effects. Dependence along the genome, say, can be accounted for by the random effects. This makes sense whenever neighboring regions are likely to exhibit more similar behavior than regions lying further apart. Even correlated random effects between different data sources, for example methylation sequencing data and copy number variation data, can be specified within the same framework without requiring novel application-specific implementations.

The advantage of a Bayesian approach is the incorporation of hyperparameter uncertainty. While the latent parameters need to follow a Gaussian distribution (in the standard `INLA`-setting; this can be relaxed in `ShrinkBayes`), arbitrary prior distributions can be designed to the hyperparameters. Here, the user can choose from different standard distributions, but also use expert-based prior distributions. These can be either defined using a function or by assigning a matrix which represent the prior distribution by specific paired x and y values. This is of interest when incorporating either expert knowledge or information available from related studies.

The flexibility of `INLA` and its extension `ShrinkBayes` comes at a price: although more efficient than most other Bayesian approaches, it is inherently computationally more demanding than frequentist methods. Typically, an analysis with `ShrinkBayes` takes a few hours, whereas p-value based methods usually render results within a couple of minutes. The comparison with edgeR and DESeq in Fig. 4.1 should also be seen in that light: while, in this setting, `ShrinkBayes` performs better, edgeR and DESeq are more convenient in use given their computational advantage. Also, performances are likely to converge when sample sizes increase due to the diminishing effect of shrinkage. Hence, in practice the method of choice will depend on a number of factors like sample sizes, complexity of the inferential problem, presence of random effects, importance of accounting for zero-inflation and the balance between performance and computational burden.

In the future, computing times may decrease by reducing the `INLA`-overhead, which stores more results than necessary sometimes. Importantly, the number of features has little to no impact on computing time for the shrinkage procedures, only for the actual fitting procedures. Therefore, a pragmatic solution for very high-dimensional data sets may be to first apply an initial screen by a fast method. Then, apply shrinkage to a large enough random subset of all features, and apply the `INLA` fitting only to those features that pass a (liberal) initial screen. If the initial screen is a non-parametric one, this may also help in terms of robustness against outliers. Note that it can be verified whether the threshold for the initial screen was set too strictly: if many features close to the screening threshold turn out to be significant with `ShrinkBayes`, then a more liberal threshold should be used.

We conclude that `INLA` is certainly a useful addition to the colorful spectrum of analysis methods of count-based sequencing data. Possibly, its largest potential lies in complex inferential problems for which its abilities to use dedicated (multivariate) shrinkage priors, and to account for (correlated) random effects can be fully exploited.

References

[1] Anders, S., Huber, W.: Differential expression analysis for sequence count data. Genome Biol. **11**, R106 (2010)

[2] Cloonan, N., Forrest, A.R., Kolle, G., Gardiner, B.B., Faulkner, G.J., Brown, M.K., Taylor, D.F., Steptoe, A.L., Wani, S., Bethel, G., et al.: Stem cell transcriptome profiling via massive-scale mrna sequencing. Nat. Meth. **5**(7), 613–619 (2008)

[3] Gelfand, A.E., Smith, A.F.M.: Sampling-based approaches to calculating marginal densities. J. Am. Stat. Assoc. **85**(410), 398–409 (1990)

[4] Martins, T.G., Simpson, D., Lindgren, F., Rue, H.: Bayesian computing with INLA: new features. Comput. Stat. Data Anal. **67**, 68–83 (2013)

[5] Paul, M., Riebler, A., Bachmann, L.M., Rue, H., Held, L.: Bayesian bivariate meta-analysis of diagnostic test studies using integrated nested Laplace approximations. Stat. Med. **29**, 1325–1339 (2010)

[6] Pickrell, J., Marioni, J., Pai, A., Degner, J., Engelhardt, B., Nkadori, E., Veyrieras, J., Stephens, M., Gilad, Y., Pritchard, J.: Understanding mechanisms underlying human gene expression variation with RNA sequencing. Nature **464**, 768–772 (2010)

[7] Robinson, M.D., Oshlack, A.: A scaling normalization method for differential expression analysis of RNA-seq data. Genome Biol. **11**(3), R25 (2010)

[8] Robinson, M., McCarthy, D., Smyth, G.: edgeR: a bioconductor package for differential expression analysis of digital gene expression data. Bioinformatics **26**, 139–140 (2010)

[9] Rue, H., Held, L.: Gaussian Markov Random Fields: Theory and Applications. Chapman & Hall/CRC Press, London (2005)

[10] Rue, H., Martino, S., Chopin, N.: Approximate Bayesian inference for latent Gaussian models by using integrated nested Laplace approximations (with discussion). J. R. Stat. Soc. Series B **71**, 319–392 (2009)

[11] Schrödle, B., Held, L., Riebler, A., Danuser, J.: Using INLA for the evaluation of veterinary surveillance data from Switzerland: a case study. J. R. Stat. Soc. Series C (Appl. Stat.) **60**(2), 261–279 (2011)

[12] Scott, J., Berger, J.: An exploration of aspects of Bayesian multiple testing. J. Stat. Plan. Inference **136**, 2144–2162 (2006)

[13] Tierney, L., Kadane, J.B.: Accurate approximations for posterior moments and marginal densities. J. Am. Stat. Assoc. **81**(393), 82–86 (1986)

[14] Tse, S.K., Chow, S.C., Lu, Q., Cosmatos, D.: Testing homogeneity of two zero-inflated Poisson populations. Biom. J. **51**(1), 159–170 (2009)

[15] van de Wiel, M.A., Leday, G., Pardo, L., Rue, H., van der Vaart, A., van Wieringen, W.: Bayesian analysis of RNA sequencing data by estimating multiple shrinkage priors. Biostatistics **14**, 113–128 (2012)

[16] Ventrucci, M., Scott, E.M., Cocchi, D.: Multiple testing on standardized mortality ratios: a Bayesian hierarchical model for FDR estimation. Biostatistics **12**, 51–67 (2011)

Chapter 5
Design of RNA Sequencing Experiments

Dan Nettleton

Abstract This chapter presents strategies for the design of RNA sequencing (RNA-seq) experiments aimed at identifying differentially expressed genes. We discuss the multiphase nature of RNA-seq experiments and point out the utility of intentionally confounding nuisance factors, that inevitably arise in different design phases, with one another. We cover the concepts of biological and technical replication. We show that experimental designs that prioritize biological replication over both technical replication and increased sequencing depth per experimental unit provide improved assessments of differential expression. Several example experimental designs are presented to illustrate the featured design principles.

5.1 Introduction

RNA sequencing (RNA-seq) has become the preferred approach for comparing the abundance of each of tens of thousands of transcribed RNA molecules across treatments of scientific interest. A prototypical RNA-seq experiment involves the following steps:

1. A total of n experimental units are selected for use in the experiment.
2. The n experimental units are assigned to t treatments of interest.
3. Each experimental unit is treated with its assigned treatment.
4. A relevant biological sample of tissue or cells of a certain type is collected from each experimental unit.
5. An RNA sample is extracted from each biological sample.

D. Nettleton (✉)
Department of Statistics, Iowa State University, Ames, IA 50011-1210, USA
e-mail: dnett@iastate.edu

S. Datta and D. Nettleton (eds.), *Statistical Analysis of Next Generation Sequencing Data*, Frontiers in Probability and the Statistical Sciences, DOI 10.1007/978-3-319-07212-8_5,

6. Each RNA sample is used to generate a sample of cDNA fragments ready for sequencing with Next Generation Sequencing (NGS) technology. (This resulting sample is often referred to as a *library*, and this step is known as *library preparation*.)
7. NGS technology is used to determine the identity of millions of short nucleotide sequences (known as *reads*) from each library of cDNA fragments.

The reads generated from steps 1–7 are processed using bioinformatic algorithms that match reads from each sample to p RNA sequences of interest. These RNA sequences of interest could be genes, specific gene transcripts, exons, microRNAs, etc. For simplicity and generality, we will refer to the sequences of interest as *features* throughout the remainder of this chapter. The results of matching reads to features are often summarized in a $p \times n$ matrix of counts. The count in row i and column j of the matrix is assumed to be positively associated with the abundance of the ith feature in the jth experimental unit. A major goal of the ensuing analysis is to determine which, if any, features show significant evidence of changes in abundance due to treatment. Features whose abundance does truly depend on treatment are often referred to as *differentially expressed* (DE).

Each of steps 1–7 can introduce variation in the count matrix. The way each step is carried out can have important implications about how the count matrix should be interpreted and DE features identified. If the steps are performed carefully with experimental design principles in mind, the matrix of counts can contain valuable information about DE features. On the other hand, DE features may be unnecessarily difficult or impossible to identify if the matrix of counts is produced from a poorly planned and executed experiment. Because RNA-seq experiments are expensive endeavors, it is especially important that they be carefully designed and executed to make the best use of available resources.

In this chapter, we present practical experimental design advice for statisticians and biological researchers who wish to use RNA-seq to identify DE features. For simplicity of terminology and presentation, we focus primarily on experiments rather than observational studies. However, most of the ideas covered in this chapter apply to observational studies as well as experiments. We assume that readers have at least a basic understanding of the fundamental experimental design principles promoted by R.A. Fisher—namely, randomization, replication, and blocking. We discuss the important role of each principle in the context of RNA-seq experiments.

In addition to fundamental experimental design principles, RNA-seq experiments require special design considerations because the complex steps required for measuring transcript abundance levels with RNA-seq technology lead to *multiphase experiments*. McIntyre [15] discussed design and analysis strategies for two-phase experiments. In the first phase of a two-phase experiment, fundamental experimental design principles are used to apply treatments to experimental units as would be done in any standard experiment. The need for a second design phase arises when measuring the response variable or variables of interest is a complex process that requires experimental design considerations. Kerr [11], Jarrett and Ruggiero [10], and Nettleton [17] have discussed the two-phase nature of microarray experiments

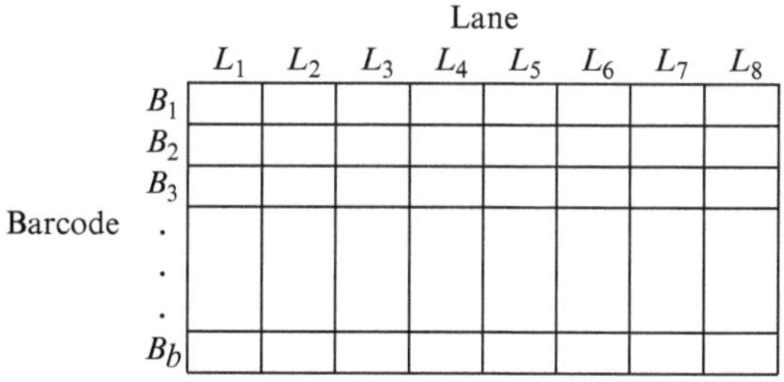

Fig. 5.1 A schematic representation of an Illumina flowcell. Each *box* in the figure represents a particular combination of barcode and lane that can be used to sequence a single library. In the final phase of an RNA-seq experimental design, libraries must be assigned to barcode × lane combinations

to measure the transcript abundance. Likewise, many RNA-seq experiments are multiphase experiments because the steps 1–7 described above can be naturally partitioned into distinct phases, with each phase requiring its own design considerations. The first phase is typically comprised of steps 1–3. The next phase or phases may naturally consist of steps 4 and 5. Finally, steps 6 and 7 can be viewed as a final phase that is often conducted by a specialized sequencing facility that produces read data from the RNA samples provided by an experimenter.

The final phase of an RNA-seq experiment is perhaps the most unique and interesting phase from a statistical perspective. The possible design options depend on the NGS sequencing platform. We will focus on the Illumina platform that has been considered by several others [1,4,5,8,14,16,21] in the RNA-seq experimental design context. Illumina sequencing occurs within the *lanes* of *flowcells*. Each flowcell has eight lanes. In the simplest scenario, eight libraries can be separately sequenced on a flowcell, with one library in each lane.[1] However, it is also possible to sequence multiple libraries together in a single lane using a technique known as *multiplexing* [4,7,9]. To enable attribution of a read to the appropriate library when multiplexing is used, a short library-specific DNA sequence known as a *barcode* is appended to the sequences in each library prior to sequencing. A given read generated from sequencing the combined and barcoded libraries in a single lane begins with the barcode that uniquely identifies the source library for that read. This allows for the feature read counts to be computed separately for each of the libraries sequenced together in a single lane.

With the structure provided by the Illumina NGS platform (see Fig. 5.1), there are many strategies that can be employed to measure feature abundance in the RNA samples extracted from experimental units. Ideally, the third-phase design strategies used to assign libraries to flowcells, lanes, and barcodes will match well

[1]Although Illumina has recommended that one of the eight lanes be used to sequence a special control sample, [5] showed that better results can be obtained by using all eight lanes to sequence libraries of direct interest.

with the design of the first and second phases of the RNA-seq experiment. Careful attention to the design of each phase and the relationship between phases will facilitate data analysis and discovery of DE features. In Sect. 5.4, we present several examples of effective RNA-seq experimental designs that account for the multiphase nature of RNA-seq experiments. Before presenting those examples, we address two issues that have important implications for the design of all phases of RNA-seq experiments: replication (Sect. 5.2) and sequencing depth (Sect. 5.3).

5.2 Replication

As stated in Sect. 5.1, a major goal of most RNA-seq experiments is to determine which features show significant evidence of changes in RNA abundance due to treatment. To determine if a treatment causes changes in the abundance of a feature, it is necessary to understand variation in feature abundance among independent experimental units treated alike. This variation in feature abundance among independent experimental units treated alike is often referred to as *biological variation*. To learn about biological variation, it is necessary to treat multiple independent experimental units alike and measure their feature abundance. This type of replication, known as *biological replication*, is an indispensable aspect of RNA-seq experiments aimed at identifying DE features. Biological replication allows for the assessment of biological variation, which permits a data analyst to determine if differences in feature abundance between experimental units treated differently are caused by the different treatments or are simply differences that can be expected by chance due to natural biological variation.

A second type of variation, known as *technical variation*, can be assessed through *technical replication*. Technical replication involves replicating the process of measuring feature abundance for one or more experimental units. Technical variation describes the variation in replicate measurements of feature abundance on a single experimental unit that is due to the measurement technology. For example, if a library derived from a single experimental unit is sequenced twice, once in each of two flowcell lanes, the two read counts obtained for a given feature often differ. This is not surprising because RNA-seq technology does not provide a complete and error free enumeration of all sequence fragments in a library. Instead, the collection of fragments that is sequenced and correctly matched to features is randomly selected from the sequence fragments in a library. Marioni et al. [14] provides empirical evidence that the read counts for a given feature obtained from repeated measurements of a single library tend to follow a Poisson distribution. However, it is reasonable to expect greater variation among measurements on a single experimental unit if the entire measurement process (steps 4 through 7 in Sect. 5.1) were replicated.

Biological and technical replication have been discussed in the context of RNA-seq experimental design by several authors. Examples include [4,6,8,16], and [21]. Auer and Doerge [4] emphasized the need for biological replication to draw conclusions about treatment effects that can be generalized beyond the experiment at hand. Busby et al. [6] pointed out that power for detecting DE features will be maximized by maximizing biological replication. Fang and Cui [8] claimed that "the most desirable replicates are biological replicates." Robles et al. [21] also favored biological over technical replicates. McIntyre et al. [16] assessed biological and technical variation in three experiments and concluded that "technical variation, while smaller than biological variation, cannot be ignored and should be accounted for in the study design."

In this chapter, we will argue that all replicates should be biological replicates when the number of measurements (i.e., number of libraries sequenced) is the limiting factor and the goal of the RNA-seq experiment is to detect features that are differentially expressed across treatments. Our claim is that, given a fixed number of measurements m, precision for estimating differential expression will be maximized by measuring m experimental units once each. This is relevant for RNA-seq experiments because the cost of each high-dimensional measurement (i.e., sequencing a library to obtain a vector of read counts) is non-trivial so that the number of measurements is the single most important factor driving experimental cost in many cases. There are, of course, exceptions. In situations where the number of experimental units is capped but budgets allow for additional measurements beyond the number of experimental units, technical replication can help reduce the impact of measurement error. However, it should be understood that repeatedly measuring the same experimental units is no substitute for measuring additional experimental units. These points are illustrated in the following subsections.

5.2.1 *A Simple Comparison of Replication Strategies*

We begin our argument in favor of exclusive biological replication with an elementary example outside the context of RNA-seq experimentation. Consider an experiment with a single treatment and a single response variable that can be measured multiple times for each experimental unit. Denote the jth measurement of the response variable for the ith experimental unit by

$$Y_{ij} = \theta + u_i + e_{ij}, \tag{5.1}$$

where θ is an unknown parameter, u_i denotes an unobserved random variable with mean zero and variance σ_u^2, and e_{ij} denotes an unobserved random variable with mean zero and variance σ_e^2. Furthermore, suppose all u_i and e_{ij} terms are mutually independent. Suppose an unlimited number of experimental units are available but we can afford to make only four measurements of the response. If our goal is to estimate θ, how shall we design the measurement process?

This is a classic measurement-error scenario where the unobserved value $\theta + u_i$ represents the true value of the response for the ith experimental unit, while Y_{ij} is the jth measurement of the response for the ith experimental unit, which differs from the

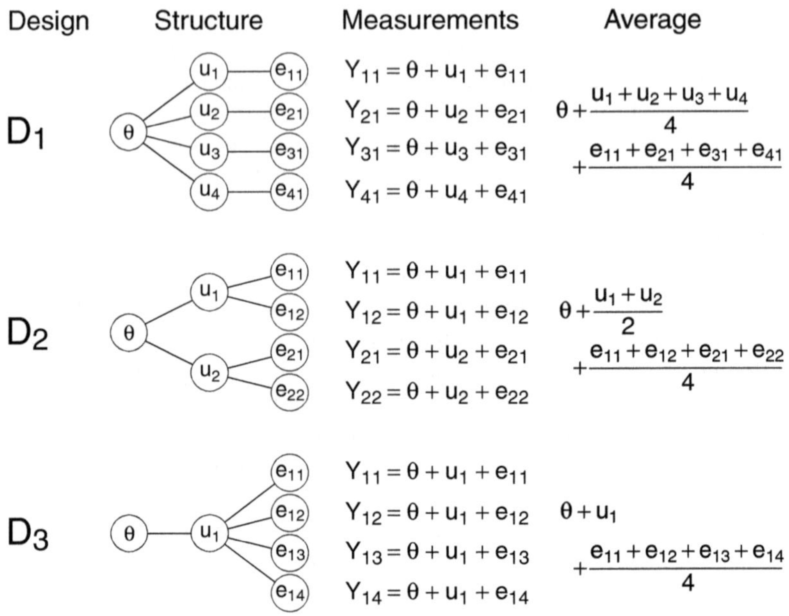

Fig. 5.2 A comparison of three replication strategies for estimating a mean parameter θ in a simple linear model with random experimental unit effects (u_i) and random measurement errors (e_{ij})

true value of the response by the measurement error e_{ij}. The variance components σ_u^2 and σ_e^2 represent variation in the true response among experimental units and variation among measurements of the response on an individual experimental unit, respectively.

Three potential designs for the measurement process are illustrated in Fig. 5.2. In design D_1, four different experimental units are each measured once. In design D_2, two experimental units are each measured twice. One experimental unit is measured four times in design D_3. For all three designs, the average of the response measurements is an unbiased estimator of θ, and the best design can be identified by determining which estimator of θ has the smallest variance. Expressions for the average of the response measurements are presented on the right side of Fig. 5.2. The variance of the average is $\sigma_u^2/4 + \sigma_e^2/4$ for design D_1, $\sigma_u^2/2 + \sigma_e^2/4$ for design D_2, and $\sigma_u^2 + \sigma_e^2/4$ for design D_3.

Note that design D_1 is guaranteed to have the smallest variance of all three designs regardless of the values of σ_u^2 and σ_e^2 (unless $\sigma_u^2 = 0$, in which case all three designs have the same variance). This is clearly a special case of a more general result that implies that each experimental unit should be measured at most once, regardless of the measurement error variance, when the number of experimental units is plentiful, the number of measurements is limited, and the goal is to estimate θ with minimum variance. The result can be easily extended to the multiple treatment case, where a design D_1 measurement strategy should be used for each treatment group if the goal is to estimate or test for differences in treatment means. Although this result is very simple from the statistical standpoint, many experimenters find it surprising. Their intuition may incorrectly suggest that measurements of the response of an experimental unit should be repeated, especially when the measurement error variance is known to be high.

There are, of course, situations where multiple measurements per experimental unit are important. For example, in order to separately estimate σ_u^2 and σ_e^2, a design like design D_2 is needed. Thus, when a new measurement technology is developed, it is quite natural to conduct some initial experiments using designs like design D_2 to obtain information about both variation in true responses of experimental units (σ_u^2) and variation introduced by the measurement process (σ_e^2). A design like design D_2 can also be useful if the number of experimental units is limited. For example, design D_2 would clearly be preferable to a design in which two experimental units are each measured just once. A design like design D_3 can be useful if the entire goal of the experiment is to estimate only the measurement error variance σ_e^2 or the response of a particular experimental unit ($\theta + u_i$) rather than the mean response of experimental units (θ). However, a design like design D_3 is a disastrous choice if the goal is to estimate θ (or a difference in treatment means in the multiple treatment case); in addition to producing an estimator that is more variable than the estimators of designs D_1 and D_2, design D_3 provides no information about σ_u^2, which makes it impossible to construct a confidence interval or test a hypothesis about θ (or a difference in treatment means in the multiple treatment case).

5.2.2 A Comparison of Replication Strategies for RNA-seq

The simple design comparison of Sect. 5.2.1 provides some support for using exclusively biological rather than technical replication. However, the linear model with additive random effects assumed for the responses in Sect. 5.2.1 is not appropriate for RNA-seq count data. Currently, the most popular model for RNA-seq data assumes that the counts for a given feature follow a negative binomial distribution with a treatment-specific mean and a dispersion parameter common across treatments (see, for example, [2, 3, 18–20], and Chaps. 2 and 3). It is well known that a negative binomial distribution can be characterized as a gamma mixture of Poisson distributions. In this subsection, we will assume that RNA-seq count data follow a lognormal mixture of Poisson distributions. Because of the similarity between lognormal and gamma distributions, a lognormal mixture of Poissons is very similar to a negative binomial distribution but has an advantage when modeling multiple sources of variability as illustrated in our formal model specification below.

Consider an RNA-seq experiment with two treatments. Suppose that one of the designs D_1, D_2, or D_3 from Sect. 5.2.1 is used to measure the experimental units

Table 5.1 Count data simulated according to model (5.2) for each design whose structure for a single treatment group is depicted in the second column of Fig. 5.2

Design	Treatment 1 data	Treatment 2 data	Obs./Exp.Unit[a]	Exp.Units/Trt[b]
D_1	$Y_{111}, Y_{121}, Y_{131}, Y_{141}$	$Y_{211}, Y_{221}, Y_{231}, Y_{241}$	1	4
D_2	$Y_{111}, Y_{112}, Y_{121}, Y_{122}$	$Y_{211}, Y_{212}, Y_{221}, Y_{222}$	2	2
D_3	$Y_{111}, Y_{112}, Y_{113}, Y_{114}$	$Y_{211}, Y_{212}, Y_{213}, Y_{214}$	4	1

[a]Number of observations per experimental unit
[b]Number of experimental units per treatment

within each treatment group. For a given feature, let Y_{ijk} be the read count for treatment i, experimental unit j, and measurement k. Suppose

$$Y_{ijk}|\lambda_{ijk} \sim \text{Poisson}(\lambda_{ijk}) \text{ and } \log(\lambda_{ijk}) = \theta + \tau_i + u_{ij} + e_{ijk}, \tag{5.2}$$

where θ, τ_1, $\tau_2 \in \mathbb{R}$ are unknown parameters, $u_{ij} \sim N(0, \sigma_u^2)$, $e_{ijk} \sim N(0, \sigma_e^2)$, all u_{ij} terms and e_{ijk} terms are mutually independent, and σ_u^2 and σ_e^2 are unknown, non-negative variance components.

In model (5.2), the u_{ij} terms are random experimental unit effects that reflect biological variation in experimental units treated alike. The e_{ijk} terms represent random measurement errors that account for technical variation in the measurement process. These e_{ijk} terms allow for multiple measurements of the same experimental unit to show variation greater than Poisson variation due to the effects of measurement steps like tissue sampling, RNA extraction, and library preparation. The Poisson mean λ_{ijk} follows a lognormal distribution with mean $\exp(\theta + \tau_i + \sigma_u^2/2 + \sigma_e^2/2)$. Thus, $E(Y_{ijk}) = \exp(\theta + \tau_i + \sigma_u^2/2 + \sigma_e^2/2)$, and the mean read count for treatment 1 divided by the mean read count for treatment 2 (often referred to as the *fold change*) is $\exp(\tau_1 - \tau_2)$. As in the case of a negative binomial distribution, the variance of a lognormal mixture of Poisson distributions is a quadratic function of the mean. In particular,

$$\text{Var}(Y_{ijk}) = E(Y_{ijk}) + \phi\{E(Y_{ijk})\}^2, \tag{5.3}$$

where $\phi = \exp(\sigma_u^2 + \sigma_e^2) - 1$.

We present a simulation study to evaluate the use of design D_1, D_2, or D_3 within each treatment group under model (5.2). The data simulated for each design are noted in Table 5.1. For each design, model (5.2) was fit to the simulated data using the function `glmer` from the `R` package `lme4`,[2] and the maximum likelihood estimate (MLE) of the log fold change $\tau_1 - \tau_2$ was recorded. The simulation was repeated 1,000 times and the mean squared error of the MLE was computed for each of the parameter settings indicated in Table 5.2.

---

[2]`R` version 2.15.2 and lme4 version 1.0–4 were used in the simulation studies of Sects. 5.2.2 and 5.3.1.

Table 5.2 Empirical estimates of mean squared error (MSE) for the maximum likelihood estimator of log fold change ($\tau_1 - \tau_2$) for designs D_1, D_2, and D_3

θ	$\tau_1 - \tau_2$	σ_u^2/σ_e^2	MSE D_1	MSE D_2	MSE D_3
2	0	4	0.098	0.121	0.170
2	0	1	0.093	0.102	0.130
2	1	4	0.107	0.136	0.164
2	1	1	0.104	0.128	0.148
2	2	4	0.144	0.149	0.201
2	2	1	0.150	0.158	0.224
4	0	4	0.034	0.057	0.091
4	0	1	0.032	0.044	0.069
4	1	4	0.037	0.061	0.100
4	1	1	0.034	0.045	0.075
4	2	4	0.038	0.063	0.101
4	2	1	0.042	0.053	0.075
8	0	4	0.025	0.045	0.090
8	0	1	0.026	0.039	0.064
8	1	4	0.024	0.042	0.084
8	1	1	0.026	0.038	0.065
8	2	4	0.026	0.044	0.083
8	2	1	0.024	0.035	0.056

For each setting, $\tau_1 + \tau_2 = 0$, and $\sigma_u^2 + \sigma_e^2 = 0.05$

Parameter values were chosen so that characteristics of the simulated read count data would match well with real RNA-seq data. The expected value of the simulated read count ranged from a low of 2.8 (when $\theta = 2$ and $\tau_2 = -1$) to a high of 8,308 (when $\theta = 8$ and $\tau_1 = 1$). To ensure a realistic mean-variance relationship in simulated read counts, the variance parameters σ_u^2 and σ_e^2 were chosen to sum to 0.05 so that the value of ϕ in (5.3) would be between the median and mean values of the gene-specific negative binomial dispersion parameter estimates in the case study of Chap. 3. The values 4 and 1 were selected for the variance ratio σ_u^2/σ_e^2 to examine behavior of the log fold change estimator in a typical situation where biological variation is higher than technical variation, and in a more extreme situation in which measurement error variance matches biological variation.

The mean squared errors in Table 5.2 show that design D_1 produced the best estimates of log fold change while design D_3 produced the worst estimates for all simulation settings. Although the results are not reported in detail here, this same ranking of designs (D_1 better than D_2 better than D_3) was seen in a second analogous study with eight observations (rather than four) per treatment group. These results indicate that researchers interested in accurately estimating the extent of differential expression due to treatment (log fold change) should use experimental designs that maximize biological replication. Simply put, adding an additional experimental unit is preferable to measuring an existing experimental unit twice. This conclusion is in complete accordance with the conclusion derived analytically in Sect. 5.2.1 for the simple linear model.

5.3 Sequencing Depth

Sequencing depth (defined formally in Chap. 2) is proportional to the number of reads obtained per library. The number of reads obtained per library is often referred to as the *library size*. If we imagine the sequencing process as a simple random sampling of the DNA fragments in a library, the number of reads Y matching any particular feature has a hypergeometric distribution well approximated by a binomial distribution with number of trials equal to the library size N and success probability p equal to the proportion of all fragments in the library that match the given feature. Because N is large and p is small, this binomial distribution is approximately Poisson with mean $\lambda = Np$. Although this characterization of the sequencing process is an oversimplification, the $Y \sim \text{Poisson}(\lambda)$ assumption is consistent with empirical data obtained by repeatedly sequencing a single library in multiple flowcell lanes [14]. The lognormal mixture of Poisson distributions detailed in model (5.2) of Sect. 5.2.2 arises because the biological and technical variation introduced in the steps leading up through library preparation cause p (and thus λ) to vary from library to library.

The coefficient of variation of the $\text{Poisson}(\lambda = Np)$ distribution is $(Np)^{-1/2}$. Thus, the greater the library size N, the lower the coefficient of variation. Because a low coefficient of variation is desirable for accurate estimation of λ, high library size N is preferred. However, sequencing cost constraints limit the total number of reads obtained across all libraries in an experiment. Thus, there is a tradeoff between high sequencing depth on the one hand and the total number of libraries sequenced on the other. In this section, we investigate this tradeoff for designs with a one-to-one correspondence between libraries and experimental units, which are the designs recommended by the arguments of Sect. 5.2.

5.3.1 *Examination of the Tradeoff Between Depth and Replication*

Suppose an experiment is conducted to compare the effect of two treatments (denoted as T_1 and T_2) on transcript abundance. Furthermore, suppose the budget allows one entire flowcell to be used for sequencing libraries associated with the experimental units. A simple design assigns the two treatments to a total of eight experimental units using a balanced and completely randomized strategy. Then the eight libraries associated with the eight experimental units can be randomly assigned to the eight lanes of the flowcell for sequencing. Such a design is depicted as design D_4 in Fig. 5.3.

Design D_5 in Fig. 5.3 shows how multiplexing can be used to sequence twice as many experimental units but at half the depth relative to design D_4. Note how four of the eight lanes use barcode B_1 for sequencing the treatment T_1 experimental units and barcode B_2 for sequencing the treatment T_2 experimental units. The other four

Design D_4

L_1	L_2	L_3	L_4	L_5	L_6	L_7	L_8
T_1	T_2	T_1	T_2	T_1	T_2	T_1	T_2

Design D_5

	L_1	L_2	L_3	L_4	L_5	L_6	L_7	L_8
B_1	T_1	T_2	T_1	T_2	T_1	T_2	T_1	T_2
B_2	T_2	T_1	T_2	T_1	T_2	T_1	T_2	T_1

Design D_6

	L_1	L_2	L_3	L_4	L_5	L_6	L_7	L_8
B_1	T_1	T_2	T_1	T_2	T_1	T_2	T_1	T_2
B_2	T_2	T_1	T_2	T_1	T_2	T_1	T_2	T_1
B_3	T_1	T_2	T_1	T_2	T_1	T_2	T_1	T_2
B_4	T_2	T_1	T_2	T_1	T_2	T_1	T_2	T_1

Fig. 5.3 Three design choices for comparing two treatments. For $k = 4, 5, 6$, design D_k uses one flowcell to sequence 2^{k-1} experimental units at depth proportional to $N/2^{k-4}$ reads per experimental unit. For $i = 1, 2$, each appearance of the term T_i represents a distinct experimental unit treated with treatment i. Treatments are arranged systematically in the designs of Fig. 5.3 to make patterns clear, but in practice, designs that randomly exchange entire rows and randomly exchange entire columns within any depicted design are recommended

lanes use the reverse assignment of barcodes to treatments to avoid any confounding between the effects of barcodes and treatments. Design D_6 in Fig. 5.3 is a design analogous to D_5 but with double the number of experimental units and half the sequencing depth.

To evaluate the merit of these designs for estimating treatment effects, we conducted a simulation study similar to the study in Sect. 5.2.2. For a given feature, let Y_{ij} be the read count for the jth experimental unit treated with treatment T_i. Let $l(i, j)$ be the lane number in which the jth experimental unit treated with treatment T_i was sequenced. For design D_5 or D_6, let $b(i,j)$ be the number of the barcode assigned to the jth experimental unit treated with treatment T_i. Then $l(i,j) \in \{1,\ldots,8\}$ for all three designs, $b(i,j) \in \{1,2\}$ for design D_5, and $b(i,j) \in \{1,2,3,4\}$ for design D_6. For design D_4, we assume

$$Y_{ij}|\lambda_{ij} \sim \text{Poisson}(\lambda_{ij}) \text{ and } \log(\lambda_{ij}) = \theta + \tau_i + \ell_{l(i,j)} + e_{ij}, \tag{5.4}$$

where terms in the model are defined as follows:

- The unknown real-valued parameters θ, τ_1, and τ_2 determine baseline expression level for each treatment group. In all our simulations, we set $\tau_1 + \tau_2$ to zero so that θ controls the overall expression level. As in Sect. 5.2.2, $\tau_1 - \tau_2$ represents the log fold change due to the effects of treatments.
- We assume $\ell_{l(i,j)} \sim N(0, \sigma_\ell^2)$, where σ_ℓ^2 is an unknown, non-negative variance component. Together, the terms $\ell_1, \ldots, \ell_8$ account for technical variation associated with the last steps of preparing eight libraries and sequencing the eight libraries in eight distinct lanes.
- We assume $e_{ij} \sim N(0, \sigma_e^2)$, where σ_e^2 is an unknown, non-negative variance component. The e_{ij} error terms account for biological variation in RNA levels between experimental units and all technical variation associated with steps in the measurement process that occur prior to the steps accounted for by the $\ell_{l(i,j)}$ terms. None of the variance components for technical or biological sources of variation can be separately estimated with design D_4, but we list the terms $\ell_{l(i,j)}$ and e_{ij} separately and explicitly in (5.4) because of the important role these terms play in the models for data from designs D_5 and D_6 that are discussed below.
- To complete the model specification, we assume all random terms are mutually independent.

For design D_5, we assume

$$Y_{ij}|\lambda_{ij} \sim \text{Poisson}(\lambda_{ij}/2) \text{ and } \log(\lambda_{ij}) = \theta + \tau_i + \beta_{b(i,j)} + \ell_{l(i,j)} + e_{ij}, \tag{5.5}$$

where the terms β_1 and β_2 are the effects of barcodes B_1 and B_2, respectively, and all other terms are as defined for model (5.4). For design D_6, we assume

$$Y_{ij}|\lambda_{ij} \sim \text{Poisson}(\lambda_{ij}/4) \text{ and } \log(\lambda_{ij}) = \theta + \tau_i + \beta_{b(i,j)} + \ell_{l(i,j)} + e_{ij}, \tag{5.6}$$

where the terms β_1, β_2, β_3, and β_4 are the effects of barcodes B_1, B_2, B_3, and B_4, respectively, and all other terms are as defined for model (5.4). Note that the Poisson means in models (5.5) and (5.6) are divided by factors two and four, respectively, to reflect the loss of sequencing depth associated with sequencing two times and four times as many experimental units as in design D_4 with the same total sequencing resources (eight lanes of a single flowcell). The motivation for introducing barcode effect terms in models (5.5) and (5.6) originates from empirical data presented by [1,7,22]. Depending on the experimental protocols and the barcoding technology used, the effects of barcodes may be substantial or

Table 5.3 Empirical estimates of mean squared error (MSE) for the maximum likelihood estimator of log fold change ($\tau_1 - \tau_2$) for designs D_4, D_5, and D_6

θ	$\tau_1 - \tau_2$	MSE D_4	MSE D_5	MSE D_6
2	0	0.099	0.086	0.076
2	1	0.105	0.097	0.092
2	2	0.143	0.128	0.121
4	0	0.032	0.021	0.013
4	1	0.037	0.022	0.017
4	2	0.037	0.024	0.017
8	0	0.025	0.010	0.004
8	1	0.024	0.009	0.005
8	2	0.025	0.010	0.005

For each setting, $\tau_1 + \tau_2 = 0$, $\sigma_\ell^2 = 0.01$, and $\sigma_e^2 = 0.05$

nearly negligible. In the simulation study of this section, we drew the barcode effects independently for each simulation run from a normal distribution with mean 0 and variance 0.05.

Using models (5.4)–(5.6), 1,000 datasets were simulated for each design and each of the parameter settings indicated in Table 5.3. As in Sect. 5.2.2, the function `glmer` from the `R` package `lme4` was used to obtain the MLE of the log fold change $\tau_1 - \tau_2$ for each dataset. The mean squared error of the MLE is provided for each combination of simulation scenario and design in Table 5.3. Figure 5.4 illustrates the distribution of log fold change estimates obtained for each design when the true log fold change $\tau_1 - \tau_2 = 1$. Although not shown here, figures for the $\tau_1 - \tau_2 = 0$ and $\tau_1 - \tau_2 = 2$ cases appear similar to Fig. 5.4.

The results in Table 5.3 and Fig. 5.4 show that design D_6 provides the best estimates of differential expression as quantified by log fold change. The mean squared error was highest for design D_4 and lowest for design D_6 for every simulation setting. These results favor designs that maximize biological replication at the sacrifice of reduced sequencing depth per experimental unit. This conclusion is consistent with the conclusions of [21], who used a different simulation strategy to show that increasing biological replication while decreasing sequencing depth led to increased power for detecting differentially expressed features. Our conclusions are also in agreement with the conclusions of [13], who used RNA-seq data and empirical analysis to determine that increased biological replication should be favored over increased depth.

Our results also show that log fold changes are more precisely estimated for features with relatively high expression levels (e.g., see the results for $\theta = 2$ vs. $\theta = 8$). This conclusion is consistent with the sample size calculations of [12] based on an exact test for differential expression [18, 19]. These sample size calculations show that power for detecting differential expression decreases as expression levels decrease. For features with very low expression levels, designs with both high sequencing depth and high biological replication may be needed to obtain precise fold change estimates and high power for detecting differential expression.

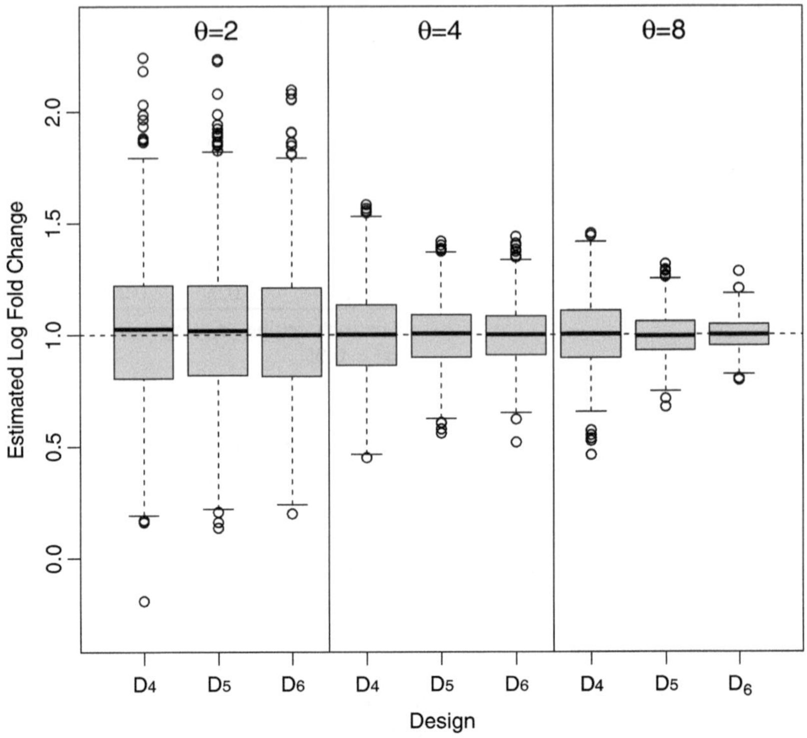

Fig. 5.4 Boxplots of 1,000 maximum likelihood estimates of log fold change ($\tau_1 - \tau_2 = 1$) produced with simulated data from designs D_4, D_5, and D_6 (depicted in Fig. 5.3) for features with low ($\theta = 2$), medium ($\theta = 4$), and high ($\theta = 8$) expression levels

5.3.2 A Comparison with Another Design for Maintaining Depth

As an alternative to a design like D_4 in Fig. 5.3, Auer and Doerge [4] proposed the use of multiplexing to sequence each of the eight samples together in each of the eight lanes.[3] Their strategy requires the creation of only one multiplex sample that is sequenced eight times. They proposed summing the eight read counts per feature obtained for each experimental unit so that the resulting dataset would have the same basic structure as a dataset generated using design D_4, i.e., one count per feature for each of the eight experimental units. Auer and Doerge pointed out that their approach provides the same sequencing depth as design D_4 but has the

[3]The example design considered by Auer and Doerge actually involved three (rather than four) experimental units for each treatment group and six (rather than eight) flowcell lanes, but we have used a natural extension of their design to match our slightly larger experimental setup.

advantage that all the presequencing preparation steps that occur after barcoding and sample combination may impact all experimental units simultaneously and equally.

Although lane effects and effects associated with post-pooling steps are balanced across treatments in Auer and Doerge's design, barcode effects are not balanced across treatments in much the same way that library preparation and lane effects are not balanced across treatments in design D_4. For this reason, we recommend designs like D_5 or D_6 that balance all known nuisance effects across treatment groups. However, it is important to acknowledge that designs like D_5 and D_6 are more labor intensive and expensive than design D_4 or the design proposed in [4] because additional experimental units and extra library preparation steps are not free. To save some sequencing expense while still maintaining balance of nuisance effects across experimental units, a subset consisting of the first two, four, or six lanes in design D_5 or D_6 may be used.

5.4 Other Example Designs

Section 5.3 presented example designs for the final phase of two-treatment RNA-seq experiments that use a balanced and completely randomized design in the initial phase. In the current section, we describe three other hypothetical but realistic RNA-seq experimental design scenarios of varying complexity.

5.4.1 An Experiment with Four Treatments

Consider an experiment to study the effects of four treatments on transcription in liver tissue of pigs. In the first phase of the experiment, 16 pigs from a population of interest are randomly assigned to the four treatments using a balanced and completely randomized design. The pigs are housed in individual pens, and the assigned treatments are imposed for a specified duration. These steps, which correspond to steps 1–3 of Sect. 5.1, comprise the first phase of the experiment.

The second phase involves collecting liver tissue from each pig (step 4 of Sect. 5.1). Because harvesting liver tissue from a pig is a non-trivial activity, the work is broken up into multiple sessions in which four pigs are processed per session. To avoid confounding potential session effects with the effects of treatments, a randomized complete block design is used to assign pigs to sessions, with one pig from each treatment in each session.

Within a given session, it is possible for the order in which pigs are processed to affect measurements of feature expression. For example, factors like room temperature or time since an animal's most recent meal could be associated with expression. Thus, processing order is assigned to pigs in sessions according to the Latin square design shown in the left panel of Fig. 5.5. The tissue samples collected in phase two are stored in ultra-low-temperature freezers until the remaining steps of the experiment can be completed.

Tissue/RNA Collection Design

	S_1	S_2	S_3	S_4
O_1	$T_1^{(1)}$	$T_2^{(2)}$	$T_3^{(3)}$	$T_4^{(4)}$
O_2	$T_2^{(1)}$	$T_3^{(2)}$	$T_4^{(3)}$	$T_1^{(4)}$
O_3	$T_3^{(1)}$	$T_4^{(2)}$	$T_1^{(3)}$	$T_2^{(4)}$
O_4	$T_4^{(1)}$	$T_1^{(2)}$	$T_2^{(3)}$	$T_3^{(4)}$

Sequencing Design

	L_1	L_2	L_3	L_4
B_1	$T_1^{(1)}$	$T_2^{(2)}$	$T_3^{(3)}$	$T_4^{(4)}$
B_2	$T_2^{(1)}$	$T_3^{(2)}$	$T_4^{(3)}$	$T_1^{(4)}$
B_3	$T_3^{(1)}$	$T_4^{(2)}$	$T_1^{(3)}$	$T_2^{(4)}$
B_4	$T_4^{(1)}$	$T_1^{(2)}$	$T_2^{(3)}$	$T_3^{(4)}$

Fig. 5.5 Designs for tissue collection, RNA extraction, and sequencing in a four-treatment experiment. For $i = 1,2,3,4$ and $j = 1,2,3,4$, the term $T_i^{(j)}$ represents the jth experimental unit treated with treatment i. The *rows* of the *left panel* define the order in which samples are processed during the tissue-collection and RNA-extraction sessions. The *columns* in the *left panel* correspond to the four sessions. The *rows* and *columns* in the *right panel* correspond to barcodes and flowcell lanes, respectively. Due to the intentional confounding of session effects and lane effects, and the intentional confounding of order effects with barcode effects, each experimental unit appears in the same relative position within both figure panels

The third experimental phase begins with thawing samples and extracting RNA. Just as in the second phase, the work of the third phase is split into manageable sessions. Because neither the session effects for tissue collection nor the session effects for RNA extraction are of scientific interest, they are intentionally confounded with each other so that the kth tissue collection session involves the same pigs as the kth RNA extraction session ($k = 1,2,3,4$). Furthermore, the order that RNA is extracted from samples within phase-three sessions is chosen to match the tissue collection order used in phase two.

In the final phase of the experiment, the 16 RNA samples from 16 experimental units are converted to libraries and sequenced on four lanes of a flowcell. The session factor is intentionally confounded with flowcell lane so that the libraries from any one session are sequenced together in a single lane using multiplexing. Furthermore, the order factor is intentionally confounded with barcode so that samples processed in the bth position are always tagged with barcode B_b ($b = 1,2,3,4$) as indicated in the right panel of Fig. 5.5. This confounding allows the effects of the nuisance factors tissue collection session, RNA-extraction session, and flowcell lane to be accounted for with a single factor in the model ultimately used for data analysis. Likewise, the effects of tissue-collection order, RNA-extraction order, and barcode can be accounted for with a single factor.

	Session/Lane 1	2	3	4
O_1/B_1	$T_{11}^{(1)}$	$T_{12}^{(3)}$	$T_{21}^{(5)}$	$T_{22}^{(7)}$
O_2/B_2	$T_{12}^{(1)}$	$T_{11}^{(3)}$	$T_{22}^{(5)}$	$T_{21}^{(7)}$
O_3/B_3	$T_{21}^{(2)}$	$T_{22}^{(4)}$	$T_{11}^{(6)}$	$T_{12}^{(8)}$
O_4/B_4	$T_{22}^{(2)}$	$T_{21}^{(4)}$	$T_{12}^{(6)}$	$T_{11}^{(8)}$

Fig. 5.6 Designs for tissue collection, RNA extraction, and sequencing in a split-plot experiment. The term $T_{ij}^{(k)}$ represents the experimental unit that received diet i and drug j in pen k. The *rows* of the table define processing order, which is intentionally confounded with barcode. The *columns* of the table define sessions, which are intentionally confounded with flowcell lanes. Note that a pair of pigs from a given pen (i.e., a whole-plot experimental unit) is always processed consecutively in either the first two or last two order positions of a session

5.4.2 A Split-Plot Experiment

Consider an experiment to study the effects of two diets and two drugs on transcription in liver tissue of pigs. Suppose 16 pigs, housed in eight pens with two pigs per pen, are available for use in the experiment. Because each pen contains a single feeder from which both pigs in a given pen eat, the two diets are randomly assigned to pens rather than individual pigs; i.e., the two pigs in any given pen are necessarily treated with the same diet. Because the drug is delivered by injection, the two drugs are randomly assigned to individual pigs within each pen. Thus, the first phase of the experiment is conducted according to a split-plot design with diet as the whole-plot factor and drug as the split-plot factor.

The four combinations of diet and drug define four treatments. Hence, the same basic strategy used to determine tissue collection, RNA extraction, and sequencing plans in Sect. 5.4.1 can be used here. However, slightly more care should be taken due to the phase-one split-plot design. Whole-plot experimental units (pairs of pigs from the same pen) should be kept together during subsequent phases of the experiment. Thus, each session should involve a pair of pens that received different diets. Furthermore, pigs from the same pen should be processed consecutively so that the ordering of the four treatments within the four sessions follows a restricted Latin square design. One example design that satisfies these principles is depicted in Fig. 5.6. Note that the processing and sequencing structure shown in the figure mimics the structure of the phase-one split-plot design.

5.4.3 A Balanced Incomplete Block Design

Consider an experiment to study the effects of four treatments on feature expression. Suppose available resources allow for sequencing of 12 experimental units in two flowcell lanes. One option would be to conduct the first phase of the experiment as a randomized complete block design with three blocks, where each block consists of one experimental unit for each of the four treatments. However, with access to only two flowcell lanes, a randomized complete block design in the first phase would not lead to attractive options for the final phase of the experiment. For example, two blocks could be sequenced in the first lane and the third block sequenced in the second lane, but this strategy would lead to a twofold difference in depth between experimental units sequenced in the first and second lanes. While normalization methods (see Chap. 9) could help adjust for this imbalance, it is much better to work with data from experimental units sequenced at a consistent depth. If processing order and barcode effects are ignorable, the strategy of [4] could be used to sequence all 12 RNA-samples together in both lanes. Because both of these approaches (and others not mentioned here) have potentially serious drawbacks, we seek a better option.

Rather than trying to make a first-phase randomized complete block design with three blocks fit with final-phase sequencing in two flowcell lanes, we recommend a balanced incomplete block design for the first phase that will allow for better control of the effects of potential nuisance factors. Experimental units should be grouped into six blocks of two experimental units each. The $\binom{4}{2} = 6$ possible treatment pairs are randomly assigned to the six blocks, with the two treatments for any one block randomly assigned to the two experimental units in the block. Subsequent processing steps are then carried out on a block-by-block basis. In final-phase sequencing, all 12 samples can be sequenced together in lane one, and sequenced a second time in lane two but with reversed barcode assignments as illustrated in Fig. 5.7. By reversing the barcode assignments within each incomplete block, confounding between barcode and treatment effects is avoided.

In contrast to the designs we have recommended previously, this balanced incomplete block design involves measuring feature expression levels twice for each experimental unit. We consider this strategy to be the best choice given the available resources (12 experimental units and two flowcell lanes) for studying expression differences across four treatments. An even better design would sequence a new set of 12 experimental units in the second flowcell lane rather than sequencing the same 12 experimental units a second time, but additional experimental units are not always affordable or available.

Fig. 5.7 Final-phase sequencing design for the balanced incomplete block design. The term $T_i^{(k)}$ represents the experimental unit that received treatment *i* in block *k*. The *rows* and *columns* of the table indicate barcode and lane assignments, respectively. Note that each of the 12 experimental units is sequenced once in lane L_1 and once in lane L_2, but with different barcodes

	L_1	L_2
B_1	$T_1^{(1)}$	$T_2^{(1)}$
B_2	$T_2^{(1)}$	$T_1^{(1)}$
B_3	$T_1^{(2)}$	$T_3^{(2)}$
B_4	$T_3^{(2)}$	$T_1^{(2)}$
B_5	$T_1^{(3)}$	$T_4^{(3)}$
B_6	$T_4^{(3)}$	$T_1^{(3)}$
B_7	$T_2^{(4)}$	$T_3^{(4)}$
B_8	$T_3^{(4)}$	$T_2^{(4)}$
B_9	$T_2^{(5)}$	$T_4^{(5)}$
B_{10}	$T_4^{(5)}$	$T_2^{(5)}$
B_{11}	$T_3^{(6)}$	$T_4^{(6)}$
B_{12}	$T_4^{(6)}$	$T_3^{(6)}$

5.5 Conclusions

Modern high-throughput DNA sequencing technologies provide powerful tools for simultaneously measuring the transcript abundance of thousands of genomic features. The RNA-seq measurement process is complex and results in the need for carefully coordinating multiphase experiments. The nuisance factors inevitably introduced in various experimental phases should be intentionally confounded with one another to simplify data analysis and to keep the focus of the analysis on factors of primary scientific interest. When the goal of an RNA-seq experiment is to detect differentially expressed features, maximizing the number of biological replications should take precedence over replicating technical steps in the measurement process or obtaining high sequencing depth for each experimental unit. RNA-seq experiments designed with these principles in mind will produce datasets rich in information about differential expression and well structured for analysis with the techniques described in Chaps. 2–4.

Acknowledgements Research reported in this chapter was supported by National Institute of General Medical Sciences (NIGMS) of the National Institutes of Health and the joint National Science Foundation/NIGMS Mathematical Biology Program under award number R01GM109458. The content is solely the responsibility of the author and does not necessarily represent the official views of the National Institutes of Health or the National Science Foundation.

References

[1] Alon, S., Vigneault, F., Eminaga, S., Christodoulou, D.C., Seidman, J.G., Church, G.M., Eisenberg, E.: Barcoding bias in high-throughput multiplex sequencing of miRNA. Genome Res. **21**(9), 1506–1511 (2011)

[2] Anders, S., Huber, W.: Differential expression analysis for sequence count data. Genome Biol. **11**(10), R106 (2010)

[3] Anders, S., McCarthy, D.J., Chen, Y., Okoniewski, M., Smyth, G.K., Huber, W., Robinson, M.D.: Count-based differential expression analysis of RNA sequencing data using R and Bioconductor. Nat. Protoc. **8**(9), 1765–1786 (2013)

[4] Auer, P.L., Doerge, R.: Statistical design and analysis of RNA sequencing data. Genetics **185**(2), 405–416 (2010)

[5] Bullard, J., Purdom, E., Hansen, K., Dudoit, S.: Evaluation of statistical methods for normalization and differential expression in mRNA-seq experiments. BMC Bioinform. **11**(1), 94 (2010)

[6] Busby, M.A., Stewart, C., Miller, C.A., Grzeda, K.R., Marth, G.T.: Scotty: a web tool for designing RNA-seq experiments to measure differential gene expression. Bioinformatics **29**(5), 656–657 (2013)

[7] Craig, D.W., Pearson, J.V., Szelinger, S., Sekar, A., Redman, M., Corneveaux, J.J., Pawlowski, T.L., Laub, T., Nunn, G., Stephan, D.A., Homer, N., Huentelman, M.J.: Identification of genetic variants using bar-coded multiplexed sequencing. Nat. Meth. **5**(10), 887–893 (2008)

[8] Fang, Z., Cui, X.: Design and validation issues in RNA-seq experiments. Brief. Bioinform. **12**(3), 280–287 (2011)

[9] Hamady, M., Walker, J.J., Harris, J.K., Gold, N.J., Knight, R.: Error-correcting barcoded primers for pyrosequencing hundreds of samples in multiplex. Nat. Meth. **5**(3), 235–237 (2008)

[10] Jarrett, R.G., Ruggiero, K.: Design and analysis of two-phase experiments for gene expression microarrays-Part I. Biometrics **64**(1), 208–216 (2008)

[11] Kathleen Kerr, M.: Design considerations for efficient and effective microarray studies. Biometrics **59**(4), 822–828 (2003)

[12] Li, C.I., Su, P.F., Shyr, Y.: Sample size calculation based on exact test for assessing differential expression analysis in RNA-seq data. BMC Bioinform. **14**(1), 357 (2013)

[13] Liu, Y., Zhou, J., White, K.P.: RNA-seq differential expression studies: more sequence or more replication? Bioinformatics **30**(3), 301–304 (2014)

[14] Marioni, J.C., Mason, C.E., Mane, S.M., Stephens, M., Gilad, Y.: RNA-seq: an assessment of technical reproducibility and comparison with gene expression arrays. Genome Res. **18**(9), 1509–1517 (2008)

[15] McIntyre, G.: Design and analysis of two phase experiments. Biometrics **11**(3), 324–334 (1955)

[16] McIntyre, L.M., Lopiano, K.K., Morse, A.M., Amin, V., Oberg, A.L., Young, L.J., Nuzhdin, S.V.: RNA-seq: technical variability and sampling. BMC Genomics **12**(1), 293 (2011)
[17] Nettleton, D.: Design of gene expression microarray experiments, Chap. 2. In: Hinkelmann K. (ed.) Design and Analysis of Experiments, vol. 3: Special Designs and Applications, pp. 73–108. Wiley, Hoboken (2012)
[18] Robinson, M.D., Smyth, G.K.: Moderated statistical tests for assessing differences in tag abundance. Bioinformatics **23**(21), 2881–2887 (2007)
[19] Robinson, M.D., Smyth, G.K.: Small-sample estimation of negative binomial dispersion, with applications to SAGE data. Biostatistics **9**(2), 321–332 (2008)
[20] Robinson, M.D., McCarthy, D.J., Smyth, G.K.: edgeR: a Bioconductor package for differential expression analysis of digital gene expression data. Bioinformatics **26**(1), 139–140 (2010)
[21] Robles, J.A., Qureshi, S.E., Stephen, S.J., Wilson, S.R., Burden, C.J., Taylor, J.M.: Efficient experimental design and analysis strategies for the detection of differential expression using RNA-sequencing. BMC Genomics **13**(1), 484 (2012)
[22] Willenbrock, H., Salomon, J., Søkilde, R., Barken, K.B., Hansen, T.N., Nielsen, F.C., Møller, S., Litman, T.: Quantitative miRNA expression analysis: comparing microarrays with next-generation sequencing. RNA **15**(11), 2028–2034 (2009)

[illegible]

Chapter 6
Measurement, Summary, and Methodological Variation in RNA-sequencing

Alyssa C. Frazee, Leonardo Collado Torres, Andrew E. Jaffe, Ben Langmead, and Jeffrey T. Leek

Abstract There has been a major shift from microarrays to RNA-sequencing (RNA-seq) for measuring gene expression as the price per measurement between these technologies has become comparable. The advantages of RNA-seq are increased measurement flexibility to detect alternative transcription, allele specific transcription, or transcription outside of known coding regions. The price of this increased flexibility is: (a) an increase in raw data size and (b) more decisions that must be made by the data analyst. Here we provide a selective review and extension of our previous work in attempting to measure variability in results due to different choices about how to summarize and analyze RNA-sequencing data. We discuss a standard model for gene expression measurements that breaks variability down into variation due to technology, biology, and measurement error. Finally, wee show the importance of gene model selection, normalization, and choice for statistical model on the ultimate results of an RNA-sequencing experiment.

A.C. Frazee • L.C. Torres • J.T. Leek (✉)
Department of Biostatistics, Johns Hopkins Bloomberg School of Public Health, 615 North Wolfe Street, Baltimore, MD 21205, USA
e-mail: jtleek@gmail.com

A.E. Jaffe
Lieber Institute for Brain Development, 855 North Wolfe Street, Baltiore, MD 21205, USA

B. Langmead (✉)
Department of Computer Science, Whiting School of Engineering, Johns Hopkins University, 3400 North Charles Street, Baltimore, MD 21218, USA
e-mail: langmea@cs.jhu.edu

S. Datta and D. Nettleton (eds.), *Statistical Analysis of Next Generation Sequencing Data*, Frontiers in Probability and the Statistical Sciences, DOI 10.1007/978-3-319-07212-8_6,

6.1 Introduction

RNA-sequencing (RNA-seq) has replaced microarray technology as the preferred assay for measuring gene expression [20]. The reasons include a dramatic drop in the price of measuring gene expression with RNA-seq [10, 33] and increased measurement flexibility. Specifically, RNA-seq experiments are more flexible than their microarray counterparts because they produce "reads" from the transcripts in the cells being sequenced rather than measuring only a pre-specified set of gene sequences [41]. A typical modern experiment in humans produces between 40 million and 100 million sequencing reads of length 50–100 base pairs (bp) from the transcripts in each sample. The reads are often paired-end, meaning that each transcript fragment is sequenced from both ends, producing a pair of reads from each fragment. These reads can then be aligned to the genome and assembled together to create estimates of the unobserved transcripts in the sample [24,45]; see Chap. 1 of this book.

The added flexibility of RNA-seq data comes at a cost: unlike microarray data, the raw data from RNA-seq experiments are huge, measuring in gigabytes or, for larger projects, terabytes. The size of the raw data makes even simple questions computationally difficult to answer. Most of the early work in methods development for RNA-seq focused on the computational challenges associated with aligning millions of reads to large genomes [18,44] and assembling transcripts for individual samples [9]. The maturity of these computational methods and the increased use of parallel processing and cloud computing have helped to address some of these computational challenges [19,42].

More recently, attention has turned to understanding sources of variation in RNA-sequencing data through statistical modeling [34]. In this chapter we discuss the sources of variation in RNA-sequencing measurements. We categorize this variation into three main types: across group variation, natural biological variation, and measurement error. We explain why some of the early over-optimism surrounding RNA-seq was due to underestimation of the effect of biological variation in RNA-seq. The analysis of microarray data provides an instructive historical lesson in this context—similar over-optimism about the measurement technology led to reproducibility and replicability problems early on in the application of this new technology due to underestimates of multiple sources of variability.

We also discuss sources of variation in the results of RNA-seq experiments. The extra flexibility afforded by RNA-seq gives researchers several options for how data should be analyzed, which in turn introduces variation in the processes used to move from raw RNA-seq reads to estimates of feature abundances and differential expression results. We discuss summarization variability, which is due partly to choices of which level of expression should be summarized (e.g., gene, transcript, or exon) and partly to choices about the model to use to obtain the selected summary measurement. We also discuss variability in results due to differences in how expression measurements are normalized and differences in the statistical models used to analyze the final, summarized, normalized expression data. Finally, we conclude with some thoughts about open questions related to modeling variability in RNA-seq data, summaries, and normalization.

6.2 Variability in RNA-Sequencing Data

Improvements in sequencing technology have not reduced the cost or effort associated with sample collection and sample preparation. Most investigator-initiated sequencing projects are still constrained to small or moderate sample sizes because of cost or logistics. Many RNA-seq experiments analyze fewer than three samples [12]. It has been noted that there is a relationship between depth of coverage (the amount of data) and power to detect regions of interest in sequencing experiments [3]. Despite this relationship, studies with limited sample sizes remain underpowered and prone to false positives [14, 30]. Small sample sizes also make it difficult for individual researchers to determine whether discoveries are due to artifacts or bugs in any one of the computational pipelines they have applied [16,28,29,37].

Here we review and extend the approach undertaken in [12]. We proposed a general form for the variability in gene expression data measured with any technology:

$$\text{Var(Expr)} = \text{Across Group Variation} + \text{Measurement Error} + \text{Biological Variation.}$$

Across group variability is the type of variability under investigation in most experiments. This type of variability may represent changes in gene expression over time, over developmental phases, or between biological groups.

Measurement error is variation due to the technology, unrelated to the biology of interest. Factors contributing to this variability could arise at various stages: during sample collection, during sample preparation, or within the sequencer itself. For example, the amount ("depth") of sequencing data produced can vary between samples. When one sample is sequenced to a greater depth than another, a normalization procedure can be applied to avoid spuriously concluding that genes are more highly expressed in the sample with greater depth. Normalization methods are discussed further later.

Other measurement errors are more subtle, requiring other statistical methods. For instance, sequencers are "biased" in favor of reporting reads from portions of the genome that have a certain favorable fraction of C (cytosine) and G (guanine) nucleotides [5, 36]. This fraction is called *GC content*. Sequencers tend to under-report reads from portions of the genome with extreme GC content (very high or low), and over-report reads from portions with intermediate GC content. The magnitude of this imbalance can vary from sample to sample, and therefore can confound results. GC-content bias and other sequence- and position-specific biases have been studied by several groups, and solutions have been proposed that involve re-weighting reads [11], integrating bias correction with isoform quantitation [39], and using conditional quantile normalization [13], among other methods [27, 38]. Other measurement errors that have been observed in sequencing data include batch effects [23] and lab effects [1].

Biological variation is natural variation in gene expression between individuals. Gene expression has been shown to be a stochastic process [6] which can only be incompletely modeled by phenotypic or technical characteristics of the samples being considered. We previously demonstrated that variability between unrelated individuals from the same population exists regardless of whether microarrays or RNA-sequencing are applied to measure gene expression [12]. Briefly, we collected RNA-seq and microarray data on n = 60 and n = 69 distinct individuals from two biological populations. We then summarized both the microarray and RNA-sequencing data at the level of genes. Figure 6.1 plots the standard deviation of the expression measurements comparing microarrays for two distinct populations (Fig. 6.1a, b). It also highlights two specific genes which show consistent levels of variability in expression regardless of measurement technology (Fig. 6.1c).

An important and often-overlooked component of biological variability involves biological factors as potential confounders [21]. Batch and other technological artifacts are well-known confounders, now commonly included in models for gene expression data [23]. But common sources of biological variation due to environmental, genetic, or demographic variables may also confound estimates of across group variation. Some of these variables might be measured in the course of a study, but some might not. Statistical methods have been developed to identify, estimate, and remove these sources of signal directly from the expression measurements using latent factor estimation techniques [8, 22, 43]. An alternative approach is to reduce the impact of confounding factors by binarizing expression measurements into two categories—expressed or not expressed. This type of binarization is called gene expression barcoding and was originally proposed for expression measurements from microarrays [32] but has recently been extended to the analysis of RNA-seq experiments [35].

Our estimates of biological variation (Fig. 6.1) also suggest two key properties about variation in gene expression data that have a bearing on experimental design. The first is that biological variation is not eliminated by the choice of technology. It has been suggested that RNA-seq data summarized at the level of counts may be modeled using a Poisson or over-dispersed Poisson model [4]. If the model is reasonable, this implies that it may be possible to estimate both the mean and variance of gene expression measurements. Our results suggest that it will be difficult to appropriately capture the potentially large inter-individual biological variability using only a small number of biological replicates. Therefore, studies using a small number of samples may be underpowered and more likely to produce spurious results. The second property is that with a small number of replicates, it is difficult to distinguish whether specific expression patterns are characteristic to the individuals in the study or can be generalized to larger populations. These ideas are well known in the statistical community, but were overlooked in early RNA-seq experiments. To appropriately and reproducibly characterize patterns of variation in expression between groups or over time, experiments must be designed to do the following:

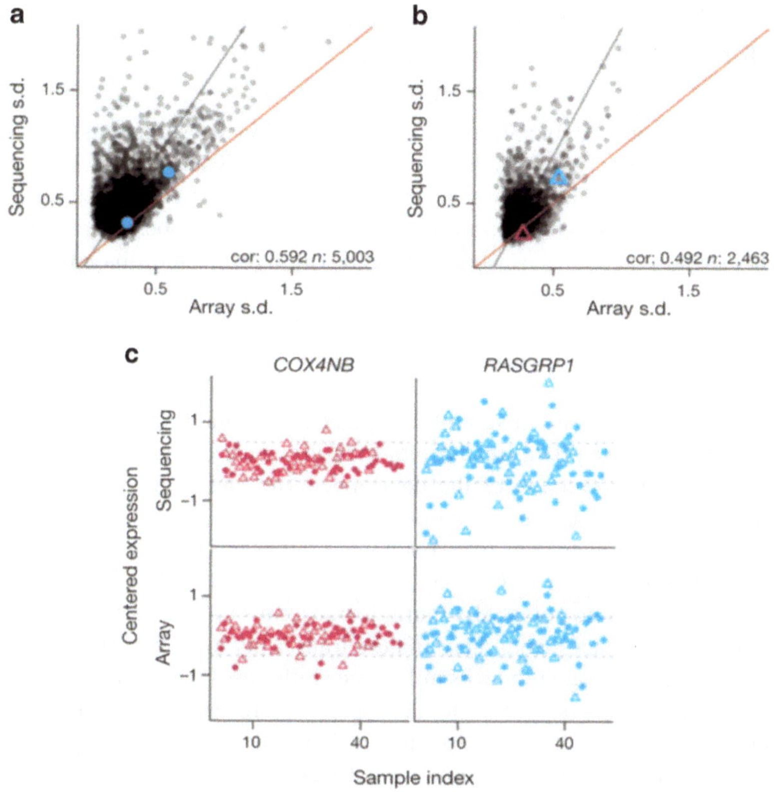

Fig. 6.1 Biological variability measured with sequencing and microarrays. (**a**) A plot of the standard deviation (s.d.) of expression values as measured with microarrays (x axis) and sequencing (y axis) in a Caucasian population. The estimates of expression variability from sequencing are similar to the estimates from microarrays. (**b**) A plot of the s.d. of expression values as measured with microarrays (x axis) and sequencing (y axis) in an African population. The estimates of expression variability from sequencing are again almost the same as estimates from microarrays. In each plot, the *black line* is the best linear fit and the *red line* is the line $y = x$. (**c**) A plot of the expression for two genes COX4NB (*left column, pink*) and RASGRP1 (*right column, blue*) as measured with sequencing (*top row*) and microarrays (*bottom row*) versus biological sample. Mean-centered measurements from the two studies are plotted as *circles* and *triangles*, respectively. The s.d. for the two genes are highlighted in **a**, **b**. The plot shows that regardless of the measurement technology or study, COX4NB expression is much less variable than RASGRP1 expression. This figure and caption reproduced from Hansen et al. [12]

1. Analyze enough biological replicates to accurately capture and model biological variation.
2. Block or randomize to reduce the role of potential confounding factors such as batch effects.

6.3 Variability in Summarization Methods for RNA-Sequencing Data

The model we proposed for variation in gene expression measurements derived from RNA-seq experiments assumes that a choice of summary has already been put into place. Microarrays again provide an instructive example: generally, each gene on a microarray is measured by more than one probe. The probes represent distinct sequences whose abundances were measured with hybridization-based technology. Translating raw microarray data—measurements of light intensity from multiple probes falling within the same gene—into an overall gene expression measurement is not straightforward. Robust estimates of gene expression levels derived from probe intensities were eventually developed [31].

There is a similar issue in summarizing RNA-seq data: how should researchers translate raw RNA-seq data (reads) into measurements of expression for specific genes? An idealized version of the data generating and counting process shows some of the potential sources of variation created when choosing how to measure expression for each gene (Fig. 6.2). Each gene represents a sequence consisting of exons (colored blocks in Fig. 6.2a) and introns (grey lines in Fig. 6.2b). The exons are spliced together into potentially multiple transcripts for each gene. The RNA-seq reads are then sequences from these mature transcripts, with the introns ostensibly removed. The technology produces millions or billions of these reads.

Given reads, our first job is to determine where each read originated with respect to the set of all gene sequences, also called the "transcriptome." This is accomplished using a read aligner: a software tool that attempts to determine each read's point of origin with respect to one or more reference sequences. For RNA sequencing data analysis, the reference sequences might consist of all the previously-observed ("annotated") transcripts for the species under study. Alternately, the reference sequences might simply consist of the chromosomes of the genome. The latter approach has the advantage that it does not pre-suppose a particular, possibly incomplete, set of possible transcripts. However, the latter approach also requires that the read aligner handle RNA sequencing reads that overlap introns. When a read spans an intron, its alignment to the genome will contain large (intron-sized) gaps (Fig. 6.2b). Such read aligners are called *spliced* read aligners and popular spliced read aligners include TopHat [45], GSNAP [47] and MapSplice [46]. Standard non-spliced read aligners include Bowtie [17, 18] and BWA [25, 26].

Whether a spliced or non-spliced read aligner is used, the aligner itself is an important source of measurement error and variability. Different aligners have different policies about which alignments are "good enough" to be reported to the user, as well as about which alignments are filtered out for the sake of speed. Also, the problem of read alignment (both spliced and unspliced) is inherently ambiguous. Because of repeated sequences in the genome, the read aligner cannot necessarily determine a read's point of origin with high confidence. The analyst therefore might instruct the read aligner to discard or otherwise down-weight evidence from reads

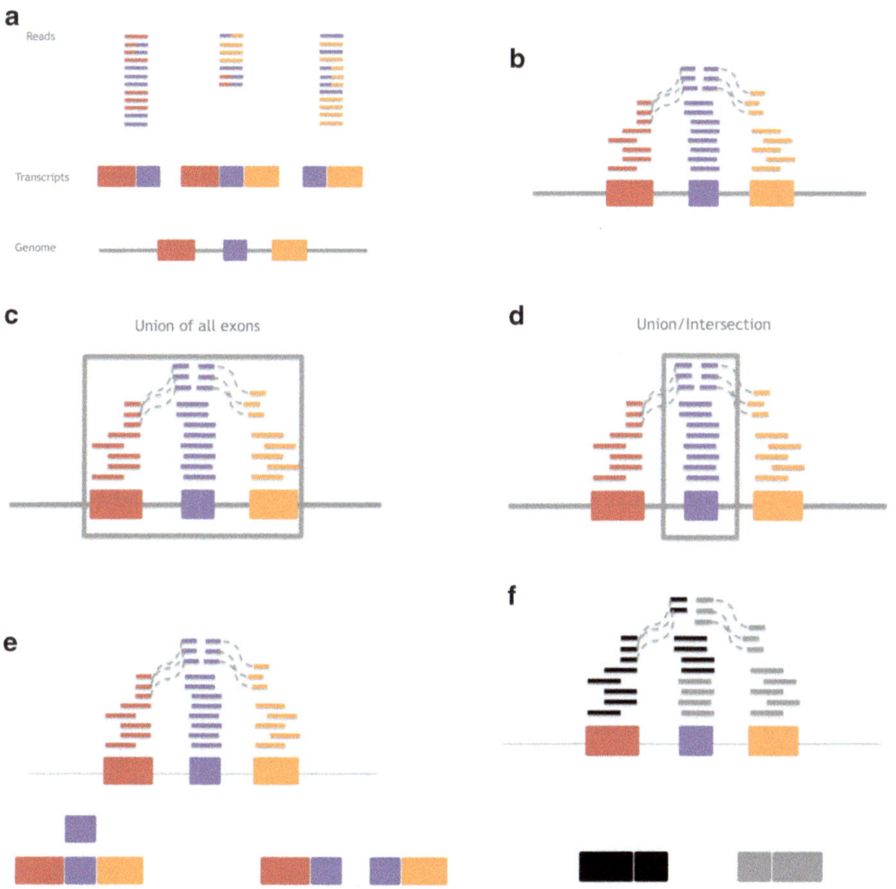

Fig. 6.2 Idealized summary of RNA-seq data generation and gene-level summarization. (**a**) The genome sequence may produce multiple transcripts—corresponding to different combinations of exons within genes. The reads represent sequences from the transcripts, which may not be contiguous sequences in the genome. (**b**) The first step in summarizing RNA-sequencing data is to identify the places in the genome or transcriptome where the reads came from—called alignment. This image shows splicing-aware alignment which is able to identify reads that do not cover contiguous sequence in the genome. Counts for each gene are obtained by adding the number of alignments that intersect a particular gene model. (**c**) In the union model, all reads touching any exon in the gene count toward the measurement for that gene. (**d**) In the union-intersection model only reads that touch exons which appear in all transcripts for that gene count toward the measurement. (**e**) Assembly pipelines first combine the aligned reads to estimate assembled transcripts. These assemblies may be unidentified or ambitious based on the read alignments. In this example, both assemblies (shown below the aligned reads) are equally supported by the observed reads. (**f**) Abundance estimates for each transcripts are obtained by assigning proportions of the reads to each transcript (*grey versus black*) using, e.g., mixture models

whose point of origin is ambiguous. The analyst has many choices when deciding how to parameterize such tools, and these choices can themselves have a major impact on the tool's ability to find correct alignments.

Once we have obtained alignments, we can observe where the alignments fall with respect to genes (or exons or transcripts), and summarize expression levels. Until recently there have been two major classes of summarization methods after alignment. The first class summarizes the reads based on previously annotated features such as exons or genes (Fig. 6.2c, d) [4]. There are multiple ways that variability may be introduced at this counting step. For example, a read may align to the genome in a position that lies entirely within the boundaries of two versions of an exon—say with alternative starting sites. In this case, a choice must be made as to whether to assign the count for that read to one exon, to both, or to neither. This variability is compounded when considering features consisting of multiple distinct genomic regions. A large percentage of variability in RNA-seq measurements at this stage is due to the choice of a gene model. The union model counts all of the reads that touch any exon within a particular gene (Fig. 6.2c). The advantage of this approach is that it does not throw away data and gives a global summary of that gene's expression. However, if the gene has multiple transcripts then the union model may not be a stable estimate of that gene's abundance. For example, consider the case of transcript switching. Imagine two transcripts for the same gene that show dramatically different expression when comparing biological groups of interest. It is entirely possible that the sum of the counts for these two transcripts may be nearly identical across samples, but the expression levels of specific exons are dramatically different. The union-intersection model is an alternative that counts only the reads that touch exons that appear in all transcripts for that gene. The union-intersection model may be more stable in the presence of alternative transcription, but may also unnecessarily throw away a large amount of data [4].

An alternative to summarizing features at the level of genes is to attempt to assemble and quantify transcripts directly from the reads themselves (Fig. 6.2e, f) [9]. In this model, the read alignments are aggregated within specific regions of the genome and estimates of the transcripts are "assembled" based on those read alignments. There are several potential sources of variability in the assembly process, including ambiguous start and end points for individual exons and junctions with a small number of supporting reads. In some cases it is not possible to unambiguously choose between different assemblies that are nearly equally supported by the observed raw data (Fig. 6.2e). There has been relatively little work to quantify the impact of assembly variation on downstream significance analyses, although hierarchical models for assembly and abundance estimation have been proposed.

Once a set of transcripts is obtained—either through estimation from the reads themselves or through previous knowledge—the abundance of each transcript must be estimated (Fig. 6.2b). The most common approach to obtaining these estimates is to form a count for each exon and potentially for each junction, then fit a mixture model for the abundance of each transcript based on these raw quantities [15].

In addition to the inherent variability that comes from estimating latent mixture models, choices about how to map reads may again introduce variability into the abundance estimates if they affect the counts for individual exons or junctions.

6.4 Variability in Statistical Methods for RNA-Sequencing Data

After abundance estimates have been obtained for the features of interest, downstream normalization and statistical modeling can be performed to identify differentially expressed features: exons, genes, or transcripts. There is a large and rapidly growing literature dedicated to developing statistical models that appropriately address the sources of variability in summarized RNA-seq data [34]. Early efforts focused primarily on developing Poisson-based models, since expression estimates obtained from RNA-seq at this time were usually counts of reads per feature [4]. These methods have been extended to account for over-dispersion in sequencing count data and to borrow strength across features to better estimate variances and differential expression effects [2, 40]. These extended methods have been extensively used and tested, and they are approaching the maturity level of well-known statistical methods for analyzing differential expression using microarray data. This maturity illustrates that many of the ideas from microarray analysis have been successfully carried over and adapted to the new technology. In the sequel we present a selective review and extension of our recent work.

A comparative disadvantage of RNA-seq at this stage is that there do not exist spike-in experiments that properly incorporate natural biological variability. Comprehensive comparisons of statistical methods for differential expression have thus largely focused on comparing how they perform when the only source of variability in the experiment is measurement error. To explore statistical properties in a more realistic setting, we analyzed a large public dataset with software we had developed, called *Myrna*, to perform differential expression analysis at scale on commercial cloud-computing resources such as Amazon EC2 [19]. We conducted an experiment comparing multiple metrologies for differential expression on this large dataset, which included 69 distinct individuals and thus had real biological variation in the gene expression measurements.

We compared three categories of statistical tests for detecting differential expression: (1) a test assuming a naive Poisson model, (2) a test using log-transformed counts and a standard linear (Gaussian) model, and (3) a permutation test after modeling the log-transformed counts using a linear (Gaussian) model. We also compared multiple normalization strategies: (1) modeling variation in sequencing depth using the 75th percentile of the count or log-count distribution as an offset and (2) allowing for a gene-specific relationship between the 75th percentile and the counts. The first normalization strategy corresponds to the usual approach of normalizing read counts by dividing by library size.

We applied these methods to a set of data from a genetically homogeneous population where the group labels for differential expression testing were assigned at random. Because groups were randomly assigned, testing for differential expression between the groups should not yield any differential expression signal. This means the experiment does not give insight into the power of each method for detecting differential expression, but it does speak to the control of type I error rates and model fit. The results suggested that the Poisson model fails to capture biological variability and, in this case, leads to spurious differential expression signal (Fig. 6.3a, b). Our results also suggested that including the library size as an offset term—equivalent to dividing by library size—may introduce differential expression in genes with low counts when library size is correlated with the groups of interest [19] (Fig. 6.3a, c, e). Under gene-specific library size normalization, both the standard linear model and the permutation procedure produce the expected null distribution of P-values (Fig. 6.3e, f).

Our results suggest that even after summarization, the choice of statistical method can have a strong impact on the results of differential expression analysis of RNA-seq data. In particular, methods that do not flexibly estimate the relationship between mean and variance in RNA-seq measurements may underestimate biological variability and lead to spurious results. This is particularly true for lowly expressed features supported by only a small number of reads that can be heavily influenced by normalization factors. In our work, we have shown that gene and feature specific normalization performs best when analyzing RNA-seq data. This observation is not unique to our analysis; it has also been pointed out that sample-specific GC content normalization may be most appropriate for RNA-seq count data [13]. The most mature statistical methods [2, 40] for count-based RNA-seq data take these principles into account.

6.5 Where Do We Go from Here?

Early tools for RNA-seq analysis focused on the computational and engineering challenges associated with the size of the raw data [42]. Statisticians primarily focused on highly summarized RNA-seq data at the level of counts for exons, genes, or transcripts. As the computational and engineering challenges have been resolved, statistical modeling has been brought closer to the raw data. This trend will likely continue as statisticians aim to model the sources of variation in RNA-seq data due to technology, biology, alignment, assembly, and summarization. Statistical models will also continue to do a better job of capturing variability across samples and due to confounding factors.

One example of how statisticians are dealing with less processed forms of RNA-seq data is an approach we have recently proposed that is intermediate between summarizing reads into feature counts and full-scale assembly [7]. The approach involves first aligning the reads to the genome using a splicing-aware alignment tool. For each base, we calculate the number of reads that overlap that base, leading

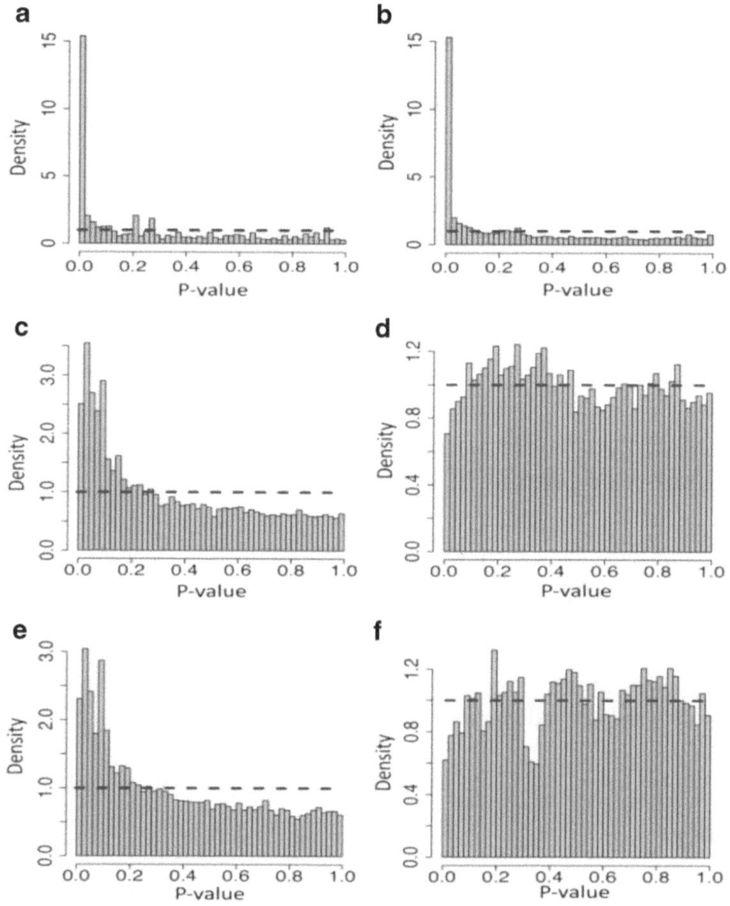

Fig. 6.3 Histograms of P-values from six different analysis strategies applied to randomly labeled samples. In each case the P-values should be uniformly distributed (*blue dotted line*) since the labels are randomly assigned. (**a**) Poisson model, 75th percentile normalization. (**b**) Poisson model, 75th percentile included as term. (**c**) Gaussian model, 75th percentile normalization. (**d**) Gaussian model, 75th percentile included as term. (**e**) Permutation model, 75th percentile normalization. (**f**) Permutation model, 75th percentile included as term. Figure and caption reproduced from Langmead et al. [19]

to an estimate of coverage. We then perform a statistical test at each base to identify bases that show a differential expression signal of interest, appropriately modeling all sources of variability. Contiguous bases showing similar differential expression signals are then grouped into candidate differentially expression regions (DERs). We then treat candidate DERs as the measurement unit of interest, estimate measures of statistical significance for these DERs, and attempt to reconcile them with previous annotation. This approach to identifying regions of differential expression from RNA-seq data does not incur the variability due to choosing gene models or due to assembly. Analyzing data at the level of base-resolution coverage can reveal novel transcribed regions and help identify mistakes in annotation.

Substantial development is still needed from the statistical community to tackle the complexity of RNA-seq data. Normalization approaches are still immature and do not fully incorporate information from genomic architecture, mappability, or summarization choices. Statistical models for assessing, comparing, and modeling assemblies are largely undeveloped. Comparisons between approaches are limited by the relatively small number of spike-in experiments and the limited set of simulation tools that can produce raw reads from differential expression experiments. Many of these areas are focuses of active development in the statistical community and promise to dramatically improve our understanding of the statistical properties of RNA-seq data.

6.6 Acknowledgement of Previous Publication

Section 6.2 of this chapter is based on paraphrasing and extending reference [12], including reproduction of the figure and figure caption from that paper in Fig. 6.1. Section 6.4 of this chapter is based on paraphrasing and extending reference [19], including reproduction of a figure and figure caption from that paper in Fig. 6.3. The reproduction of these figures is permitted under the publishing agreements for *Nature Biotechnology* and *Genome Biology*.

References

[1] A C't Hoen, P., Friedländer, M.R., Almlöf, J., Sammeth, M., Pulyakhina, I., Anvar, S.Y., Laros, J.F., Buermans, H.P., Karlberg, O., Brännvall, M., et al.: Reproducibility of high-throughput mrna and small rna sequencing across laboratories. Nat. Biotechnol. **31**, 1015–1022 (2013)

[2] Anders, S., Huber, W.: Differential expression analysis for sequence count data. Genome Biol. **11**, R106 (2010). doi:10.1186/gb-2010-11-10-r106. http://genomebiology.com/2010/11/10/R106/

[3] Auer, P.L., Doerge, R.W.: Statistical design and analysis of RNA sequencing data. Genetics **185**(2), 405–416 (2010)

[4] Bullard, J., Purdom, E., Hansen, K.D., Dudoit, S.: Evaluation of statistical methods for normalization and differential expression in mrna-seq experiments. BMC Bioinform. **11**, 94 (2010). R package version 1.10.0

[5] Dohm, J.C., Lottaz, C., Borodina, T., Himmelbauer, H.: Substantial biases in ultra-short read data sets from high-throughput dna sequencing. Nucleic Acids Res. **36**(16), e105–e105 (2008)

[6] Elowitz, M., Levine, A., Siggia, E., Swain, P.: Stochastic gene expression in a single cell. Science **297**(5584), 1183 (2002)

[7] Frazee, A., Sabunciyan, S., Hansen, K., Irizarry, R., Leek, J.: Differential expression analysis 362 of RNA-seq data at single-base resolution. Biostatistics doi: 10.1093/biostatistics/kxt053 (2014)

[8] Friguet, C., Kloareg, M., Causer, D.: A factor model approach to multiple testing under dependence. J. Am. Stat. Assoc., 104:488, 1406–1415 (2009)
[9] Garber, M., Grabherr, M., Guttman, M., Trapnell, C.: Computational methods for transcriptome annotation and quantification using rna-seq. Nat. Meth. **8**(6), 469–477 (2011)
[10] Glenn, T.C.: Field guide to next-generation dna sequencers. Mol. Ecol. Resour. **11**(5), 759–769 (2011)
[11] Hansen, K.D., Brenner, S.E., Dudoit, S.: Biases in illumina transcriptome sequencing caused by random hexamer priming. Nucleic Acids Res. **38**(12), e131 (2010)
[12] Hansen, K.D., Wu, Z., Irizarry, R.A., Leek, J.T.: Sequencing technology does not eliminate biological variability. Nat. Biotechnol. **29**(7), 572–573 (2011)
[13] Hansen, K.D., Irizarry, R.A., Wu, Z.: Removing technical variability in rna-seq data using conditional quantile normalization. Biostatistics **13**(2), 204–216 (2012)
[14] Ioannidis, J.P.: Why most published research findings are false. PLoS Med. **2**(8), e124 (2005)
[15] Jiang, H., Wong, W.: Statistical inferences for isoform expression in rna-seq. Bioinformatics **25**(8), 1026–1032 (2009)
[16] Kleinman, C.L., Majewski, J.: Comment on "widespread RNA and DNA sequence differences in the human transcriptome". Science **335**(6074), 1302; author reply 1302 (2012)
[17] Langmead, B., Salzberg, S.L.: Fast gapped-read alignment with bowtie 2. Nat. Meth. **9**(4), 357–359 (2012)
[18] Langmead, B., Trapnell, C., Pop, M., Salzberg, S.L.: Ultrafast and memory-efficient alignment of short DNA sequences to the human genome. Genome Biol. **10**(3), R25 (2009)
[19] Langmead, B., Hansen, K.D., Leek, J.T.: Cloud-scale RNA-sequencing differential expression analysis with Myrna. Genome Biol. **11**(8), R83 (2010)
[20] Ledford, H.: The death of microarrays? Nature **455**(7215), 847 (2008)
[21] Leek, J., Storey, J.: Capturing heterogeneity in gene expression studies by 'surrogate variable analysis'. PLoS Genet. **3**, e161 (2007)
[22] Leek, J., Storey, J.: A general framework for multiple testing dependence. PNAS **105**, 18,718–18,723 (2008)
[23] Leek, J.T., Scharpf, R.B., Bravo, H.C., Simcha, D., Langmead, B., Johnson, W.E., Geman, D., Baggerly, K., Irizarry, R.A.: Tackling the widespread and critical impact of batch effects in high-throughput data. Nat. Rev. Genet. **11**, 733–739 (2010)
[24] Li, B., Dewey, C.: Rsem: accurate transcript quantification from rna-seq data with or without a reference genome. BMC Bioinform. **12**(1), 323 (2011)
[25] Li, H., Durbin, R.: Fast and accurate short read alignment with burrows–wheeler transform. Bioinformatics **25**(14), 1754–1760 (2009)
[26] Li, H., Durbin, R.: Fast and accurate long-read alignment with burrows–wheeler transform. Bioinformatics **26**(5), 589–595 (2010)
[27] Li, J., Jiang, H., Wong, W.: Modeling non-uniformity in short-read rates in rna-seq data. Genome Biol. **11**(5), R25 (2010)
[28] Li, M., Wang, I.X., Li, Y., Bruzel, A., Richards, A.L., Toung, J.M., Cheung, V.G.: Widespread rna and dna sequence differences in the human transcriptome. Science **333**(6038), 53–58 (2011)
[29] Lin, W., Piskol, R., Tan, M.H., Li, J.B.: Comment on "widespread RNA and DNA sequence differences in the human transcriptome". Science **335**(6074), 1302; author reply 1302 (2012)
[30] MacArthur, D.: Methods: face up to false positives. Nature **487**(7408), 427–428 (2012)
[31] McCall, M.N., Bolstad, B.M., Irizarry, R.A.: Frozen robust multiarray analysis (frma). Biostatistics **11**(2), 242–253 (2010)
[32] McCall, M.N., Uppal, K., Jaffee, H.A., Zilliox, M.J., Irizarry, R.A.: The gene expression barcode: leveraging public data repositories to begin cataloging the human and murine transcriptomes. Nucleic Acids Res. **39**(Suppl 1), D1011–D1015 (2011)
[33] NHGRI: DNA sequencing costs. http://www.genome.gov/sequencingcosts/
[34] Oshlack, A., Robinson, M.D., Young, M.D., et al.: From rna-seq reads to differential expression results. Genome Biol. **11**(12), 220 (2010)

[35] Piccolo, S.R., Withers, M.R., Francis, O.E., Bild, A.H., Johnson, W.E.: Multiplatform single-sample estimates of transcriptional activation. Proc. Natl. Acad. Sci. **110**(44), 17,778–17,783 (2013)

[36] Pickrell, J., Marioni, J., Pai, A., Degner, J., Engelhardt, B., Nkadori, E., Veyrieras, J., Stephens, M., Gilad, Y., Pritchard, J.: Understanding mechanisms underlying human gene expression variation with rna sequencing. Nature **464**(7289), 768–772 (2010)

[37] Pickrell, J.K., Gilad, Y., Pritchard, J.K.: Comment on "widespread RNA and DNA sequence differences in the human transcriptome". Science **335**(6074), 1302; author reply 1302 (2012)

[38] Risso, D., Schwartz, K., Sherlock, G., Dudoit, S.: Gc-content normalization for rna-seq data. BMC Bioinform. **12**(1), 480 (2011)

[39] Roberts, A., Trapnell, C., Donaghey, J., Rinn, J., Pachter, L., et al.: Improving rna-seq expression estimates by correcting for fragment bias. Genome Biol. **12**(3), R22 (2011)

[40] Robinson, M., McCarthy, D., Smyth, G.: edgeR: a bioconductor package for differential expression analysis of digital gene expression data. Bioinformatics **26**(1), 139–140 (2010)

[41] Shendure, J., Ji, H.: Next-generation DNA sequencing. Nat. Biotechnol. **26**, 1135–1145 (2008)

[42] Stein, L.D.: The case for cloud computing in genome informatics. Genome Biol. **11**(5), 207 (2010)

[43] Teschendorff, A.E., Zhuang, J., Widschwendter, M.: Independent surrogate variable analysis to deconvolve confounding factors in large-scale microarray profiling studies. Bioinformatics **27**, 1496–1505 (2011)

[44] Trapnell, C., Pachter, L., Salzberg, S.L.: TopHat: discovering splice junctions with RNA-Seq. Bioinformatics **25**(9), 1105–1111 (2009)

[45] Trapnell, C., Williams, B.A., Pertea, G., Mortazavi, A., Kwan, G., van Baren, M.J., Salzberg, S.L., Wold, B.J., Pachter, L.: Transcript assembly and quantification by RNA-Seq reveals unannotated transcripts and isoform switching during cell differentiation. Nat. Biotechnol. **28**(5), 511–515 (2010)

[46] Wang, K., Singh, D., Zeng, Z., Coleman, S.J., Huang, Y., Savich, G.L., He, X., Mieczkowski, P., Grimm, S.A., Perou, C.M., et al.: Mapsplice: accurate mapping of rna-seq reads for splice junction discovery. Nucleic Acids Res. **38**(18), e178 (2010)

[47] Wu, T.D., Nacu, S.: Fast and snp-tolerant detection of complex variants and splicing in short reads. Bioinformatics **26**(7), 873–881 (2010)

Chapter 7
DE-FPCA: Testing Gene Differential Expression and Exon Usage Through Functional Principal Component Analysis

Hao Xiong, James Bentley Brown, Nathan Boley, Peter J. Bickel, and Haiyan Huang

Abstract RNA-seq, next-generation sequencing (NGS) applied to RNA, is rapidly becoming the platform of choice for gene expression profiling. Existing methods, mostly parametric, describe the expression level of a gene or transcript by a single number that summarizes all reads mapped to that gene or transcript. However, assay noise often makes such parametric models unwieldy, non-intuitive, and difficult to implement. To overcome these limitations, we have developed a nonparametric approach, based on functional principal component analysis (FPCA), to differential expression estimation. Our approach, named DE-FPCA, represents the expression profile of a gene by a random curve, which is modeled as a linear combination of orthogonal functional principal components (FPCs), and tests differential expression between two groups of samples using a statistic defined by the FPC scores. We applied our method to 26 RNA-seq samples collected in *Drosophila melanogaster*. This application demonstrates that our new FPCA-based test statistic has substantial power to detect differential usage of exons and isoforms, in addition to gene-level differential expression, and is robust to random fluctuations in the RNA-seq data.

H. Xiong
Microbiology, University of Washington, Seattle, WA, USA
e-mail: haoxiong@uw.edu

J.B. Brown
Genome Dynamics, Lawrence Berkeley National Laboratory, Berkeley, CA, USA
e-mail: benofberkeley@gmail.com

N. Boley
Biostatistics, University of California, Berkeley, CA, USA
e-mail: npboley@gmail.com

P.J. Bickel • H. Huang (✉)
Statistics, University of California, Berkeley, CA, USA
e-mail: bickel@stat.berkeley.edu; hhuang@stat.berkeley.edu

S. Datta and D. Nettleton (eds.), *Statistical Analysis of Next Generation Sequencing Data*, Frontiers in Probability and the Statistical Sciences, DOI 10.1007/978-3-319-07212-8_7,

7.1 Introduction

RNA-seq, the application of next-generation sequencing (NGS) technology to transcriptomes, is rapidly the platform of choice for gene expression profiling [43]. RNA-seq has the ability to measure isoform and allele-specific expressions [2, 40], and identify SNP variants, small or large indels, alternative splicing, alternative transcript start and end sites, post-transcriptional RNA editing, gene fusions, gene inversions, and chromosome rearrangements [29]. RNA-seq gene expression estimates have less background noise and a wider dynamic range than the microarray estimates [11, 43]. However, the analysis of RNA-seq data can be challenging due to bias and noise introduced during data generation [14, 39].

One of the main goals of RNA-seq data analysis is to discover genes that are differentially expressed between two groups of samples [29]. The primary task in such analysis is to model and distinguish between different types of expression variability, which arise from three primary sources: (i) real biological differences in different experimental groups or conditions, (ii) measurement errors and (iii) random biological and/or technical variation [15]. The first type of variability is of real biological interest; however, it is confounded with measurement errors and random biological/technical variation. Properly accounting for the latter two types of variability is the fundamental question in differential expression analysis of RNA-seq data [18] and the topic of this paper.

The most widely used software tools for differential expression analysis include Cuffdiff [42], edgeR [34], DESeq [1], PoissonSeq [22], baySeq [16], and limma [37]. Each of these takes a parametric approach, assuming that read counts follow a Poisson distribution [26], or a negative binomial distribution [1,24] or other variants [4,21]. The Poisson-based approaches assume that the number of reads overlapping a gene is independently sampled from a multinomial distribution which can be approximated by a Poisson distribution [23,26,38,44]. However, the Poisson model is too restrictive to model all sources of variability in RNA-seq data; in particular it tends to underestimate the variance [1, 24]. Attempts to model the overdispersion have been approached with two-parameter models, typically based on negative binomial distributions or generalized Poisson distributions [1, 5, 33, 34]. In these models, an additional parameter is introduced to model the over-dispersion, and it appears that this can help reduce bias [21, 24].

All the above models assume that reads are sampled uniformly from a transcript. However, empirically, the coverage of reads across transcripts varies greatly even within an exon. Some of this variance has been characterized. It is well known that base-composition bias (e.g., GC-content [45]) and position-specific bias (e.g. 5' end, 3' end depletion [46]) exist. Methods have been developed that model GC-content bias [31], fragment size bias [32], and 5'/3' end effects [41]. PennSeq [19] improves upon Cufflinks2's [42] approach for quantification by allowing non-uniform sampling probabilities at each base. Li et al. (2010) and Hu et al. (2012) incorporate base-specific variation into their models which tend improves the accuracy of quantification. These models are useful but also computationally intensive and involve a large number of parameters. Furthermore, these methods cannot fully describe observed read inhomogeneity, suggesting that uncharacterized sources of bias still exist.

Genes can express multiple transcript isoforms. In mammals it has been estimated that 95 % of genes express multiple isoforms [25]; in *Drosophila* half of all genes express multiple transcript isoforms. The estimation of expression at isoform level relies on the correct identification of all expressed isoforms within a gene, a challenging problem which is not solvable in all gene loci [21].

In summary, attempting to identify differentially expressed loci using single-value representations of gene expression introduces many (known or unknown) sources of bias and variability, which can lead to decisions with high and often unknown type-I or type-II errors.

DEXSeq [2] attempts to overcome these problems by identifying differential expression at the exon level. The method uses generalized linear models to estimate an overall expression level for each exon or bin (merged exons), and then infers differential gene expression by assessing the changes of expression/usages at the exon/bin level between genes. This is an improvement over early methods such as DESeq [1] that models the expression level of a gene with a single number, but it still does not permit read variation within exons. In addition, multiple testing issues are exacerbated due to testing each exon in every gene. To overcome these limitations, coverage functions were proposed by Okoniewski et al. (2012) [27], where read counts at each base along a gene sequence are viewed as a function of genomic position. However, that method is hampered by ad hoc comparison methods and a lack of formal test statistics for differential expression analysis.

We developed a statistically grounded method for testing gene differential expression with expression values considered at single-base resolution. Our method represents the expression profile of a gene by a functional curve, called a "gene expression function". We use the Karhunen-Loève decomposition [28] to decompose the random gene expression function into orthogonal FPCs. Then, we test for differential gene expression by comparing FPC coefficients between two groups of samples. Our method, named DE-FPCA, makes minimum assumptions about the data generation procedure and is sensitive to subtle changes in the expression level of genes. However, our method requires multiple replicates in each treatment group. Because our approach is sensitive even in low read-coverage regimes, we propose that sequencing additional biological replicates in multiplexed sequencing runs may be more advantageous than deep sequencing of particular replicates when differential gene expression is the primary analysis goal.

We applied our method DE-FPCA to the modENCODE *Drosophila* RNA-seq dataset [13]. The results demonstrate that our FPCA-based method is especially powerful in detecting alternative splicing, while remaining robust to assay noise.

7.2 Method

7.2.1 *Definition of a Gene Expression or Coverage Function*

We define a gene expression function as follows. Let t be a genomic position within a genomic region and T be the length of the genomic region being considered. We consider two conditions (or two groups of samples): case and control. Assume that n_A random case samples and n_B random control samples are sequenced. Let $x_i(t)$ denote the number of reads, covering the genomic position t, from the ith case sample. We similarly define $y_i(t)$ for the ith control sample. The functions $x_i(t)$ and $y_i(t)$ are the empirical gene expression or coverage functions.

7.2.2 *Differential Analysis of RNA-seq Data Using Functional Principal Component Analysis*

7.2.2.1 Review on Functional Principal Component Analysis (FPCA)

We first give a brief review of FPCA [28]. Let $X(t)$ be a centered, square-integrable function, in our case describing read coverage over a gene region, and therefore gene expression. Let $\phi_1, \phi_2, \dots$ be the orthonormal eigenfunctions. By the Karhunen-Loève theorem, one can express the centered process in the eigenbasis functions as $X(t) = \sum_{k=1}^{\infty} \xi_k \phi_k(t)$, where $\xi_k = \int X(t)\phi_k(t)dt$ is the principal component coefficient associated with the kth eigenfunction $\phi_k(t)$, and has the property that $E(\xi_k) = 0$ and $E(\xi_k \xi_l) = 0$ for $k \neq l$. Furthermore, the covariance function $R(s,t)$ can be written as $R(s,t) = Cov(X(s), X(t)) = \sum_{k=1}^{\infty} \lambda_k \phi_k(s)\phi_k(t)$ with $\lambda_k = Var(\xi_k)$.

The first eigenfunction $\phi_1(t)$ depicts the dominant mode of variation in $X(t)$. That is, $\phi_1(t)$ is the ϕ that maximizes the variance of $\xi = \int X(t)\phi(t)dt$ [17]:

$$var(\xi) = var\left[\int X(t)\phi(t)dt\right] = \int\int \phi(s)R(s,t)\phi(t)dsdt.$$

Similarly, ϕ_k is the function that maximizes $var(\xi)$ in the functional space that is orthogonal to $\phi_1, \dots, \phi_{k-1}$.

Since $X(t) = \sum_{k=1}^{\infty} \xi_k \phi_k(t)$, the centered process $X(t)$ is equivalent to the vector $(\xi_1, \xi_2, \dots)$. Also note that since $R(s,t) = Cov(X(s), X(t)) = \sum_{k=1}^{\infty} \lambda_k \phi_k(s)\phi_k(t)$, the eigenfunctions $\phi_1, \phi_2, \dots$ should satisfy, with $\lambda_1 \geq \lambda_2 \geq \lambda_3 \geq \dots$,

$$\int R(s,t)\phi_k(s)ds = \lambda_k \phi_k(t), \tag{7.1}$$

for any integer $k \geq 1$. Solving this equation provides a way to find $\phi_1, \phi_2, \dots$ [28].

7.2.2.2 Performing FPCA on RNA-seq Data

We name our method DE-FPCA. In the context of our study, we can perform FPCA and find the eigenfunctions and corresponding principal components as follows. Let $X(t) = [X_1(t), X_2(t), \ldots]^T$ be a vector-valued function with $X_i(t)$ denoting the gene expression function for the ith sample. There are N replicate samples. We select an orthonormal basis (e.g., a Fourier basis) with P functions $\Delta(t) = [\delta_1(t), \delta_2(t), \ldots, \delta_p(t)]$ and assume that the gene expression functions $X_1(t), \cdots, X_N(t)$ and the eigenfunctions $\phi_1, \phi_2, \cdots$ can be expressed as a linear combination of $\delta_1(t), \delta_2(t), \ldots, \delta_p(t)$ (Note that this is a function approximation). That is, for the N replicate gene expression profiles, we have $X(t) = C\Delta(t)$, where the ijth element in the matrix C is $C_{ij} = \int X_i(t)\delta_j(t)dt$, with $i = 1, \ldots, N$ and $j = 1, \ldots, P$. Similarly, we can express $\phi(t)$ as $\phi(t) = \Delta^T(t)\beta$, where $\beta = [\beta_1, \ldots, \beta_p]^T$ with $\beta_j = \int \phi(t)\delta_j(t)dt$. To find eigenfunctions, or equivalently, to determine β, we make use of (7.1), which has the following equivalent expressions under the current context:

$$E\left[\begin{pmatrix} \xi_1 \\ \vdots \\ \xi_p \end{pmatrix} (\xi_1 \ldots \xi_p)\right] \begin{pmatrix} \beta_1 \\ \vdots \\ \beta_p \end{pmatrix} = \lambda \begin{pmatrix} \beta_1 \\ \vdots \\ \beta_p \end{pmatrix}. \tag{7.2}$$

Next we replace $E(\xi_i \xi_j)$ by its empirical estimate from the sample gene expression functions $X_1(t), \cdots, X_N(t)$ to obtain an empirical version of equation (7.2):

$$\frac{1}{N} C^T C\beta = \lambda\beta. \tag{7.3}$$

The eigenfunctions can be found by solving the above multivariate eigenvalue (λ) and eigenvector (β) problem. The number of eigenfunctions can be chosen based on percentage of variance explained. We used 90 % in the following applications. Different values other than 90 % could be used depending on how accurate the function approximation is needed.

7.2.3 *Test Statistic*

We pooled the expression functions $x_i(t)$'s (for case samples) and $y_i(t)$'s (for control samples) together to estimate the orthonormal eigenfunctions $\phi_j(t)$, $j = 1, \ldots, k$ following the procedure described in Sect. 7.2.2. We expanded $x_i(t)$'s and $y_i(t)$'s on the obtained eigenfunctions. Let the corresponding principal components associated with $\phi_j(t)$ be ξ_{ij} and η_{ij}, for $x_i(t)$ and $y_i(t)$ respectively. We next use the Hotelling T^2 statistic defined through the ξ_{ij}'s and η_{ij}'s to assess the difference between the case and control samples. In more details, we denote the average vector of functional principal components in the case and control groups by $\bar{\xi} = [\bar{\xi}_1, \ldots, \bar{\xi}_k]^T$ and $\bar{\eta} = [\bar{\eta}_1, \ldots, \bar{\eta}_k]^T$ respectively, where $\bar{\xi}_j = \frac{1}{n_A}\sum_{i=1}^{n_A} \xi_{ij}$, and $\bar{\eta}_j = \frac{1}{n_B}\sum_{i=1}^{n_B} \eta_{ij}$, $j = 1, \ldots, k$. Then the pooled covariance matrix is $S = \frac{1}{n_A+n_B-2}\left(\sum_{i=1}^{n_A}(\xi_i - \bar{\xi})(\xi_i - \bar{\xi})^T + \sum_{i=1}^{n_B}(\eta_i - \bar{\eta})(\eta_i - \bar{\eta})^T\right)$, where $\xi_i = [\xi_{i1}, \ldots, \xi_{ik}]^T$, $\eta_i = [\eta_{i1}, \ldots, \eta_{ik}]^T$. Let $\Lambda = (\frac{1}{n_A} + \frac{1}{n_B})S$. Now we define our test statistic as $T^2 = (\bar{\xi} - \bar{\eta})^T \Lambda^{-1}(\bar{\xi} - \bar{\eta})$. Under the null hypothesis of no differential expression between the case and control groups, the statistic T^2 asymptotically follows a central $\chi^2_{(k)}$ distribution, where k is the number of functional principal components used for expansion of expression functions. Alternatively, we can also empirically estimate the null distribution of T^2 and then obtain p-values by randomly shuffling the case and control samples. To get an accurate estimation of p-values, a large number of replicates in each treatment group is preferred. We note that when there are enough samples, statistics other than T^2, such as the maximum of the absolute difference between principal curves, $max|\bar{\xi} - \bar{\eta}|$ which focus on particular aspects of the curves, could also be used as a test statistic.

7.3 Results

7.3.1 Data Set

We used the following dataset to validate our method. Libraries (consisting of poly-A(+) RNAs) extracted from adult fly heads (18 samples) and adult fly carcasses (8 samples) were sequenced on Illumina Analyzer IIx or HiSeq 2000 platforms, generating paired-end reads of length 76 or 100 bases. A total of 14,125 genes were analyzed after filtering out genes with zero read coverage and genes in heterochromatin.

7.3.2 Assessment of Expression Representation by FPCA

We used DE-FPCA to fit nonparametric functional gene expression curves at base resolution: we first derived the principal eigenfunctions based on 401 Fourier basis functions, and then expanded the observed gene expression function in terms of the derived eigenfunctions, as described in Sect. 7.2.2. To demonstrate how well the estimated curves approximate the observed read counts along a gene sequence, we plotted the estimated curve and the observed average expression profile (over the 26 samples) for the gene Dscam1 in Fig. 7.1. *Dscam* is about 25,000 bp long, and contains 95 alternative exons giving it the potential to encode 38,016 distinct proteins [7]. Figure 7.1 shows that the estimated curve nicely approximates the

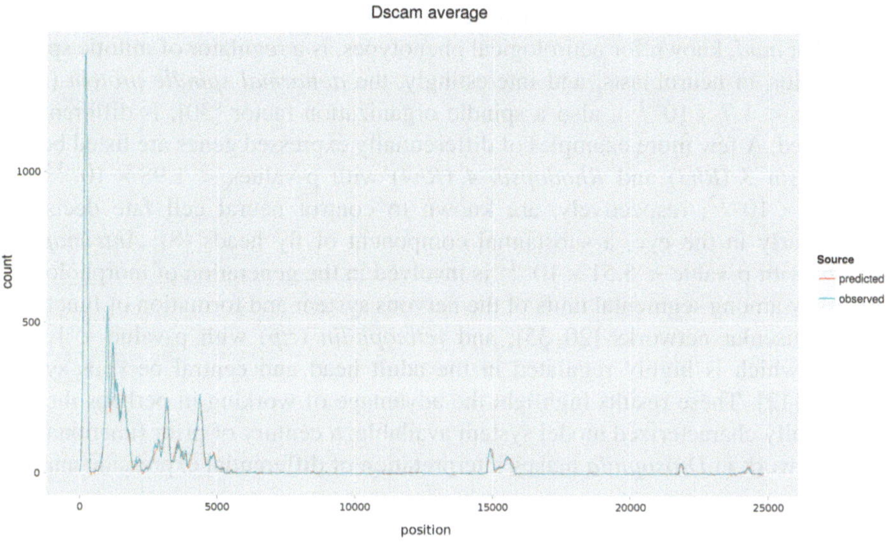

Fig. 7.1 The observed and predicted read counts of gene Dscam where curves in *red* and *blue* represent predicted and observed read count curves, respectively

observed read-count curve even though the expressions of the first several exons in the gene are dramatically fluctuating across the bases. Since Dscam has a large number of alternative isoforms, we observe large read count variation between the alternate exons. Figure 7.1 demonstrates that FPCA can capture expression variation at the levels of genes, exons and single bases.

7.3.3 *Differential Expression Analysis*

We next used the DE-FPCA statistic (Sect. 7.2.3) to identify gene regions differentially expressed between the head and carcass samples, testing 14,125 genes. We identified 588 genes that were differentially expressed (significance level 0.05 after Bonferroni correction). The most significantly differentially expressed gene between heads and carcass is *Abdominal B* (*Abd-B*) with p-value $< 1.0 \times 10^{-17}$ (this is before Bonferroni correction, the same as the p-values listed below). The Hox protein *Abd-B* is critical for early body patterning during embryogenesis [6]. Another Hox protein *Ultrabithorax* (*Ubx*) was also identified as highly differentially expressed between heads and carcass (p-value $< 1.11 \times 10^{-16}$). *Ubx* encodes a transcription factor that regulates many levels of developmental pathways and specifies morphological traits [20]. As expected, key early body patterning factors are expressed both at different relative levels and in different isoforms in adult heads vs. carcass. Similarly, the gene *mushroom body defect* (*mud*) (p-value $< 6.9 \times 10^{-15}$), the critical brain-development factor is differentially expressed in heads [3]. We note that *mud*, known for neurological phenotypes, is a regulator of mitotic spindle orientation in neuroblasts, and interestingly, the *abnormal spindle protein* (*ASP*) (p-value $< 1.7 \times 10^{-12}$), also a spindle organization factor [30], is differentially expressed. A few more examples of differentially expressed genes are listed below: *Rhodopsin 3* (*Rh3*) and *Rhodopsin 4* (*Rh4*) with p-values $< 1.98 \times 10^{-12}$ and $< 2.89 \times 10^{-12}$, respectively, are known to control neural cell fate decisions, particularly in the eye, a substantial component of fly heads [8]; *Antennapedia* (*Antp*) with p-value $< 5.51 \times 10^{-12}$ is involved in the generation of morphological diversity among segmental units of the nervous system and formation of functional neuromuscular networks [20, 35]; and *retinophilin* (*rtp*) with p-value $< 1.21 \times 10^{-11}$, which is highly regulated in the adult head and central nervous system tissues [12]. These results highlight the advantage of working in perhaps the best genetically characterized model system available: a century of prior functional and genetic work in *Drosophila* makes interpretation of differential expression analysis tenable.

7.3.4 Differentially Expressed Isoforms and Differential Exon Usage

One strength of the DE-FPCA statistic is its ability to detect differential expression at the isoform and exon levels. Figures 7.2, 7.3, 7.4 illustrate this.

Figure 7.2 shows the expression profiles of the gene *shaggy* (*sgg*) in heads and carcasses. Our method found that *sgg* was differentially expressed between heads and carcasses with p-value $< 3.13 \times 10^{-8}$. However, DEseq [1] analysis did not find sgg differentially expressed (p-value < 0.42). It is estimated that *sgg* has at least 17 isoforms [10]. The plot in Fig. 7.2 reveals complex sample-to-sample expression variation among head samples and among carcass samples. It is interesting to see that the expression curves, particularly near the end of *sgg*, are significantly different between heads and carcasses while the difference in the mean number of reads for the entire gene can be ignored.

Another differentially expressed gene detected by DE-FPCA is gene *Moe* with p-value $< 4.13 \times 10^{-6}$. *Moe* plays a role in synaptic development at the neuromuscular junction [36]. We plotted the expression profiles of gene *Moesin* (*Moe*) as a function of genome position in Fig. 7.3. The mean average numbers of reads in the heads and carcasses are 11,994.83 and 11,119.22, respectively. Due to these close values in the mean number of reads, as expected, DESeq did not detect a significant difference between the heads and carcasses (p-value ≈ 0.866). This detection difference between DE-FPCA and DEseq is that *Moe* exhibits differential expression at the exon level: Some exons have higher expression levels in carcass samples while other exons have the opposite property, and these regional differences cancel each other out. As a result, overall read counts are similar but the difference is apparent in the graph of gene expression functions.

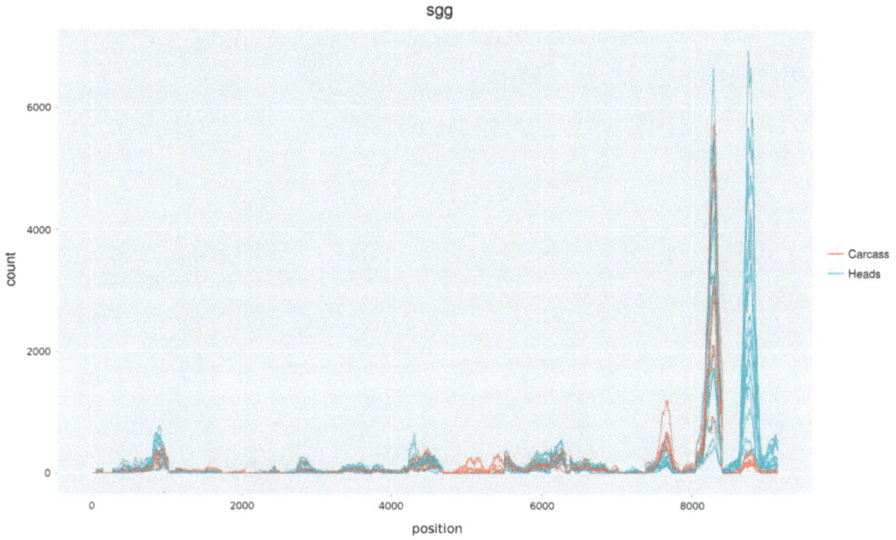

Fig. 7.2 RNA-seq expression profiles for gene shaggy (sgg) where curves in *red* and *blue* represent expression profiles for sgg in carcass and heads, respectively

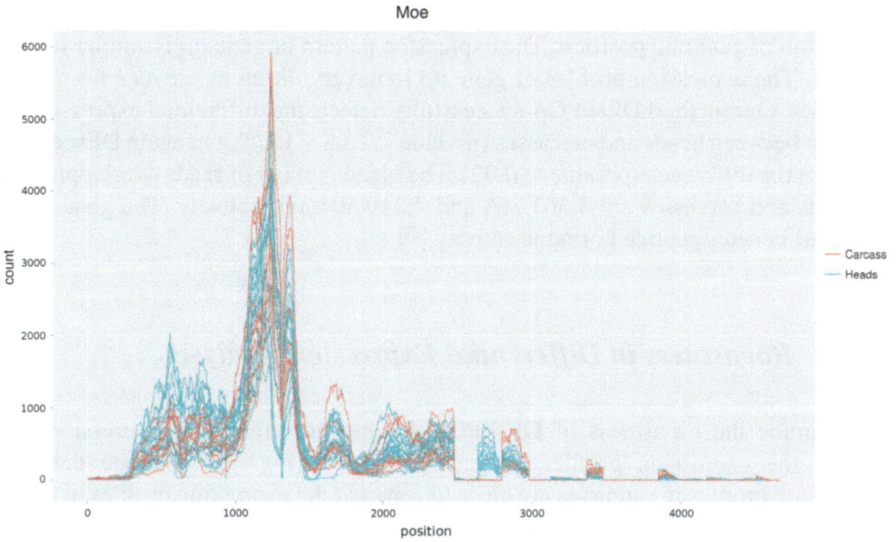

Fig. 7.3 RNA-seq expression profiles for gene Moesin (Moe) where curves in *red* and *blue* represent expression profiles for Moe in carcass and heads, respectively

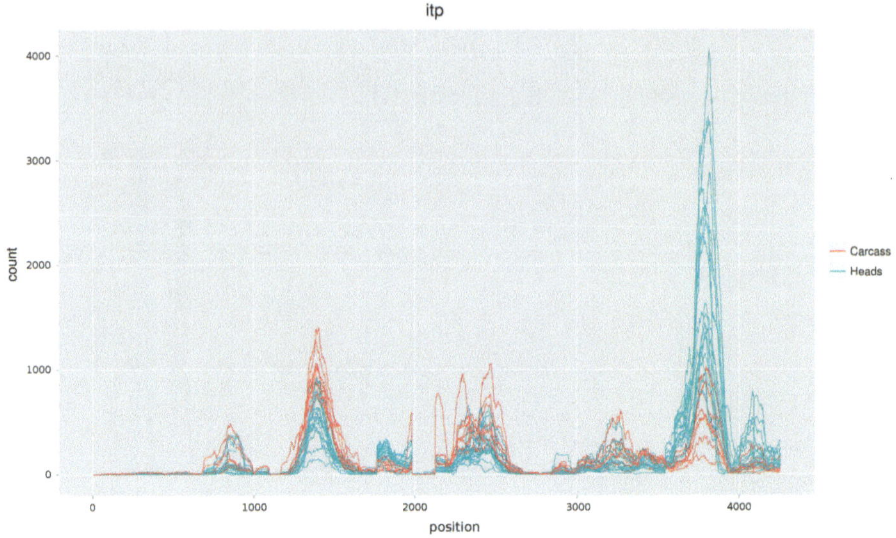

Fig. 7.4 RNA-seq expression profiles for gene ion transport peptide (itp) where curves in *red* and *blue* represent expression profiles for itp in carcass and heads, respectively

Figure 7.4 plotted the expression profiles of gene *ion transport peptide* (*itp*) as a function of genomic position. The expression pattern of gene *itp* is similar to that of *Moe*. The expression profiles of gene *itp* look very different between heads and carcasses. Our method DE-FPCA successfully detects the differential expression of gene *itp* between heads and carcasses (p-value $< 1.68 \times 10^{-6}$), but again DEseq fails to detect the difference (p-value ≈ 0.92). The mean number of reads overlapping *itp* in heads and carcasses are 4,367.865 and 4,249.504, respectively. The gene *itp* is involved in neuropeptide hormone activity [9].

7.3.5 Robustness in Differential Expression Analysis

To examine the robustness of DE-FPCA for testing differential expression, we plotted the expression profiles of gene *CG32553* (Fig. 7.5). We see that the expression profiles in carcasses are close to zero and the expression profiles in heads are randomly fluctuating across the entire gene with large positive overall expression values. Accordingly, DEseq concluded that gene *CG32553* is significantly differentially expressed between heads and carcasses (p-value $< 2.02 \times 10^{-10}$), but our method DE-FPCA did not find a significant differential expression for *CG32553* (p-value < 0.423). To illustrate that the significant results of *CG32553* identified by DEseq was likely to be false positive, we further plotted Fig. 7.6 showing expression profiles of *CG32553* in the adult heads samples corresponding to 1, 4 and 20 day old adults, respectively. A couple of samples displayed large peaks in one or two

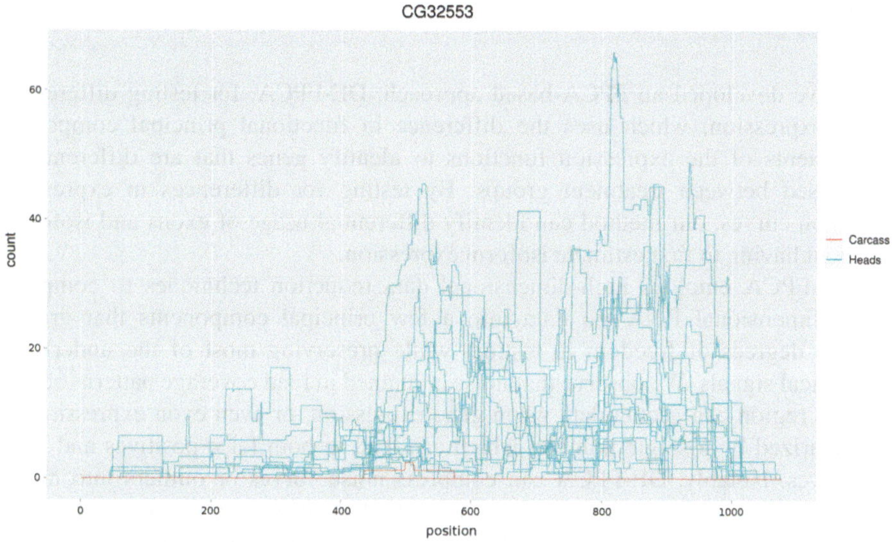

Fig. 7.5 RNA-seq expression profiles for gene CG32553 where curves in *red* and *blue* represent expression profiles for CG32553 in carcass and heads, respectively

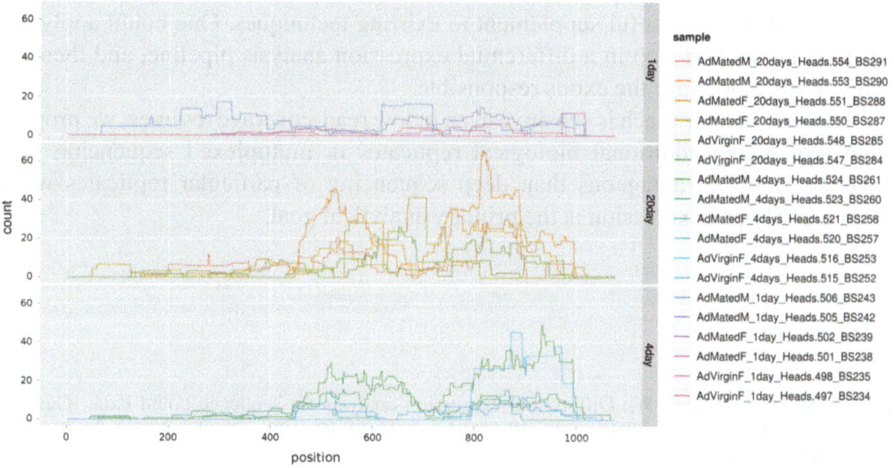

Fig. 7.6 RNA-seq expression profiles for gene CG32553 of six head samples at 1, 4 and 20 days, respectively

regions, but most head samples had low read counts throughout the gene with no discernible pattern. The random pattern of read counts in Fig. 7.6 suggests that the observed peaks in read coverage are likely either unrelated to the underlying gene model, or that they arise from another source of noise in the assay, and thus the detection of significant differential expression for *CG32553* may be spurious.

7.4 Discussion

We have developed an FPCA-based approach, DE-FPCA, for testing differential gene expression, which uses the difference in functional principal component coefficients of the expression functions to identify genes that are differentially expressed between treatment groups. By testing for differences in expression function curves, our method can identify differential usage of exons and isoforms without having to first estimate isoform expression.

DE-FPCA employs high-dimensional data reduction techniques to compress high dimensional RNA-seq data into a few principal components that greatly reduce degrees of freedom in testing, while preserving most of the underlying biological signals. We observe that noise contained in read coverage patterns across a gene region can accumulate when gene expression, or even exon expression, is summarized by a single number, which can lead to both false positives and false negatives. Notably, DE-FPCA can compress noise curves or outliers into minor principal components and hence help mitigate the impact of sequencing errors on tests. Hence, FPCA-based tests for differential expression are more robust than those based simply on counts.

Since the expression curves contain enriched information on isoforms that are due to varying usage of splice sites and transcription start and end sites, our method should constitute a useful supplement to existing techniques. One could apply our method as the first step in a differential expression analysis pipeline, and then use other tools to identify the exons responsible.

Because our approach is sensitive even in low read-coverage regimes, we propose that sequencing additional biological replicates in multiplexed sequencing runs may be more advantageous than deep sequencing of particular replicates when differential gene expression is the primary analytical goal.

References

[1] Anders, S., Huber, W.: Differential expression analysis for sequence count data. Genome Biol. **11**(10), R106 (2010). doi:10.1186/gb-2010-11-10-r106. PMID: 20979621 PMCID: PMC3218662

[2] Anders, S., Reyes, A., Huber, W.: Detecting differential usage of exons from RNA-seq data. Genome Res. **22**(10), 2008–2017 (2012). doi:10.1101/gr.133744.111. PMID: 22722343 PMCID: PMC3460195

[3] de Belle, J.S., Heisenberg, M.: Expression of drosophila mushroom body mutations in alternative genetic backgrounds: a case study of the mushroom body miniature gene (mbm). Proc. Natl. Acad. Sci. U.S.A. **93**(18), 9875–9880 (1996). PMID: 8790424 PMCID: PMC38522

[4] Bi, Y., Davuluri, R.V.: NPEBseq: nonparametric empirical bayesian-based procedure for differential expression analysis of RNA-seq data. BMC Bioinform. **14**, 262 (2013). doi:10.1186/1471-2105-14-262. PMID: 23981227 PMCID: PMC3765716

[5] Bullard, J.H., Purdom, E., Hansen, K.D., Dudoit, S.: Evaluation of statistical methods for normalization and differential expression in mRNA-Seq experiments. BMC Bioinform. **11**, 94 (2010). doi:10.1186/1471-2105-11-94. PMID: 20167110 PMCID: PMC2838869

[6] Celniker, S.E., Keelan, D.J., Lewis, E.B.: The molecular genetics of the bithorax complex of *Drosophila*: characterization of the products of the abdominal-b domain. Genes Dev. **3**(9), 1424–1436 (1989). PMID: 2575066

[7] Celotto, A.M., Graveley, B.R.: Alternative splicing of the drosophila dscam pre-mRNA is both temporally and spatially regulated. Genetics **159**(2), 599–608 (2001). PMID: 11606537 PMCID: PMC1461822

[8] Charlton-Perkins, M., Whitaker, S.L., Fei, Y., Xie, B., Li-Kroeger, D., Gebelein, B., Cook, T.: Prospero and pax2 combinatorially control neural cell fate decisions by modulating ras- and notch-dependent signaling. Neural Dev. **6**, 20 (2011). doi:10.1186/1749-8104-6-20. PMID: 21539742 PMCID: PMC3123624

[9] Dircksen, H., Tesfai, L.K., Albus, C., Nässel, D.R.: Ion transport peptide splice forms in central and peripheral neurons throughout postembryogenesis of *Drosophila melanogaster*. J. Comp. Neurol. **509**(1), 23–41 (2008). doi:10.1002/cne.21715. PMID: 18418898

[10] Flicek, P., Amode, M.R., Barrell, D., Beal, K., Brent, S., Chen, Y., Clapham, P., Coates, G., Fairley, S., Fitzgerald, S., Gordon, L., Hendrix, M., Hourlier, T., Johnson, N., Kähäri, A., Keefe, D., Keenan, S., Kinsella, R., Kokocinski, F., Kulesha, E., Larsson, P., Longden, I., McLaren, W., Overduin, B., Pritchard, B., Riat, H.S., Rios, D., Ritchie, G.R.S., Ruffier, M., Schuster, M., Sobral, D., Spudich, G., Tang, Y.A., Trevanion, S., Vandrovcova, J., Vilella, A.J., White, S., Wilder, S.P., Zadissa, A., Zamora, J., Aken, B.L., Birney, E., Cunningham, F., Dunham, I., Durbin, R., Fernández-Suarez, X.M., Herrero, J., Hubbard, T.J.P., Parker, A., Proctor, G., Vogel, J., Searle, S.M.J.: Ensembl 2011. Nucleic Acids Res. **39**(Suppl 1), D800–D806 (2011). doi:10.1093/nar/gkq1064. PMID: 21045057

[11] Garber, M., Grabherr, M.G., Guttman, M., Trapnell, C.: Computational methods for transcriptome annotation and quantification using RNA-seq. Nat. Meth. **8**(6), 469–477 (2011). doi:10.1038/nmeth.1613. PMID: 21623353

[12] Goldman, T.D., Arbeitman, M.N.: Genomic and functional studies of *Drosophila* sex hierarchy regulated gene expression in adult head and nervous system tissues. PLoS Genet. **3**(11), e216 (2007). doi:10.1371/journal.pgen.0030216. PMID: 18039034 PMCID: PMC2082469

[13] Graveley, B.R., Brooks, A.N., Carlson, J.W., Duff, M.O., Landolin, J.M., Yang, L., Artieri, C.G., van Baren, M.J., Boley, N., Booth, B.W., Brown, J.B., Cherbas, L., Davis, C.A., Dobin, A., Li, R., Lin, W., Malone, J.H., Mattiuzzo, N.R., Miller, D., Sturgill, D., Tuch, B.B., Zaleski, C., Zhang, D., Blanchette, M., Dudoit, S., Eads, B., Green, R.E., Hammonds, A., Jiang, L., Kapranov, P., Langton, L., Perrimon, N., Sandler, J.E., Wan, K.H., Willingham, A., Zhang, Y., Zou, Y., Andrews, J., Bickel, P.J., Brenner, S.E., Brent, M.R., Cherbas, P., Gingeras, T.R., Hoskins, R.A., Kaufman, T.C., Oliver, B., Celniker, S.E.: The developmental transcriptome of *Drosophila melanogaster*. Nature **471**(7339), 473–479 (2011). doi:10.1038/nature09715. http://www.nature.com/nature/journal/v471/n7339/full/nature09715.html

[14] Hansen, K.D., Brenner, S.E., Dudoit, S.: Biases in illumina transcriptome sequencing caused by random hexamer priming. Nucleic Acids Res. **38**(12), e131 (2010). doi:10.1093/nar/gkq224. PMID: 20395217 PMCID: PMC2896536

[15] Hansen, K.D., Wu, Z., Irizarry, R.A., Leek, J.T.: Sequencing technology does not eliminate biological variability. Nat. Biotechnol. **29**(7), 572–573 (2011). doi:10.1038/nbt.1910. PMID: 21747377 PMCID: PMC3137276

[16] Hardcastle, T.J., Kelly, K.A.: baySeq: empirical bayesian methods for identifying differential expression in sequence count data. BMC Bioinform. **11**, 422 (2010). doi:10.1186/1471-2105-11-422. PMID: 20698981 PMCID: PMC2928208

[17] Henderson, D., Plaschko, P.: Stochastic Differential Equations in Science and Engineering. World Scientific, New Jersey (2006)

[18] Hu, M., Zhu, Y., Taylor, J.M.G., Liu, J.S., Qin, Z.S.: Using poisson mixed-effects model to quantify transcript-level gene expression in RNA-Seq. Bioinformatics **28**(1), 63–68 (2012). doi:10.1093/bioinformatics/btr616. PMID: 22072384 PMCID: PMC3244770

[19] Hu, Y., Liu, Y., Mao, X., Jia, C., Ferguson, J.F., Xue, C., Reilly, M.P., Li, H., Li, M.: PennSeq: accurate isoform-specific gene expression quantification in RNA-Seq by modeling non-uniform read distribution. Nucleic Acids Res. **42**(3), e20 (2014). doi:10.1093/nar/gkt1304. PMID: 24362841 PMCID: PMC3919567

[20] Lewis, E.B.: A gene complex controlling segmentation in *Drosophila*. Nature **276**(5688), 565–570 (1978). PMID: 103000

[21] Li, J., Jiang, H., Wong, W.H.: Modeling non-uniformity in short-read rates in RNA-Seq data. Genome Biol. **11**(5), R50 (2010). doi:10.1186/gb-2010-11-5-r50. PMID: 20459815 PMCID: PMC2898062

[22] Li, J., Witten, D.M., Johnstone, I.M., Tibshirani, R.: Normalization, testing, and false discovery rate estimation for RNA-sequencing data. Biostatistics **13**(3), 523–538 (2012). doi:10.1093/biostatistics/kxr031. PMID: 22003245 PMCID: PMC3372940

[23] Marioni, J.C., Mason, C.E., Mane, S.M., Stephens, M., Gilad, Y.: RNA-seq: an assessment of technical reproducibility and comparison with gene expression arrays. Genome Res. **18**(9), 1509–1517 (2008). doi:10.1101/gr.079558.108. PMID: 18550803 PMCID: PMC2527709

[24] McCarthy, D.J., Chen, Y., Smyth, G.K.: Differential expression analysis of multifactor RNA-Seq experiments with respect to biological variation. Nucleic Acids Res. **40**(10), 4288–4297 (2012). doi:10.1093/nar/gks042. PMID: 22287627 PMCID: PMC3378882

[25] Merkin, J., Russell, C., Chen, P., Burge, C.B.: Evolutionary dynamics of gene and isoform regulation in mammalian tissues. Science **338**(6114), 1593–1599 (2012). doi: 10.1126/science.1228186.PMID:23258891PMCID:PMC3568499

[26] Mortazavi, A., Williams, B.A., McCue, K., Schaeffer, L., Wold, B.: Mapping and quantifying mammalian transcriptomes by RNA-Seq. Nat. Meth. **5**(7), 621–628 (2008). doi: 10.1038/nmeth.1226.PMID:18516045

[27] Okoniewski, M.J., Leśniewska, A., Szabelska, A., Zyprych-Walczak, J., Ryan, M., Wachtel, M., Morzy, T., Schäfer, B., Schlapbach, R.: Preferred analysis methods for single genomic regions in RNA sequencing revealed by processing the shape of coverage. Nucleic Acids Res. **40**(9), e63 (2012). doi:10.1093/nar/gkr1249. PMID: 22210855 PMCID: PMC3351146

[28] Ramsay, J.O., Silverman, B.W.: Functional Data Analysis. Springer, New York (2005)

[29] Rapaport, F., Khanin, R., Liang, Y., Pirun, M., Krek, A., Zumbo, P., Mason, C.E., Socci, N.D., Betel, D.: Comprehensive evaluation of differential gene expression analysis methods for RNA-seq data. Genome Biol. **14**(9), R95 (2013). doi:10.1186/gb-2013-14-9-r95. PMID: 24020486

[30] Riparbelli, M.G., Massarelli, C., Robbins, L.G., Callaini, G.: The abnormal spindle protein is required for germ cell mitosis and oocyte differentiation during *Drosophila* oogenesis. Exp. Cell Res. **298**(1), 96–106 (2004). doi:10.1016/j.yexcr.2004.03.054. PMID: 15242765

[31] Risso, D., Schwartz, K., Sherlock, G., Dudoit, S.: GC-content normalization for RNA-Seq data. BMC Bioinform. **12**, 480 (2011). doi:10.1186/1471-2105-12-480. PMID: 22177264 PMCID: PMC3315510

[32] Roberts, A., Trapnell, C., Donaghey, J., Rinn, J.L., Pachter, L.: Improving RNA-Seq expression estimates by correcting for fragment bias. Genome Biol. **12**(3), R22 (2011). doi:10.1186/gb-2011-12-3-r22. PMID: 21410973 PMCID: PMC3129672

[33] Robinson, M.D., Smyth, G.K.: Small-sample estimation of negative binomial dispersion, with applications to SAGE data. Biostatistics **9**(2), 321–332 (2008). doi: 10.1093/biostatistics/kxm030.PMID:17728317

[34] Robinson, M.D., McCarthy, D.J., Smyth, G.K.: edgeR: a bioconductor package for differential expression analysis of digital gene expression data. Bioinformatics **26**(1), 139–140 (2010). doi:10.1093/bioinformatics/btp616. PMID: 19910308 PMCID: PMC2796818

[35] Rogulja-Ortmann, A., Renner, S., Technau, G.M.: Antagonistic roles for ultrabithorax and antennapedia in regulating segment-specific apoptosis of differentiated motoneurons in the drosophila embryonic central nervous system. Development **135**(20), 3435–3445 (2008). doi:10.1242/dev.023986. PMID: 18799545

[36] Seabrooke, S., Stewart, B.A.: Moesin helps to restrain synaptic growth at the *Drosophila* neuromuscular junction. Dev. Neurobiol. **68**(3), 379–391 (2008). doi:10.1002/dneu.20595. PMID: 18161855

[37] Smyth, G.K.: Linear models and empirical bayes methods for assessing differential expression in microarray experiments. Stat. Appl. Genet. Mol. Biol. **3**, Article3 (2004). doi:10.2202/1544-6115.1027.PMID:16646809

[38] Srivastava, S., Chen, L.: A two-parameter generalized poisson model to improve the analysis of RNA-seq data. Nucleic Acids Res. **38**(17), e170 (2010). doi:10.1093/nar/gkq670.PMID: 20671027PMCID:PMC2943596

[39] Sun, Z., Zhu, Y.: Systematic comparison of RNA-Seq normalization methods using measurement error models. Bioinformatics **28**(20), 2584–2591 (2012). doi:10.1093/bioinformatics/bts497. PMID: 22914217

[40] Suo, C., Calza, S., Salim, A., Pawitan, Y.: Joint estimation of isoform expression and isoform-specific read distribution using multisample RNA-Seq data. Bioinformatics **30**(4), 506–513 (2014). doi:10.1093/bioinformatics/btt704. PMID: 24307704

[41] Trapnell, C., Roberts, A., Goff, L., Pertea, G., Kim, D., Kelley, D.R., Pimentel, H., Salzberg, S.L., Rinn, J.L., Pachter, L.: Differential gene and transcript expression analysis of RNA-seq experiments with TopHat and cufflinks. Nat. Protoc. **7**(3), 562–578 (2012). doi:10.1038/nprot.2012.016. PMID: 22383036 PMCID: PMC3334321

[42] Trapnell, C., Hendrickson, D.G., Sauvageau, M., Goff, L., Rinn, J.L., Pachter, L.: Differential analysis of gene regulation at transcript resolution with RNA-seq. Nat. Biotechnol. **31**(1), 46–53 (2013). doi:10.1038/nbt.2450. PMID: 23222703 PMCID: PMC3869392

[43] Wang, Z., Gerstein, M., Snyder, M.: RNA-seq: a revolutionary tool for transcriptomics. Nat. Rev. Genet. **10**(1), 57–63 (2009). doi:10.1038/nrg2484. PMID: 19015660 PMCID: PMC2949280

[44] Wang, L., Feng, Z., Wang, X., Wang, X., Zhang, X.: DEGseq: an r package for identifying differentially expressed genes from RNA-seq data. Bioinformatics **26**(1), 136–138 (2010). doi:10.1093/bioinformatics/btp612. PMID: 19855105

[45] Wu, Z., Wang, X., Zhang, X.: Using non-uniform read distribution models to improve isoform expression inference in RNA-seq. Bioinformatics **27**(4), 502–508 (2011). doi:10.1093/bioinformatics/btq696.PMID:21169371

[46] Zheng, W., Chung, L.M., Zhao, H.: Bias detection and correction in RNA-sequencing data. BMC Bioinform. **12**, 290 (2011). doi:10.1186/1471-2105-12-290. PMID: 21771300 PMCID: PMC3149584

Chapter 8
Mapping of Expression Quantitative Trait Loci Using RNA-seq Data

Wei Sun and Yijuan Hu

Abstract RNA sequencing (RNA-seq) is replacing expression microarrays for genome-wide assessment of gene expression abundance. Many sophisticated statistical methods have been developed to map gene expression quantitative trait loci (eQTL) using microarray data. These methods can potentially be applied to RNA-seq data with minor modifications. However, they fail to exploit two types of novel information that are available from RNA-seq but not from microarrays: the allele-specific expression (ASE) and the isoform-specific expression (ISE). This chapter gives an overview of the statistical methods that are specifically designed for eQTL mapping using RNA-seq data, as well as the challenges and some future directions.

8.1 Introduction

In most living organisms, the DNA information stored in a cell is transcribed into messenger RNA (mRNA) and then translated into protein, which is the working force of the cell. The amount of mRNA produced by a gene is generally referred to as gene expression. Since mid 1990s, gene expression microarrays have been widely employed to assess mRNA abundance genome-wide. The huge amount of data produced by expression microarrays have not only greatly improved our understanding of cell biology, but also provided invaluable resources to guide the diagnosis and treatment of human diseases. For example, gene expression profiles have been used to dissect cancer subtypes [45] and to predict drug sensitivities [20].

W. Sun (✉)
Department of Biostatistics, UNC Chapel Hill, Chapel Hill, NC, USA
e-mail: weisun@email.unc.edu

Y. Hu
Department of Biostatistics and Bioinformatics, Emory University, Atlanta, GA, USA
e-mail: yijuan.hu@emory.edu

S. Datta and D. Nettleton (eds.), *Statistical Analysis of Next Generation Sequencing Data*, Frontiers in Probability and the Statistical Sciences, DOI 10.1007/978-3-319-07212-8_8,

The mRNA abundance of a gene may be associated with the genotype of one or more genetic loci, which are referred to as expression quantitative trait loci (eQTL). In most eQTL studies, genome-wide gene expression data and DNA genotype data of genetic markers such as single nucleotide polymorphisms (SNPs) are collected in a common set of samples. Then eQTLs are identified by linkage/association analysis in which the expression of each gene is treated as a quantitative trait. We refer the readers to [10, 51] for reviews on eQTL studies and their potential impacts on understanding the genomic basis of human complex traits, and to [33, 68] for reviews on statistical methods and computational tools for eQTL studies using gene expression from microarrays.

In this chapter, we will focus on eQTL mapping using RNA-seq data. RNA-seq, i.e., high-throughput RNA sequencing, is replacing expression microarrays for transcriptome studies. To explain the motivations of designing statistical methods specifically for RNA-seq data, it is helpful to first describe the differences between the microarray and RNA-seq platforms. In microarray experiments, the abundance of gene expression is measured by fluorescent signals on a set of probes, where each probe contains a specific short piece of DNA sequence (e.g., 25 base pairs for most Affymetrix arrays). The amount of information that can be obtained is limited by the design of the microarray:

- The quantification of gene expression is confined to the regions where the probes are placed. The probes are pre-selected to cover known genes, and in most array platforms, the probes are located at the 3' ends of the transcripts instead of being uniformly distributed across exonic regions. Therefore, previously unknown transcripts cannot be measured for expression and the measurements at known transcripts may be biased by the signals at the 3' ends.
- The same probe sequences are used for all samples and do not accommodate the genetic differences across samples or the differences between the paternal and maternal alleles of a sample. Therefore, the gene expression from the paternal and maternal alleles cannot be distinguished.

In RNA-seq experiments, the expression of a gene is measured by the number of sequence reads mapped to that gene [18, 42]. RNA-seq overcomes the two limitations of microarrays. First, RNA-seq objectively quantifies the genome-wide transcript abundance without relying on pre-selected probes. Second, an RNA-seq read delivers allele-specific information if it overlaps with at least one heterozygous SNP/indel (i.e., a SNP or an insertion or deletion that is heterozygous between the paternal and maternal alleles).

Figure 8.1 illustrates the data generated by the two platforms. In particular, microarray data take continuous values and RNA-seq data are discrete counts. If that is all the difference between the two platforms, then there is no need to develop novel statistical methods for RNA-seq data because one can simply replace the linear regression model for continuous microarray data with the generalized linear regression model (with Poisson or negative binomial distribution assumption) for count data. In fact, the raw sequence data from RNA-seq contain much more information than a single count as shown in Fig. 8.1. First, in a diploid genome such

a

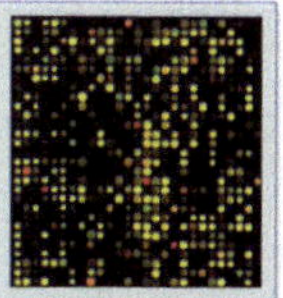

	Sample 1	Sample 2	...	Sample n
Gene 1	10.45	10.52		10.40
Gene 2	4.63	4.76		4.70
Gene 3	8.80	8.96		8.82
...				
...				
Gene p	8.30	8.44		8.27

b

```
@HWI-EAS134:1:1:0:1446#0/1
GGGACTTAATCAACGCAAGCTTATGACCCGCACTT
+
BB<2A81(66B<2>BB@>?B=BA6=8@BCBB@?3/
```

	Sample 1	Sample 2	...	Sample n
Gene 1	512	339		286
Gene 2	1043	1212		888
Gene 3	20	12		10
...				
...				
Gene p	78	65		42

Fig. 8.1 **(a)** Gene expression data from a microarray. Each sample is measured by an array with tens of thousands of pre-selected probes. The expression of one gene is estimated by combining the fluorescent signals of multiple probes. **(b)** Gene expression data from RNA-seq. The data of each sample is stored in a text file, usually in the FASTQ format. An FASTQ file contains millions of records and each record corresponds to an RNA-seq read with four lines: the sequence identifier, the actual DNA sequence, a separator, and the sequencing quality scores for every base pair of the sequence

as the genome of human or mouse, there are two sets of chromosomes, one from the father and one from the mother. Thus most genes (e.g., autosomal genes and X-linked genes in females) have two copies and each copy is called an allele of this gene. The expression of each allele of a gene, i.e., **allele-specific expression** (ASE), can be extracted from the raw RNA-seq data. Second, in a higher organism such as a human or mouse, one gene often comprises of several exons and the exons can be grouped in different ways to produce different proteins or non-coding RNA molecules. Each combination of the exons of a gene is called a transcript or an RNA isoform. The expression of each isoform, i.e., **isoform-specific expression** (ISE), can also be inferred from the raw RNA-seq data. In summary, the RNA-seq platform delivers much more information than the microarrays and thus warrants the development of novel statistical methods to fully exploit the new features.

The remainder of this chapter is organized as follows. Sections 8.2 and 8.3 will introduce eQTL mapping using ASE and ISE, respectively. Section 8.4 will discuss some challenges and future directions.

8.2 eQTL Mapping Using ASE

We will first describe the quantification of ASE and show how the ASE enables the detection of *cis*-regulatory eQTLs. Then we will introduce statistical methods for eQTL mapping using ASE under two scenarios, namely, with and without known haplotypes between the candidate eQTL and the gene of interest.

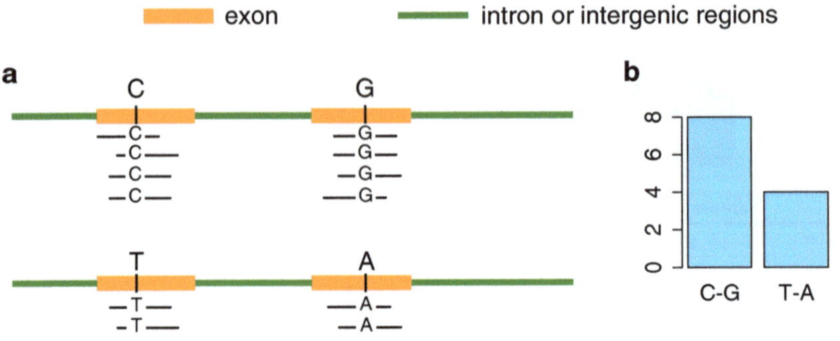

Fig. 8.2 An example of ASE abundance quantification using RNA-seq, for a hypothetical gene with two exons and one heterozygous SNP within each exon. **(a)** Two haplotypes of this gene. **(b)** The number of allele-specific reads from these two haplotypes

8.2.1 Quantification of ASE Using RNA-seq

ASE can be measured by the number of RNA-seq reads that are mapped to the gene and overlapped with at least one SNP or indel with heterozygous genotype. Figure 8.2 illustrates the quantification of ASE for a hypothetical gene with two exons. There are two SNPs with heterozygous genotypes on the exonic regions of this gene, one SNP for each exon. Given the genotype at each SNP, allele-specific read count (**ASReC**) can be obtained by counting the number of reads harboring a particular SNP allele. For example, there are 6 reads overlapping with the first SNP with genotype CT, and the ASReCs are 4 and 2 for SNP alleles C and T, respectively. Then, the ASE of this gene can be estimated by combining ASReCs across multiple SNPs if the haplotype information is available. In the example shown in Fig. 8.2a, the genotypes of the two SNPs are CT and GA and the possible haplotype pairs are (C-G, T-A) and (C-A, T-G). If we knew that the underlying haplotype pair is (C-G, T-A), we could obtain the gene-level ASReCs as shown in Fig. 8.2b.

Next we discuss a few issues related to ASE quantification: haplotype phasing, sequence mapping bias, and expected ASReC.

8.2.1.1 Haplotype Phasing

Many algorithms (e.g., [8, 12, 36]) have been developed to infer the haplotype phases from the genotypes of unrelated individuals. It is well known that the phasing accuracy deteriorates as the length of the haplotype increases. However, it is still reasonable to assume that the phasing is accurate within the exonic regions of a gene because those regions are relatively short (∼90 % of the annotated genes are shorter than 100 kb [16]) and tend to undergo less recombination [62]. In addition, the switch errors (i.e., mistaken swapping from one haplotype to the other) in exonic regions can be captured and corrected by RNA-seq reads (either single or paired-end reads) that overlap with two or more heterozygous SNPs (i.e., SNPs with heterozygous genotypes) and thus provide direct information on the haplotype phase. Some reads may even span over non-adjacent exons due to alternative splicing and thus provide information on long-range phase.

8.2.1.2 Sequence Mapping Bias

A common practice in RNA-seq studies is to map the reads of all samples against the same reference genome. This may induce mapping bias because the reads harboring reference alleles tend to be mapped more accurately than those harboring alternative alleles. There are several solutions to this problem.

1. Identify and remove SNPs that may cause mapping bias by mapping simulated reads to the reference genome [46].
2. Employ an allele-aware sequence aligner [70] that uses both the reference genome and alternate alleles to map reads.
3. Construct the two haploid genomes for each diploid individual and map the reads against the two genomes separately [26, 30].

The third approach is the most unbiased and most comprehensive one, although it requires more information, i.e., the complete haploid genomes, and more computational time. Such an effort can be well justified for certain diploid samples with two very different haploid genomes, e.g., F1 mice from a cross of two inbred mouse strains with different genome backgrounds.

8.2.1.3 Expected ASReC

What proportion of RNA-seq reads are allele-specific? The answer depends on two factors, the density of DNA polymorphisms (usually SNPs or indels) with heterozygous genotypes and the read length. Clearly, the more different are the two haploid genomes, the more reads are allele-specific; the longer the reads are, the more likely they overlap with heterozygous DNA polymorphisms. The expected proportion of allele-specific reads can vary from 0.5 % in a human study with short reads [46, 55] (Fig. 8.3a) to 35 % in an F1 mouse study with longer reads [11] (Fig. 8.3b). To be specific, the human study [46,55] adopted an RNA-seq experiment with 35 bp single-end reads and used $\sim$1.4 million HapMap SNPs to extract allele-specific reads. The number of heterozygous SNPs for an individual ranges from 392,800 to 415,500 with a median of 409,100. In another on-going study involving 550 breast cancer patients from The Cancer Genome Atlas (TCGA) using 2$\times$50 bp paired-end reads and $\sim$30 million 1000G SNPs, we identified 3.4 % reads as allele-specific. The number of heterozygous SNPs across these TCGA samples ranges from 1.91 million to 2.02 million with a median of 1.97 million. The increase of the proportion of allele-specific reads from 0.5 % to 3.4 % in the two human studies can be attributed to both the longer reads and the larger number of heterozygous

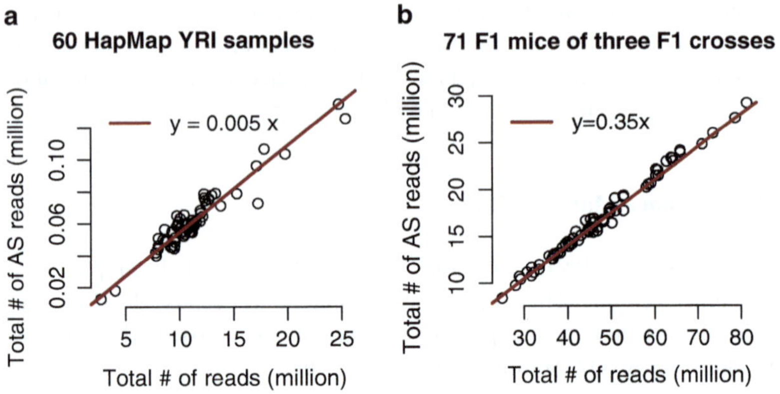

Fig. 8.3 Scatter plot of the total number of RNA-seq reads versus the total number of allele-specific reads for all the samples in (**a**) a human study of unrelated individuals of African population (HapMap YRI samples) [55] and (**b**) a mouse study of three reciprocal F1 crosses of three mouse inbred strains (CAST/EiJ, PWK/PhJ and WSB/EiJ) representative of three subspecies within the *Mus musculus* species group (*M. m. castaneus*, *M. m. musculus* and *M. m. domesticus*, respectively)

SNPs. By contrast, the mouse study [11] collected 2×100 bp paired-end RNA-seq reads from F1 mice with around 17.5 million heterozygous SNPs/indels per sample, making it possible to harvest 35 % of RNA-seq reads as allele-specific.

8.2.2 ASE for cis-eQTL Mapping

Given ASE, we can assess whether there is allelic imbalance of gene expression. In some publications, the terms ASE and allelic imbalance are used exchangeably. In this book chapter, however, ASE indicates the expression measurement from a particular allele. ASE is available for a gene if it has exonic SNPs/indels with heterozygous genotypes, and thus having ASE does not imply allelic balance. A number of pioneering studies have shown that allelic imbalance in gene expression exists and may be associated with disease susceptibility [17, 27, 35, 40, 60, 73]. For example, the reduction in the expression of one allele at the TGFBR1 gene in blood cells (germline) leads to an elevated risk of colorectal cancer [60]. In addition, effective treatments can be developed by silencing the disease allele while sparing the expression of the wild type allele [41]. Here, we focus on mapping the DNA polymorphism that leads to allelic imbalance of gene expression, which is called a *cis*-eQTL and is a main mechanism of allelic imbalance.

To better understand *cis*-eQTLs, it is helpful to introduce the concept of *trans*-eQTL and clarify their differences. *Cis*-eQTL and *trans*-eQTL have been widely used to refer to eQTLs that are close to the associated genes and eQTLs that are distant, respectively. An arbitrary distance, such as 200 kb or 1 Mb, is often used to distinguish local and distant eQTLs. It has been pointed out before [51] and is worthwhile to be emphasized again: it is misleading to refer to a local or distant eQTL as a *cis*- or *trans*-eQTL as the latter have their own biological meanings.

The Latin words *cis* and *trans* mean "on the same side" and "across", respectively. A *cis*-eQTL is located on the same chromosome as its target gene and influences the gene expression in an allele-specific manner. Specifically, a mutation in the maternal allele only changes the gene expression from the maternal allele but does not affect the expression from the paternal allele (Fig. 8.4a). A plausible scenario is that a *cis*-eQTL is located at the transcriptional factor binding site of a gene and thus interferes with the transcriptional factor binding in the allele-specific manner. A *cis*-eQTL is likely to be a local eQTL, though this is not always true. By contrast, a *trans*-eQTL of a gene can be located anywhere in the genome and it influences the gene expression of both alleles to the same extent. One possible mechanism is that a *trans*-eQTL modifies the activity or abundance of a protein that regulates the gene and such regulation does not distinguish the two alleles of the gene [67] (Fig. 8.4b). Therefore, *cis*- and *trans*-eQTLs should be distinguished by ASE (Fig. 8.4a, b) [14, 52] rather than their physical distance to the target gene. Note that *cis*- and *trans*-eQTLs cannot be distinguished by the total expression of the gene, which shows the same pattern at the population level (Fig. 8.4c, d).

From the above discussions, it is clear that ASE is informative for *cis*-eQTL mapping. Figure 8.5a–d shows a hypothetical example of *cis*-eQTL mapping using ASE. Assume that the gene of interest has two exons with one SNP for each. We wish to test whether a candidate eQTL, displayed on the left of the gene in Fig. 8.5a, *cis*-regulates the gene expression. First, we count the number of allele-specific reads. As mentioned in Sect. 8.2.1, an RNA-seq read is allele-specific if it can be assigned to one of the two alleles of the gene without ambiguity. As illustrated in Fig. 8.5a, individuals (i) and (ii) have heterozygous genotypes for at least one exonic SNP, and thus their ASE can be measured by the number of reads that overlap with the heterozygous SNPs. Haplotype information is required to combine ASE measured at individual exonic SNPs into the gene-level ASE. For example, for individual (i), we count the number of allele-specific reads mapped to the haplotypes A-A and T-G. Next, we associate ASE with the candidate eQTL. For individual (i) in Fig. 8.5a, given the longer haplotypes C-A-A and T-T-G that span over the gene as well as the candidate eQTL, we can link ASE of the A-A and T-G haplotypes of the gene to the C and T alleles of the candidate eQTL, respectively (Fig. 8.5c). The association testing seeks to answer the question whether one allele of the candidate eQTL is associated with a higher or lower ASE of the gene. If the answer is yes (and assuming there is no other factor inducing the allelic imbalance), then we expect allelic imbalanced expression when the genotype of the candidate eQTL is heterozygous and allelic balanced expression when the genotype is homozygous; in other words, the candidate eQTL is a *cis*-eQTL. For example, individual (i) has a heterozygous genotype C/T at the candidate eQTL and has a higher ASE corresponding to the C allele than the T allele (Fig. 8.5c). Individual (ii) has a homozygous genotype C/C at the candidate eQTL, each C allele corresponding to the same ASE (Fig. 8.5d). A real data example of 65 HapMap samples is shown in Fig. 8.5f.

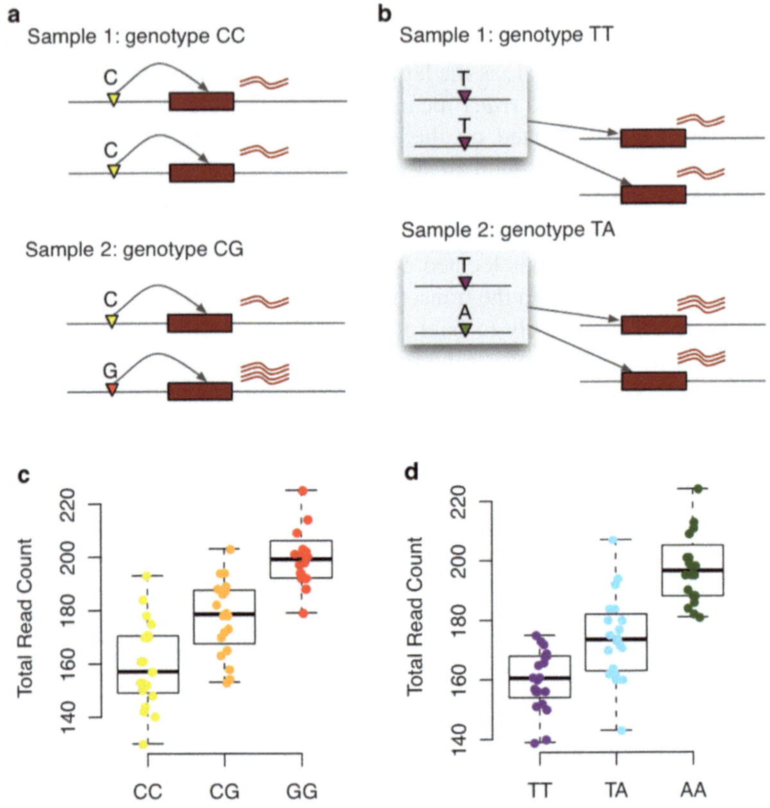

Fig. 8.4 **(a)** An example of a *cis*-eQTL in two samples. In sample 2 where the candidate eQTL (the SNP for which we test association) has a heterozygous genotype CG, the expression of the two alleles are different. **(b)** An example of a *trans*-eQTL in two samples. In sample 2 where the candidate eQTL has a heterozygous genotype TA, the expression of the two alleles are the same. **(c)** A simulated data for a *cis*-eQTL across 60 samples with 20 samples within each genotype class. **(d)** A simulated data for a *trans*-eQTL across 60 samples with 20 samples within each genotype class. This figure is adapted from Fig. 1 in our earlier paper Sun and Hu (2013) [56]

The total read count (**TReC**) is also informative for *cis*-eQTL mapping, which is similar to the traditional eQTL mapping using gene expression measured by microarrays. While ASE provides information at the allele level, TReC contributes at the individual level and in a way that is consistent with the allele level. In Fig. 8.5a–d, the C allele of the candidate eQTL is associated with a higher ASE, which is manifested at the allele level (Fig. 8.5c, d) and at the individual level (Fig. 8.5b). In general, the TReC of a gene is much greater than the sum of the two ASReCs in that TReC includes many reads that do not overlap with any heterozygous SNPs/indels.

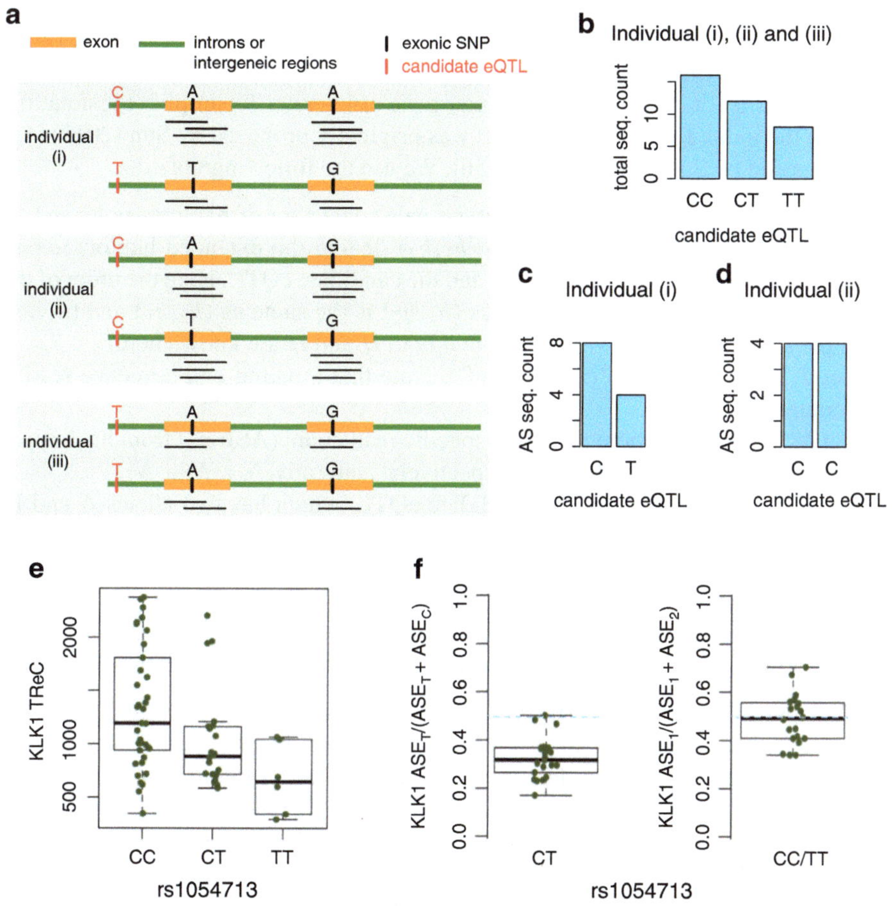

Fig. 8.5 **(a)–(d)** A hypothetical example of *cis*-eQTL mapping. **(a)** RNA-seq measurements of a gene with two exons in three individuals. **(b)** TReC (total read count) for the three individuals. **(c–d)** ASE for individual (i) and (ii). **(e)–(f)** A real data example of *cis*-eQTL mapping between gene KLK1 and SNP rs1054713. **(e)** Association between the genotypes and TReC. The y-axis is the total number of reads mapped to the gene KLK1 and each point corresponds to one of the 65 samples. **(f)** Association between the genotypes and ASE. When the genotype of rs1054713 is heterozygous, the ASE of the two alleles of this gene can be associated with the two alleles of rs1054713. ASE_T and ASE_C denote the ASReC corresponding to the T and C allele of rs1054713, respectively. When the genotype of rs1054713 is homozygous, we denote the ASReC of the two alleles of this gene by ASE_1 and ASE_2, respectively. This figure is a modified version of Figs. 2 and 4 of the earlier paper by Sun and Hu (2013) [56]

8.2.3 eQTL Mapping Using ASE with Known Haplotypes

While the haplotypes across the exonic regions of a gene can be accurately phased, those extending from the gene to a candidate eQTL may not be reliably phased because the candidate eQTL may be far away from the gene. In this section, we assume that the extended haplotypes are known and defer the scenario with unknown haplotypes to the next section.

Our statistical model is for a particular gene of interest. To simplify the notation, we skip the index for gene. The model was originally proposed by Sun (2012) [55] and reviewed by Sun and Hu (2013) [56]. We use the following notation.

- Let $H = (h_1, h_2)$ denote the haplotype pair consisting of haplotypes h_1 and h_2 across the exonic SNPs. Let $\tilde{H} = (\tilde{h}_1, \tilde{h}_2)$ denote the extended haplotype pair consisting of both the exonic SNPs and the candidate eQTL. Here the order of the two haplotypes is arbitrary and thus (h_1, h_2) is the same as (h_2, h_1) and $(\tilde{h}_1, \tilde{h}_2)$ is the same as $(\tilde{h}_2, \tilde{h}_1)$. We assume that both H and $\tilde{H}$ are known here.
- Let T be the total read count (TReC). Note that a paired-end sequence read is counted as one read.
- Let N_1, N_2 and N denote the allele-specific read count (ASReC) from haplotypes h_1 and h_2 and the total ASReC, respectively. Naturally, $N = N_1 + N_2$.
- Let G be the genotype of the candidate eQTL, which has two alleles A and B. Under the additive genetic effect, $G = 0$, 1, and 2 for genotypes AA, AB and BB, respectively. Dominant, recessive, and co-dominant effects can also be modeled using appropriate coding for genotypes.
- Let $\mathbf{X}$ be the relevant covariates including an intercept. Typically, $\mathbf{X}$ include the log form of the total read count per sample reflecting the read depth.

We model the probability of T given G and $\mathbf{X}$ by a negative binomial distribution indexed by parameters $(\boldsymbol{\gamma}, \beta_T, \phi)$, which is denoted by $P_{\text{TReC}}(T|G, \mathbf{X}; \boldsymbol{\gamma}, \beta_T, \phi)$. A negative binomial distribution can be considered as an infinite gamma mixture of Poisson distributions. It allows over-dispersion in the read counts, a phenomenon that is often observed in sequencing data across biological replicates. Thus the negative binomial distribution has been commonly used for RNA-seq data analysis [5]. In particular, we assume that T follows the negative binomial distribution with mean μ and a dispersion parameter ϕ:

$$P_{\text{TReC}}(T|G, \mathbf{X}; \boldsymbol{\gamma}, \beta_T, \phi) = \frac{\Gamma(T + 1/\phi)}{T!\Gamma(1/\phi)} \left(\frac{1}{1 + \phi\mu} \right)^{1/\phi} \left(\frac{\phi\mu}{1 + \phi\mu} \right)^{T},$$

where

$$\log(\mu) = \boldsymbol{\gamma}^{\mathrm{T}} \mathbf{X} + w(G, \beta_T),$$

and

$$w(G, \beta_T) = \begin{cases} 0 & \text{if } G = 0 \\ \log\left[1 + \exp(\beta_T)\right] - \log 2 & \text{if } G = 1 \\ \beta_T & \text{if } G = 2. \end{cases}$$

The functional form of $w(G, \beta_T)$ reflects the additive genetic effect. To see this, we write the means of T given $\mathbf{X}$ and $G = 0, 1, 2$ by $\mu_{AA,\mathbf{X}}$, $\mu_{AB,\mathbf{X}}$ and $\mu_{BB,\mathbf{X}}$, respectively, where

$$\mu_{AA,\mathbf{X}} = \exp(\boldsymbol{\gamma}^{\mathrm{T}}\mathbf{X}),$$
$$\mu_{AB,\mathbf{X}} = \exp(\boldsymbol{\gamma}^{\mathrm{T}}\mathbf{X} + \log[1 + \exp(\beta_T)] - \log 2)$$
$$\mu_{BB,\mathbf{X}} = \exp(\boldsymbol{\gamma}^{\mathrm{T}}\mathbf{X} + \beta_T).$$

We can see that β_T characterizes the difference between $\log(\mu_{AA,\mathbf{X}})$ and $\log(\mu_{BB,\mathbf{X}})$ and $\mu_{AB,\mathbf{X}}$ is at the mid point between $\mu_{AA\mathbf{X}}$ and $\mu_{BB,\mathbf{X}}$, i.e., $\mu_{AB,\mathbf{X}} = (\mu_{AA,\mathbf{X}} + \mu_{BB,\mathbf{X}})/2$.

We model the probability of N_1 given N, $\tilde{H}$ and $\mathbf{X}$ assuming that N_1 follows a beta-binomial distribution indexed by parameters (β_A, ψ) and denote the model by $P_{\texttt{ASReC}}(N_1|N, \tilde{H}, \mathbf{X}; \beta_A, \psi)$. A beta-binomial distribution extends a binomial distribution to allow over-dispersion. In particular, we assume that N_1 follows a beta-binomial distribution with mean p and a dispersion parameter ψ:

$$P_{\texttt{ASReC}}(N_1|N, \tilde{H}, \mathbf{X}; \beta_A, \psi) = \binom{N}{N_1} \frac{\prod_{k=0}^{N_1-1}(p + k\psi)\prod_{k=0}^{N-N_1-1}(1 - p + k\psi)}{\prod_{k=1}^{N-1}(1 + k\psi)},$$

where

$$p = \begin{cases} 0.5 & \text{if the candidate eQTL has a homozygous genotype AA or BB,} \\ q & \text{if } \tilde{H} \text{ indicates haplotype configuration B-}h_1 \text{ and A-}h_2\text{, respectively,} \\ 1-q & \text{if } \tilde{H} \text{ indicates haplotype configuration A-}h_1 \text{ and B-}h_2\text{, respectively.} \end{cases}$$

Thus q characterizes the proportion of ASReC corresponding to the B allele among the total ASReC corresponding to the heterozygous genotype AB. We further express q as $e^{\beta_A}/(1 + e^{\beta_A})$. Note that the covariate effects are ignored here because they are expected to be the same on the two alleles of a gene within an individual. When the candidate eQTL *cis*-regulates the expression of the gene, we have $\beta_A = \beta_T$. To see this, we first define μ_A and μ_B as the mean ASReC corresponding to the A and B alleles, respectively, at the baseline of $\mathbf{X}$. Then, $\beta_A = \log[q/(1-q)] = \log(\mu_B/\mu_A)$. On the other hand, $\beta_T = \log(\mu_{BB,\mathbf{X}}/\mu_{AA,\mathbf{X}}) = \log\{(2\mu_B)/(2\mu_A)\}$, where the second equation follows from the additive genetic effect and from canceling out the individual-specific covariate effects. By contrast, when the candidate eQTL *trans*-regulates the gene expression, we have $\beta_T \neq 0$ but $\beta_A = 0$.

The likelihood based on the TReC and ASReC data of n unrelated individuals takes the form

$$L(\boldsymbol{\Theta}) = \prod_{i=1}^{n} P_{\texttt{TReC}}(T_i|G_i, \mathbf{X}_i; \boldsymbol{\gamma}, \beta_T, \phi) P_{\texttt{ASReC}}(N_{i1}|N_i, \tilde{H}_i, \mathbf{X}_i; \beta_A, \psi), \quad (8.1)$$

where $\boldsymbol{\Theta} = (\boldsymbol{\gamma}, \beta_T, \phi, \beta_A, \psi)$. We refer to (8.1) as the **TReCASE** model, which is the novel model for *cis*-eQTL mapping using RNA-seq data. For *trans*-eQTL mapping, since ASE data are uninformative, the likelihood is only based on the TReC data: $L(\boldsymbol{\gamma}, \beta_T, \phi) = \prod_{i=1}^{n} P_{\texttt{TReC}}(T_i | G_i, \mathbf{X}_i; \boldsymbol{\gamma}, \beta_T, \phi)$. A hypothesis testing method has been developed to distinguish whether an eQTL is *cis*- or *trans*- by testing H_0: $\beta_T = \beta_A$ [55].

8.2.4 eQTL Mapping Using ASE with Unknown Haplotypes

When the haplotypes connecting the candidate eQTL and the gene of interest are unknown, we consider all possible haplotype pairs $(\tilde{h}_k, \tilde{h}_l)$ that are compatible with the known haplotypes in the gene body (H) and the genotype at the candidate eQTL (G). We denote these haplotype pairs as $(\tilde{h}_k, \tilde{h}_l) \sim (G, H)$. Then the likelihood function is a weighted summation of the probabilities, each corresponding to a possible haplotype pair and given by (8.1), i.e.,

$$L(\boldsymbol{\Theta}) = \prod_{i=1}^{n} P_{\texttt{TReC}}(T_i | G_i, \mathbf{X}_i; \boldsymbol{\gamma}, \beta_T, \phi) \times \sum_{(\tilde{h}_k, \tilde{h}_l) \sim (G_i, H_i)} P_{\texttt{ASReC}}(N_{i1} | N_i, \tilde{h}_k, \tilde{h}_l, \mathbf{X}_i; \beta_A, \psi) P(\tilde{h}_k, \tilde{h}_l; \boldsymbol{\pi}) f_{kl}(\mathbf{X}_i), \quad (8.2)$$

where $\boldsymbol{\Theta} = (\boldsymbol{\gamma}, \beta_T, \phi, \beta_A, \psi, \boldsymbol{\pi}, \{f_{kl}(.)\}_{k,l})$. We explain the terms that are not in (8.1) as follows.

Suppose there are K possible haplotypes across the exonic SNPs and the candidate eQTL. Write the frequency of the kth haplotype by $\pi_k = \Pr(\tilde{h} = \tilde{h}_k)$ and $\boldsymbol{\pi} = (\pi_1, \ldots, \pi_K)$. We denote the model for the probability of $\tilde{H} = (\tilde{h}_k, \tilde{h}_l)$ indexed by $\boldsymbol{\pi}$ by $P(\tilde{h}_k, \tilde{h}_l; \boldsymbol{\pi})$. Under the assumption of Hardy-Weinberg equilibrium, $P(\tilde{h}_k, \tilde{h}_l; \boldsymbol{\pi}) = \pi_k \pi_l$.

The density function of $\mathbf{X}$ given $\tilde{H} = (\tilde{h}_k, \tilde{h}_l)$ is denoted by $f_{kl}(\mathbf{X})$. Under the assumption of gene-environment independence, $f_{kl}(\mathbf{X})$ reduces to the marginal density function of $\mathbf{X}$ and will drop out from (8.2). In some applications, $\tilde{H}$ and $\mathbf{X}$ are correlated. One important example is when $\mathbf{X}$ represent the principal components for ancestry. Another example is when the gene influences both the environmental exposure (e.g., cigarette smoking) and the disease occurrence (e.g., lung cancer) [3]. In such cases, $f_{kl}(\mathbf{X})$ can be specified using a generalized odds-ratio function [28].

8.3 Isoform-Specific eQTL Mapping

More than 90 % of human multi-exon genes can be alternatively spliced, resulting in RNA isoforms [44, 64]. Alternative splicing may directly cause a disease or modify certain disease susceptibility [19, 61, 63]. Although several methods have been proposed for detecting the event of alternative splicing and estimating the RNA-isoform abundance [2, 4, 21, 23, 31, 34, 38, 39, 50, 59, 65], only a few have been developed for testing the differential RNA-isoform usage between two groups of samples (e.g., cases vs. controls) [22, 54, 59]. Differential isoform usage refers to the changes of RNA-isoform expression relative to the total expression of the corresponding gene. The purpose of isoform-specific eQTL mapping is to dissect the genetic basis of the differential isoform usage. There are a few points worth mentioning from the statistical perspective on isoform-specific eQTL mapping.

- Because the isoform structure or abundance cannot be directly measured, transcriptome reconstruction and abundance estimation are necessary steps of isoform-specific eQTL mapping. The uncertainty of the transcriptome reconstruction and the abundance estimation should be incorporated into isoform-specific eQTL mapping.
- In most eQTL studies or genome-wide association studies, SNP genotype effects are assumed to be additive. Thus the SNP genotype is essentially a quantitative covariate. However, most existing methods assess the differential isoform usage between two groups of samples (e.g., cases vs. controls) and few methods can test the association between the isoform usage and a quantitative covariate.
- One gene may be differentially expressed with respect to a covariate, both in terms of the total expression and the isoform usage. It will be useful to jointly test for differential expression and differential isoform usage.

8.3.1 Transcriptome Reconstruction and Isoform Abundance Estimation

A gene usually occupies a consecutive segment of the DNA sequence and it is often composed of several exons that are separated by introns. A subset of the exons may be employed by the cell to construct alternatively spliced messenger RNAs (mRNAs). These mRNAs may be translated to different proteins. Each RNA isoform is often referred to as a transcript and thus each gene can be considered as a transcript cluster. In some organism such as a human or a mouse, there are existing annotations on the kinds of transcripts a gene may encode. Such annotations are often incomplete or inaccurate, for example, some transcripts may be express in a particular tissue and/or developmental stage. In some other organisms, such as those without complete reference genomes, such transcriptome annotations are not available at all. Therefore, one may need to reconstruct the transcriptome from the observed RNA-seq data. This task can be achieved with or without a reference genome [18]. The reference genome-guided reconstruction is often more accurate and computationally more efficient than the de novo transcriptome construction without a reference genome. Thus the former approach is more popular for organisms that have reference genomes. Given the transcriptome annotation, the

$$
\mathbf{X}^{\mathsf{T}} = \begin{pmatrix}
\eta_{\{1\}} & 0 & 0 & 0 & 0 & 0 & 0 \\
0 & \eta_{\{2\}} & 0 & 0 & 0 & 0 & 0 \\
0 & 0 & \eta_{\{3\}} & 0 & 0 & 0 & 0 \\
\eta_{\{1\}} & \eta_{\{2\}} & 0 & \eta_{\{1,2\}} & 0 & 0 & 0 \\
0 & \eta_{\{2\}} & \eta_{\{3\}} & 0 & \eta_{\{2,3\}} & 0 & 0 \\
\eta_{\{1\}} & 0 & \eta_{\{3\}} & 0 & 0 & \eta_{\{1,3\}} & 0 \\
\eta_{\{1\}} & \eta_{\{2\}} & \eta_{\{3\}} & \eta_{\{1,2\}} & \eta_{\{2,3\}} & \eta_{\{1,(2),3\}} & \eta_{\{1,2,3\}}
\end{pmatrix}
$$

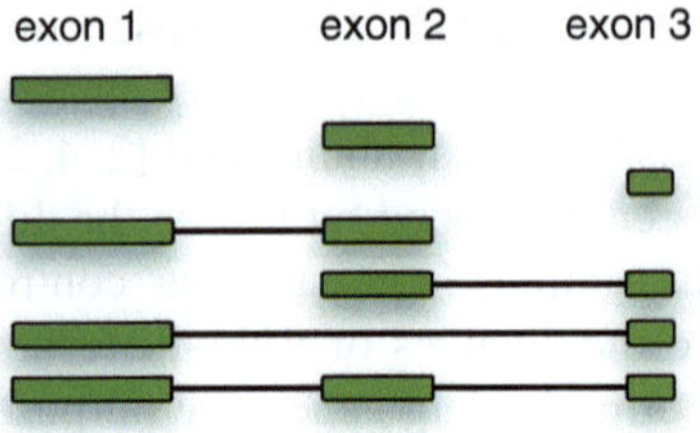

Fig. 8.6 All possible isoforms of a gene with three exons and the corresponding design matrix $\mathbf{X}^{\mathsf{T}}$

abundance of each transcript can be estimated by the number of RNA-seq reads aligned to that transcript. However, most RNA-seq fragments cannot be uniquely assigned to a specific transcript. To estimate transcript abundance in the presence of such alignment ambiguity is the focus of many existing works [31, 32, 37, 43, 48, 49, 53, 59, 72]. Penalized regression methods have been developed to simultaneously reconstruct transcriptome and estimate transcript/isoform abundance [6, 38, 39, 71]. The method we will describe next is an example of such penalized regression methods.

8.3.2 *Isoform-Specific eQTL Mapping*

The method presented here is based on Sun et al. (2013) [58]. We first illustrate the statistical model by a hypothetical gene with three exons (Fig. 8.6). An RNA-seq read may overlap with one or more exons. Thus we count the number of RNA-seq reads per exon set. For this simple gene, there are seven possible exon sets, denoted by $\{1\}$, $\{2\}$, $\{3\}$, $\{1,2\}$, $\{2,3\}$, $\{1,3\}$, and $\{1,2,3\}$. Note that each RNA-seq read is only counted once. For example, if an RNA-seq read overlaps with both exon 1 and 2, it will be counted for exon set $\{1,2\}$ instead of exon set $\{1\}$ or $\{2\}$. There are seven possible isoforms (right panel of Fig. 8.6). We code each isoform as a covariate, which corresponds to one row of the design matrix $\mathbf{X}^{\mathsf{T}}$ (left panel), where ^{T} denotes matrix transpose. The seven columns of matrix $\mathbf{X}^{\mathsf{T}}$ correspond to exon sets $\{1\}$, $\{2\}$, $\{3\}$, $\{1,2\}$, $\{2,3\}$, $\{1,3\}$, and $\{1,2,3\}$. Each element in $\mathbf{X}^{\mathsf{T}}$ is the effective length of the column-specific exon set within the row-specific isoform. Intuitively, the effective length of an exon set A, denoted by η_A, is the number of unique locations within A, where a randomly selected sequence fragment can be sampled. We defer the details of effective length calculation to the next section, but would like to point out that there are special exon sets that consist of non-contiguous exons in the specific isoform. For example, the exons in set $\{1,3\}$ is non-contiguous with respective to isoform 1-2-3 and the effective length of $\{1,3\}$ is denoted by $\eta_{\{1,(2),3\}}$. Our effective length calculation accurately reflects the fact that sequence reads of exon set $\{1,3\}$ are more likely from isoform 1-3 rather than isoform 1-2-3.

In this example, the gene expression in the ith sample is denoted by a vector: $\mathbf{y}_i = (y_{i\{1\}}, y_{i\{2\}}, y_{i\{3\}}, y_{i\{1,2\}}, y_{i\{2,3\}}, y_{i\{1,3\}}, y_{i\{1,2,3\}})^\mathsf{T}$, where y_{iA} indicates the TReC at the exon set A. As in Sect. 8.2.3, we model the probability of a TReC via a negative binomial distribution. Let $f_{NB}(\mu,\phi)$ be a negative binomial distribution with mean μ and a dispersion parameter ϕ. We assume that $y_{iA} \sim f_{NB}(\mu_{iA},\phi)$. Assuming independence of y_{iA}'s given the underlying RNA isoforms, then $\mathbf{y}_i \sim f_{NB}(\boldsymbol{\mu}_i,\phi) \equiv \prod_A f_{NB}(\mu_{iA},\phi)$ where $\boldsymbol{\mu}_i = (\mu_{i\{1\}}, \mu_{i\{2\}}, \ldots, \mu_{i\{1,2,3\}})^\mathsf{T}$. By the definition of the design matrix $\mathbf{X}$, we transform the problem of isoform deconvolution to a regression problem: $\mathbf{y}_i \sim f_{NB}(\boldsymbol{\mu}_i,\phi)$, $\boldsymbol{\mu}_i = T_i \mathbf{X}\boldsymbol{\gamma} = T_i \sum_{u=1}^{7} \mathbf{x}_u b_u$, where T_i is TReC of this gene in sample i, $\mathbf{X} = (\mathbf{x}_1, \ldots, \mathbf{x}_7)$, $\boldsymbol{\gamma} = (b_1, \ldots, b_7)^\mathsf{T}$, and $b_u \geq 0$ is the expression rate of the uth isoform. Note that b_u quantifies the relative expression abundance with respect to the total expression T_i.

Next, we present the general method. Suppose that we study the isoform-specific expression of a gene with m exon sets and p possible isoforms across n individuals, and we are particularly interested in whether a covariate G has an influence on the isoform-specific expression of this gene. We assess this hypothesis by a likelihood ratio test. Under the null hypothesis, we solve the problems of isoform selection and abundance estimation by assuming that the isoform usage is the same for all samples. Thus we use a negative binomial regression with the link function $\boldsymbol{\mu}_i = T_i \mathbf{X}\boldsymbol{\gamma}$. Note that a linear link function instead of commonly used log link function is used to reflect the fact that the total number of reads is the summation of the number of reads from all the isoforms. Under the alternative, we model the effect of G as follows. Let g_i be the value of G in the ith sample. Without loss of generality, we restrict the range of g_i to be [0,1]. For example, if G is genotype of a SNP, we set g_i = 0, 1/2, and 1 for genotypes AA, AB, and BB, respectively. Provided $\boldsymbol{\mu}_i = T_i \mathbf{X}\boldsymbol{\gamma}$, we model the influence of G on b_u $(1 \leq u \leq p)$ by a linear model: $b_u = \gamma_u(1-g_i) + \gamma_{u+p} g_i$, where $\gamma_j \geq 0$ for $1 \leq j \leq 2p$. Therefore, we have two negative binomial problems, with p and $2p$ covariates, under null and alternative, respectively.

The major difficulty of this problem comes from the high dimensionality of the possible isoforms [25]. We address this difficulty by two sequential steps. First we identify the candidate isoforms for a gene using a modified connectivity graph approach [23, 38]. Next we select among the candidate isoforms using a penalized negative binomial regression problem. For example, under the alternative, the objective function becomes $f(\boldsymbol{\gamma},\phi) = \sum_{i=1}^{n} \log\left[f_{NB}(\boldsymbol{\mu}_i,\phi)\right] - \sum_{j=1}^{2p} \lambda \log(\gamma_j + \tau)$, where λ and τ are two tuning parameters that can be selected by BIC or extended BIC [57]. We use the log penalty $\lambda \log(\gamma_j + \tau)$ because of its superior theoretical and empirical advantages over other penalties [9,15,57]. Given λ and τ, the parameters $\boldsymbol{\gamma}$ and ϕ can be estimated by a coordinate descent algorithm [57]. The above model is formulated when the isoform usage is associated with one quantitative covariate; it is straightforward to extend it to include multiple quantitative covariates. For a categorical covariate (e.g., under the dominant or recessive effect of a SNP), we can simply code it as a number of dummy variables, which can be treated as multiple quantitative covariates.

Due to the variable selection (i.e., selecting expressed RNA isoforms) under both the null and the alternative hypotheses, the asymptotic distribution of the likelihood ratio statistic is unknown. Thus we estimate the null distribution of the statistic by parametric bootstrap. Specifically, we generate the vth bootstrap sample, denoted by $\tilde{\mathbf{y}}^{(v)}$ (a vector of length nm), by sampling from a negative binomial distribution with mean $\hat{\boldsymbol{\mu}}_0$ and a dispersion parameter $\hat{\phi}_0$, where $\hat{\boldsymbol{\mu}}_0$ (a vector of length nm) and $\hat{\phi}_0$ are estimated under the null. Then using this bootstrap sample, we apply the penalized regression approach under the null and the alternative to obtain a likelihood ratio statistic LR_v. Repeat the parametric bootstrap for a large number of times (e.g. 10,000 times) and pool the LR_v's, we obtain the null distribution for the observed statistic LR. The final p-value is the proportion of LR_v's that are equal to or larger than the likelihood ratio statistic from original data.

The above solution only tests differential isoform usage, which is the difference of relative abundance of an isoform with respect to the total expression of the gene for different values of G. If we are interested in testing both the differential expression and the differential isoform usage of a gene, the original link function $\boldsymbol{\mu}_i = T_i\mathbf{X}\boldsymbol{\gamma}$ can be changed to be $\boldsymbol{\mu}_i = R_i\mathbf{X}\boldsymbol{\gamma}$, where R_i is the total number of RNA-seq reads of the ith sample across all genes. The reason is as follows. The original link function can be written as $\boldsymbol{\mu}_i = T_i\mathbf{X}\boldsymbol{\gamma} = R_i(T_i/R_i)\mathbf{X}\boldsymbol{\gamma}$, where (T_i/R_i) measures the total expression of the gene in the ith sample. Then skipping the ratio (T_i/R_i) in the original link function leads to the new link function, which is equivalent to assuming this gene has a constant expression rate across samples.

8.3.3 Calculation of Effective Length

An RNA-seq fragment is a segment of RNA to be sequenced. Usually only part of an RNA-seq fragment is sequenced: one end or both ends, hence single-end sequencing or paired-end sequencing. All the discussions in this section are for paired-end reads, though the extension to single-end reads is straightforward. The minimum fragment size is the read length, denoted by d. This happens when the two reads of a fragment completely overlap. We impose an upper bound for the fragment length based on prior knowledge of the experimental procedure and denote the upper bound by l_M. Then the fragment length l satisfies $d \le l \le l_M$. We denote the distribution of the fragment length for sample i by $\varphi_i(l)$, which can be calculated using observed read alignment information. The fragment length distribution is incorporated in our model to allow across-sample variations due to the differences in fragment length distribution.

For the ith sample, the effective length of exon j of r_j base pairs (bps) is

$$\eta_{i,\{j\}} = f(r_j, d, l_M, \varphi_i) = \begin{cases} 0 & \text{if } r_j < d \\ \sum_{l=d}^{\min(r_j, l_M)} \varphi_i(l)(r_j + 1 - l) & \text{if } r_j \ge d \end{cases}.$$

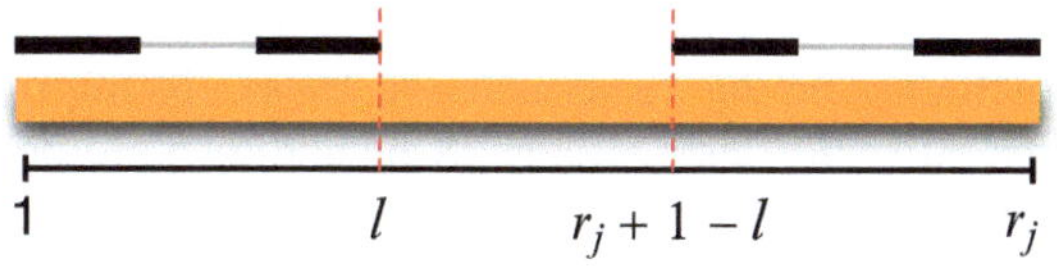

Fig. 8.7 An illustration of effective length calculation for an exon of r_j bps and RNA-seq fragment of l bps. The *orange box* indicates the exon, and the *black lines* above the *orange box* indicate two RNA-seq fragments, while each RNA-seq fragment is sequenced by a paired-end read. There are r_j+1-l distinct choices to select an RNA-seq fragment of l bps from this exon, and thus the effective length is r_j+1-l

If $r_j < d$, the exon is shorter than the shortest fragment length, and thus the effective length of this exon is 0. In other words, no RNA-seq fragment is expected to overlap and only overlap with this exon. If $r_j \geq d$, the effective length is r_j+1-l, i.e., there are r_j+1-l distinct RNA-seq fragments that can be sequenced from this exon (Fig. 8.7). Then $\sum_{l=d}^{\min(r_j,l_M)} \varphi_i(l)(r_j+1-l)$ is summation across all likely fragment lengths, weighted by the probability of having fragment length l.

In the following discussions, to simplify the notation, we skip the subscript of i. For two exons j and k $(j < k)$ of lengths r_j and r_k, which are adjacent in the transcript, the effective length for the fragments that cover both exons is

$$\eta_{\{j,k\}} = f(r_j+r_k,d,l_M,\varphi) - \eta_{\{j\}} - \eta_{\{k\}}. \tag{8.3}$$

For three exons j, h, and k $(j < h < k)$ of lengths r_j, r_h and r_k, which are adjacent in the transcript, the effective length for the fragments that cover all three exons is

$$\eta_{\{j,h,k\}} = f(r_j+r_h+r_k,d,l_M) - \eta_{\{j,h\}} - \eta_{\{h,k\}} - \eta_{\{j,(h),k\}} - \eta_{\{j\}} - \eta_{\{h\}} - \eta_{\{k\}},$$

where $\eta_{\{j,(h),k\}}$ is the effective length in the scenario that the transcript covers consecutive exons j, h, and k, whereas the observed paired-end read only covers exons j and k.

$$\eta_{\{j,(h),k\}} = \begin{cases} 0 & \text{if } (r_j,r_h,r_k) \in R_1 \\ \displaystyle\sum_{l=2d+r_h}^{\min(r_j+r_h+r_k,l_M)} \varphi(l)\delta_l & \text{otherwise} \end{cases}$$

where $R_1 = \{(r_j,r_h,r_k) : \; r_j < d \text{ or } r_k < d \text{ or } r_h+2d > l_M\}$, and $\delta_l = \min(r_j,l-r_h-d) - \max(d,l-r_h-r_k)+1$. The above formula is derived by the following arguments. Let l_j and l_k be the lengths of the parts of the fragment that overlaps with exon j and k, respectively. Given l, the restriction of l_j and l_k are $l = l_j+l_k+r_h$, $d \leq l_j \leq r_j$, and $d \leq l_k \leq r_k$, and thus the range of l_j is $\max(d,l-r_h-r_k) \leq l_j \leq \min(r_j,l-r_h-d)$. For more than three consecutive exons, the effective lengths can be calculated using recursive calls to the above equations.

In practice, a few sequence fragments may be observed even when the effective length is zero, which may be due to sequencing errors. To improve the robustness of our method, we modify the design matrix **X** by adding a pre-determined constant `eLenMin` to each element of **X**.

8.4 Discussion

We conclude this chapter by a few discussion points.

8.4.1 eQTL Mapping Using Both ASE and ISE

We have introduced statistical methods of using ASE or ISE for eQTL mapping. A natural extension is to use both ASE and ISE for eQTL mapping. The likelihood can be similar to the one for eQTL mapping using ASE, but using count data from exon sets intend of genes. Such a model can explain more subtle changes in the gene expression data. For example, one isoform is used in one allele, but not in the other allele, i.e., allele-specific isoform usage. A major challenge would be computational feasibility. Thus a more computationally efficient implementation is needed for such an effort.

8.4.2 cis-eQTL and Imprinting

Allelic imbalance of gene expressions may be due to factors other than *cis*-eQTL. Arguably, the second most likely factor causing allelic imbalance, after *cis*-eQTL, is imprinting. Imprinted genes are differentially expressed on maternal and paternal alleles. Thus imprinting is also referred to as the parent-of-origin effect [47]. An important lesson we learned from our recent study of ASE in F1 mice [11] is that "imprinting is incomplete for most genes and *cis*-acting mutations can modify the strength of imprinting". Usually imprinting effect is much more subtle than *cis*-eQTL effects. Therefore, to obtain more sensitive and more accurate estimates of imprinting effects, it is crucial to jointly study imprinting and *cis*-eQTL.

8.4.3 Quality Control and Possible Non-genetic Factors

Quality control (QC) is a necessary step for eQTL mapping using RNA-seq data. Low quality samples may be detected by checking the sequencing quality scores, mapping quality, percentage of uniquely mapped reads, percentage of reads mapped to exonic regions, percentage of rRNA reads, and the distribution of insert size for paired-end reads [1, 13, 66]. Sample identity check is a very important step in genome-wide genomic studies. Between sample contamination may be detected by the percentage of heterozygous SNPs, sex-mismatch (recorded sex from demographic information vs. sex inferred from genomic data), or the D-statistic that measures the median correlation of gene expression between one sample versus each of the other samples [1, 69]. Sample swap will seriously reduce the power of eQTL analysis. Fortunately, checking for sample swap is relatively easy using RNA-seq data than using microarray data [29]. A QC step that is crucial for ASE data is the mapping bias toward reference alleles, which has been discussed at Sect. 8.2. For ISE data, checking the coverage of the whole gene body is important because there may be a trend of increasing read depth towards the 3' end of a gene. The method described in Sect. 8.3 assumes a uniform distribution of read depth, though the hypothesis testing method is not sensitive to this assumption due to the resampling nature of the test [58].

The effect of non-genetic factors can be accounted for by including them (or an appropriate transformation of them) as covariates in eQTL mapping. First, the overall read depth per sample is one factor that should always be included. In addition, GC content and dinucleotide frequencies may influence gene expression in a sample-specific manner. For example, gene expression and GC content may be positively correlated in some samples, but negatively correlated in other samples [74]. A conditional quantile normalization method has been proposed to model such sample-specific effects from sequence contents within the framework of generalized linear regression models [24]. This approach can be employed in the eQTL-mapping framework described in this book chapter.

8.4.4 The Genetic Architecture of Gene Expression

Figure 8.8 shows the results of two genome-wide eQTL studies: a yeast study of ~6,000 genes and ~1,000 SNPs in 112 yeast segregants (offspring) (Fig. 8.8a) and a human study of ~18,000 genes and ~1,000,000 SNPs (germline genotype) in 550 breast cancer patients. Gene expression abundance was measured by microarrays in the yeast study and by RNA-seq in the human study. The difference in the genetic architecture of gene expression between the two studies is remarkable. In both studies, the eQTL plots have a diagonal pattern, which corresponds to a large number of local eQTLs. In the yeast study, there are several vertical bands, each corresponding to an eQTL hotspot, i.e., a genetic locus that is eQTL of many genes. In contrast, there is no such eQTL hotspot in the human study. The two studies are representative for experimental cross and human studies. In experimental cross, usually two strains with very different genetic backgrounds are crossed and thus some loci may have large and broad effects on many genes. For example, in the yeast study, several eQTL hotspots arise because one strain has several genes deleted. In human studies, the genetic differences across humans are much smaller than

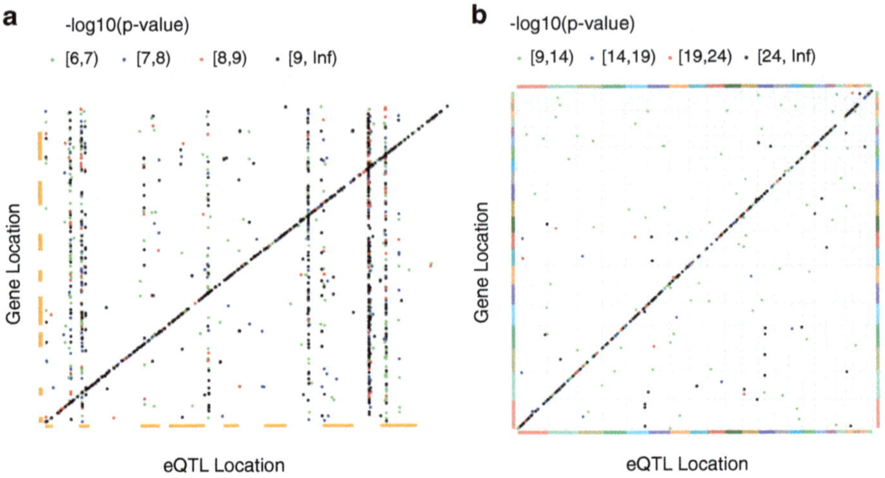

Fig. 8.8 The results of eQTL studies in **(a)** 112 yeast sergeants of two yeast strains [7] and **(b)** 550 breast cancer patients of an on-going study. Each point represents a genome-wide significant association. The *color* indicates certain range of the p-value. More liberal p-values are used for the yeast study because there is a smaller number of genes and SNPs and hence less burden of multiple testing correction

in experimental crosses and generally no single locus can substantially alter the expression of many genes. We have reported similar findings in a recent human eQTL studies with 2,494 twins and a validation data set of 1,895 independent subjects [69]. The conclusion is that, for human studies, the vast majority of genetic effects on gene expression are through local eQTL and most of the local eQTL are likely to be *cis*-eQTL [55]. This implies that the identification of distant eQTLs may be as difficult as or even more difficult than genome-wide association studies for complex traits.

References

[1] A C't Hoen, P., Friedländer, M.R., Almlöf, J., Sammeth, M., Pulyakhina, I., Anvar, S.Y., Laros, J.F., Buermans, H.P., Karlberg, O., Brännvall, M., et al.: Reproducibility of high-throughput mrna and small rna sequencing across laboratories. Nat. Biotechnol. **31**, 1015–1022 (2013)

[2] Ameur, A., Wetterbom, A., Feuk, L., Gyllensten, U.: Global and unbiased detection of splice junctions from RNA-seq data. Genome Biol. **11**(3), R34 (2010)

[3] Amos, C.I., Wu, X., Broderick, P., Gorlov, I.P., Gu, J., Eisen, T., Dong, Q., Zhang, Q., Gu, X., Vijayakrishnan, J., et al.: Genome-wide association scan of tag snps identifies a susceptibility locus for lung cancer at 15q25. 1. Nature Genet. **40**(5), 616–622 (2008)

[4] Au, K., Jiang, H., Lin, L., Xing, Y., Wong, W.: Detection of splice junctions from paired-end RNA-seq data by splicemap. Nucleic Acids Res. **38**(14), 4570–4578 (2010)

[5] Auer, P.L., Doerge, R.: Statistical design and analysis of rna sequencing data. Genetics **185**(2), 405–416 (2010)

[6] Bohnert, R., Rätsch, G.: rquant. web: a tool for RNA-seq-based transcript quantitation. Nucleic Acids Res. **38**(Suppl 2), W348–W351 (2010)

[7] Brem, R.B., Storey, J.D., Whittle, J., Kruglyak, L.: Genetic interactions between polymorphisms that affect gene expression in yeast. Nature **436**(7051), 701–703 (2005)

[8] Browning, S., Browning, B.: Rapid and accurate haplotype phasing and missing-data inference for whole-genome association studies by use of localized haplotype clustering. Am. J. Hum. Genet. **81**(5), 1084–1097 (2007)

[9] Chen, T., Sun, W., Fine, J.: Designing penalty functions in high dimensional problems: the role of tuning parameters. Technical Report, UNC Chapel Hill (2011)

[10] Cookson, W., Liang, L., Abecasis, G., Moffatt, M., Lathrop, M.: Mapping complex disease traits with global gene expression. Nat. Rev. Genet. **10**(3), 184–194 (2009)

[11] Crowley, J.J., Zhabotynsky, V., Sun, W., Huang, S., Pakatci, I.K., Kim, Y., Wang, J.R., Morgan, A.P., Calaway, J.D., Aylor, D.L., Yun, Z., Bell, T.A., Buus, R.J., Calaway, M.E., Didion, J.P., Gooch, T.J., Hansen, S.D., Robinson, N.N., Shaw, G.D., Spence, J.S., Quackenbush, C.R., Barrick, C.J., Xie, Y., Valdar, W., Lenarcic, A.B., Wang, W., Welsh, C.E., Fu, C.P., Zhang, Z., Holt, J., Guo, Z., Threadgill, D.W., Tarantino, L.M., Miller, D., R., Zou, F., McMillan, L., Sullivan, P.F., Pardo-Manuel de Villena, F.: Pervasive allelic imbalance revealed by allele-specific gene expression in highly divergent mouse crosses. Nat. Genet. (2013, in revision)

[12] Delaneau, O., Zagury, J., Marchini, J., et al.: Improved whole-chromosome phasing for disease and population genetic studies. Nat. Meth. **10**(1), 5–6 (2013)

[13] DeLuca, D.S., Levin, J.Z., Sivachenko, A., Fennell, T., Nazaire, M.D., Williams, C., Reich, M., Winckler, W., Getz, G.: Rna-seqc: Rna-seq metrics for quality control and process optimization. Bioinformatics **28**(11), 1530–1532 (2012)

[14] Doss, S., Schadt, E., Drake, T., Lusis, A.: Cis-acting expression quantitative trait loci in mice. Genome Res. **15**(5), 681 (2005)

[15] Fan, J., Lv, J.: Non-concave penalized likelihood with np-dimensionality. IEEE Trans. Inf. Theory **57**(8), 5468–5484 (2011)

[16] Flicek, P., Amode, M., Barrell, D., Beal, K., Brent, S., Chen, Y., Clapham, P., Coates, G., Fairley, S., Fitzgerald, S., et al.: Ensembl 2011. Nucleic Acids Res. **39**(Suppl 1), D800 (2011)

[17] Fogarty, M., Xiao, R., Prokunina-Olsson, L., Scott, L., Mohlke, K.: Allelic expression imbalance at high-density lipoprotein cholesterol locus mmab-mvk. Hum. Mol. Genet. **19**(10), 1921–1929 (2010)

[18] Garber, M., Grabherr, M., Guttman, M., Trapnell, C.: Computational methods for transcriptome annotation and quantification using RNA-seq. Nat. Meth. **8**(6), 469–477 (2011)

[19] Garcia-Blanco, M., Baraniak, A., Lasda, E.: Alternative splicing in disease and therapy. Nat. Biotechnol. **22**(5), 535–546 (2004)

[20] Garnett, M., Edelman, E., Heidorn, S., Greenman, C., Dastur, A., Lau, K., Greninger, P., Thompson, I., Luo, X., Soares, J., et al.: Systematic identification of genomic markers of drug sensitivity in cancer cells. Nature **483**(7391), 570–575 (2012)

[21] Grabherr, M., Haas, B., Yassour, M., Levin, J., Thompson, D., Amit, I., Adiconis, X., Fan, L., Raychowdhury, R., Zeng, Q., et al.: Full-length transcriptome assembly from RNA-seq data without a reference genome. Nat. Biotechnol. **29**(7), 644–652 (2011)

[22] Griffith, M., Griffith, O., Mwenifumbo, J., Goya, R., Morrissy, A., Morin, R., Corbett, R., Tang, M., Hou, Y., Pugh, T., et al.: Alternative expression analysis by RNA sequencing. Nat. Meth. **7**(10), 843–847 (2010)

[23] Guttman, M., Garber, M., Levin, J., Donaghey, J., Robinson, J., Adiconis, X., Fan, L., Koziol, M., Gnirke, A., Nusbaum, C., et al.: Ab initio reconstruction of cell type-specific transcriptomes in mouse reveals the conserved multi-exonic structure of lincRNAs. Nat. Biotechnol. **28**(5), 503–510 (2010)

[24] Hansen, K.D., Irizarry, R.A., Zhijin, W.: Removing technical variability in rna-seq data using conditional quantile normalization. Biostatistics **13**(2), 204–216 (2012)

[25] Hiller, D., Jiang, H., Xu, W., Wong, W.: Identifiability of isoform deconvolution from junction arrays and RNA-Seq. Bioinformatics **25**(23), 3056 (2009)

[26] Holt, J., Huang, S., McMillan, L., Wang, W.: Read annotation pipeline for high-throughput sequencing data. In: Proceedings of the International Conference on Bioinformatics, Computational Biology and Biomedical Informatics, p. 605. ACM (2013)
[27] Hosokawa, Y., Arnold, A.: Mechanism of cyclin d1 (ccnd1, prad1) overexpression in human cancer cells: analysis of allele-specific expression. Genes Chrom. Cancer **22**(1), 66–71 (1998)
[28] Hu, Y., Lin, D., Zeng, D.: A general framework for studying genetic effects and gene–environment interactions with missing data. Biostatistics **11**(4), 583–598 (2010)
[29] Huang, J., Chen, J., Lathrop, M., Liang, L.: A tool for rna sequencing sample identity check. Bioinformatics **29**(11), 1463–1464 (2013)
[30] Huang, S., Kao, C.Y., McMillan, L., Wang, W.: Transforming genomes using mod files with applications. In: Proceedings of the International Conference on Bioinformatics, Computational Biology and Biomedical Informatics, p. 595. ACM (2013)
[31] Jiang, H., Wong, W.: Statistical inferences for isoform expression in RNA-seq. Bioinformatics **25**(8), 1026 (2009)
[32] Katz, Y., Wang, E., Airoldi, E., Burge, C.: Analysis and design of RNA sequencing experiments for identifying isoform regulation. Nat. Meth. **7**(12), 1009–1015 (2010)
[33] Kendziorski, C., Wang, P.: A review of statistical methods for expression quantitative trait loci mapping. Mamm. Genome **17**(6), 509–517 (2006)
[34] Li, B., Dewey, C.: Rsem: accurate transcript quantification from RNA-seq data with or without a reference genome. BMC Bioinform. **12**(1), 323 (2011)
[35] Li, Y., Grupe, A., Rowland, C., Nowotny, P., Kauwe, J., Smemo, S., Hinrichs, A., Tacey, K., Toombs, T., Kwok, S., et al.: Dapk1 variants are associated with Alzheimer's disease and allele-specific expression. Hum. Mol. Genet. **15**(17), 2560–2568 (2006)
[36] Li, Y., Willer, C., Ding, J., Scheet, P., Abecasis, G.: Mach: using sequence and genotype data to estimate haplotypes and unobserved genotypes. Genetic Epidemiol. **34**(8), 816–834 (2010)
[37] Li, B., Ruotti, V., Stewart, R., Thomson, J., Dewey, C.: RNA-seq gene expression estimation with read mapping uncertainty. Bioinformatics **26**(4), 493–500 (2010)
[38] Li, W., Feng, J., Jiang, T.: IsoLasso: a LASSO regression approach to RNA-seq based transcriptome assembly. J. Comput. Biol. **18**(11), 1693–1707 (2011)
[39] Li, J., Jiang, C., Hu, Y., Brown, B., Huang, H., Bickel, P.: Sparse linear modeling of RNA-seq data for isoform discovery and abundance estimation. Proceedings of the National Academy of Sciences **108**(50), 19867–19872 (2011)
[40] Meyer, K., Maia, A., O'Reilly, M., Teschendorff, A., Chin, S., Caldas, C., Ponder, B.: Allele-specific up-regulation of fgfr2 increases susceptibility to breast cancer. PLoS Biol. **6**(5), e108 (2008)
[41] Miller, V., Xia, H., Marrs, G., Gouvion, C., Lee, G., Davidson, B., Paulson, H.: Allele-specific silencing of dominant disease genes. Proc. Natl. Acad. Sci. U.S.A. **100**(12), 7195 (2003)
[42] Ozsolak, F., Milos, P.: RNA sequencing: advances, challenges and opportunities. Nat. Rev. Genet. **12**(2), 87–98 (2010)
[43] Pachter, L.: Models for transcript quantification from RNA-seq. Arxiv preprint arXiv:1104.3889 (2011)
[44] Pan, Q., Shai, O., Lee, L., Frey, B., Blencowe, B.: Deep surveying of alternative splicing complexity in the human transcriptome by high-throughput sequencing. Nat. Genet. **40**(12), 1413–1415 (2008)
[45] Perou, C., Sørlie, T., Eisen, M., van de Rijn, M., Jeffrey, S., Rees, C., Pollack, J., Ross, D., Johnsen, H., Akslen, L., et al.: Molecular portraits of human breast tumours. Nature **406**(6797), 747–752 (2000)
[46] Pickrell, J., Marioni, J., Pai, A., Degner, J., Engelhardt, B., Nkadori, E., Veyrieras, J., Stephens, M., Gilad, Y., Pritchard, J.: Understanding mechanisms underlying human gene expression variation with RNA sequencing. Nature **464**(7289), 768–772 (2010)
[47] Reik, W., Walter, J., et al.: Genomic imprinting: parental influence on the genome. Nat. Rev. Genet. **2**(1), 21–32 (2001)

[48] Richard, H., Schulz, M., Sultan, M., Nürnberger, A., Schrinner, S., Balzereit, D., Dagand, E., Rasche, A., Lehrach, H., Vingron, M., et al.: Prediction of alternative isoforms from exon expression levels in RNA-seq experiments. Nucleic Acids Res. **38**(10), e112 (2010)
[49] Roberts, A., Trapnell, C., Donaghey, J., Rinn, J., Pachter, L., et al.: Improving RNA-seq expression estimates by correcting for fragment bias. Genome Biol. **12**(3), R22 (2011)
[50] Robertson, G., Schein, J., Chiu, R., Corbett, R., Field, M., Jackman, S., Mungall, K., Lee, S., Okada, H., Qian, J., et al.: De novo assembly and analysis of RNA-seq data. Nat. Meth. **7**(11), 909–912 (2010)
[51] Rockman, M.V., Kruglyak, L.: Genetics of global gene expression. Nat. Rev. Genet. **7**(11), 862–872 (2006)
[52] Ronald, J., Brem, R., Whittle, J., Kruglyak, L.: Local regulatory variation in Saccharomyces cerevisiae. PLoS Genet. **1**(2), e25 (2005)
[53] Salzman, J., Jiang, H., Wong, W.: Statistical modeling of RNA-seq data. Stat. Sci. **26**(1), 62–83 (2011)
[54] Singh, D., Orellana, C., Hu, Y., Jones, C., Liu, Y., Chiang, D., Liu, J., Prins, J.: Fdm: a graph-based statistical method to detect differential transcription using RNA-seq data. Bioinformatics **27**(19), 2633–2640 (2011)
[55] Sun, W.: A statistical framework for eQTL mapping using RNA-seq data. Biometrics **68**(1), 1–11 (2012)
[56] Sun, W., Hu, Y.: eQTL mapping using RNA-seq data. Stat. Biosci. **5**(1), 198–219 (2013)
[57] Sun, W., Ibrahim, J., Zou, F.: Genomewide multiple-loci mapping in experimental crosses by iterative adaptive penalized regression. Genetics **185**(1), 349 (2010)
[58] Sun, W., Liu, Y., Crowley, J.J., Chen, T.H., Zhou, H., Chu, H., Huang, S., Kuan, P.F., Li, Y., Miller, D., Shaw, G., Wu, Y., Zhabotynsky, V., McMillan, L., Zou, F., Sullivan, P.F., Pardo-Manuel de Villena, F.: IsoDOT detects differential RNA-isoform usage with respect to a categorical or continuous covariate with high sensitivity and specificity. arXiv preprint arXiv:1402.0136 (2014)
[59] Trapnell, C., Williams, B., Pertea, G., Mortazavi, A., Kwan, G., van Baren, M., Salzberg, S., Wold, B., Pachter, L.: Transcript assembly and quantification by RNA-Seq reveals unannotated transcripts and isoform switching during cell differentiation. Nat. Biotechnol. **28**(5), 511–515 (2010)
[60] Valle, L., Serena-Acedo, T., Liyanarachchi, S., Hampel, H., Comeras, I., Li, Z., Zeng, Q., Zhang, H., Pennison, M., Sadim, M., et al.: Germline allele-specific expression of tgfbr1 confers an increased risk of colorectal cancer. Science **321**(5894), 1361 (2008)
[61] Venables, J.: Aberrant and alternative splicing in cancer. Cancer Res. **64**(21), 7647 (2004)
[62] Wahls, W.P., Davidson, M.K.: Dna sequence-mediated, evolutionarily rapid redistribution of meiotic recombination hotspots commentary on genetics 182: 459–469 and genetics 187: 385–396. Genetics **189**(3), 685–694 (2011)
[63] Wang, G., Cooper, T.: Splicing in disease: disruption of the splicing code and the decoding machinery. Nat. Rev. Genet. **8**(10), 749–761 (2007)
[64] Wang, E., Sandberg, R., Luo, S., Khrebtukova, I., Zhang, L., Mayr, C., Kingsmore, S., Schroth, G., Burge, C.: Alternative isoform regulation in human tissue transcriptomes. Nature **456**(7221), 470–476 (2008)
[65] Wang, K., Singh, D., Zeng, Z., Coleman, S., Huang, Y., Savich, G., He, X., Mieczkowski, P., Grimm, S., Perou, C., et al.: Mapsplice: accurate mapping of RNA-seq reads for splice junction discovery. Nucleic Acids Res. **38**(18), e178 (2010)
[66] Wang, L., Wang, S., Li, W.: Rseqc: quality control of rna-seq experiments. Bioinformatics **28**(16), 2184–2185 (2012)
[67] Wittkopp, P., Haerum, B., Clark, A.: Evolutionary changes in cis and trans gene regulation. Nature **430**(6995), 85–88 (2004)
[68] Wright, F.A., Shabalin, A.A., Rusyn, I.: Computational tools for discovery and interpretation of expression quantitative trait loci. Pharmacogenomics **13**(3), 343–352 (2012)

[69] Wright, F., Sullivan, P., Brooks, A., Zou, F., Sun, W., Xia, K., Madar, V., Abdellaoui, A., Batista, S., Butler, C., Chen, G., Chen, T., W., C., et al.: Heritability and genomics of gene expression in peripheral blood. Nature Genet. **46**(5), 430–437 (2014)
[70] Wu, T.D., Nacu, S.: Fast and snp-tolerant detection of complex variants and splicing in short reads. Bioinformatics **26**(7), 873–881 (2010)
[71] Xia, Z., Wen, J., Chang, C., Zhou, X.: Nsmap: A method for spliced isoforms identification and quantification from RNA-seq. BMC Bioinform. **12**(1), 162 (2011)
[72] Xing, Y., Yu, T., Wu, Y., Roy, M., Kim, J., Lee, C.: An expectation-maximization algorithm for probabilistic reconstructions of full-length isoforms from splice graphs. Nucleic Acids Res. **34**(10), 3150 (2006)
[73] Zhao, Q., Kirkness, E., Caballero, O., Galante, P., Parmigiani, R., Edsall, L., Kuan, S., Ye, Z., Levy, S., Vasconcelos, A., et al.: Systematic detection of putative tumor suppressor genes through the combined use of exome and transcriptome sequencing. Genome Biol. **11**(11), R114 (2010)
[74] Zheng, W., Chung, L.M., Zhao, H.: Bias detection and correction in rna-sequencing data. BMC Bioinform. **12**(1), 290 (2011)

Chapter 9
The Role of Spike-In Standards in the Normalization of RNA-seq

Davide Risso, John Ngai, Terence P. Speed, and Sandrine Dudoit

Abstract Normalization of RNA-seq data is essential to ensure accurate inference of expression levels, by adjusting for sequencing depth and other more complex nuisance effects, both within and between samples. Recently, the External RNA Control Consortium (ERCC) developed a set of 92 synthetic spike-in standards that are commercially available and relatively easy to add to a typical library preparation. In this chapter, we compare the performance of several state-of-the-art normalization methods, including adaptations that directly use spike-in sequences as controls. We show that although the ERCC spike-ins could in principle be valuable for assessing accuracy in RNA-seq experiments, their read counts are not stable enough to be used for normalization purposes. We propose a novel approach to normalization that can successfully make use of control sequences to remove unwanted effects and lead to accurate estimation of expression fold-changes and tests of differential expression.

D. Risso (✉) • T.P. Speed
Department of Statistics, University of California, Berkeley, CA, USA
e-mail: davide.risso@berkeley.edu

J. Ngai
Department of Molecular and Cell Biology, Helen Wills Neuroscience Institute, and Functional Genomics Laboratory, University of California, Berkeley, CA, USA
e-mail: jngai@berkeley.edu

T.P. Speed
Bioinformatics Division, Walter and Eliza Hall Institute, Melbourne, VIC, Australia

Department of Mathematics and Statistics, The University of Melbourne, Victoria, Australia
e-mail: terry@wehi.edu.au

S. Dudoit
Division of Biostatistics and Department of Statistics, University of California, Berkeley, CA, USA
e-mail: sandrine@stat.berkeley.edu

S. Datta and D. Nettleton (eds.), *Statistical Analysis of Next Generation Sequencing Data*, Frontiers in Probability and the Statistical Sciences, DOI 10.1007/978-3-319-07212-8_9,

9.1 Introduction

RNA-seq has become the assay of choice for measuring gene expression levels, and its routine use has grown exponentially in the last few years. Despite initial optimistic claims [37], *normalization* remains a crucial, yet often overlooked step that can have a large impact on subsequent analyses, such as differential expression (DE) or clustering [7, 10]. Normalization is essential to ensure that observed differences in expression measures between samples and/or genomic regions are truly due to differential expression and not nuisance experimental/technical effects.

A variety of normalization approaches have been proposed to correct for *between-sample* nuisance effects, e.g., differences in sequencing depths, library preparation effects [7, 32], as well as *within-sample* gene-specific effects, e.g., due to gene length or GC-content [17, 28].

The simplest and most intuitive normalization method adjusts each sample for *sequencing depth* (i.e., total number of mapped reads) by dividing gene-level read counts by the total read count. This is the approach used in the multiplicative Poisson model of [20] and in the Reads Per Kilobase of exon model per Million mapped reads (RPKM) of [22]. Several authors have shown that the total number of reads in a sample is influenced by very few highly-expressed genes, making total-count (TC) normalization less effective and sensitive to outliers [7, 32]. In addition to sequencing depth, the RPKM method adjusts for gene length, assuming uniform read coverage within genic regions. Since this assumption is hardly ever met in practice, RPKM-normalized measures are often still dependent on gene length [7,23,34].

In addition to length effects, other gene-specific biases have been documented. Risso et al. [28] proposed within-sample normalization methods to account for library-specific GC-content effects; Hansen et al. [17] proposed a conditional quantile normalization (CQN) procedure to account for GC-content and length effects, as well as between-sample effects. Finally, several authors have proposed normalization methods to account for sequence composition biases, such as random hexamer priming biases [16], non-uniform cDNA fragment distribution [30], and nucleotide composition [40].

In this chapter, we focus on between-sample normalization methods. Bullard et al. [7] demonstrated the large impact of normalization on differential expression results; in some contexts, sensitivity varies more between normalization procedures than between testing methods. They showed that upper-quartile (UQ) and full-quantile (FQ) normalization procedures are more robust than TC normalization and improve sensitivity without loss of specificity. Robinson and Oshlack [32] and Anders and Huber [1] independently proposed two pairwise global-scaling normalization procedures that scale read counts by a robust measure of "global expression fold-change" between each sample and a reference.

In the context of differential expression analysis, and as with microarrays, all between-sample normalization methods mentioned thus far work properly when the majority of the genes are not DE between the conditions under study, a reasonable assumption in most applications. In practice, most procedures continue to work well even when a high proportion of genes are DE, provided that they are roughly equi-distributed between up- and down-regulation (see [27] for a discussion in the context of microarrays). However, in case of a global shift in expression, usual between-sample normalization approaches will fail [19]. Consider a simple example where there are two samples, a treated and a control sample, and the treatment is so strong that it causes more than half of the genes to be up-regulated. Then, all standard normalization procedures will wrongly scale down the counts of the treated sample, causing an increase in both false negatives (undetected up-regulated genes) and false positives (unaffected genes declared down-regulated). In this case, normalization based on control sequences may be the only option. Lovén et al. [19] proposed a three-step normalization procedure for such a situation: (i) count the number of cells in each sample; (ii) spike in control sequences to each sample in proportion to the number of cells; (iii) normalize read counts based on cyclic loess regression [5] only on the spike-in counts. The spike-in sequences used in [19] were developed by the External RNA Control Consortium (ERCC) as a set of RNA standards for RNA-seq experiments [3, 18].

The aims of this chapter are threefold: (i) to assess the performance of the ERCC spike-in standards; (ii) to evaluate and compare normalization methods; (iii) to explore the possibility of using the ERCC spike-ins as controls in the normalization procedure. In particular, we extend several normalization methods proposed in the literature for the latter purpose. Our study is based on two datasets that differ greatly in terms of organisms, experimental designs, and biological and technical effects: the "real" Zebrafish dataset [11] and the "artificial" SEQC benchmarking dataset [33].

9.2 Methods

9.2.1 Datasets

9.2.1.1 Zebrafish Dataset

Olfactory sensory neurons were isolated from three pairs of gallein-treated and control embryonic zebrafish pools and purified by fluorescence activated cell sorting (FACS) [11]. Each RNA sample was enriched in poly(A)+ RNA from 10–30 ng total RNA and 1 μL (1:1000 dilution) of Ambion ERCC ExFold RNA Spike-in Control Mix 1 was added to 30 ng of total RNA before mRNA isolation. cDNA libraries were prepared according to manufacturer's protocol. The six libraries were sequenced in two multiplex runs on an Illumina HiSeq2000 sequencer, yielding approximately 50 million 100 base pair (bp) paired-end reads per library. See Fig. 9.1a for a schema of the experimental design. We made use of a custom reference sequence, defined as the union of the zebrafish reference genome (Zv9, downloaded from Ensembl [12], v. 67) and the ERCC spike-in sequences (http://tools.invitrogen.com/downloads/

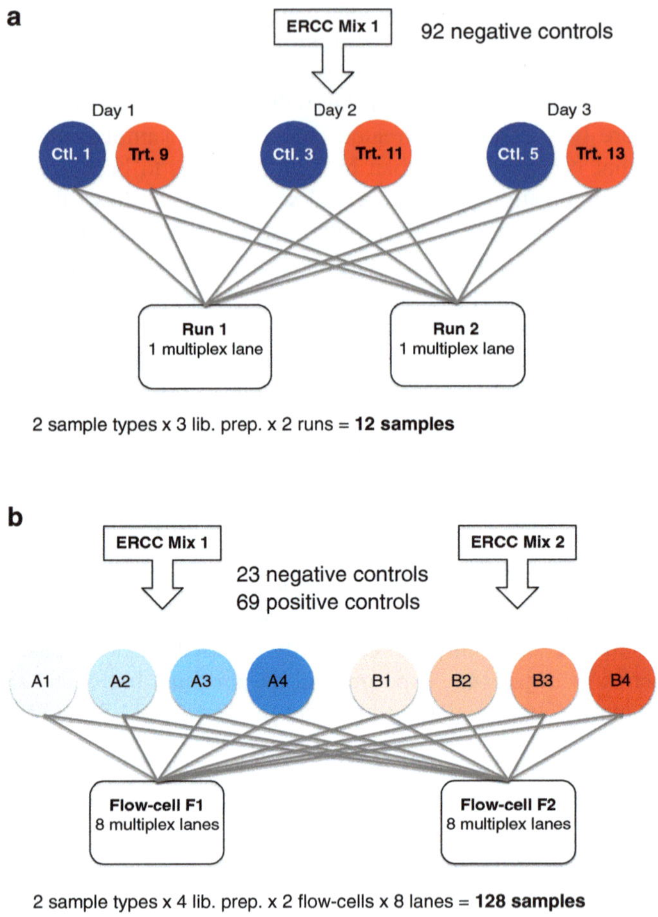

Fig. 9.1 *Experimental design.* Schematic representation of the design of the two experiments considered in this chapter (**a**) Zebrafish dataset (**b**) SEQC dataset

ERCC92.fa). Reads were mapped with TopHat [36] (v. 2.0.4, default parameters and supplying the Ensembl GTF annotation through the -G option). Gene-level read counts were obtained using the htseq-count python script [2] in the "union" mode and Ensembl (v. 67) gene annotation. After verifying that there were no run-specific biases (data not shown), we used the sums of the counts of the two runs as the expression measures for each library. Genes/spike-ins with more than five reads in at least two libraries were retained, resulting in a total of 20,806 (out of 32,561) expressed genes and 59 (out of 92) "present" spike-ins. The FASTQ files containing the raw data are publicly available in GEO with the accession number GSE53334.

9.2.1.2 SEQC Dataset

The third phase of the MicroArray Quality Control (MAQC) Project, also known as SEquencing Quality Control (SEQC) Project [33], aims to assess the technical performance of high-throughput sequencing platforms by generating benchmarking datasets. The design includes four different sample types, namely Samples A, B, C, and D. Sample A is Stratagene's universal human reference (UHR) RNA; Sample B is Ambion's human brain reference RNA; Samples C and D are mixes of Samples A and B, in a 3:1 and 1:3 ratio, respectively. The four types of reference RNA samples were sent to several sequencing centers around the world and sequenced using different platforms. We focus on the Illumina HiSeq2000 data. Each center prepared 4 libraries for each sample type and multiplex pools of the resulting 16 barcoded libraries were sequenced in 8 lanes of 2 flow-cells, yielding a total of 16 (technical) replicates per library and 64 replicates per sample type. Prior to library preparation, Ambion ERCC ExFold RNA Spike-in Control Mix 1 and Mix 2 were added to Sample A and Sample B RNA, respectively, in a proportion of 50 μL per 2,500 μL of total RNA (nominal proportion of 0.02). Here, we consider only Sample A and Sample B sequenced at the Australian Genome Research Facility (AGRF). See Fig. 9.1b for a schema of the experimental design.

The data consist of an average of ten million 100 bp paired-end reads per sample.[1] We made use of a custom reference sequence, defined as the union of the human reference genome (GRCh37, downloaded from Ensembl, v. 69) and the ERCC spike-in sequences. Reads were mapped with TopHat (v. 2.0.6, default parameters and supplying the Ensembl GTF annotation through the -G option). Gene-level read counts were obtained using the htseq-count python script in the "union" mode and Ensembl (v. 69) gene annotation. Genes/Spike-ins with more than five reads in at least ten samples were retained, resulting in a total of 21,559 (out of 55,933) expressed genes and 59 (out of 92) present spike-ins. The FASTQ files containing the raw data will be made publicly available in GEO upon publication of the main SEQC report.

In addition to the *internal* ERCC spike-in positive and negative controls, we use *external* qRT-PCR positive and negative controls from the original MAQC study [8]. As in our previous work [7,28], among the genes assayed by qRT-PCR, we consider only those that match a unique Ensembl gene, are called present in at least three out of each of the four Sample A and Sample B qRT-PCR runs, and have standard errors across the 8 runs not exceeding 0.25. We found 698 qRT-PCR genes in common with the RNA-seq filtered genes and use this subset to compare expression measures between the two assays.

[1]Throughout this chapter, we shall use the term *sample* to refer to an observational unit of interest, i.e., a set of reads from a given lane for a particular library. Thus, as indicated in Fig. 9.1b, there are 128 samples in total for the SEQC dataset, 64 of the reference Sample A type and 64 of the reference Sample B type.

9.2.1.3 ERCC Spike-In Standards

The External RNA Control Consortium (ERCC) [3] developed a set of 92 polyadenylated transcripts that mimic natural eukaryotic mRNAs. These standards are designed to have a wide range of lengths (250–2,000 nucleotides) and GC-contents (5–51 %) and can be spiked into RNA samples prior to library preparation at various concentrations. Ambion commercializes two ERCC spike-in mixes, ERCC ExFold RNA Spike-in Control Mix 1 and Mix 2. The two mixes contain the same set of 92 spike-in standards, but at different concentrations. This allows the design of experiments in which the spike-ins can be used both as positive and negative controls. In particular, the spike-ins are divided into four groups of 23 transcripts each, spanning a 10^6-fold concentration range, with approximately the same length and GC-content distributions. The first group has an expected fold-change (cf. molar ratio) of 4:1 between the two mixes (Mix1:Mix2); the second group has an expected fold-change of 1:1 (negative controls); the third and fourth groups have expected fold-changes of 2:3 and 1:2, respectively. (See the white paper at http://tools.invitrogen.com/content/sfs/manuals/cms_086340.pdf for additional details.)

In the Zebrafish dataset, Mix 1 was added to all samples, so that all spike-ins can be used as negative controls. In the SEQC dataset, Mix 1 was added to Sample A and Mix 2 to Sample B, so that 23 spike-ins can be used as negative controls and 69 as positive controls (23 over-represented and 46 under-represented in Sample A compared to Sample B).

9.2.2 Normalization Methods

In the remainder of the chapter, in order to avoid problems with the log transformation of zero counts, we will adopt the convention (typical in the RNA-seq literature) of adding a small offset to the counts. Hence, $\log(x)$ should be interpreted as $\log(x+\varepsilon)$. In the data analysis, we set $\varepsilon = 1$.

Recently-proposed between-sample normalization procedures to be compared in our study fall into two main groups, global-scaling and non-linear approaches.

9.2.2.1 Global-Scaling Normalization

Most normalization approaches proposed thus far scale gene level read counts by a single *normalization factor* per sample, implicitly assuming linear nuisance effects on gene expression measures. Such *global-scaling normalization* methods differ in their choice of summary statistic for the per-sample read count distribution to be used as scaling factor.

Using notation similar to that of [32], one can present all global-scaling normalization procedures within the following common framework. Define Y_{ij} as the observed read count for gene j ($j = 1,\dots,J$) in sample i ($i = 1,\dots,n$) and μ_{ij} as the corresponding unknown expected expression level, i.e., the number of transcripts for gene j in sample i, times the length of gene j. The expected value of Y_{ij} can then be modeled as

$$E[Y_{ij}|N_i] = \mu_{ij} \cdot S_i, \tag{9.1}$$

where $N_i = Y_{i\cdot} = \sum_j Y_{ij}$ is the total number of mapped reads for sample i and S_i the unknown normalization factor for sample i.

Total-count (TC) normalization estimates S_i by $\hat{S}_i = N_i$, implicitly assuming that all samples have the same total RNA output. This assumption is problematic, as total RNA output can vary drastically from sample to sample depending on RNA composition, and explains why TC normalization can lead to biased inference of differential expression [7, 32].

Upper-quartile (UQ) normalization estimates S_i by the upper-quartile of the distribution of gene-level read counts for sample i [7].

The *Trimmed Mean of M values* or *TMM* method of [32] takes a slightly different, *pairwise* approach to global-scaling normalization. Since relative RNA abundance in two samples is easier to estimate than absolute abundance in one sample, the normalization factors S_i are estimated based on measures of expression fold-changes between each sample and a reference, such as the sample for which the upper-quartile of the read count distribution is closest to the average upper-quartile across samples. Specifically, given a reference sample r, define, for each gene j and sample i, measures of expression log-fold-change (M_{ij}^r) and absolute expression (A_{ij}^r) in the following way:

$$M_{ij}^r = \log_2 \frac{Y_{ij}/N_i}{Y_{rj}/N_r},$$

$$A_{ij}^r = \frac{1}{2} \log_2 \left((Y_{ij}/N_i) \cdot (Y_{rj}/N_r) \right), \quad \text{for } Y_{ij} > 0, Y_{rj} > 0. \tag{9.2}$$

The gene-level read counts for sample i are then scaled by $\hat{S}_i^r$, defined in terms of a variance-weighted trimmed average of M values,

$$\log_2(N_i/\hat{S}_i^r) = \frac{\sum_{j \in \mathscr{J}^r} w_{ij}^r M_{ij}^r}{\sum_{j \in \mathscr{J}^r} w_{ij}^r},$$

where the set $\mathscr{J}^r$ corresponds to the genes that remain after discarding those with M values in the top and bottom 30th percentiles and A values in the top and bottom fifth percentiles (trimming percentages could be tailored to the data at hand) and the weights

$$w_{ij}^r = \frac{N_i - Y_{ij}}{N_i Y_{ij}} + \frac{N_r - Y_{rj}}{N_r Y_{rj}}$$

reflect the variance of the read counts, as estimated by the delta method. The resulting estimator of S_i can hence be seen as a robust estimator of the global expression fold-change between sample i and the reference r.

Anders and Huber [1] independently proposed a very similar pairwise global-scaling normalization procedure, where S_i is estimated by the median fold-change between the counts of sample i and a synthetic reference sample, whose counts are defined as the geometric means of the counts across samples. We refer to this method as AH, after the authors' initials.

9.2.2.2 Non-Linear Normalization

Bullard et al. [7] proposed to normalize RNA-seq gene-level read counts using a *full-quantile* (FQ) approach inspired from the microarray literature [5]. Briefly, the procedure consists in specifying a common reference distribution defined in terms of a function of the sorted counts (e.g., median) across samples and in projecting the quantiles of the count distribution of each sample onto that reference. This is equivalent to matching all quantiles of the read count distributions across samples.

Also borrowed from the microarray literature, *loess* normalization involves robust local regression fits for mean-difference plots (MD-plots) of log counts for pairs of samples [5, 9]. Specifically, *cyclic loess* (CL) normalization considers all possible pairs of samples (i,i'), regresses log-fold-change measures $\log Y_{i'j} - \log Y_{ij}$ on overall expression measures $(\log Y_{ij} + \log Y_{i'j})/2$ using loess, and defines normalized expression measures based on residuals from the regression. Loess normalization can also be performed for each sample paired with a synthetic reference obtained, for example, by averaging counts across samples. The loess fits can be based on either all genes or only a set of controls. For instance, Lovén et al. [19] applied CL normalization to RNA-seq with ERCC spike-in sequences, by performing loess fits only on the spike-ins and then interpolating/extrapolating the fits and computing residuals for all of the genes.

Building on the microarray normalization method of [13], we developed a novel normalization strategy for RNA-seq, coined *RUV* for *remove unwanted variation* [29]. Briefly, RUV works as follows. Consider a generalized linear model, where the observed RNA-seq read counts are regressed on both the known covariates of interest (e.g., treatment) and unknown factors of unwanted variation (e.g., library preparation). RUV makes use of a subset of the data (e.g., negative control genes) to estimate the unwanted factors and adjusts for these in the model for differential expression analysis (see Sect. 9.2.3 for details).

9.2.2.3 Using Control Sequences for Normalization

Suppose one has a set of negative control sequences known a priori not to be differentially expressed between samples, e.g., ERCC spike-ins. Global-scaling normalization methods can be readily adapted to make use of such controls, by simply computing the normalization factors based only on the controls.

Likewise, in loess normalization, the regression can be fit using only the control sequences and then interpolated/extrapolated to normalize all of the genes. In RUV, the factors of unwanted variation can be estimated using only the controls. Full-quantile normalization, however, cannot be readily adapted to make use of controls.

Note that although Lovén et al. [19] used cyclic loess normalization based on the ERCC spike-ins, our conclusions with CL normalization are not fully applicable to their entire method, as the ERCC sequences were not spiked in proportion to the number of cells in either the Zebrafish or SEQC dataset.

9.2.3 A General Framework for Normalization in the Context of Differential Expression

In several RNA-seq applications, it is inevitable to separate normalization from subsequent analyses. For instance, when dealing with class prediction or clustering, the obvious pipeline is to first normalize gene-level counts into pseudo-counts and then apply the classifier or clustering algorithm to the normalized expression measures.

In differential expression (DE) analysis, however, it might be sensible and convenient to directly model the original read counts and include normalization as part of the model. Specifically, in the context of DE, the normalization model of (9.1) can be combined with a Poisson or negative binomial generalized linear model (GLM) [21] for read counts, such as the one used by *edgeR* [31] and *DESeq* [1]. Considering all J genes and n samples at once, this leads to the log-linear model

$$E[Y|X] = \exp(X\beta), \tag{9.3}$$

where Y is the $n \times J$ matrix of observed read counts, X an $n \times p$ design matrix corresponding to covariates of interest/factors of "wanted variation" (e.g., treatment effects), and β a $p \times J$ matrix of parameters of interest. In the typical context of multiple class comparison, X is an ANOVA-like design matrix of indicator dummy variables and the parameters in β represent expression log-fold-changes between pairs of classes. Combining the models in (9.1) and (9.3), it is easy to show that

$$\log \mu = X\beta - \log S, \tag{9.4}$$

where μ is the $n \times J$ matrix of unknown expected expression levels and S is an $n \times J$ matrix of normalization factors. The term $O = -\log S$ is referred to as an *offset* in the usual GLM terminology [21]. For global-scaling normalization, the offsets are constant within samples/rows, i.e., $S_{ij} = S_i$ for each gene j. By allowing gene-specific offsets, one can generalize the model to non-linear between-sample normalization, as well as within-sample normalization, such as in [17, 28] for GC-content effects.

Remove unwanted variation (RUV) normalization is a somewhat different approach, that includes in the model a term representing the nuisance factors. This leads to the general model,

$$\log \mu = X\beta + W\alpha + O, \tag{9.5}$$

where W is an $n \times k$ matrix with the unknown factors of "unwanted variation", α a $k \times J$ matrix of corresponding parameters, and O an $n \times J$ matrix of (possibly gene-specific) offsets. This specification includes as special cases all normalization methods of interest in this study. In particular, by forcing $W\alpha = 0$, one recovers all the usual normalization methods of Sect. 9.2.2.

In order to estimate W for a given k, RUV assumes that one can identify a set of J_c negative control sequences, i.e., a set of non-DE genes, for which $\beta_c = 0$ and $\log \mu_c = W\alpha_c + O_c$, where the subscript c denotes the restriction of matrices to the set of J_c control sequences. The procedure works as follows.

- Either set the offset O to zero or estimate O from some other normalization procedure such as upper-quartile normalization.
- Perform the singular value decomposition (SVD) of $\log Y_c - O_c$, that is, write $\log Y_c - O_c = U\Lambda V^T$, where U is an $n \times n$ orthogonal matrix with columns the left singular vectors of $\log Y_c - O_c$, V a $J_c \times J_c$ orthogonal matrix with columns the right singular vectors, and Λ an $n \times J_c$ rectangular diagonal matrix of singular values (at most $\min(n, J_c)$ distinct non-zero singular values).
- For a given k, estimate $W\alpha_c$ by $\widehat{W\alpha_c} = U\Lambda_k V^T$ and W by $\hat{W} = U\Lambda_k$, where Λ_k is the $n \times J_c$ rectangular diagonal matrix obtained from Λ by retaining only the k largest singular values and setting other diagonal entries to zero (drop null columns to obtain W).
- Plug $\hat{W}$ into (9.5), for the full set of J genes, and estimate both α and β by GLM regression.

Note that for RUV normalization, one may or may not include an offset O in the model. In practice, we found that estimating O via upper-quartile normalization or forcing $O = 0$ in (9.5) lead to very similar results (data not shown).

The two main tuning parameters of RUV are the set of negative control sequences and the number k of factors of unwanted variation. Here, we consider two types of negative controls: a set of "in silico" *empirical controls*, defined as all but the 5,000 most DE genes (found prior to RUV normalization, e.g., based on UQ-normalized counts) and the set of *ERCC spike-in controls*. The choice of k should be guided by considerations that include sample size, extent of technical effects captured by the first k factors, and extent of differential expression [13, 14]. For instance, the small sample size ($n = 6$) for the Zebrafish dataset only allows one or two factors of unwanted variation. Here, we set $k = 1$. The SEQC dataset has a much greater sample size ($n = 128$) and more factors can be considered. Here, we drop the first factor, as it captures the biological factor of interest, and retain the next $k = 3$ factors. We observed that RUV is robust to the choice of k for the SEQC dataset [29].

9.2.4 Evaluation Criteria

9.2.4.1 Relative Log Expression

A particularly useful transformation of read counts is the relative log expression (RLE), defined, for each gene, as the log-ratio of a read count to the median read count across samples. Comparable samples should have similar RLE distributions, that are centered around zero. Unusual RLE distributions could reveal suspicious samples (e.g., problematic library) or batch effects.

9.2.4.2 Differential Expression Analysis

To compare normalization procedures in terms of their impact on differential expression results, we consider the negative binomial GLM analysis of *edgeR* [31], with tag-wise dispersion. Likelihood ratio tests of DE are performed for the following effects: for the Zebrafish dataset, treatment effect, and for the SEQC dataset, Sample A vs. B effect and, in the "null" experiment of Fig. 9.6, library preparation effects for samples of type A. A gene is declared DE if the associated null hypothesis is rejected at a false discovery rate (FDR) [4] of 0.05.

9.2.4.3 Bias in Expression Fold-Change Estimation

Expression log-fold-changes are estimated by log-ratios of average normalized read counts between two sets of samples corresponding to the two conditions of interest. In order to compute bias, one needs to know the true value of an expression fold-change.

Since qRT-PCR is typically viewed as producing accurate estimates of expression levels, we use the qRT-PCR expression measures as a gold standard for the SEQC dataset. Specifically, we define as "true" Sample A/Sample B expression log-fold-change the log-ratio between the average of the 4 qRT-PCR measures of Sample A and the average of the 4 measures of Sample B. The corresponding RNA-seq estimate is the log-ratio between the average of the normalized counts for the 64 samples of type A and the average of the normalized counts for the 64 samples of type B. For a given gene, bias is then estimated as the difference between the estimated log-fold-changes from the two technologies.

9.2.5 Software Implementation

All normalization methods examined in this study are implemented in open-source R [26] packages released through the Bioconductor Project (http://www.biocondutor.org) [15]: *affy* (CL), *DESeq* (AH), *EDASeq* (UQ, FQ), *edgeR* (TMM), *RUVSeq* (RUV).

9.3 Results

9.3.1 *Behavior of the ERCC Spike-In Controls*

One of the aims of this chapter is to evaluate the performance of the ERCC spike-in standards, with particular focus on the possibility of using them as controls in the normalization procedure. In order for the spike-ins to be trusted for normalization, two conditions must be satisfied: (i) spike-in read counts are not affected by the biological factors of interest and (ii) the unwanted variation affects spike-in and gene read counts similarly.

Figures 9.2, 9.3, and 9.4 show that, in our datasets, neither condition is met. In Fig. 9.2, while the proportion of reads mapping to the ERCC spike-in sequences is similar for samples from the same library, it can vary substantially between libraries and can deviate markedly from the nominal value. This result confirms the findings of [25] that suggest that poly(A) selection may play a role in spike-in detection. Even more troubling is the observed dependence of spike-in counts on the biological condition: for the SEQC dataset, the proportion of reads mapping to the spike-ins is consistently greater in Sample B than in Sample A (Fig. 9.2a); for the Zebrafish dataset, the proportion is consistently greater in treated than in control samples (Fig. 9.2b).

Figure 9.3 illustrates the variability across samples of individual spike-in read counts for the SEQC dataset (similar findings for the Zebrafish dataset are not shown). It is clear that the variability of read counts is very large, especially for

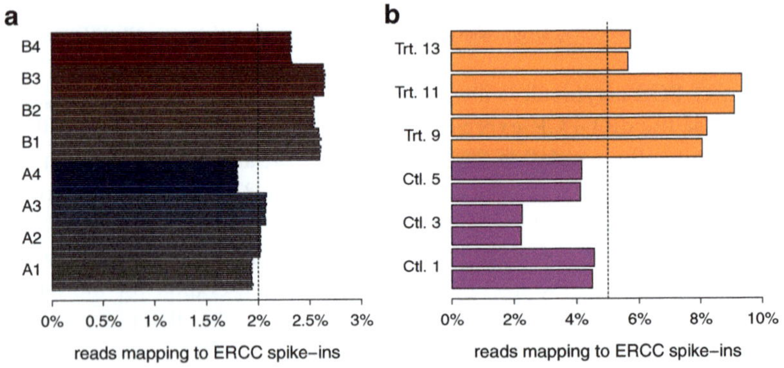

Fig. 9.2 *Behavior of the ERCC spike-in controls*. Proportion of reads mapping to ERCC spike-in sequences out of total number of mapped reads; vertical dashed line indicates nominal proportion. (**a**): SEQC dataset, each shade of blue represents one of the four Sample A libraries and each shade of red represents one of the four Sample B libraries (16 replicates per library). (**b**): Zebrafish dataset, treated libraries are displayed in orange and control libraries in purple. There is an evident library preparation effect and, more disturbingly, a treatment effect, on the proportion of reads mapping to the spike-ins. This may lead to confounding when identifying differentially expressed genes

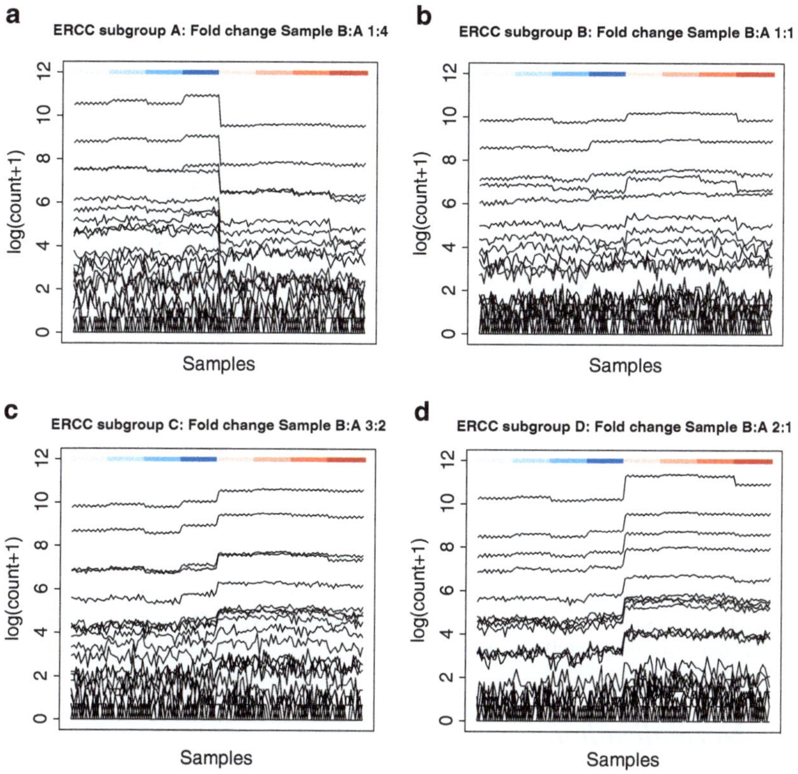

Fig. 9.3 *Behavior of the ERCC spike-in controls, SEQC dataset.* Unnormalized log(count+1) for individual ERCC spike-in sequences across the 128 samples, by ERCC control group. Sample A: Sample B nominal fold-change: (**a**) Group A: 4:1; (**b**) Group B: 1:1; (**c**) Group C: 2:3; (**d**) Group D: 1:2. The color bar at the top of each plot indicates the library corresponding to each sample, using the same color code as in Fig. 9.2a

sequences spiked in at low concentrations. Out of the full set of 92 spike-ins, only the ~ 40 most abundant have a reasonable signal-to-noise ratio, while counts for the least abundant are practically noise. Even though the variability in read counts is expected to depend on sequencing depth, we found that the same set of 59 spike-ins were present in our two datasets, despite much deeper sequencing for the Zebrafish dataset (see Sect. 9.2.1). Not surprisingly, these were the 59 spike-ins with highest concentration and, hence, highest read counts.

Finally, systematic library and/or treatment effects on the ERCC spike-ins have important downstream effects. For instance, when comparing two control libraries in the Zebrafish dataset, the distribution of log-ratios of read counts for the spike-ins is markedly different from the corresponding distribution for the ensemble of genes (Fig. 9.4a). This is most likely an artifact, as one does not expect any treatment effects when comparing two control samples.

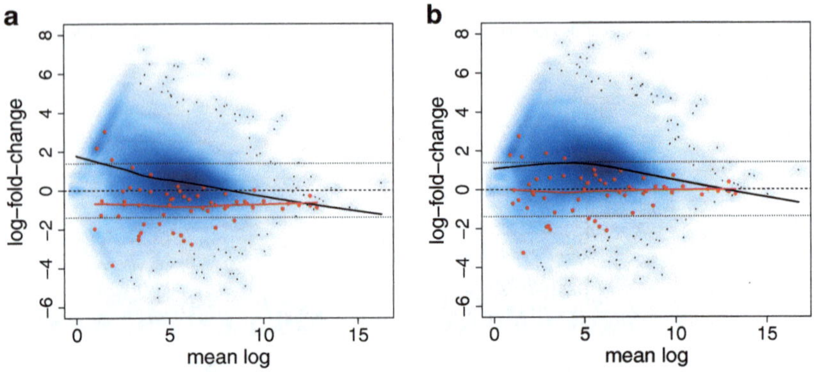

Fig. 9.4 *Behavior of the ERCC spike-in controls, Zebrafish dataset.* Mean-difference plots (MD-plots) of log counts for two control samples (Library 3 vs. Library 1): red points correspond to ERCC spike-ins; black and red lines are lowess fits for all genes and only the spike-ins, respectively. (**a**): Unnormalized counts. As expected, count log-ratios are scattered around the zero line, indicating that most genes are equally expressed in the two control samples. The negative slope of the black line indicates the need for normalization. The very different behavior of the spike-ins with respect to the bulk of the genes is highlighted by the difference in the two lowess fits (red and black lines). (**b**): Cyclic loess normalization on ERCC spike-ins. By normalizing all genes based on the spike-in loess fit, CL normalization wrongly shifts upward gene expression log-fold-changes for the ensemble of genes

9.3.2 Normalization

9.3.2.1 Normalization Based on All Genes

In this section, we evaluate the performance of several normalization methods, applied to the full set of genes.

Figure 9.5 displays boxplots of relative log expression for normalized counts for the Zebrafish dataset. Properly normalized counts should have RLE distributions centered around zero and as similar as possible across samples. All methods lead to reasonable normalized counts, with the exception of TC normalization which leads to RLE distributions similar to those of unnormalized counts. The excessive variability of Library 11 is not removed by global-scaling normalization and only partially reduced by FQ and CL normalization. By contrast, RUV has the extreme effect of shrinking the expression measures of Library 11 towards zero.

To assess the ability of each normalization method to remove unwanted technical effects in the SEQC dataset, we test the null hypothesis of equal mean expression for the four libraries of Sample A (i.e., libraries A1–4), considering the 16 samples within each library as replicates. Given the same starting RNA sample, none of the genes are expected to be DE between the four libraries and p-values are expected to follow a uniform distribution. Figure 9.6 displays the empirical cumulative distribution function (ECDF) of *edgeR* p-values for tests of differential expression

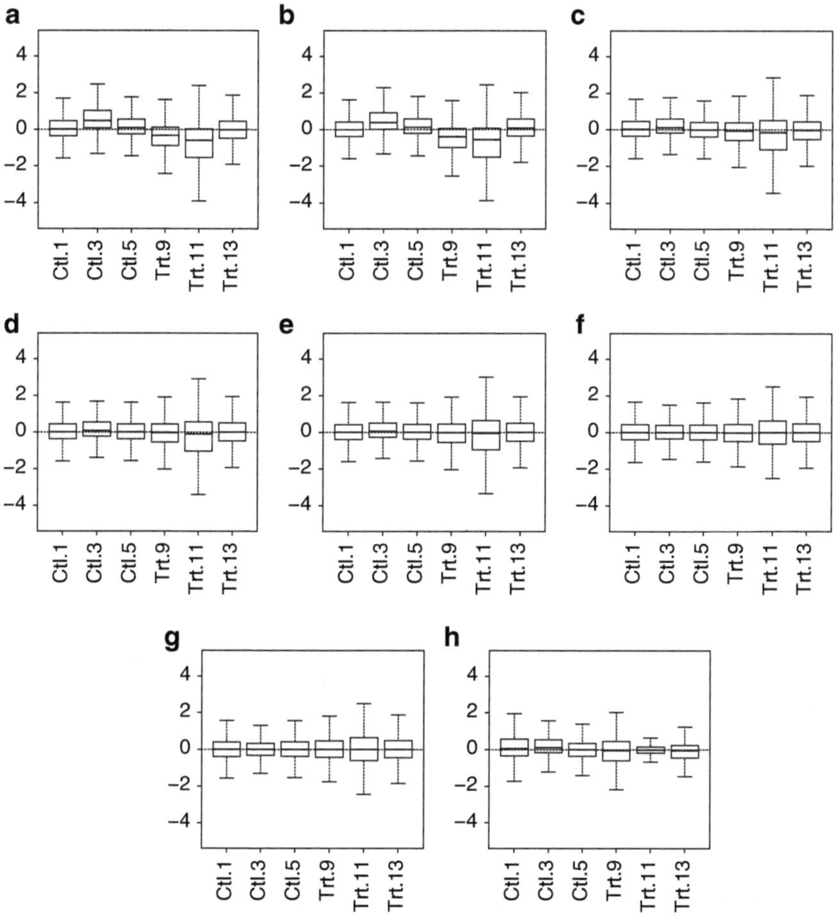

Fig. 9.5 *RLE distributions for normalized counts, Zebrafish dataset.* Boxplots of relative log expression (RLE) for normalized counts. RLE distributions are expected to be centered around zero and as similar as possible across samples (**a**) Unnormalized (**b**) TC (**c**) UQ (**d**) TMM (**e**) AH (**f**) FQ (**g**) CL (**h**) RUV

between replicate libraries. The significant library preparation effect observed for unnormalized counts is only attenuated by almost all normalization methods; only RUV is able to remove most of this unwanted effect.

9.3.2.2 Normalization Based on ERCC Spike-In Controls

One largely unexplored direction is the use of control sequences in the RNA-seq normalization procedure itself. Control sequences have been successfully employed in microarray normalization, for mRNA arrays [24, 39] and, more recently, microRNA

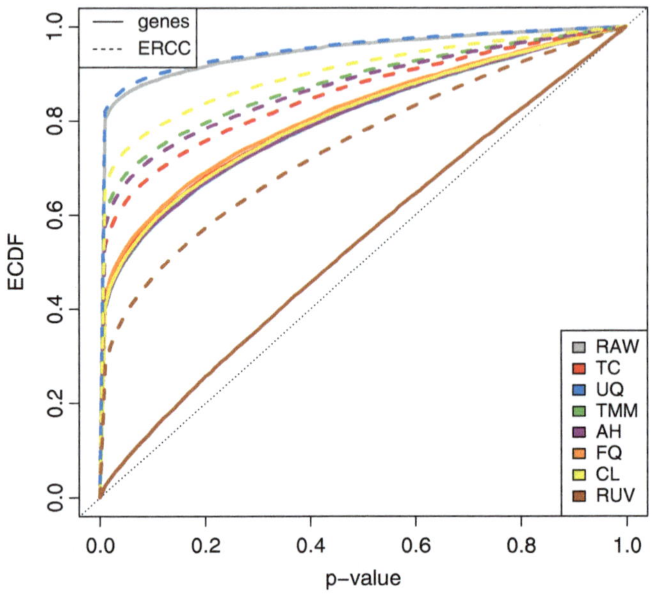

Fig. 9.6 *p-value distributions for tests of library preparation effects based on normalized counts, SEQC dataset.* Empirical cumulative distribution function (ECDF) of p-values for tests of differential expression between Sample A replicate libraries. Solid and dashed lines correspond, respectively, to normalization procedures based on all genes and only the ERCC spike-in sequences. We expect no DE and hence p-values to follow a uniform distribution, with ECDF as close as possible to the identity line. Only with RUV based on all genes do p-values behave close to expectation

arrays [38]. One of the advantages of using negative controls for normalization is the possibility of relaxing the common assumption that the majority of the genes are not differentially expressed between the conditions under study.

For the Zebrafish dataset, the RLE boxplots of Fig. 9.7 indicate that none of the normalization procedures, except for RUV, lead to reliable read count distributions when based only on the ERCC spike-ins. The poor performance of control-based normalization is likely due to the misbehavior of the spike-ins, as noted above and illustrated, in particular, in Fig. 9.2, which suggests that the proportion of reads mapping to the spike-ins could be confounded with sequencing depth. The MD-plots of log counts for a pair of control libraries in Fig. 9.4 further demonstrate the poor results of control-based CL normalization: normalizing based on a loess fit only on the spike-ins (red line) wrongly shifts upward all count log-ratios in a situation where one does not expect any global shift in expression.

The poor performance of normalization based on the ERCC spike-ins is confirmed with the SEQC dataset. The ECDF of p-values in Fig. 9.6 indicate that, for each normalization procedure, the version based on all of the genes (solid lines) is always more successful in reducing library preparation effects than that based only

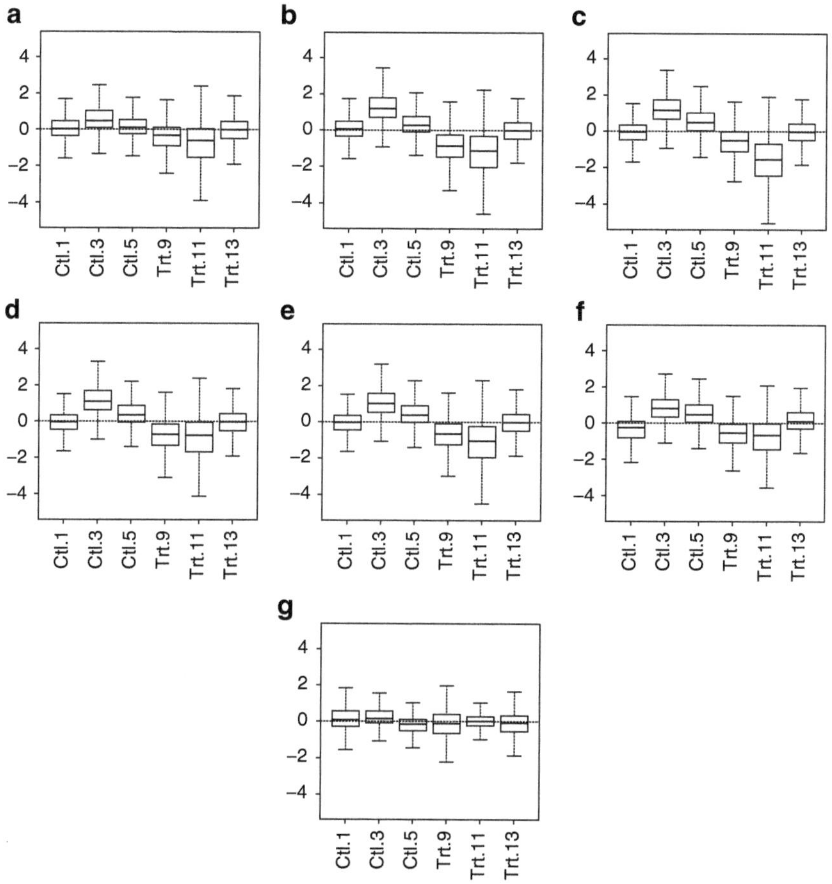

Fig. 9.7 *RLE distributions for spike-in-normalized counts, Zebrafish dataset.* Boxplots of relative log expression for normalized counts based on the ERCC spike-in controls. RLE distributions are expected to be centered around zero and as similar as possible across samples (**a**) Unnormalized (**b**) TC (**c**) UQ (**d**) TMM (**e**) AH (**f**) CL (**g**) RUV

on the spike-ins (dashed lines). This is true for RUV as well; although its spike-in-based version still outperforms all other approaches.

9.3.2.3 Impact on Differential Expression

Normalization has been shown to have a strong impact on the inference of differential expression [7, 10, 28]. Here, we exploit the availability of external qRT-PCR controls for the SEQC dataset (see Sect. 9.2.1.2) to compare normalization methods in terms of DE results.

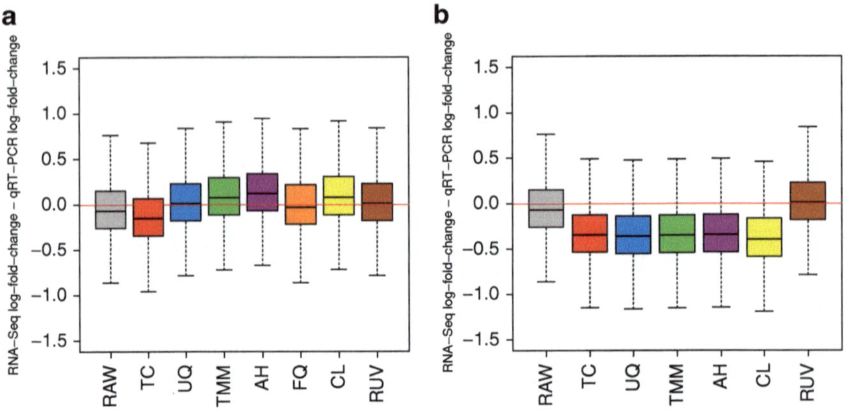

Fig. 9.8 *Impact of normalization on differential expression results, SEQC dataset.* Difference between RNA-seq and qRT-PCR estimates of Sample A/Sample B expression log-fold-changes, i.e., bias in RNA-seq when viewing qRT-PCR as a gold standard. (**a**): Normalization based on all genes. (**b**): Normalization based only on the ERCC spike-ins

Figure 9.8 displays boxplots of differences between RNA-seq and qRT-PCR estimated Sample A/Sample B expression log-fold-changes for various normalization approaches. By viewing qRT-PCR as a gold standard, one can interpret these differences as estimating bias in RNA-seq. When using all genes to normalize the data (Fig. 9.8a), all methods lead to good results with respect to qRT-PCR. Thanks to the large sample size and balanced design of the SEQC dataset, even without normalization, there is only a slight bias in log-fold-change estimation. TC-normalized log-fold-changes are more biased than unnormalized log-fold-changes; all other methods improve upon no normalization, with UQ, FQ, and RUV leading to unbiased estimates. The results are much worse when normalization is based only on the ERCC spike-ins. All spike-in-based methods, but RUV, introduce more bias than no normalization and only RUV leads to unbiased estimates.

The absence of biological replication and the extreme difference between Sample A and Sample B make the SEQC dataset rather artificial and a more realistic and biologically meaningful dataset is required to confirm our findings. However, the Zebrafish dataset lacks external controls to validate differential expression results. One can nonetheless examine the distribution of p-values for tests of DE between treated and control samples. Under the assumption that most genes are non DE, one expects this distribution to be a mixture between a uniform distribution (for the majority of non-DE genes) and a point mass at zero (for the few DE genes). The ECDF of p-values in Fig. 9.9 show that this is indeed the case for normalization procedures based on all genes (solid lines). However, using only the ERCC spike-ins leads to unrealistic distributions, very far from uniform (dashed lines). Once again, the notable exception is RUV, which leads to good results even when using the spike-ins as negative controls.

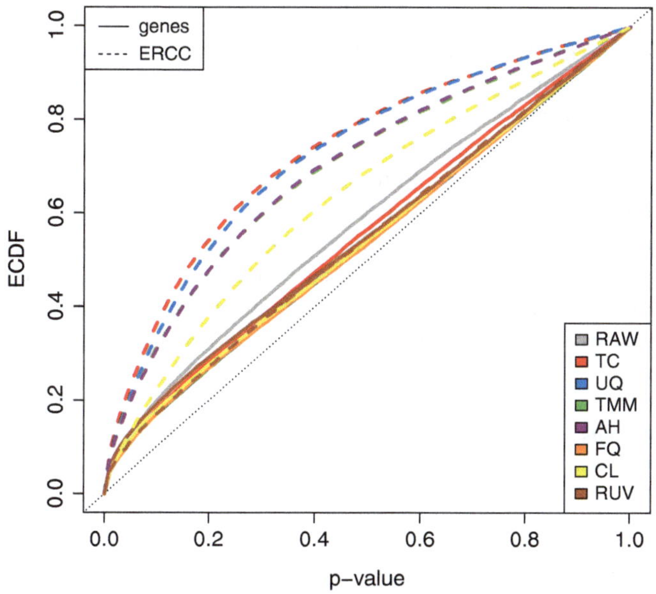

Fig. 9.9 *Impact of normalization on differential expression results, Zebrafish dataset.* Empirical cumulative distribution functions of *p*-values for tests of differential expression between treated and control samples. Solid and dashed lines correspond, respectively, to normalization procedures based on all genes and only the ERCC spike-in sequences. We expect a uniform distribution for the bulk of non-DE genes, with a spike at zero corresponding to a few DE genes. This is the case for normalization procedures based on all genes (solid lines), but, with the exception of RUV, not for procedures based only on the ERCC spike-ins (dashed lines)

9.4 Conclusions

The possibility of using spike-in sequences as controls to normalize RNA-seq data is appealing, as it allows us, among other things, to relax the usual assumption that only a small proportion of genes are differentially expressed. Here, we have seen that the ERCC spike-in standards are not reliable or stable enough to be used in standard global-scaling or regression-based normalization procedures. In particular, for both the Zebrafish and the SEQC datasets, spike-in read count distributions vary substantially between technical replicate libraries (Fig. 9.2). Moreover, signal-to-noise ratio is disturbingly low, especially for low-concentration spike-ins (Fig. 9.3).

All normalization methods compared here perform similarly in terms of relative log expression and impact on differential expression results: they lead to satisfactory results when based on all genes and to very poor results when based only on the ERCC spike-ins (Figs. 9.5, 9.7, 9.8, and 9.9). This is not surprising, given the previously-noted misbehavior of the spike-ins. One notable exception is RUV, which leads to good results whether based on all genes or only the spike-ins. For the SEQC dataset, RUV is the only normalization procedure able to fully remove library preparation effects (Fig. 9.6); for the Zebrafish dataset, its effect is to down-weight outlying Library 11 (Fig. 9.5h), thereby adding robustness to subsequent tests of differential expression.

Internal and external controls are essential for the analysis of high-throughput data and spike-in sequences have the potential to help researchers better adjust for unwanted technical effects. With the advent of single-cell sequencing [35], the role of spike-in standards should become even more important, both to account for technical variability [6] and to allow the move from relative to absolute RNA expression quantification. It is therefore essential to ensure that spike-in standards behave as expected and to develop a set of controls that are stable enough across replicate libraries and robust to both differences in library composition and library preparation protocols.

Acknowledgements We thank Leming Shi for providing the SEQC pilot data and Laurent Jacob for his help with the software implementation of the RUV method.

References

[1] Anders, S., Huber, W.: Differential expression analysis for sequence count data. Genome Biol. **11**(10), R106 (2010)

[2] Anders, S., Pyl, P.T., Huber, W.: HTSeq: a Python framework to work with high-throughput sequencing data. Technical Report, bioRxiv preprint (2014). doi:10.1101/002824

[3] Baker, S.C., Bauer, S.R., Beyer, R.P., Brenton, J.D., Bromley, B., Burrill, J., Causton, H., Conley, M.P., Elespuru, R., Fero, M., et al.: The external RNA controls consortium: a progress report. Nat. Meth. **2**(10), 731–734 (2005)

[4] Benjamini, Y., Hochberg, Y.: Controlling the false discovery rate: a practical and powerful approach to multiple testing. J. R. Stat. Soc. Series B **57**, 289–300 (1995)

[5] Bolstad, B.M., Irizarry, R.A., Åstrand, M., Speed, T.P.: A comparison of normalization methods for high density oligonucleotide array data based on variance and bias. Bioinformatics **19**(2), 185–193 (2003)

[6] Brennecke, P., Anders, S., Kim, J.K., Kołodziejczyk, A.A., Zhang, X., Proserpio, V., Baying, B., Benes, V., Teichmann, S.A., Marioni, J.C., Heisler, M.G.: Accounting for technical noise in single-cell RNA-seq experiments. Nat. Meth. **10**, 1093–1095 (2013)

[7] Bullard, J., Purdom, E., Hansen, K., Dudoit, S.: Evaluation of statistical methods for normalization and differential expression in mRNA-Seq experiments. BMC Bioinform. **11**(1), 94 (2010)

[8] Canales, R.D., Luo, Y., Willey, J.C., Austermiller, B., Barbacioru, C.C., Boysen, C., Hunkapiller, K., Jensen, R.V., Knight, C.R., Lee, K.Y., et al.: Evaluation of DNA microarray results with quantitative gene expression platforms. Nat. Biotechnol. **24**(9), 1115–1122 (2006)

[9] Cleveland, W.S.: Robust locally weighted regression and smoothing scatterplots. J. Am. Stat. Assoc. **74**(368), 829–836 (1979)

[10] Dillies, M.A., Rau, A., Aubert, J., Hennequet-Antier, C., Jeanmougin, M., Servant, N., Keime, C., Marot, G., Castel, D., Estelle, J., et al.: A comprehensive evaluation of normalization methods for Illumina high-throughput RNA sequencing data analysis. Brief. Bioinform. **14**(6), 671–683 (2013)

[11] Ferreira, T., Wilson, S.R., Choi, Y.G., Risso, D., Dudoit, S., Speed, T.P., Ngai, J.: Silencing of odorant receptor genes by G Protein $\beta\gamma$ signaling ensures the expression of one odorant receptor per olfactory sensory neuron. Neuron **81**, 847–859 (2014)

[12] Flicek, P., Amode, M.R., Barrell, D., Beal, K., Brent, S., Carvalho-Silva, D., Clapham, P., Coates, G., Fairley, S., Fitzgerald, S., et al.: Ensembl 2012. Nucleic Acids Res. **40**(D1), D84–D90 (2012)

[13] Gagnon-Bartsch, J., Speed, T.: Using control genes to correct for unwanted variation in microarray data. Biostatistics **13**(3), 539–552 (2012)

[14] Gagnon-Bartsch, J., Jacob, L., Speed, T.P.: Removing unwanted variation from high dimensional data with negative controls. Technical Report 820, Department of Statistics, University of California, Berkeley (2013)

[15] Gentleman, R.C., Carey, V.J., Bates, D.M., Bolstad, B., Dettling, M., Dudoit, S., Ellis, B., Gautier, L., Ge, Y., Gentry, J., Hornik, K., Hothorn, T., Huber, W., Iacus, S., Irizarry, R.A., Leisch, F., Li, C., Maechler, M., Rossini, A.J., Sawitzki, G., Smith, C., Smyth, G.K., Tierney, L., Yang, Y.H., Zhang, J.: Bioconductor: open software development for computational biology and bioinformatics. Genome Biol. **5**(10), R80 (2004)

[16] Hansen, K.D., Brenner, S.E., Dudoit, S.: Biases in Illumina transcriptome sequencing caused by random hexamer priming. Nucleic Acids Res. **38**(12), e131 (2010)

[17] Hansen, K.D., Irizarry, R.A., Zhijin, W.: Removing technical variability in RNA-seq data using conditional quantile normalization. Biostatistics **13**(2), 204–216 (2012)

[18] Jiang, L., Schlesinger, F., Davis, C.A., Zhang, Y., Li, R., Salit, M., Gingeras, T.R., Oliver, B.: Synthetic spike-in standards for RNA-seq experiments. Genome Res. **21**(9), 1543–1551 (2011)

[19] Lovén, J., Orlando, D., Sigova, A., Lin, C., Rahl, P., Burge, C., Levens, D., Lee, T., Young, R.: Revisiting global gene expression analysis. Cell **151**(3), 476–482 (2012)

[20] Marioni, J., Mason, C., Mane, S., Stephens, M., Gilad, Y.: RNA-seq: an assessment of technical reproducibility and comparison with gene expression arrays. Genome Res. **18**(9), 1509 (2008)

[21] McCullagh, P., Nelder, J.: Generalized Linear Models. Chapman and Hall, New York (1989)

[22] Mortazavi, A., Williams, B.A., McCue, K., Schaeffer, L., Wold, B.: Mapping and quantifying mammalian transcriptomes by RNA-Seq. Nat. Meth. **5**(7), 621–628 (2008)

[23] Oshlack, A., Wakefield, M.: Transcript length bias in RNA-seq data confounds systems biology. Biol. Direct **4**(1), 14 (2009)

[24] Oshlack, A., Emslie, D., Corcoran, L.M., Smyth, G.K.: Normalization of boutique two-color microarrays with a high proportion of differentially expressed probes. Genome Biol. **8**(1), R2 (2007)

[25] Qing, T., Yu, Y., Du, T., Shi, L.: mRNA enrichment protocols determine the quantification characteristics of external RNA spike-in controls in RNA-Seq studies. Sci. China Life Sci. **56**(2), 134–142 (2013)

[26] R Core Team: R: a language and environment for statistical computing. R Foundation for Statistical Computing, Vienna, Austria (2013). http://www.R-project.org

[27] Risso, D., Massa, M.S., Chiogna, M., Romualdi, C.: A modified LOESS normalization applied to microRNA arrays: a comparative evaluation. Bioinformatics **25**(20), 2685–2691 (2009)

[28] Risso, D., Schwartz, K., Sherlock, G., Dudoit, S.: GC-content normalization for RNA-Seq data. BMC Bioinform. **12**(1), 480 (2011)

[29] Risso, D., Ngai, J., Speed, T., Dudoit, S.: Normalization of RNA-seq data using factor analysis of control genes or samples. Nat. Biotechnol. (2014, in press).

[30] Roberts, A., Trapnell, C., Donaghey, J., Rinn, J.L., Pachter, L.: Improving RNA-Seq expression estimates by correcting for fragment bias. Genome Biol. **12**(3), R22 (2011)

[31] Robinson, M.D., McCarthy, D.J., Smyth, G.K.: edgeR: a bioconductor package for differential expression analysis of digital gene expression data. Bioinformatics **26**(1), 139–140 (2010)

[32] Robinson, M.D., Oshlack, A.: A scaling normalization method for differential expression analysis of RNA-seq data. Genome Biol. **11**(3), R25 (2010)

[33] Su, Z., Labaj, P., Li, S., Thierry-Mieg, J., Thierry-Mieg, D., Shi, W., et al.: Power and limitations of RNA-Seq. Nat. Biotechnol. (2014, in press)

[34] Sun, Z., Zhu, Y.: Systematic comparison of RNA-Seq normalization methods using measurement error models. Bioinformatics **28**(20), 2584–2591 (2012)
[35] Tang, F., Barbacioru, C., Wang, Y., Nordman, E., Lee, C., Xu, N., Wang, X., Bodeau, J., Tuch, B.B., Siddiqui, A., et al.: mRNA-Seq whole-transcriptome analysis of a single cell. Nat. Meth. **6**(5), 377–382 (2009)
[36] Trapnell, C., Pachter, L., Salzberg, S.L.: TopHat: discovering splice junctions with RNA-Seq. Bioinformatics **25**(9), 1105–1111 (2009)
[37] Wang, Z., Gerstein, M., Snyder, M.: RNA-Seq: a revolutionary tool for transcriptomics. Nat. Rev. Genet. **10**(1), 57–63 (2009)
[38] Wu, D., Hu, Y., Tong, S., Williams, B.R., Smyth, G.K., Gantier, M.P.: The use of miRNA microarrays for the analysis of cancer samples with global miRNA decrease. RNA **19**(7), 876–888 (2013)
[39] Yang, Y.H., Dudoit, S., Luu, P., Lin, D.M., Peng, V., Ngai, J., Speed, T.P.: Normalization for cDNA microarray data: a robust composite method addressing single and multiple slide systematic variation. Nucleic Acids Res. **30**(4), e15 (2002)
[40] Zheng, W., Chung, L.M., Zhao, H.: Bias detection and correction in RNA-sequencing data. BMC Bioinform. **12**(1), 290 (2011)

Chapter 10
Cluster Analysis of RNA-Sequencing Data

Peng Liu and Yaqing Si

Abstract RNA-seq technology has been widely adopted to study global gene expression. By grouping genes with similar expression profiles across treatments, cluster analysis enables us to organize and visualize results from RNA-seq experiments. Such analysis often provides insights into gene functions and gene networks, and hence it is a useful technique that has been routinely applied in gene expression studies. In this chapter, we describe several clustering algorithms that have been applied to RNA-seq data analysis: *K*-means clustering, hierarchical clustering, model-based clustering, and hybrid-hierarchical clustering algorithms. In addition, we illustrate the applications of these clustering algorithms in a maize dataset and discuss some remaining challenges in cluster analysis of RNA-seq data.

10.1 Introduction

Next-generation sequencing (NGS) technologies can be used to measure the abundance of messenger RNA (mRNA) in a sample, and this resulting technology is called RNA-sequencing (RNA-seq). In the pioneering studies using RNA-seq, only two treatment groups were analyzed [22, 40]. In such experiments, detecting genes that are differentially expressed between treatment groups is the major goal of the analysis. With the increasing popularity of RNA-seq experiments, there have been more experiments with larger scale that involve multiple treatment groups or many samples. Cluster analysis is a useful tool for learning information from such data.

P. Liu (✉)
Department of Statistics, Iowa State University, Ames, IA, USA
e-mail: pliu@iastate.edu

Y. Si
School of Statistics, Southwestern University of Finance and Economics,
Chengdu, Sichuan, China
e-mail: sy@swufe.edu.cn

S. Datta and D. Nettleton (eds.), *Statistical Analysis of Next Generation Sequencing Data*, Frontiers in Probability and the Statistical Sciences, DOI 10.1007/978-3-319-07212-8_10,

Table 10.1 An RNA-seq data matrix [19]. A total of G genes are measured for each of the n samples, and the count of sequence reads mapped to each gene measures the expression level of the corresponding gene

	Sample				
Gene	1	2	3	...	n
1	16	35	27	...	22
2	313	306	300	...	279
3	8	8	6	...	30
4	226	156	231	...	350
...	...	...	...	...	...
G	11	18	17	...	28

Cluster analysis is the process of "grouping similar objects". In other words, cluster analysis searches for groups (clusters) in the data, in such a way that objects belonging to the same cluster share similar features whereas objects in different clusters are dissimilar. In the context of gene expression studies, either gene-based or sample-based clustering may be applied. Gene-based clustering refers to the process of grouping genes based on their expression values across different treatment groups. Such cluster analysis helps identify groups of genes that are co-expressed, and co-expressed genes tend to be functionally related. For example, such genes might be co-regulated by the same transcription factor or be involved in the same biological pathway. Hence gene-based clustering may shed light on functions of genes that may not be easy to discover by going through the list of differentially expressed genes one by one. Cluster analysis may also facilitate gene network analysis and gene annotation. Hence, gene-based clustering has been widely used in interpreting gene expression data. Alternatively, sample-based clustering refers to the process of grouping samples based on a long list of features that correspond to the expression levels of genes. Such sample-based clustering might reveal relationships among the samples. For example, patients with similar clinical phenotype might have different molecular profiles. Clustering samples may be used for the detection of sub-groups which are difficult to identify based on traditional morphology-based methods [16]. Many of the clustering algorithms can be applied to both gene-based and sample-based clustering. In this chapter, we focus on gene-based clustering algorithms that cluster genes according to their expression profiles. We will comment on sample-based clustering in Sect. 10.3.

In a typical RNA-seq experiment, mRNA molecules of each sample are extracted, converted to a library of complementary DNA fragments, and then sequenced with a high-throughput sequencing platform, such as Illumina's Genome Analyzer. Millions of short sequences, or *reads*, are obtained for each sample and then mapped to a reference genome. The count of reads mapped to a given gene measures the expression level of this gene. Note that we use the term gene loosely in this chapter, and it may refer to any interesting genomic feature which might correspond to an exon or a gene isoform. Table 10.1 presents partial RNA-seq data from a maize study [19], where each row corresponds to a gene and each column corresponds to a sample.

By the nature of the technology, RNA-seq data are counts, different from microarray data that are typically treated as continuous and modeled with normal distributions after a log transformation. In addition, there exists an apparent mean-variance relationship for RNA-seq data [1], and the majority of genes have low counts [4]. Based on these characteristics, normal distributions are not appropriate for directly modeling of such data. Methods for microarray data are mostly based on normal distributions and cannot be directly applied to the count data generated by RNA-seq. Two types of methods have been applied for cluster analysis of RNA-seq data. The first type of methods are based on transformed data where the transformations aim to result in distributions that are closer to normal. Then methods for microarray analysis can be borrowed to analyze the transformed RNA-seq data. The other type of methods work with the count data directly by using appropriate discrete distributions.

Another characteristic of RNA-seq data is its high-dimensionality. RNA-seq technology measures an enormous number of genes simultaneously for each sample. For example, the number of measured genes is more than 30,000 for human beings [27], and more than 50,000 for maize [19]. In many biological systems, it is often believed that only a small proportion of genes are differentially expressed between different conditions or treatments. Including genes that are equivalently expressed across conditions would increase the background noise and computational cost of cluster analysis. Therefore, we recommend cluster analysis only for genes that are identified as differentially expressed with proper multiple testing error control. This typically leads to clustering a much smaller number of genes and results in more meaningful interpretations.

This chapter is organized as follows. In Sect. 10.2, we describe the models that could be used to fit RNA-seq data. In Sect. 10.3, we present several clustering algorithms that have been applied to RNA-seq data analysis. Section 10.4 presents an example study, and Sect. 10.5 describes the implementation of different algorithms using available R packages. This chapter concludes with some discussion in Sect. 10.6.

10.2 Models for RNA-seq Data

In this chapter, we let Y_{gij} denote the count of reads mapped to gene g for replicate j of treatment i for $g = 1, \cdots, G, i = 1, \cdots, I, j = 1, \cdots, n_i$, where G is the total number of genes of interest, I is the number of treatment groups, and n_i is the number of replicates for treatment i. In this section, we first introduce some transformation methods that have been applied to RNA-seq data, and then we discuss two models used to fit the count data directly.

10.2.1 Transformations Applied to RNA-seq Data

Gene expression measurements using RNA-seq technology are discrete counts. Typically, the majority of genes have low counts, and the distributions of data are right-skewed. Commonly used dissimilarity measures in cluster analysis, such as the Euclidean distance and correlation based distance (to be introduced in Sect. 10.3) may not work well for the count data because these dissimilarity measures are sensitive to outliers and skewness. There were several published cluster analyses of RNA-seq data applied to the log-transformed RPKM values [19, 24, 34] where RPKM stands for reads per kilobase of exon model per million mapped reads and is defined [23] as

$$RPKM_{gij} = \frac{10^9 \cdot Y_{gij}}{C_{ij} \cdot L_g} \tag{10.1}$$

where $C_{ij} = \sum_{g=1}^{G} Y_{gij}$ is the total count for the ijth sample, and L_g is the length of gene g in number of bases. RPKM was introduced by Mortazavi et al. [23] as a normalization method to make RNA-seq data comparable across samples and across genes because the read count Y_{gij} is influenced by the sequencing depth (which is measured by the total number of reads for each sample in RPKM) and the length of each gene. The RPKM values are no longer discrete counts, and log-transformation of RPKM values reduces the skewness of the data. The log-transformed RPKM values are closer to normally distributed than the original count data. Methods that work well for normally distributed data, such as hierarchical clustering and K-means clustering using Euclidean distance, have been applied to the log-transformed RPKM values [19, 24, 34]. Note that the original definition of RPKM uses the total number of reads for each sample, C_{ij}, to measure the sequencing depth [23]. Because this measure is affected by the most highly expressed genes that are only a small proportion of all genes [4], it may not be a good quantification of the sequencing depth. Several different estimates of the relative sequencing depth (the normalization factor) have been proposed and will be reviewed in Sect. 10.2.3. These normalization factors (s_{gij}) can be used as a substitute for C_{ij} in (10.1) to calculate normalized expression values. The average of transformed data can be calculated for each gene and treatment, and the resulting data can be subjected to gene-based clustering.

Another characteristic of RNA-seq data is that there exists an apparent mean-variance relationship [1]. It was noted that "it is more important to model the mean-variance relationship correctly than it is to specify the exact probabilistic distribution of the counts" [18]. Several efforts have been made to transform RNA-seq data to reduce or eliminate the mean-variance relationship. The `R` package, `DESeq` [1], provides a function that performs variance stabilizing transformation, and the `R` package, `limma` [35], includes a function `voom` that estimates the mean-variance relationship and generates a precision weight for each individual observation. It has been shown in differential expression analysis that both transformations followed by the normal-distribution based empirical Bayes method (`limma`, [35]) provide reasonably good results [38]. Such transformed data may also be used for cluster analysis.

10.2.2 Discrete Distributions Proposed for RNA-seq Data

Other than making efforts to transform the count data generated by RNA-seq experiments, there are also methods proposed to directly model the count data using discrete probability distributions. The Poisson distribution has been shown to be appropriate when only technical replicates are included [4, 22]. When there are biological replicates, RNA-seq data may exhibit more variability than expected by a Poisson distribution, which is known as the overdispersion phenomenon [1]. The negative binomial model has been proposed to handle the overdispersion problem [1,32]. Methods based on Poisson and negative binomial models have been developed recently by Witten [43] and Si et al. [37] for cluster analysis of RNA-seq data. In this subsection, we describe our parameterization of the two models and normalization for RNA-seq data.

10.2.2.1 Poisson Distribution

Suppose that Y_{gij} follows a Poisson distribution with mean λ_{gij} that is parameterized as

$$\log \lambda_{gij} = \log(s_{gij}) + \alpha_g + \beta_{gi} \quad (10.2)$$

with $\sum_{i=1}^{I} \beta_{gi} = 0$. The offset term s_{gij} is a normalization factor to adjust for varying sequencing depths and other technical effects across replicates and across genes. Several methods have been offered to estimate the normalization factor, and we will review them in Sect. 10.2.3. These reviewed methods estimate one normalization factor for each sample, i.e., $s_{gij} = s_{ij}$ for $g = 1, 2, \cdots, G$. To be more general, we still denote the normalization factor by s_{gij} just in case gene-specific features, such as gene length, are also involved in estimating the normalization factor.

10.2.2.2 Negative Binomial Distribution

The mean of the negative binomial distribution, λ_{gij}, is modeled the same way as in (10.2) for the Poisson distribution. As in Robinson and Smyth [31], the variance of the negative binomial distribution is parameterized as

$$\text{Var}(Y_{gij}) = \lambda_{gij} + \phi_g \lambda_{gij}^2. \quad (10.3)$$

Compared with the Poisson model, an extra dispersion parameter, ϕ_g, is introduced for each gene and this allows the negative binomial model to handle extra variability in the data as compared with the Poisson model. As $\phi_g \to 0$, the negative binomial distribution converges to the Poisson distribution [5]. Because the negative binomial model allows more flexibility in modeling the variance, and has been shown to fit RNA-seq data well [1], it has been used often in differential expression analysis of RNA-seq data [1, 15, 32, 36].

10.2.3 Normalization for RNA-seq Data

As in microarray data analysis, normalization needs to be done to remove or reduce systematic biases, such as sequencing depth, in order to make different samples comparable. Several methods have been proposed to normalize RNA-seq data through estimating the offset term s_{gij} in both the Poisson and the negative binomial models. A simple minded estimate of s_{gij} is the total number of reads for the ijth sample [25], C_{ij}. Because the total read count is largely influenced by those highly expressed genes that may only be a small proportion of all genes, Bullard et al. [4] proposed the upper quartile of gene counts within the ijth sample as a more robust estimate for s_{gij} where the upper quartile is calculated only using genes with non-zero reads in at least one sample. Similar to the upper quartile normalization method, two popularly applied `R` packages for RNA-seq data analysis, `DESeq` and `edgeR`, employ normalization methods that are also robust. The `DESeq` package uses a method that estimates s_{gij} by first calculating the ratio of the gene count in a sample to the geometric mean of this gene across samples and then obtaining the median of those ratios across genes for each sample [1]:

$$s_{gij} = s_{ij} = \texttt{median}_g \left(\frac{Y_{gij}}{(\prod_{i,j} Y_{gij})^{1/\sum_i n_i}} \right).$$

The Trimmed Mean of M Values (TMM) method [30] implemented in `edgeR` computes each normalization factor from a trimmed mean of the gene-wise log fold changes of the current sample to a reference sample. The reference [9] provides a comprehensive evaluation of normalization methods using several datasets, and the authors found that the median method of `DESeq` and the TMM method performed better than other methods, including the total count and the upper quartile methods. This observation is consistent with our experience in differential expression analysis [36]. Once estimated from data, the normalization factor is often treated as known in subsequent analysis [4, 22, 30].

10.3 Clustering Methods

Cluster analysis has been routinely applied in multivariate analysis to discover interesting patterns in a dataset. Many clustering methods tend to work well for data that follow the normal distribution, either exactly or approximately, in each true cluster. Examples of such algorithms include the K-means clustering [11,17] and the model-based clustering [12]. The hierarchical clustering [11,17] using the Euclidean distance is also a popular method for clustering continuous data. Such methods have been applied to transformed RNA-seq data [19, 34], and they will be introduced in Sect. 10.3.1. There have been two methods [37, 43] proposed for cluster analysis using RNA-seq count data directly; these will be described in Sect. 10.3.2.

10.3.1 Clustering Methods for Transformed RNA-seq Data

Cluster analysis has been applied to transformed RNA-seq data such as the log-transformed RPKM values [19, 34]. A variance stabilizing transformation or other functions that transform count data closer to normality could also be applied with the clustering methods discussed in this subsection. To cluster gene expression profiles, we first take the average transformed data over replicates for each treatment and each gene. The resulting vector of sample treatment means for each gene g, $\mathbf{X}_g$, is of length I and stores the features for gene (object) g to be used in cluster analysis.

10.3.1.1 Dissimilarity Measures

When clustering objects, we aim to put similar objects in the same cluster and dissimilar objects in different clusters. The dissimilarity measure determines how different two objects are. There are many choices of dissimilarity measures. Here, we discuss two that have been widely used in gene expression data analysis.

Suppose $\mathbf{X}_g$ and $\mathbf{X}_h$ are the I-dimensional data vectors for the gth and hth objects, respectively. Euclidean distance is one of the most commonly used dissimilarity measure in a cluster analysis:

$$d_E(\mathbf{X}_g, \mathbf{X}_h) = ||\mathbf{X}_g - \mathbf{X}_h|| = \sqrt{\sum_{i=1}^{I} (X_{gi} - X_{hi})^2}. \quad (10.4)$$

Euclidean distance measures the geometric distance between two objects, and it is sensitive to mean-shifting and scale-changing because the magnitude of each individual feature enters the calculation of this distance. In gene expression data analysis, biologists are often more interested in grouping the patterns of gene expression profiles, not necessarily requiring the expression magnitudes to be similar for genes in the same cluster. Thus, the Euclidean distance is often applied to mean-centered data, where the mean ($\bar{X}_g = \sum_{i=1}^{I} X_{gi}/I$) is subtracted from each element of the vector $\mathbf{X}_g$. It is also common to apply the Euclidean distance to standardized data, where standardization is done to each object by subtracting the mean and dividing by the standard deviation of the elements in $\mathbf{X}_g$.

Another commonly used dissimilarity measure is based on the Pearson correlation. The 1 minus correlation dissimilarity is defined as:

$$d_{corr}(\mathbf{X}_g, \mathbf{X}_h) = 1 - r_{\mathbf{X}_g,\mathbf{X}_h} = 1 - \frac{\sum_{i=1}^{I} (X_{gi} - \bar{X}_g)(X_{hi} - \bar{X}_h)}{\sqrt{\sum_{i=1}^{I} (X_{gi} - \bar{X}_g)^2} \sqrt{\sum_{i=1}^{I} (X_{hi} - \bar{X}_h)^2}}. \quad (10.5)$$

Using the dissimilarity defined in (10.5), genes with similar expression profiles tend to be clustered together, and hence it is one of the most popularly used dissimilarity measures in gene expression analysis. In fact, the Euclidean distance of the standardized data for $\mathbf{X}_g$ and $\mathbf{X}_h$ is equal to $\sqrt{2(I-1)}\sqrt{1 - r_{\mathbf{X}_g,\mathbf{X}_h}}$. So the relative rankings of similarity using the two distance measures are the same.

Although we use gene-based clustering to illustrate the dissimilarity measure, both the Euclidean distance and the Pearson-correlation based dissimilarity can be applied to sample-based clustering too. In sample-based clustering, the objects are the samples and each object has G features, one for each gene.

Both Euclidean distance and 1 minus correlation dissimilarity tend to be influenced by skewness or outliers in the data and may not work well for non-normally distributed data [16]. Although non-parametric measures of dissimilarity, such as distance based on the Spearman's rank correlation, may be used, such rank-based measures use substantially less information [16]. Therefore, the transformations discussed in Sect. 10.2.1 are recommended when using these Euclidean and 1 minus correlation dissimilarities.

10.3.1.2 *K*-Means Clustering

The K-means algorithm partitions the whole set of objects into K groups. Let $\mathbf{X}_1, \mathbf{X}_2, \ldots, \mathbf{X}_G$ denote the objects to be clustered (each $\mathbf{X}_g$ is an I-dimensional vector). Let $C(g)$ denote the cluster assignment for the gth object. For a given K, the K-means algorithm attempts to find a clustering of objects that minimizes

$$\sum_{k=1}^{K} \sum_{C(g)=k} ||\mathbf{X}_g - \bar{\mathbf{X}}_k||^2, \quad (10.6)$$

where $\bar{\mathbf{X}}_k = \sum_{C(i)=k} \mathbf{X}_i / n_k$ and n_k is the number of objects in the kth cluster.

The K-means algorithm starts from an initial partition of the objects (genes) and proceeds by iteratively calculating the centers (means) of clusters and reassigning each object to the closest cluster center. This iteration continues until no more reassignments take place. The initial clustering is often random, and several different runs of the K-means algorithm starting from different random initializations are recommended. The top K-means clustering solution used in practice is obtained by selecting the one that results in the smallest within-cluster sum of distances. The number of runs that results in the solution may be used to check the reliability of the solution. If it is only found in a few runs, better K-means clustering solutions may exist, and more runs may be needed [8].

The K-means algorithm is fast and has been included in many standard statistical software like R or popular tools for analyzing gene expression data such as "Gene Cluster" [8] or "Cluster and TreeView" [10]. Hence, it has been widely applied in gene expression data analysis. However, it requires a pre-specified K which is typically unknown in practice. Determining the correct number of clusters, K, is an interesting statistical problem and many solutions have been offered in the cluster analysis literature, such as using the silhouette width [33], the gap statistic [41] or the weighted rank aggregation method [28]. Ideally, different algorithms should arrive at the same K, the optimal number of clusters. However, in the analysis of real data that may consist of groups of genes whose expression profiles are not easily distinguished from each other, different algorithms may lead to different choices of K. Practitioners of K-means algorithm may use different choices of K in the hopes that the results will provide a more complete view of the data. Alternatively, hierarchical clustering can be applied.

10.3.1.3 Hierarchical Clustering

Hierarchical clustering algorithms use a different philosophy from the partitioning methods such as the K-means algorithm. Instead of partitioning all objects into a pre-defined number of groups, hierarchical clustering methods build a nested sequence of clusters that can be displayed using a dendrogram. On the dendrogram, the height of a node represents the dissimilarity between the two clusters merged together at the node (Fig. 10.1). The tree can be cut at different levels to generate the partition of objects into different numbers of clusters. Consequently, hierarchical clustering allows the flexibility of choosing different K's. In addition, the resulting dendrogram or tree structure reveals the relationships among clusters and such relationships are often interesting in studying functions of genes. Because of these advantages, hierarchical clustering has been popularly applied to both microarray and RNA-seq data analysis.

Two different algorithms can be used to generate the hierarchical tree. The agglomerative (bottom–up) method starts with each object in its own little cluster, and then successively merges clusters until only one large cluster remains which is the whole dataset. On the other hand, the divisive (top–down) method works in an opposite direction. It starts by considering the whole dataset as one cluster, and then splits up clusters until each object is separate. The divisive method requires higher computational cost, and thus it is not so popular in analysis of gene expression data that often include thousands of objects (genes) to be clustered. Here, we focus on

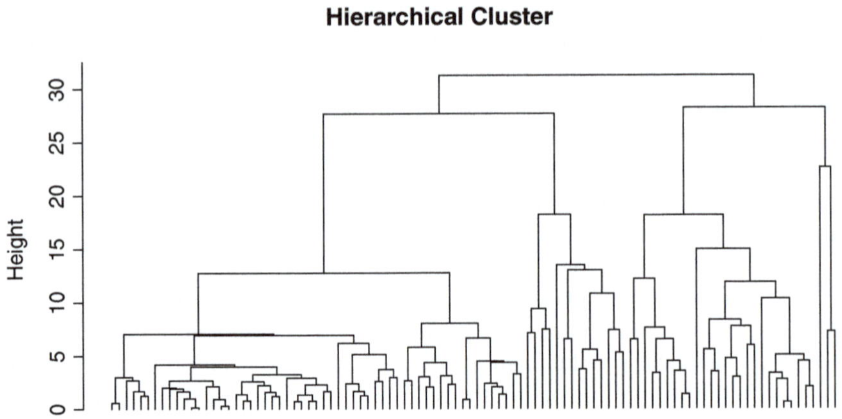

Fig. 10.1 The dendrogram of hierarchical clusters. The height of a node represents the dissimilarity between the two clusters merged together at the node

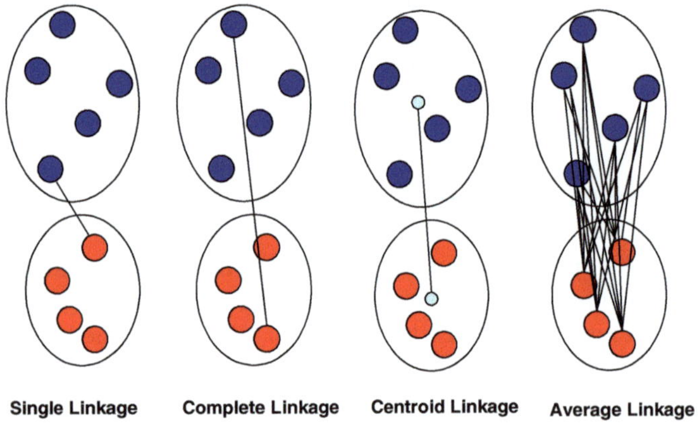

Fig. 10.2 The linkage methods that determine between-cluster dissimilarity

the agglomerative algorithm; readers are referred to [11] for more discussion on divisive algorithms.

Once a dissimilarity measure is determined, the first step of agglomerative method is to merge the pair of objects that have the smallest dissimilarity. When we have clusters with several objects, we need to define the dissimilarity between different clusters using linkage methods. Several linkage methods have been applied in hierarchical agglomerative clustering (Fig. 10.2). Suppose that we want to measure the dissimilarity between cluster A (shown in blue in Fig. 10.2) and cluster B (shown in red in Fig. 10.2).

- Single linkage uses the minimum of all dissimilarities from an object in cluster A to an object in cluster B to measure the dissimilarity between clusters A and B. It is also called the nearest neighbor approach.

- Complete linkage uses the maximum of all dissimilarities from an object in cluster A to an object in cluster B to measure the dissimilarity between clusters A and B. It is also called the farthest neighbor approach.
- Centroid linkage uses the dissimilarity between the centers of the clusters to measure the dissimilarity between them.
- The average linkage uses the average of all dissimilarities from an object in cluster A to an object in cluster B to measure the dissimilarity between clusters A and B.

Different linkage methods often result in different hierarchical trees. Discussion on properties and problems of these linkage methods can be found in [11, 17]. Single linkage method tends to result in long chains and cannot discern poorly separated clusters. On the other hand, complete linkage may produce distinct clusters when they do not really exist. Average linkage tends to produce the most appealing trees. When using single linkage and complete linkage, the dissimilarity between two clusters is measured by one pair of objects. Thus, different dissimilarity measures that have the same relative ranking of dissimilarities will result in the same clusters. For example, the Euclidean distance of standardized data and the 1 minus correlation dissimilarity will result in the same results when single linkage or complete linkage is used. However, when using average linkage method, two dissimilarity measures that have the same relative ranking may result in different final configurations of clusters.

10.3.2 Clustering Methods for Count Data

The K-means algorithm and the hierarchical algorithms using the dissimilarity measures described in previous subsections have been widely applied to gene expression data including transformed RNA-seq data. These methods provide reasonable results if data are normally distributed, and many studies of the properties of these methods also focus on normally distributed data [11]. However, the count data that arise from RNA-seq experiments are far from normally distributed for most genes. Another problem is that many genes may have low counts or zero counts in some treatment groups, and this introduces problems in the log transformation. Alternatively, model-based approaches using Poisson or negative binomial models can handle this problem easily. Recently, there have been methods proposed for clustering count data based on Poisson and negative binomial models [37, 43], and these methods will be described in this subsection.

10.3.2.1 Hierarchical Clustering Using Poisson Dissimilarity

The Euclidean distance works well for normal distributions. In fact, the likelihood ratio test statistic for testing the equality of two mean vectors of multivariate normal random variables [43] is a monotonic function of the Euclidean distance. In a similar fashion, Witten [43] defined the Poisson dissimilarity measure for RNA-seq data under the assumptions of Poisson distribution and independence among genes and replicates. Basically, the Poisson dissimilarity is calculated as the approximated log likelihood ratio statistic for testing whether two objects have the same mean vector or not (with appropriate normalization adjustment for RNA-seq data). The paper [43] focuses on sample-based clustering, not gene-based clustering. The Poisson dissimilarity expression in [43] is proposed specifically for clustering samples. Here, we follow the same idea and derive the Poisson dissimilarity for gene-based clustering.

Assuming gene g and gene h have Poisson read counts $Y_{gij} \sim Pois(s_{gij}\lambda_{gi})$ and $Y_{hij} \sim Pois(s_{hij}\lambda_{hi})$, respectively, where s_{gij}, s_{hij} are normalization factors and $\lambda_{gi}, \lambda_{hi}$ are the normalized gene expression levels under treatment i. Using our notation in (10.2), $\log(\lambda_{gi}) = \alpha_g + \beta_{gi}$, and $\log(\lambda_{hi}) = \alpha_h + \beta_{hi}$. The Poisson dissimilarity can be derived as the log likelihood ratio statistic to test the null hypothesis $H_0 : \lambda_{gi} = \lambda_{hi}$ for all $i = 1, 2, \cdots, I$:

$$d_{pois}(\mathbf{Y}_g, \mathbf{Y}_h) = \sum_i \left[Y_{gi} \log\left(\frac{Y_{gi}(S_{gi}+S_{hi})}{S_{gi}(Y_{gi}+Y_{hi})}\right) + Y_{hi} \log\left(\frac{Y_{hi}(S_{gi}+S_{hi})}{S_{hi}(Y_{gi}+Y_{hi})}\right)\right], \quad (10.7)$$

where $Y_{gi} = \sum_j Y_{gij}, S_{gi} = \sum_j s_{gij}$ and $Y_{hi} = \sum_j Y_{hij}, S_{hi} = \sum_j s_{hij}$. Here, we use the additive property of independent Poisson distributions which implies that $Y_{gi} \sim Pois(S_{gi}\lambda_{gi})$ and $Y_{hi} \sim Pois(S_{hi}\lambda_{hi})$. When the normalization factors are only sample-specific but not gene specific, i.e., $S_{gi} = S_{hi} = S_i$, (10.7) can be simplified to

$$d_{pois}(\mathbf{Y}_g, \mathbf{Y}_h) = \sum_i \left[Y_{gi} \log\left(\frac{2Y_{gi}}{Y_{gi}+Y_{hi}}\right) + Y_{hi} \log\left(\frac{2Y_{hi}}{Y_{gi}+Y_{hi}}\right)\right]. \quad (10.8)$$

Note that Y_{gi} or Y_{hi} can equal zero for genes that are not detected in ith treatment. However, the log of zero is not defined. Such problem exists both in gene-based and sample-based clustering. To avoid this problem, [43] replaces the maximum likelihood estimators (MLEs) of the mean parameters by the posterior means under a *Gamma* prior. Similar strategy could be applied for λ_{gi} and λ_{hi} here. More specifically, assuming that λ_{gi} and λ_{hi} follow a $Gamma(a, a)$ distribution with mean 1 as in [43], the posterior means for of λ_{gi} and λ_{hi} can be derived as $(Y_{gi}+a)/(S_{gi}+a)$ and $(Y_{hi}+a)/(S_{hi}+a)$. The author of [43] used $a = 1$ in all examples of that paper. These posterior means can be used to get a modified log likelihood ratio statistic that does not suffer from the problem of having to take the logarithm of zero.

As mentioned in Sect. 10.2, RNA-seq data with biological replicates often exhibit over-dispersion compared to the Poisson model. To use the Poisson dissimilarity, [43] proposed a power transformation introduced in [20]: $Y_{gi}^a \to Y'_{gi}$ where $a \in (0, 1]$ so that

$$\sum_{i=1}^{I}\sum_{g=1}^{G}\frac{(Y'_{gi}-Y'_{g\cdot}Y'_{\cdot i}/Y'_{\cdot\cdot})^2}{Y'_{g\cdot}Y'_{\cdot i}/Y'_{\cdot\cdot}} \approx (I-1)(G-1).$$

Such transformed data approximately follow a Poisson distribution, and formulas (10.7) and (10.8) still apply although the transformed data do not take on integer values.

Note that with the Poisson dissimilarity defined as (10.7) and (10.8), the goal is to cluster $\boldsymbol{\lambda}_g = (\lambda_{g1}, \lambda_{g2}, \cdots, \lambda_{gI})'$. If clustering gene expression profiles $\beta_g = (\beta_{g1}, \cdots, \beta_{gI})'$ is of interest, then the likelihood ratio statistics should be derived for testing this null hypothesis: $H_0 : \beta_{gi} = \beta_{hi}$ for all $i = 1, 2, \cdots, I$, where gene-specific α_g needs to be estimated under null hypothesis. There is no closed-form expression of the maximum-likelihood estimator of the parameters in this case, and the Poisson dissimilarity needs to be evaluated numerically.

Given the Poisson dissimilarity, hierarchical clustering may be applied with any linkage method described in Sect. 10.3.1.3. In [43], the author applied complete linkage with the Poisson dissimilarity derived for sample-based clustering, and found that the proposed Poisson dissimilarity outperformed the Euclidean distance whether applied directly to count data or applied to data transformed to stabilize variance using the `DESeq` method [1].

10.3.2.2 Model-Based Clustering

Different from the hierarchical clustering approach using a Poisson dissimilarity measure, Si et al. [37] proposed a partitioning method: model-based cluster analysis for count data generated by RNA-seq experiments. Studies of clustering algorithms with microarray data suggested that model-based algorithms perform better than heuristic algorithms such as the K-means method [45]. Extensive research has been done in model-based clustering with multivariate normal mixture distributions. See, for example, [13], for an excellent review among others. As described in Sect. 10.2, RNA-seq data are counts, and are typically fitted with Poisson or negative binomial distributions. In [37], a model-based method based on Poisson and negative binomial models is introduced for clustering gene expression profiles of RNA-seq data.

The mean of Poisson and negative binomial models are both expressed as λ_{gij} where $\log(\lambda_{gij}) = \log(s_{gij}) + \alpha_g + \beta_{gi}$ [model (10.2)]. As in [37], we focus on clustering gene expression profiles according to gene expression changes along different treatment groups, i.e., $\beta_g = (\beta_{g1}, \cdots, \beta_{gI})'$, because this is often more interesting than clustering the actual gene expression levels. The reference [37] explains how to change the clustering algorithm if the goal is to cluster on the basis of $\boldsymbol{\lambda}_g$ which define gene expression levels across I treatments.

Model-based clustering methods assume that data are generated by a mixture of probability distributions where each component corresponds to one cluster. We will first present the model-based clustering algorithm using a Poisson mixture model.

Then we discuss the algorithm for negative binomial mixture models. Suppose there are K clusters, and let $\mu_k = (\mu_{k1}, \cdots, \mu_{kI})'$ denote the center of cluster k with $\sum_{i=1}^{I} \mu_{ki} = 0$ for $k = 1, \cdots, K$. The likelihood of the Poisson mixture model for gene g is $\sum_k p_k f(\mathbf{Y}_g | \alpha_g, \beta_g = \mu_k)$, where $\mathbf{Y}_g = \{Y_{gij}\}$, $f(\mathbf{Y}_g | \alpha_g, \beta_g = \mu_k)$ is the likelihood if gene g belongs to the kth cluster and p_k is the mixing proportion with $p_k \geq 0$ and $\sum_{k=1}^{K} p_k = 1$. Taking all genes together, the likelihood is:

$$L = \prod_g \sum_k p_k f(\mathbf{Y}_g | \alpha_g, \beta_g = \mu_k). \tag{10.9}$$

Note that [37] assumed independence among genes as in other studies of gene expression data analysis [43, 45]. In reality, it is likely that some subsets of genes are dependent on each other. Currently there is little prior knowledge about the relationship among genes, and reliable estimation of the correlation structure cannot be done for tens of thousands of genes with only several replicates. For now, we proceed with the assumption of independence among genes in calculating the likelihood. Under this assumption, [37] proposed an Expectation-Maximization (EM) algorithm, as described below, to estimate the model parameters and cluster genes.

The MB-EM Algorithm

This entire algorithm effectively consists of the following two parts. Let the cluster indicator, Z_{gk}, be 1 if gene g belongs to the kth cluster and $Z_{gk} = 0$ otherwise. Taking all indicator variables $\mathbf{Z} = \{Z_{gk} : g = 1, \cdots, G; k = 1, \cdots, K\}$ as missing data and the EM algorithm proceeds by iteratively calculating the conditional expectations of $\mathbf{Z}$ and updating the estimates for model parameters until convergence. Here is the EM algorithm proposed in [37].

Part 1: The EM Algorithm

(i) *Initialization*: Set $p_k^{(1)}$ according to prior knowledge about the cluster size. If no such information is available, let $p_k^{(1)} = 1/K$ for $k = 1, \cdots, K$. Choose K vectors $\mu_1^{(1)}, \cdots, \mu_K^{(1)}$, with $\sum_{i=1}^{I} \mu_{ki}^{(1)} = 0$, for $k = 1, \cdots, K$, as the initial set of cluster centers. See Part 2 for one recommended way to choose these $\mu_k^{(1)}$. Obtain the initial values of $\boldsymbol{\alpha}^{(1)} = \{\alpha_{gk}^{(1)} : g = 1, \cdots, G; k = 1, \cdots, K\}$ by maximizing $f(\mathbf{Y}_g | \alpha_{gk}, \mu_k^{(1)})$ with respect to α_{gk} for each combination of gene g and cluster k.

(ii) *E-step*: Calculate the conditional expectation of Z_{gk} given data and parameters estimated from the mth step $(\mu^{(m)}, \mathbf{p}^{(m)}, \boldsymbol{\alpha}^{(m)})$, where $\mu^{(m)} = \{\mu_k^{(m)} : k = 1, \cdots, K\}, \mathbf{p}^{(m)} = \{p_k^{(m)} : k = 1, \cdots, K\}$, and $\boldsymbol{\alpha}^{(m)} = \{\alpha_{gk}^{(m)} : g = 1, \cdots, G; k =$

$1, \cdots, K\}$. To simplify the notation, we use $\hat{Z}_{gk}^{(m)}$ to denote the conditional expectation $\mathrm{E}(Z_{gk}|\mathbf{Y}_g, \mu^{(m)}, \mathbf{p}^{(m)}, \boldsymbol{\alpha}^{(m)})$ given by

$$\hat{Z}_{gk}^{(m)} = \frac{p_k^{(m)} f(\mathbf{Y}_g|\alpha_{gk}^{(m)}, \mu_k^{(m)})}{\sum_l p_l^{(m)} f(\mathbf{Y}_g|\alpha_{gl}^{(m)}, \mu_l^{(m)})}. \tag{10.10}$$

(iii) *M-step*: Update the parameter estimates by

$$\mu_k^{(m+1)} = \underset{\{\sum_i \mu_{ki}=0\}}{\operatorname{argmax}} \sum_g \hat{Z}_{gk}^{(m)} \log f\left(\mathbf{Y}_g \middle| \alpha_{gk}^{(m)}, \mu_k\right),$$

$$p_k^{(m+1)} = \sum_g \hat{Z}_{gk}^{(m)} / G,$$

and

$$\alpha_{gk}^{(m+1)} = \underset{\alpha_{gk}}{\operatorname{argmax}} f\left(\mathbf{Y}_g|\alpha_{gk}, \mu_k^{(m+1)}\right).$$

(iv) Return to step (ii) or stop the iteration if change of the total log-likelihood is small.
(v) For each $g = 1, \cdots, G$, assign gene g to cluster $k = \operatorname{argmax}_l \hat{Z}_{gl}$, where $\hat{Z}_{gl}$ is obtained by (10.10) after the convergence of above steps.

Note the EM algorithm not only assigns each gene g to a cluster k but also provides a measure of the uncertainty in the assignment by $1 - \hat{Z}_{gk}$. This measure provides additional information compared to hierarchical clustering and K-means clustering.

It is well known that the initialization of EM algorithm impacts both the speed of convergence and the final results [13, 14, 26]. Through simulation studies, the authors of [37] compared the initialization by random sampling from the whole set of objects with a model-based initialization of cluster centers for the EM algorithm, and reported that the model-based initialization dramatically improved the performance of the EM algorithm. The idea of the model-based initialization is to select initial cluster centers in a specific way such that they are well separated from each other with respect to a likelihood-based dissimilarity measure. In the context of gene-based clustering with a Poisson mixture model, the model-based initialization algorithm [37] is described below:

Part 2: Model-based Initialization for Cluster Centers

Initialization Step. Choose one gene randomly from all genes, and set the initial center for cluster 1, $\mu_1^{(1)}$, to be the maximum likelihood estimate (MLE) of β_g of the selected gene.

Addition Step. In the mth step, where $m = 2, 3, \cdots, K$, one additional cluster center is selected by the following procedures conditional on the current set of $(m-1)$ cluster centers.

(1) Calculate a measure of distance, $d_{g,l}$, between each gene g and each selected cluster center $\mu_l^{(1)}$ by

$$d_{g,l} = \log \frac{\max_{\alpha_g \in \mathscr{R}, \Sigma \beta_{gi}=0} f(\mathbf{Y}_g | \alpha_g, \beta_g)}{\max_{\alpha_g \in \mathscr{R}} f(\mathbf{Y}_g | \alpha_g, \beta_g = \mu_l^{(1)})},$$

for $g = 1, \cdots, G$ and $l = 1, \cdots, m-1$.

(2) Randomly select a gene from the multinomial distribution with probabilities $q_g = d_g^2 / \sum_{g'=1}^{G} d_{g'}^2$ for $g = 1, 2, \cdots, G$ and $d_g = \min\{d_{g,1}, \cdots, d_{g,(m-1)}\}$.

(3) Set the new cluster center, $\mu_m^{(1)}$, at $\hat{\beta}_g$, where $\hat{\beta}_g$ is the maximum likelihood estimate of β_g for the selected gene in (2).

Using Part 2, a total of K steps select the K initial cluster centers. Only the first cluster center is chosen uniformly at random, and the additional centers are selected one at a time based on the distance between each gene and each of the selected centers. The selection is done through the distance d_g and the selection probability q_g, which are defined such that a gene is more likely to be selected if it is far away from all existing centers. Hence the K centers chosen by this algorithm are expected to be separated better than a set of centers that are randomly selected.

In [37], the authors also evaluated two stochastic versions of the EM algorithm, the simulated annealing (SA) and the deterministic annealing (DA) algorithms. In their simulation study, these stochastic versions did not show further improvement over the EM algorithm initialized by the model-based initialization.

Model-Based Clustering Using the Negative Binomial Mixture Model

The model-based clustering method using the MB-EM algorithm can be applied to a negative binomial mixture model too. Compared with the Poisson model, an extra parameter, ϕ_g, is introduced for each gene in the negative binomial model, and this dispersion parameter needs to be estimated for each gene. The reference [31] describes several methods to estimate ϕ_g. The paper [37] recommends first estimating ϕ_g for each gene g using methods such as the quasi-likelihood (QL) method, and then treating the estimated ϕ_g as known when calculating the likelihood for negative binomial model. By this means, the unknown parameters are the same for the Poisson and the negative binomial models. Then the MB-EM algorithm can be directly applied to the negative binomial mixture model, where the likelihood function $f(\mathbf{Y}_g | \alpha_g, \beta_g)$ is evaluated using the negative binomial probability mass function with estimated ϕ_g for gene g. Through simulation studies, Si et al. [37] found that the performance of model-based clustering using such estimated dispersion was very close to the performance when true dispersion parameter values were used to calculate the likelihood in the EM-algorithm.

Choosing the Number of Clusters

As in K-means clustering, the model-based clustering algorithm is also a partitioning algorithm and requires the specification of the number of partitions, K. The silhouette width [33] or the gap statistic [41] can both be applied here. Because the model-based clustering is a likelihood-based approach, criteria used for model-selection can also be applied. The authors of [37] recommended using the Akaike information criterion (AIC) defined by $-2(\log L - n_p)$, where L is the likelihood and is calculated by (10.9) and n_p is the number of parameters in the model. A low value of AIC indicates a better model. In [37], it was shown that AIC identified the true number of clusters in their simulation studies, and AIC also provided a reasonable number of clusters in their real data analysis.

10.3.2.3 Hybrid-Hierarchical Clustering Algorithm

The partitioning algorithms such as a model-based method or the K-means method group objects into a pre-specified number of groups. Such algorithms are fast, but the relationships between clusters are not revealed. In addition, choosing the number of clusters, K, is challenging, and different algorithms might result in different K values. On the other hand, the results of hierarchical clustering provide information about the relationships of clusters and allows flexibility of obtaining different number of clusters by cutting the tree at different levels. However, the computational cost is high because the number of genes included in the cluster analysis is often large. To borrow strength from both hierarchical clustering and partitioning methods, hybrid-hierarchical (HH) clustering algorithms have been introduced [42, 46]. Following such an idea, a model-based HH clustering has been proposed by combining the model-based and hierarchical clustering methods [37].

The model-based HH clustering algorithm proposed in [37] uses an agglomerative (bottom-up) strategy starting with K_0 clusters, where K_0 is a number relatively large to allow enough resolution but far less than the number of genes, G. The model-based clustering algorithm is used to obtain the initial set of K_0 clusters. Then two clusters with the smallest 'distance' among all possible pairs are merged at each of the following steps. After $K_0 - 1$ steps, all genes belong to a single cluster and the hierarchical tree is built up. Here, the term 'hybrid' is used to point out that the HH algorithm combines the starting steps that obtain K_0 clusters using non-hierarchical methods and the merging steps that are similar to ordinary hierarchical clustering.

In [37], the distance between two clusters is measured by the reduction of total log-likelihood from before to after the merger of two clusters. Hence, merging clusters with the minimal distance aims to achieve the maximum log-likelihood in each step. The hybrid hierarchical clustering algorithm may also combine the K-means method and hierarchical clustering methods as introduced in Sect. 10.3.2.

10.4 Case Studies

In this section, we apply the clustering methods described in Sect. 10.3 to a maize RNA-seq dataset generated by Li et al. [19]. This dataset quantifies transcript abundance of four segments along a leaf developmental gradient, with two biological replicates for each segment. The total number of genes detected in this experiment is more than 50,000. For the purpose of cluster analysis, we only use the list of 12,631 genes that were identified as differentially expressed among the four segments using generalized linear model analysis based on the negative binomial distribution [19, 37].

10.4.1 Results of Clustering the Count Data

We first the applied the MB-EM algorithm to the count data based on mixture of negative binomial models as described in Sect. 10.3.2.2. The algorithm was run for $K = 2, 3, \cdots, 50$, and AIC reached its minimum at $K = 15$ (Fig. 10.3a). When $K = 15$, the resulting clusters from the model-based algorithm are presented in Fig. 10.3b where each grey line plots the expression profile across the four segments of a gene, and the red line corresponds to the cluster center.

We also applied the hybrid-hierarchical approach as described in Sect. 10.3.2.3. The list of 12,631 differentially expressed genes were first grouped into 100 small clusters using the MB-EM algorithm. Genes in each cluster had almost identical expression profiles at $K = 100$ based on visual inspection. This indicates that clustering the 100 small clusters likely leads to similar results to those from clustering all the genes. Of course, one may increase $K = 100$ to a bigger number, which may give better resolution. With the 100 small clusters, the hierarchical clustering algorithm was applied with dissimilarity between clusters defined by the likelihood reduction. Figure 10.4 depicts the dendrogram and the heatmap of gene expression profiles. The cluster tree may be cut at different levels to obtain partitions of the genes. Based on the heatmap, there are three major clusters depending on whether the genes were more abundantly expressed in segment 4 (the major cluster on the left), or in the middle segments (the middle cluster) or in the basal segments, segments 1 and 2 (the major cluster on the right).

10.4.2 Results of Clustering Transformed Data

We also applied the K-means and hybrid hierarchical methods to transformed data. Let X_{gi} be the average of log-transformed RPKM values across replicates for gene g and treatment i. To avoid taking logarithm of zero, we added a small constant to those RPKM values at zero before log transformation. Note that the

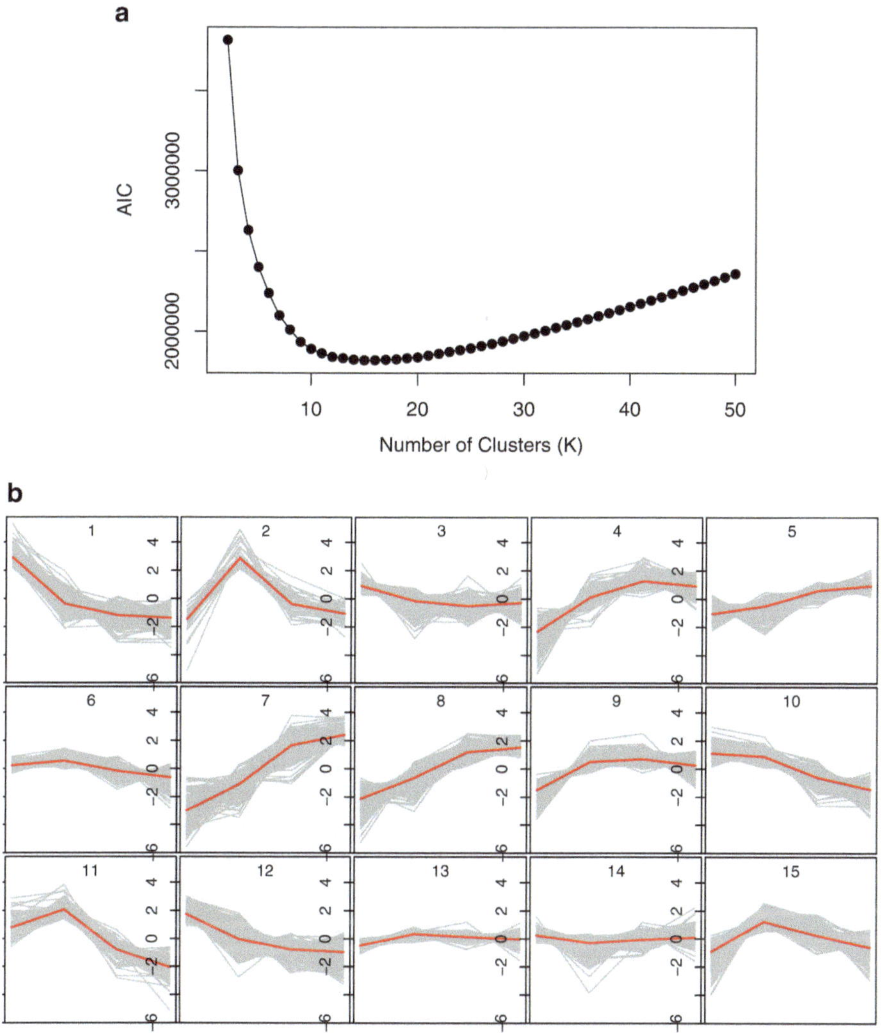

Fig. 10.3 Model-based clustering using the negative binomial model and EM algorithm. (**a**) AIC scores based on the negative binomial model for different number of clusters. (**b**) Clustering results from model-based methods using the negative binomial model and $K = 15$. Each *grey line* corresponds to the expression profile measured using mean-centered, replicate-averaged, and log-transformed normalized expression values. The *red line* corresponds to the profile defined by each cluster center

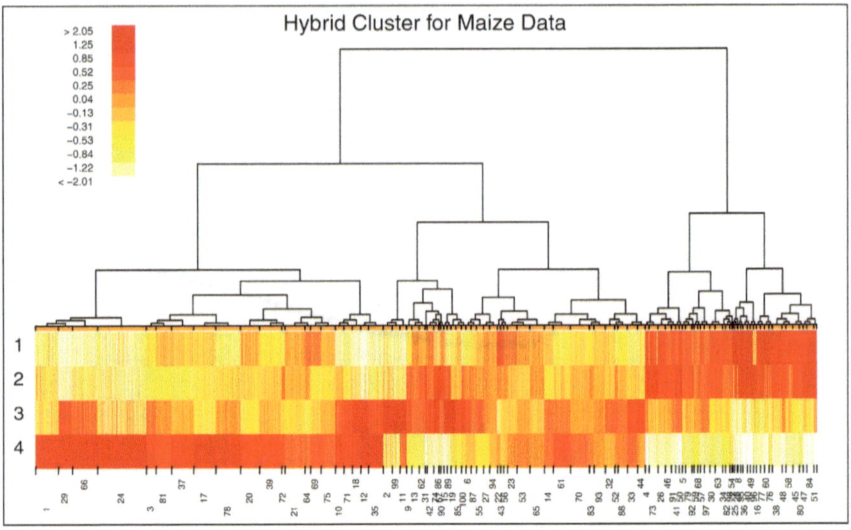

Fig. 10.4 Hybrid-Hierarchical clustering results using the model-based method based on the negative binomial model. The hierarchical structure starts from 100 small clusters. Each number labeled at the bottom of the graph represents one of the 100 small clusters

model-based method did not require such arbitrary modification of data. We also subtracted the mean $\bar{X}_g = \sum_{i=1}^{I} X_{gi}$ from each X_{gi} before clustering in order to group the patterns of gene expression changes along the leaf segments. Then, we applied the K-means method to partition the list of genes into $K = 15$ groups using Euclidean distance. A plot similar to Fig. 10.3b may be presented. One thing to keep in mind is that the results depend on the initial partition, and different initialization often ends up with different clustering results using the K-means method. So in practice, we recommend running the K-means algorithm several times using different initializations. The Part 2 of the MB-EM algorithm described in Sect. 10.3.2.2 may be applied with appropriate dissimilarity measures in selecting the initial objects.

A hierarchical clustering algorithm may also be applied to this dataset. Considering the dimension of objects (genes), we instead performed a hybrid-hierarchical clustering. First, we used the K-means method with Euclidean distance to cluster the transformed data into 100 small clusters. Visual inspection indicated that genes in each cluster shared very similar patterns. Then we applied hierarchical clustering with Euclidean distance and complete linkage. The results could be plotted as in Fig. 10.4, and the tree could be truncated at some level to obtain a partitioning of genes. The results obtained from the model-based method, K-means, and hierarchical clustering are presented in the next subsection.

10.4.3 Comparison Between Different Methods

In simulation studies, the clustering results have been evaluated based on the comparison between the true partition of objects used to simulate data and the resulting partition obtained by a clustering algorithm. Better performance is indicated by more agreement between the two partitions. The following criteria have been used to evaluate the agreement. For all four statistics listed below, higher values indicate better performance.

- *Pairwise Sensitivity*: the proportion of pairs of genes (objects) that are clustered together among all pairs that had the same original assignment [2, 37, 44].
- *Pairwise Specificity*: the proportion of pairs of genes (objects) that are clustered to different groups among all pairs that had different original assignment [2,37,44].
- *Rand Index*: the proportion of pairs of genes (objects) that are correctly clustered together and that are correctly clustered into different clusters out of all possible pairs of objects. Let M be the number of pairs of genes (objects) that are in the same cluster in the true partition and in the resulting partition, and let N be the number of pairs of genes (objects) that are in different clusters in the true partition and in the resulting partition. Then the Rand Index is defined to be $(M+N)/\binom{G}{2}$ where G is the total number of objects. The author of [43] used the clustering error rate to evaluate clustering results, which is just one minus the rand index.
- *Normalized Mutual Information (NMI)*: Mutual information (MI) is used in information theory to measure the amount of information one random variable contains about another, or equivalently, the reduction in the uncertainty of one due to the knowledge of the other. Here, MI is used to quantify the shared information between the true partition and the clustering result. Let u be a cluster given by the true partition $\mathscr{U}$, v be a cluster given by the clustering result $\mathscr{V}$, and p_u, p_v, p_{uv} be the proportions of genes in u, v, and their intersection, respectively. The MI between $\mathscr{U}$ and $\mathscr{V}$ is defined as $MI = \sum_{u,v} p_{uv} \log \frac{p_{uv}}{p_u p_v}$. The MI value is high if there is strong dependence (more shared information) between the two partitions, and is close to zero otherwise. Since there is no upper bound for MI, it is normalized to a range between 0 and 1 by dividing by the total entropy. The normalized MI (NMI) between two partitions $\mathscr{U}$ and $\mathscr{V}$ is $NMI = 1 - \frac{\sum_{u,v} p_{uv} \log p_{uv}}{\sum_u p_u \log p_u + \sum_v p_v \log p_v}$, where $0 \times \log 0 \equiv 0$ if encountered. NMI is often desirable for easier comparison [39].

These four criteria are presented to evaluate how one clustering algorithm performs by comparing its result with the true partition that is used to generate data. They could also be used to compare the concordance of two clustering results. In real data analysis, there is no knowledge of true partition. To evaluate the performance of different clustering algorithms, [6, 7] proposed measures for statistical consistency (stability) of the clustering results and measures for the biological congruence of the clustering results. Similar to the idea of using biological functions to validate clustering results, [37] compared clustering results with the gene functional categories. Their rationale is that genes within the same functional category tend to

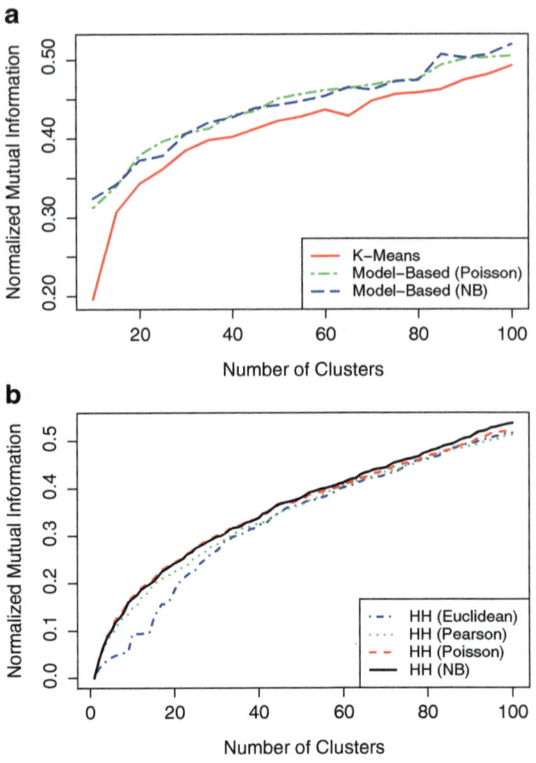

Fig. 10.5 NMI between clustering results and gene annotations. (**a**) *K*-means vs. model-based clustering. (**b**) Comparison of different hybrid-hierarchical clustering methods

have correlated expression patterns and thus are more likely to be grouped together. Here we use the same strategy by comparing gene annotations from Mapman as described in [19] and clustering results. Excluding categories that contain less than five or more than 500 genes, we have 306 non-overlapping categories with a total of 5,002 genes. We quantify the concordance between clustering results and gene functional categories for these genes using NMI.

We first compare the results of partitioning methods. For K ranging from 10 to 100 in steps of 5, we performed the K-means method using Euclidean distance, and performed model-based clustering methods based on Poisson and negative binomial distributions, respectively. Figure 10.5a shows that the NMI scores of the model-based algorithms are higher than the NMI score of K-means method for all K. This indicates that the groups partitioned by model-based methods agree more with the gene categories than the K-means method. We also applied hybrid-hierarchical clustering by first partitioning the 5,002 genes into 100 small clusters using K-means and then using average linkage in based on Euclidean or 1 minus correlation distance, and applied the HH algorithm introduced in Sect. 10.3.2.3 based on Poisson and negative binomial likelihood functions. Figure 10.5b shows that hybrid-hierarchical methods perform better in terms of NMI when the distances are based on likelihoods. The figures also show that the scores for model-based methods using

Poisson and negative binomial models are similar. This is likely because that this dataset only has two replicates for each condition, and the difference between the fitting of the Poisson model and that of the negative binomial distribution is not big.

10.5 Implementation of Clustering Methods

All methods introduced in this chapter may be implemented using the statistical software `R`. K-means and hierarchical clustering are implemented in functions `kmeans()` and `hclust()`, respectively, included with the base distribution of `R`. Several packages, such as the package `amap` by [21], can also be applied to perform K-means and hierarchical clustering. The package `MBCluster.Seq` is designed by [37] to perform the model-based clustering algorithms described in Sect. 10.3.2.2. Packages `clValid` [3] and `RankAggreg` [29] contain functions for validating the results of a clustering analysis. All the above-mentioned packages can be downloaded from The Comprehensive R Archive Network (CRAN).

The functions `Kmeans` and `hcluster` in the R package `amap` are straightforward to apply in order to perform the K-means method and the hierarchical clustering method, respectively. One could also perform hybrid hierarchical clustering by first partitioning all objects into K groups using `Kmeans` and then building the hierarchical tree using the `hcluster` function.

The main functions in the package `MBCluster.Seq` to perform model-based clustering include the following.

- `RNASeq.Data()` organizes the data into the format used by this package. For example:

  ```
  > mydata=RNASeq.Data(Count=counts,Normalize=log(scalar),
   Treatment=treats,GeneID=geneID)
  ```

  where the matrix `counts` stores the RNA-seq data with columns corresponding to treatments specified by the vector `treats` and rows corresponding to genes specified by the vector `geneID`. The argument `scalar` stores the normalization factors, one for each column, that may be estimated by the methods introduced in Sect. 10.2.3.
- `KmeansPlus.RNASeq()` selects the objects in the initialization step by part 2 of the MB-EM algorithm. For example, to select 100 initial cluster centers:

  ```
  > c0=KmeansPlus.RNASeq(mydata,nK=100)$centers
  ```

- `Cluster.RNASeq()` performs model-based clustering algorithm using the EM or other stochastic versions of the EM algorithm given the number of clusters. For example, to obtain $K = 15$ clusters:

  ```
  > cls=Cluster.RNASeq(data=mydata,model="nbinom",
   centers=c0[1:15,],method="EM")$cluster
  ```

  The output stores the cluster IDs for each gene.

- `lglk.cluster()` calculates the log-likelihood given the clustering results from `Cluster.RNASeq`. For example:

```
> lglk=lglk.cluster(mydata,model="nbinom",cluster=cls)
```

  The resulting log-likelihood can be used to calculate AIC values to select the number of clusters, K.
- For model-based hybrid-hierarchical clustering, `Hybrid.Tree()` performs the hierarchical clustering after obtaining the results from `Cluster.RNASeq()`. For example:

```
> tr=Hybrid.Tree(data=mydata,cluster=cls,model="nbinom")
```

  where `cls` stores the initial set of clusters obtained by `Cluster.RNASeq()`.
- The results of model-based hybrid-hierarchical clustering can be plotted using `plotHybrid.Tree()`. For example:

```
> plotHybrid.Tree(merge=tr,cluster=cls,logFC=mydata$logFC,
tree.title="Hybrid Cluster for Maize Data",colorful=TRUE)
```

  where `mydata$logFC` stores the normalized data $(X_{gi} - \bar{X}_g)$ as discussed in Sect. 10.4.2.

10.6 Discussion

Cluster analysis is a multivariate analysis technique that has been routinely applied in gene expression studies. Although this chapter focuses on the gene-based clustering, the methods described here can be applied to sample-based clustering with the appropriate switch of dimension and model specification. In both applications, the dimension of genes is high in RNA-seq datasets, and the genes are typically not interesting if their expression levels do not changes across treatment groups. Hence, we recommend to use only the subset of differentially expressed genes in cluster analysis to reduce computational cost and noise in the dataset.

This chapter introduces several algorithms for cluster analysis of RNA-seq data. Unless the dataset includes very distinct clusters, different algorithms usually lead to different groupings. For the partitioning methods that depend on the initialization, different initial sets may also result in different groupings. Because of such uncertainties in the results, cluster analysis is not used to draw inference but is more of a way to organize and visualize data. In practice, it is reasonable to apply different algorithms to the same dataset. If all algorithms generate similar clusters, the results are more reliable and worth further investigating. Widely different groupings suggest that no distinct cluster structure exists in the dataset.

In this chapter, we have described the K-means method, hierarchical clustering algorithms, and finite mixture model-based methods. Model-based methods offer a more unified approach compared with the other two. Model-based methods rely on a model-selection criterion to select the number of clusters and provide uncertainties of the clustering results. In cluster analysis of the maize dataset illustrated in

Sect. 10.4, the model-based method provided good performance. Still, there is room to improve this method. First, the estimation of parameters may be improved using some shrinkage method. RNA-seq data is one case of "large p, small n" data, where there are a large number of variables and a small number of replicates. Shrinkage methods have been shown to improve the parameter estimation and performance of hypothesis tests that identify differentially expressed genes. Better estimation of the parameters might improve the performance of model-based clustering methods. Second, current models do not consider random effects. Some experimental designs, such as split-plot designs, result in random effects that should be accounted for. Also, some experiments might have correlated samples, such as experiments with repeated measures. It is a challenging question to estimate the parameters for mixtures of log-linear models with random effects. More statistical research in this area is warranted.

References

[1] Anders, S., Huber, W.: Differential expression analysis for sequence count data. Genome Biol. **11**, R106 (2010)

[2] Booth, J., Casella, G., Hobert, J.: Clustering using objective functions and stochastic search. J. Roy. Stat. Soc. Ser. B Stat. Meth. **70**, 119–139 (2008)

[3] Brock, G., Pihur, V., Datta, S., Datta, S.: clValid, an R package for cluster validation. J. Stat. Software **25**, 4 (2008)

[4] Bullard, J., Purdom, E., Hansen, K., Dudoit, S.: Evaluation of statistical methods for normalization and differential expression in mrna-seq experiments. BMC Bioinform. **11**, 94 (2010)

[5] Cameron, A.C., Trivedi, P.K.: Regression Analysis of Count Data. Cambridge University Press, Cambridge (1998)

[6] Datta, S., Datta, S.: Comparisons and validation of statistical clustering techniques for microarray gene expression data. Bioinformatics **19**(4), 459–466 (2003)

[7] Datta, S., Datta, S.: Evaluation of clustering algorithms for gene expression data. BMC Bioinform. **7**(Suppl 4), S17 (2006)

[8] de Hoon, M.J.L., Imoto, S., Nolan, J., Miyano, S.: Open source clustering software. Bioinformatics **20**(9), 1453–1454 (2004)

[9] Dillies, M.A., Rau, A., Aubert, J., Hennequet-Antier, C., Jeanmougin, M., Servant, N., Keime, C., Marot, G., Castel, D., Estelle, J., et al.: A comprehensive evaluation of normalization methods for Illumina high-throughput RNA sequencing data analysis. Brief. Bioinformatics **14**(6), 671–683 (2013)

[10] Eisen, M.: Software: Cluster and TreeView (2002). http://rana.lbl.gov/EisenSoftware.htm

[11] Everitt, B.S.: Cluster Analysis, 3rd edn. Edward Arnold, London (1993)

[12] Fraley, C.: Algorithms for model-based gaussian hierarchical clustering. SIAM J. Sci. Comput. **20**(1), 270–281 (1999)

[13] Fraley, C., Raftery, A.: Model-based clustering, discriminant analysis, and density estimation. J. Am. Stat. Assoc. **97**, 611–631 (2002)

[14] Hall, L., Özyurt, I., Bezdek, J.: Clustering with a genetically optimized approach. IEEE Trans. Evol. Comput. **3**, 103–112 (1999)

[15] Hardcastle, T.J., Kelly, K.A.: baySeq: empirical Bayesian methods for identifying differential gene expression in sequence count data. BMC Bioinform. **11**, 422 (2010)

[16] Jiang, D., Tang, C., Zhang, L.: Cluster analysis for gene expression data: a survey. IEEE Trans. Knowledge Data Eng. **16**(11), 1370–1386 (2004)
[17] Johnson, R.A., Wichern, D.W.: Applied Multivariate Statistical Analysis, 5th edn. Prentice Hall, Englewood Cliffs (2002)
[18] Law, C.W., Chen, Y., Shi, W., Smyth, G.K.: Voom! precision weights unlock linear model analysis tools for RNA-seq read counts. Genome Biol. **15**(2), R29 (2014)
[19] Li, P., Ponnala, L., Gandotra, N., Wang, L., Si, Y., Tausta, S., Kebrom, T., Provart, N., Patel, R., Myers, C., Reidel, E., Turgeon, R., Liu, P., Sun, Q., Nelson, T., Brutnell, T.: The developmental dynamics of the maize leaf transcriptome. Nat. Genet. **42**, 1060–1067 (2010)
[20] Li, J., Witten, D.M., Johnstone, I.M., Tibshirani, R.: Normalization, testing, and false discovery rate estimation for RNA-sequencing data. Biostatistics **13**(3), 523–538 (2012)
[21] Lucas, A.: amap: another multidimensional analysis package. Available from the Comprehensive R Archive Network http://cran.r-project.org/web/packages/amap/index.html (2013)
[22] Marioni, J.C., Mason, C.E., Mane, S.M., Stephens, M., Gilad, Y.: RNA-seq: an assessment of technical reproducibility and comparison with gene expression arrays. Genome Res. **18**, 1509–1517 (2008)
[23] Mortazavi, A., Williams, B.A., McCue, K., Schaeffer, L., Wold, B.: Mapping and quantifying mammalian transcriptomes by RNA-seq. Nat. Meth. **5**, 621–628 (2008)
[24] O'Rourke, J.A., Yang, S.S., Miller, S.S., Bucciarelli, B., Liu, J., Rydeen, A., Bozsoki, Z., Uhde-Stone, C., Tu, Z.J., Allan, D., Gronwald, J.W., Vance, C.P.: An RNA-Seq transcriptome analysis of orthophosphate-deficient white lupin reveals novel insights into phosphorus acclimation in plants. Plant Physiol. **161**(2) 705–724 (2013)
[25] Oshlack, A., Robinson, M.D., Young, M.D.: From RNA-seq reads to differential expression results. Genome Biol. **11**, 220 (2010)
[26] Park, H., Yoo, S., Cho, S.: Evolutionary fuzzy clustering algorithm with knowledge-based evaluation and applications for gene expression profiling. J. Comput. Theor. Nanosci. **2**, 1–10 (2005)
[27] Pickrell, J., Marioni, J., Pai, A., Degner, J., Engelhardt, B., Nkadori, E., Veyrieras, J., Stephens, M., Gilad, Y., Pritchard, J.K.: Understanding mechanisms underlying human gene expression variation with RNA sequencing. Nature **464**, 768–772 (2010)
[28] Pihur, V., Datta, S., Datta, S.: Weighted rank aggregation of cluster validation measures: a Monte Carlo cross-entropy approach. Bioinformatics **23**, 1607–1615 (2007)
[29] Pihur, V., Datta, S., Datta, S.: RankAggreg, an R package for weighted rank aggregation. BMC Bioinform. **10**, 62 (2009)
[30] Robinson, M.D., Oshlack, A.: A scaling normalization method for differential expression analysis of RNA-seq data. Genome Biol. **11**, R25 (2010)
[31] Robinson, M.D., Smyth, G.K.: Small-sample estimation of negative binomial dispersion, with applications to sage data. Biostatistics **9**, 321–332 (2008)
[32] Robinson, M.D., McCarthy, D.J., Smyth, G.K.: edgeR: a bioconductor package for differential expression analysis of digital gene expression data. Bioinformatics **26**(1), 139–140 (2010)
[33] Rousseeuw, P.J.: Silhouettes: a graphical aid to the interpretation and validation of cluster analysis. J. Comput. Appl. Math. **20**, 53–65 (1987)
[34] Severin, A.J., Woody, J.L., Bolon, Y-T, Joseph, B., Diers, B.W., Farmer, A.D., Muehlbauer, G.J., Nelson, R.T., Grant, D., Specht, J.E., Graham, M.A., Cannon, S.B., May, G.M., Vance, C.P., Shoemaker, R.C.: RNA-seq atlas of glycine max: a guide to the soybean transcriptome. BMC Plant Biol. **10**, 160 (2010)
[35] Smyth, G.K.: Linear models and empirical Bayes methods for assessing differential expression in microarray experiments. Stat. Appl. Genet. Mol. Biol. **3**(1), Article 3 (2004)
[36] Si, Y., Liu, P.: An optimal test with maximum average power while controlling FDR with application to RNA-seq data. Biometrics **69**, 594–605 (2013)
[37] Si, Y., Liu, P., Li, P., Brutnell, T.: Model-based clustering of RNA-seq data. Bioinformatics **30**(2), 197–205 (2014)
[38] Soneson, C., Delorenzi, M.: A comparison of methods for differential expression analysis of RNA-seq data. BMC Bioinform. **14**, 91 (2013)

[39] Strehl, A., Ghosh, J.: Cluster ensembles - a knowledge reuse framework for combining partitions. J. Mach. Learn. Res. **3**, 583–617 (2002)

[40] Sultan, M., Schulz, M.H., Richard, H.: A global view of gene activity and alternative splicing by deep sequencing of the human transcriptome. Science **321**, 956–960 (2008)

[41] Tibshirani, R., Walther, G., Hastie, T.: Estimating the number of clusters in a data set via the gap statistic. J. Roy. Stat. Soc. Ser. B Stat. Meth. **63**, 411–423 (2001)

[42] Vaithyanathan, S., Dom, B.: Model-based hierarchical clustering. In: Proceedings of the 16th Conference on Uncertainty in Artificial Intelligence, pp. 599–608 (2000)

[43] Witten, D.M.: Classification and clustering of sequencing data using a poisson model. Ann. Appl. Stat. **5**(4), 2493–2518 (2011)

[44] Woodard, D., Goldszmidt, M.: Model-based clustering for online crisis identification in distributed computing. J. Am. Stat. Assoc. **106**(493), 49–60 (2011)

[45] Yeung, K., Fraley, C., Murua, A., Faftery, A., Ruzzo, W.: Model-based clustering and data transformations for gene expression data. Bioinformatics **17**(10), 977–987 (2001)

[46] Zhong, S., Ghosh, J.: A unified framework for model-based clustering. J. Mach. Learn. Res. **4**, 1001–1037 (2003)

Chapter 11
Classification of RNA-seq Data

Kean Ming Tan*, Ashley Petersen*, and Daniela Witten

Abstract Next-generation sequencing technologies have made it possible to obtain, at a relatively low cost, a detailed snapshot of the RNA transcripts present in a tissue sample. The resulting reads are usually binned by gene, exon, or other region of interest; thus the data typically amount to read counts for tens of thousands of features, on no more than dozens or hundreds of observations. It is often of interest to use these data to develop a classifier in order to assign an observation to one of several pre-defined classes. However, the high dimensionality of the data poses statistical challenges: because there are far more features than observations, many existing classification techniques cannot be directly applied. In recent years, a number of proposals have been made to extend existing classification approaches to the high-dimensional setting. In this chapter, we discuss the use of, and modifications to, logistic regression, linear discriminant analysis, principal components analysis, partial least squares, and the support vector machine in the high-dimensional setting. We illustrate these methods on two RNA-sequencing data sets.

11.1 Introduction

In the past 15 years, much effort has focused on characterizing the *transcriptome*—the identity and quantity of all transcripts in a cell or population of cells—under a variety of conditions and disease states. The first technique that was widely-used to quantify the transcriptome was the microarray, which uses hybridization probes in order to measure the relative expression of transcripts in a cell. This technology is now relatively inexpensive and widespread, but it suffers from some weaknesses. First, the hybridization probes must be selected a priori, so only known

* indicates joint first authorship.

K.M. Tan (✉) • A. Petersen (✉) • D. Witten
Department of Biostatistics, University of Washington, Seattle, WA 98195-7232, USA
e-mail: keanming@uw.edu; ajpete@uw.edu; dwitten@uw.edu

S. Datta and D. Nettleton (eds.), *Statistical Analysis of Next Generation Sequencing Data*, Frontiers in Probability and the Statistical Sciences, DOI 10.1007/978-3-319-07212-8_11,

transcripts can be measured, and novel transcripts cannot be discovered. Second, cross-hybridization can occur, in which an unintended molecule binds to the probe; this results in an imprecise measure of expression of each transcript. Today, with the advent of *RNA-sequencing* (RNA-seq), a more sensitive and complete quantification of the transcriptome is possible [32, 38, 58, 66, 71, 81, 90].

Transcriptomic data (often referred to as *gene expression* data) can be used for inference and prediction. For instance, suppose that RNA-sequencing has been performed on a set of n tissue samples (observations), each of which belongs to one of K pre-specified classes (such as cancer versus normal, or Disease A versus Disease B versus Disease C). A common inferential goal is to identify the features (e.g. genes or exons) that are *differentially expressed* across the classes—that is, those that have higher or lower mean expression among the tissue samples belonging to a certain class, as compared to the baseline. However, in this chapter, our goal concerns prediction—specifically, classification. Given the class labels and feature measurements for a set of observations, how can we predict the class membership for a new observation for which we know the gene expression measurements but not the class membership?

11.1.1 RNA-Sequencing Data

Here we briefly describe the steps involved in obtaining RNA-seq data. First, the RNA from a tissue sample is isolated and converted into cDNA. This cDNA is then fragmented and directly sequenced using next-generation sequencing. Once the sequencing is complete, the reads are mapped to a reference genome, if available, or otherwise aligned using de novo assembly. Regions of interest (for instance, genes or exons) are then identified on the mapping, and the number of reads per region is quantified. In what follows, we will refer to these regions as *features*. We refer the interested reader to [38, 66, 71, 81, 90] for a much more detailed discussion.

We now introduce some notation that will be used throughout the chapter. Let $\mathbf{X}$ be a $n \times p$ matrix, where n is the number of observations, and p is the number of features. Each element x_{ij} contains the number of reads of the jth feature (e.g. gene or exon) in the ith observation (e.g. tissue sample). Thus, the elements of $\mathbf{X}$ are nonnegative and integer-valued. We let $\mathbf{x}_i$ denote the ith row of $\mathbf{X}$. Additionally, each observation is assumed to belong to one of K classes. We define $\mathbf{y}$ to be an n-vector that contains the class labels for the n observations: $y_i \in \{1, \ldots, K\}$. Furthermore, we let C_k denote the set of indices of observations belonging to the kth class: $C_k = \{i : y_i = k\}$. Lastly, we let $\mathbf{x}_*$ denote a vector of feature measurements for a new observation, which we would like to classify into one of the K classes. That is, we wish to predict the unknown class label y_* corresponding to $\mathbf{x}_*$.

11.1.2 Statistical Issues in Classification of High-Dimensional Data

RNA-seq data are high-dimensional, in the sense that the number of features, p, typically greatly exceeds the number of observations, n. As we will see, most standard approaches for classification, such as logistic regression and linear discriminant analysis, cannot be applied directly in this setting. Even when standard techniques can be applied, the resulting models are usually too complex given the number of observations in the training set.[1] This leads to *overfitting*. Furthermore, direct application of standard techniques for classification can lead to *difficulty in interpretation* in high dimensions. We now briefly discuss both of these concepts.

Overfitting occurs when a fitted classifier models not only the signal, but also the noise, in the training set. This is a grave concern particularly in high dimensions, in which direct application of a standard classification technique will typically yield a classifier that perfectly classifies all training set observations. However, application of this classifier to an independent set of observations not used in model training will yield very poor results. In general, classification techniques that are well-suited for the high-dimensional setting perform *dimension reduction* or *regularization* in order to reduce the complexity of the model fit to the training data, thereby reducing the risk and extent of overfitting. Such techniques typically involve a user-specified tuning parameter that controls the amount of regularization or dimension reduction. That is, the tuning parameter controls the trade-off between an overly complex model given the number of training observations (overfitting) and an overly simplistic model that does not capture the underlying signal (underfitting). Tuning parameter selection is typically performed via cross-validation [3, 27, 34, 79, 82].

In the context of high-dimensional data, we often believe that only a subset of the features is associated with the response. Hence, we may be interested in *sparse* classifiers that make use of only a subset of the features in the classification rule. Even if we do not truly believe that only a small subset of features is associated with the response, we may prefer a sparse classifier for practical reasons, since classifying a new observation using a sparse classifier requires measurement of only a subset of the features. Furthermore, a sparse classifier can have advantages in terms of interpretability: it is much easier to obtain intuition about the underlying biological rationale for a classifier that involves only a small subset of the features.

11.1.3 Organization of This Chapter

The rest of this chapter is organized as follows. In Sects. 11.2–11.6, we consider various approaches for high-dimensional classification. In Sect. 11.7, we briefly

[1] The *training set* is the set of observations used to fit the classifier.

discuss how to normalize RNA-seq data prior to classification. We illustrate the performance of the methods on two RNA-seq data sets, and provide guidance on software, in Sect. 11.8. The Discussion is in Sect. 11.9.

11.2 Logistic Regression

11.2.1 Logistic Regression in Low Dimensions

We first consider the task of performing logistic regression in the low-dimensional setting, in which $n > p$. Though logistic regression can be applied in the presence of $K > 2$ classes, it is most often used when $K = 2$; for simplicity, here we restrict the discussion to the latter setting. The probability that the ith observation belongs to either class 1 or class 2 is modeled as

$$\Pr(y_i = 1|\mathbf{x}_i) = \frac{1}{1+e^{\alpha+\beta^T\mathbf{x}_i}}; \quad \Pr(y_i = 2|\mathbf{x}_i) = \frac{e^{\alpha+\beta^T\mathbf{x}_i}}{1+e^{\alpha+\beta^T\mathbf{x}_i}}, \tag{11.1}$$

where α is an unknown scalar and β is an unknown vector of length p. The corresponding log-likelihood takes the form

$$l(\alpha,\beta) = \sum_{i=1}^{n} \left\{ (y_i - 1)\alpha + (y_i - 1)\beta^T\mathbf{x}_i - \log\left(1 + e^{\alpha+\beta^T\mathbf{x}_i}\right) \right\}. \tag{11.2}$$

Then maximum likelihood estimates of α and β—denoted as $\hat{\alpha}$ and $\hat{\beta}$—can be obtained by maximizing $l(\alpha,\beta)$ using iteratively reweighted least squares [1,42,64].

Logistic regression yields estimates of the probability that a test observation belongs to a particular class: that is,

$$\widehat{\Pr}(y_* = 1|\mathbf{x}_*) = \frac{1}{1+e^{\hat{\alpha}+\hat{\beta}^T\mathbf{x}_*}}.$$

In practice, in order to classify the test observation, we must decide upon a cutoff point, $0 \le t \le 1$, such that we classify to class 1 if $\widehat{\Pr}(y_* = 1|\mathbf{x}_*) > t$, and to class 2 otherwise. (This results in a linear decision boundary, in the sense that it amounts to classifying the test observation based on the value of $\hat{\alpha} + \hat{\beta}^T\mathbf{x}_*$.) Typically, either a cutoff of $t = 0.5$ is used, or a cutoff is chosen based on other considerations, such as the intended application for the classifier.

11.2.2 Logistic Regression in High Dimensions

In the high-dimensional setting when $p > n$, the classes are typically *linearly separable*—that is, it is possible to perfectly separate the n training observations using a linear decision boundary. Hence, the estimate of β obtained using logistic regression is unstable, is not unique, and may contain elements that are infinite [99]. Consequently, in high dimensions, the logistic regression classifier is not suitable; some form of regularization is required to reduce the feature space. Even if n is slightly larger than p, so that the logistic regression solution is unique, it is likely that logistic regression will overfit the data.

To overcome this problem, one option is to regularize the log-likelihood (11.2) by applying a penalty to the coefficient vector β. If we let $P(\beta)$ denote a convex penalty function and λ a nonnegative tuning parameter, then maximization of the penalized log-likelihood

$$l(\alpha,\beta) = \sum_{i=1}^{n}\left\{(y_i-1)\alpha+(y_i-1)\beta^T\mathbf{x}_i-\log\left(1+e^{\alpha+\beta^T\mathbf{x}_i}\right)\right\}-\lambda P(\beta) \quad (11.3)$$

is a convex optimization problem. In general, the larger the value of λ in (11.3), the less prone the model is to overfitting.

In (11.3), two commonly used penalties are $P(\beta) = ||\beta||^2$ and $P(\beta) = ||\beta||_1$. The former is a ridge (ℓ_2) penalty [43], and the latter is a lasso (ℓ_1) penalty [84]. Both penalties can successfully regularize the logistic regression problem in order to yield stable, unique, and well-defined coefficient estimates in high dimensions. However, the ℓ_1 penalty has a particularly attractive feature: it encourages the estimated coefficients to be sparse (equal to zero) when λ is sufficiently large. Hence, it yields results that are more interpretable.

In (11.3), the tuning parameter λ can be chosen via cross-validation. We typically center and scale each feature to have mean zero and standard deviation one before solving (11.3). Many authors have proposed efficient algorithms for solving (11.3) when $P(\beta)$ is an ℓ_1 or an ℓ_2 penalty [30,51,72]. We note that other types of penalties in (11.3) are also possible; we refer the reader to [65,98,99,101] for some examples.

When there are more than two classes ($K > 2$), the penalized logistic regression approach (11.3) can be easily extended to penalized multinomial logistic regression. The details can be found in [42].

11.3 Linear Discriminant Analysis

11.3.1 Linear Discriminant Analysis in Low Dimensions

Linear discriminant analysis (LDA) is one of the most commonly-used approaches for classification [42, 59]. While there are several ways to motivate the LDA classifier, here we motivate it as the Bayes decision boundary under a set of assumptions on the distribution of the observations.

Let $\mathscr{X}$ and $\mathscr{Y}$ denote random variables for the expression data and the class label, respectively. Let $\pi_k = P(\mathscr{Y} = k)$ denote the prior probability that an observation belongs to the kth class; note that $\sum_{k=1}^{K} \pi_k = 1$. Also, let $p_k(\mathbf{x}) = \mathrm{f}(\mathscr{X} = \mathbf{x}|\mathscr{Y} = k)$ be the density function of an observation that belongs to the kth class. By Bayes' theorem, the posterior probability that an observation belongs to the kth class is

$$\Pr(\mathscr{Y} = k|\mathscr{X} = \mathbf{x}) = \frac{\pi_k p_k(\mathbf{x})}{\sum_{l=1}^{K} \pi_l p_l(\mathbf{x})}. \tag{11.4}$$

If p_k and π_k were known for all $k \in \{1,\ldots,K\}$, then it would be natural to classify a test observation $\mathbf{x}_*$ to the class for which the posterior probability is largest: that is, to $\mathrm{argmax}_k \Pr(\mathscr{Y} = k|\mathscr{X} = \mathbf{x}_*)$. This is called the *Bayes classifier*.

In LDA, we assume that each observation in the kth class is drawn i.i.d. from a p-dimensional Gaussian distribution $N(\mu_k, \Sigma)$: that is, each observation has a class-specific mean vector, and a common within-class covariance matrix Σ. In this case, the density for an observation in the kth class is

$$p_k(\mathbf{x}) = \frac{1}{(2\pi)^{p/2}}|\Sigma|^{-1/2} \exp\left(-\frac{1}{2}(\mathbf{x}-\mu_k)^T \Sigma^{-1} (\mathbf{x}-\mu_k)\right). \tag{11.5}$$

Substituting (11.5) into (11.4) and performing some algebra, we see that a new observation $\mathbf{x}_*$ is assigned to the class for which

$$\delta_k(\mathbf{x}_*) = \mathbf{x}_*^T \Sigma^{-1} \mu_k - \frac{1}{2}\mu_k^T \Sigma^{-1} \mu_k + \log \pi_k \tag{11.6}$$

is largest. We refer to (11.6) as the *linear discriminant function*, since it is linear in $\mathbf{x}_*$.

In practice, the parameters μ_k, Σ, and π_k are unknown, and must be estimated based on the data. Let n_k denote the number of training observations in the kth class. Then we typically use the following estimates:

- $\hat{\pi}_k = n_k/n$, the proportion of observations from the kth class;
- $\hat{\mu}_k = \sum_{i \in C_k} \mathbf{x}_i/n_k$, the mean of the observations from the kth class;
- $\hat{\Sigma} = \frac{1}{n}\sum_{k=1}^{K} \sum_{i \in C_k} (\mathbf{x}_i - \hat{\mu}_k)(\mathbf{x}_i - \hat{\mu}_k)^T$, the pooled within-class empirical covariance.

These parameter estimates can be directly plugged into (11.6) in order to obtain the LDA classifier. Figure 11.1 displays the performance of the LDA classifier on a simple simulated example.

A simple extension of LDA, called *quadratic discriminant analysis* (QDA), results from assuming that the observations in the kth class are i.i.d. from a $N(\mu_k, \Sigma_k)$ distribution, where Σ_k denotes a class-specific covariance matrix for the kth class. This results in a quadratic, rather than a linear, decision rule. Other extensions of LDA and QDA can be found in [31, 39–41].

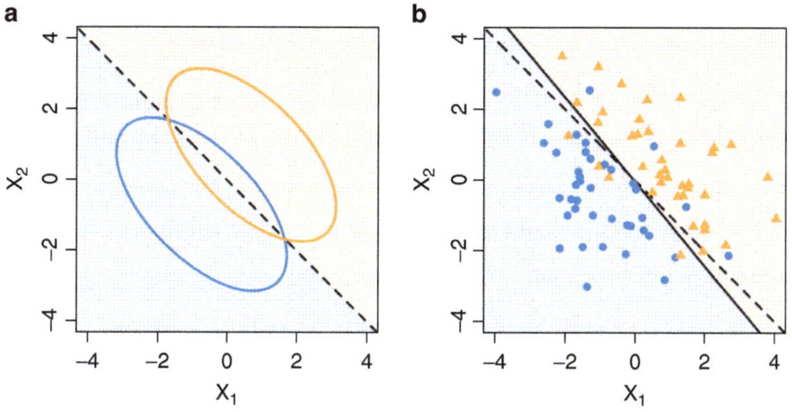

Fig. 11.1 Simulated data with $K = 2$ classes. Each observation is drawn from a two-dimensional multivariate normal distribution, with a class-specific mean vector and a common covariance matrix. (**a**) Most observations from a given class are contained within the corresponding ellipse. The Bayes decision boundary is shown (*dashed line*). (**b**) Forty observations generated from each class are displayed, as are the LDA (*solid line*) and Bayes (*dashed line*) decision boundaries

11.3.2 Linear Discriminant Analysis in High Dimensions

In high-dimensional data where $p > n$, LDA cannot be directly applied due to the singularity of the estimated within-class covariance matrix whose inverse is required in (11.6). We now consider a few proposals for performing LDA in high dimensions. Many others can be found in the literature [15, 35, 36, 54, 57, 85, 86, 88, 94].

11.3.2.1 Diagonal LDA

As mentioned earlier, in high dimensions, the standard estimate of the within-class covariance matrix is singular. To overcome this problem, several authors have considered a modification of LDA in which it is assumed that the features are independent (i.e. an observation in the kth class has a $N(\mu_k, \Sigma)$ distribution, where Σ is diagonal). We refer to this as *diagonal LDA*. Some authors have shown that the diagonal LDA classifier performs well in high-dimensional problems [8,26]. The decision rule for diagonal LDA is simple: we assign the observation $\mathbf{x}_*$ to the class for which

$$\delta_k'(\mathbf{x}_*) = -\frac{1}{2}\sum_{j=1}^{p}(x_{*j} - \hat{\mu}_{kj})^2/\hat{s}_j^2 + \log\hat{\pi}_k \quad (11.7)$$

is largest. Here, $\hat{\mu}_{kj}$ is the mean for the jth feature in the kth class, and $\hat{s}_j^2 = \frac{1}{n}\sum_{k=1}^{K}\sum_{i\in C_k}(x_{ij} - \hat{\mu}_{kj})^2$ is the pooled within-class variance for the jth feature.

Though the diagonal LDA classifier can be applied in high dimensions, it leaves something to be desired, because it yields a decision boundary involving all p features. When p is large, we may prefer a sparse classifier that involves only a subset of the features, in the interest of simplicity, interpretability, and reduced variance. Next we consider an extension of diagonal LDA that achieves sparsity.

11.3.2.2 Nearest Shrunken Centroids

The *nearest shrunken centroids* (NSC) proposal [85, 86] is an extension of diagonal LDA that yields a decision boundary involving only a subset of the features. It derives its name from the *nearest centroids* classifier, a simplified version of diagonal LDA obtained by assuming that $\pi_1 = \ldots = \pi_K$ and $s_1 = \ldots = s_p$ in (11.7). NSC involves modifying (11.7) so that $\hat{\mu}_{kj} = \frac{1}{n_k} \sum_{i \in C_k} x_{ij}$ is replaced with an estimate $\hat{\mu}'_{kj}$ that satisfies

$$\hat{\mu}'_{1j} = \ldots = \hat{\mu}'_{Kj} \tag{11.8}$$

for some features. The motivation is that if (11.8) holds, then the jth feature is not involved in the classification rule given by (11.7). This leads to a sparse classifier.

We now describe the NSC procedure in detail. Let

$$d_{kj} = \frac{\hat{\mu}_{kj} - \hat{\mu}_{.j}}{m_k(\hat{s}_j + s_0)}, \tag{11.9}$$

where $\hat{\mu}_{.j} = \frac{1}{n} \sum_{i=1}^{n} x_{ij}$ is the overall mean of the jth feature, $m_k^2 = 1/n_k - 1/n$, and s_0 is some positive constant. Then d_{kj} can be interpreted as a t-like statistic that measures the difference between the mean of the kth class for the jth feature and the overall mean for the jth feature. Note that (11.9) can be rewritten as

$$\hat{\mu}_{kj} = \hat{\mu}_{.j} + m_k(\hat{s}_j + s_0) d_{kj}. \tag{11.10}$$

In order to encourage (11.8) to hold, we *soft-threshold* d_{kj} [25], to yield

$$d'_{kj} = \text{sign}(d_{kj}) \max(|d_{kj}| - \lambda, 0), \tag{11.11}$$

where λ is a tuning parameter, typically chosen via cross-validation. We then define

$$\hat{\mu}'_{kj} = \hat{\mu}_{.j} + m_k(\hat{s}_j + s_0) d'_{kj}, \tag{11.12}$$

and we plug $\hat{\mu}'_{kj}$ into the decision rule given in (11.7). It is clear that if λ is sufficiently large, then (11.8) will hold for some of the features, which will consequently not be involved in the decision rule. There is an interesting connection between nearest shrunken centroids and an ℓ_1 minimization problem; see Exercise 18.2 in [42].

11.3.2.3 Poisson LDA

As was mentioned in the previous sections, LDA, diagonal LDA, and NSC are based on the assumption that the observations in the kth class are i.i.d. from a $N(\mu_k, \Sigma)$ distribution (where the latter two approaches further assume that Σ is diagonal). However, recall that RNA-seq data involve counts: the elements of $\mathbf{X}$ are nonnegative and integer-valued. Therefore, a number of authors have considered modeling RNA-seq data under the assumption that the counts are drawn from Poisson or negative binomial distributions rather than Gaussian distributions [4, 12, 55, 60, 76, 92].

Recently, [93] proposed the *Poisson LDA* method for classification on the basis of RNA-seq data. This approach mirrors the NSC proposal, except that it is assumed that the elements of the data matrix $\mathbf{X}$ are independently drawn from Poisson distributions rather than from Gaussian distributions. In greater detail, it is assumed that

$$x_{ij} \mid y_i = k \sim \text{Poisson}(s_i g_j e_{kj}), \tag{11.13}$$

where s_i allows for variability in the number of counts per sample, g_j allows for variability in the number of counts per feature, and e_{kj} is a measure of differential expression for the jth gene in the kth class. We can use (11.13) to write out the density for an observation in the kth class, which can then be plugged into (11.4). After performing some algebra and estimating unknown parameters, we obtain the decision rule that assigns a test observation $\mathbf{x}_*$ to the class for which the quantity

$$\delta_k''(\mathbf{x}_*) = \sum_{j=1}^{p} x_{*j} \log \hat{e}_{kj} - \hat{s}_* \sum_{j=1}^{p} \hat{e}_{kj} \hat{g}_j + \log \hat{\pi}_k \tag{11.14}$$

is largest. Note that (11.14) is linear in $\mathbf{x}_*$ and involves the jth feature in $\mathbf{x}_*$, unless $\hat{e}_{kj} = 1$ for all k. To encourage a sparse decision rule, [93] employs soft-thresholding of $\hat{e}_{kj}$ towards 1, using a tuning parameter value chosen via cross-validation.

11.4 Principal Components Classification

In this section, we discuss a two-stage approach for building a classifier for high-dimensional data. We first use principal components analysis (PCA) to obtain a lower-dimensional feature set, and then build a classifier based on this new set of features [2, 62]. Together, these two steps are known as *principal components classification* (PCC).

The motivation for PCC is that in many settings, most of the signal in the data is contained in a low-dimensional subspace. By transforming the p original features into a set of m *denoised* features where $m \ll p$, we can often obtain a classifier that has lower variance, greater accuracy, and a reduced risk of overfitting.

Unlike logistic regression and LDA, which must be modified in order to be applicable in high dimensions (Sects. 11.2.2 and 11.3.2), PCC can be performed without modification when $p > n$. This is because regardless of the value of p in the original data, the classifier built in the second step of PCC is low-dimensional: it involves only a small set of denoised features.

We now discuss the two steps that make up the PCC approach.

11.4.1 Step 1: Principal Components Analysis

11.4.1.1 Introduction to PCA

PCA has long been used as a technique for dimension reduction [46]; we provide a brief overview here. We assume that the columns of $\mathbf{X}$ are centered to have mean zero. PCA seeks the linear combinations of the columns of $\mathbf{X}$ that have the highest possible variance, subject to a constraint of orthogonality.

In greater detail, for $M = \min(n-1, p)$, the principal component *score vectors* $\mathbf{z}_1, \ldots, \mathbf{z}_M$ are n-vectors, defined as $\mathbf{z}_k = \mathbf{X}\mathbf{v}_k$, where $\mathbf{v}_k$ is the kth principal component *loading vector*. The first loading vector $\mathbf{v}_1$ is obtained by solving

$$\underset{\mathbf{v}_1 \in R^p}{\text{maximize}} \quad \mathbf{v}_1^T \mathbf{X}^T \mathbf{X} \mathbf{v}_1 \qquad \text{subject to} \quad \mathbf{v}_1^T \mathbf{v}_1 = 1. \tag{11.15}$$

In other words, the first loading vector $\mathbf{v}_1$ is chosen so that the resulting linear combination of the columns of $\mathbf{X}$ has the highest possible variance. The subsequent loading vectors $\mathbf{v}_2, \ldots, \mathbf{v}_M$ can be found by solving

$$\underset{\mathbf{v}_k \in R^p}{\text{maximize}} \quad \mathbf{v}_k^T \mathbf{X}^T \mathbf{X} \mathbf{v}_k \qquad \text{subject to} \quad \mathbf{v}_k^T \mathbf{v}_k = 1, \mathbf{v}_k^T \mathbf{v}_j = 0 \quad \forall j < k. \tag{11.16}$$

Hence, $\mathbf{v}_k$ maximizes the variance of the kth score vector, subject to the constraint that $\mathbf{v}_k$ must be orthogonal to the previous loading vectors.

As mentioned earlier, we assume in (11.15) and (11.16) that the features are centered to have mean zero. Typically, before performing PCA, we also scale the features to have standard deviation one. Otherwise, a high-variance feature (perhaps resulting from the fact that features are measured on different scales) could drive the vast majority of the variance in the data, and hence play an outsize role in the first (or first few) loading vectors.

Though there exist $M = \min(n-1, p)$ score vectors, for many purposes only the first few score vectors are of interest. The first m score vectors, where $m \ll M$, typically capture a large portion of the variation in the data. Hence, they can be used instead of the original features in downstream analyses, such as data visualization (examples will be shown in Sects. 11.8.4.1 and 11.8.4.2) or classification (to be described in Sect. 11.4.2). The problem of how to choose the value of m is outside of the scope of this chapter; we refer the reader to Sect. 6 in [46] for a detailed discussion.

There is a close connection between the *singular value decomposition* (SVD) and PCA. The SVD of the matrix $\mathbf{X}$ is

$$\mathbf{X} = \mathbf{U}\mathbf{D}\mathbf{V}^T, \qquad \mathbf{U}^T\mathbf{U} = \mathbf{I}_p, \qquad \mathbf{V}^T\mathbf{V} = \mathbf{I}_p. \tag{11.17}$$

Here, $\mathbf{U}$ is a $n \times p$ orthogonal matrix whose columns contain the left singular vectors, $\mathbf{V}$ is a $p \times p$ orthogonal matrix whose columns contain the right singular vectors, and $\mathbf{D}$ is a $p \times p$ matrix that contains the (non-increasing, nonnegative) singular values. (Here we are assuming that $n \geq p$; if instead $p > n$, the dimensions of $\mathbf{U}$, $\mathbf{D}$, and $\mathbf{V}$ must be adjusted.) One can show that the kth principal component loading vector in PCA is exactly equal to the kth right singular vector. It follows that the kth principal component score vector is proportional to the kth left singular vector; the constant of proportionality is equal to the kth singular value.

11.4.1.2 Sparse Principal Components Analysis

The principal component loading vectors are typically non-sparse, so that each score vector is a linear combination of all p features. This renders the interpretation of the loading vectors difficult. To remedy this, *sparse PCA* can be used to obtain loading vectors for which most of the elements are zero. Here, we consider two different approaches for obtaining sparse loading vectors.

We first consider the proposal of [47]. In order to obtain the first sparse loading vector, [47] proposed solving the optimization problem

$$\underset{\mathbf{v}_1 \in R^p}{\text{maximize}} \quad \mathbf{v}_1^T \mathbf{X}^T \mathbf{X} \mathbf{v}_1 \qquad \text{subject to} \quad \mathbf{v}_1^T \mathbf{v}_1 = 1, \quad \|\mathbf{v}_1\|_1 \leq c. \tag{11.18}$$

The constraint in (11.18) is roughly equivalent to placing an ℓ_1 penalty on $\mathbf{v}_1$, and hence most of its elements are zero when the nonnegative tuning parameter c is sufficiently small. Subsequent sparse loading vectors can be obtained by solving (11.16) with an additional orthogonality constraint on the kth loading vector. The optimization problem (11.18) is not convex, and the computations are difficult. In recent years, [87] proposed a projected gradient algorithm for (11.18), and [95] provided a quick iterative algorithm.

We now discuss a different approach for sparse PCA, which involves a low-rank approximation of $\mathbf{X}$ using the connection between PCA and the SVD [80, 95]. The first sparse loading vector $\mathbf{v}_1$ can be found by solving the optimization problem

$$\underset{d_1 \in R^+,\, \mathbf{u}_1 \in R^n,\, \mathbf{v}_1 \in R^p}{\text{minimize}} \quad \|\mathbf{X} - d_1 \mathbf{u}_1 \mathbf{v}_1^T\|_F^2 \quad \text{subject to} \quad \mathbf{u}_1^T \mathbf{u}_1 = \mathbf{v}_1^T \mathbf{v}_1 = 1, \quad \|\mathbf{v}_1\|_1 \leq c. \tag{11.19}$$

The kth sparse loading vector can be obtained by solving optimization problem (11.19) with the data matrix $\mathbf{X}$ replaced by the residual matrix $\mathbf{X} - \sum_{l=1}^{k-1} d_l \mathbf{u}_l \mathbf{v}_l^T$. The problem (11.19) is bi-convex in $\mathbf{u}_1$ and $\mathbf{v}_1$, and can be efficiently solved via an iterative algorithm. Note that the sparse loading vectors from this approach differ from those of [47], since in (11.19), orthogonality is not imposed on the successive right singular vectors. Other proposals for sparse PCA can be found in [18, 19, 48, 56, 102].

11.4.1.3 Other Extensions of PCA

Recall that the first few principal component score vectors explain most of the variation in the data. However, there is no guarantee that they will perform well when used as predictors in a classifier (as described in Sect. 11.4.2) [5, 6], since it may be that they provide poor summaries of the aspects of the data that differ among the classes. To overcome this problem, the *supervised PCA* proposal of [5, 6] involves first selecting features that are associated with the outcome, and then performing PCA on this reduced set of features. In greater detail: (1) Calculate test statistics that quantify the association between the response and each of the p features, (2) form a new data matrix $\mathbf{X}'$ that only includes features with test statistics that exceed a user-specified threshold in absolute value, (3) compute the principal components of $\mathbf{X}'$, and (4) use the first few score vectors to build a classifier to predict the outcome, as in Sect. 11.4.2. An extension of supervised PCA can be found in [7]. Note that Step 1 above can be tailored to RNA-seq data by using test statistics that are specifically intended to detect differential expression on the basis of RNA-seq data [4, 12, 55, 60, 63, 77].

The standard approach for PCA assumes that the values of $\mathbf{X}$ are continuous. A few proposals in the recent literature extend PCA to the case where the entries of $\mathbf{X}$ are distributed under some exponential family distribution [16, 21, 50, 78]. Just as [93] proposed an extension of LDA to the Poisson setting, one could develop an extension of PCA that is better-suited for Poisson (or negative binomial) data.

11.4.2 Step 2: Build a Classifier

In Sect. 11.4.1, we described how PCA (or a variant, such as sparse PCA or supervised PCA) can be performed on $\mathbf{X}$ in order to obtain $M = \min(n-1, p)$ score vectors, $\mathbf{z}_1, \ldots, \mathbf{z}_M$, each of length n. We now use the first $m \ll M$ of these score vectors as features in a classifier to predict the response $\mathbf{y}$. Since m is substantially smaller than n, the classifier is constructed based on low-dimensional data. Therefore, traditional classification approaches such as logistic regression or LDA can be used. Typically, a small value of m (no greater than 5 or 10) is used; this value can be chosen by cross-validation.

11.5 Partial Least Squares

Similar to PCC, *partial least squares* (PLS) is a two-stage procedure in which a reduced set of features is constructed, and a model is then fit using this new set of features [96]. PLS can be applied without modification even in high dimensions. Unlike PCC, PLS uses the outcome variable to construct the features. We first discuss PLS in the regression context (Sect. 11.5.1), and then in the classification context (Sect. 11.5.2).

11.5.1 Partial Least Squares for Regression

Though this chapter focuses on classification, we first discuss PLS in the regression setting for which it was originally developed. We begin by mean-centering the continuous response $\mathbf{y}$, and mean-centering and standardizing each feature to have variance one.

PLS yields a transformed set of features (or *components*), $\mathbf{z}_1, \ldots, \mathbf{z}_M$, where $\mathbf{z}_k = \mathbf{X}\mathbf{w}_k$. The direction vectors $\mathbf{w}_1, \ldots, \mathbf{w}_M$ can be obtained as follows:

$$\underset{\mathbf{w}_k \in R^p}{\text{maximize}} \quad \text{Var}(\mathbf{X}\mathbf{w}_k)\text{Cor}^2(\mathbf{X}\mathbf{w}_k, \mathbf{y}) \quad \text{subject to} \quad \mathbf{w}_k^T\mathbf{w}_k = 1, \quad \mathbf{w}_k^T\mathbf{X}^T\mathbf{X}\mathbf{w}_j = 0 \quad \forall j < k, \tag{11.20}$$

where we use $\text{Var}(\mathbf{X}\mathbf{w}_k)$ to denote the sample variance of the elements of $\mathbf{X}\mathbf{w}_k$ and $\text{Cor}(\mathbf{X}\mathbf{w}_k, \mathbf{y})$ to denote the sample correlation between the elements of $\mathbf{X}\mathbf{w}_k$ and those of $\mathbf{y}$. Therefore, PLS seeks components that not only explain much of the variability of the original features, but also are highly correlated with the outcome. Once the PLS components have been obtained, the first $m < M$ are used as features in a linear regression model to predict the response $\mathbf{y}$. Here m is a tuning parameter that can be selected using cross-validation [97]. PLS and PCA are closely connected: the former can be seen as a supervised version of the latter [29].

11.5.2 PLS for Classification

PLS has been extended to the classification setting by a number of authors. The simplest extension involves some slight modifications to (11.20) [9, 67, 68]:

- If the response is binary, then it is coded as a vector of 0's and 1's before solving (11.20).
- If there are $K > 2$ classes, then the response is coded using a matrix of dummy variables, and a problem closely related to (11.20) is solved.

Once the PLS components have been obtained, the first $m < M$ can be used to fit a classifier to predict the response $\mathbf{y}$, using logistic regression, LDA, or another approach suitable for low dimensions.

Several authors have proposed *generalized PLS*, an alternative framework for extending PLS to the classification setting, using an approach analogous to the extension of linear models to generalized linear models [24, 28, 61].

Just as the PCA score vectors involve all p features, each PLS component is a linear combination of all p features. Recently, [13, 14] reformulated (11.20) using an ℓ_1 penalty on the direction vectors, in order to achieve sparsity.

11.6 Support Vector Machine

The *support vector machine* (SVM) has become a very popular classification technique in the computer science community during the past two decades, and has been extensively used to classify gene expression data measured on microarrays [11, 17, 33, 37, 89]. Like PCC and PLS, SVM can be applied without modification even when $p > n$. Here we present a simple motivation of SVM. Since the standard SVM is intended for binary classification, we assume that $K = 2$. Furthermore, in this section only, suppose that the two classes are coded as $y_i = 1$ or $y_i = -1$.

First, recall that a *hyperplane* is defined as

$$\{\mathbf{x} : \mathbf{x}^T\beta + \beta_0 = 0\}, \tag{11.21}$$

where β is a p-vector. Suppose that the training observations are *linearly separable*—that is, the training observations in class 1 and those in class -1 can be separated by a hyperplane. This means that there exist β and β_0 such that an observation $\mathbf{x}_i$ for which $y_i = 1$ satisfies $\mathbf{x}_i^T\beta + \beta_0 > 0$, and an observation $\mathbf{x}_i$ for which $y_i = -1$ has $\mathbf{x}_i^T\beta + \beta_0 < 0$. Such a *separating hyperplane* can be used to build a very simple classifier: we classify a test observation $\mathbf{x}_*$ by assigning it to class

$$\begin{cases} 1 & \text{if} \quad \mathbf{x}_*^T\beta + \beta_0 > 0 \\ -1 & \text{if} \quad \mathbf{x}_*^T\beta + \beta_0 < 0 \end{cases}. \tag{11.22}$$

Some examples of separating hyperplanes are shown in Fig. 11.2a.

The approach of using a separating hyperplane to build a classifier seems simple enough, but in general there are two problems.

1. If a separating hyperplane does exist, then typically it is not unique, and there exist an infinite number of such hyperplanes (Fig. 11.2a). In practice, we must choose among them. Typically, we choose the separating hyperplane that is farthest from any training observation (as measured by the perpendicular distance between the separating hyperplane and each of the observations). This is equivalent to choosing the separating hyperplane that maximizes the *margin*

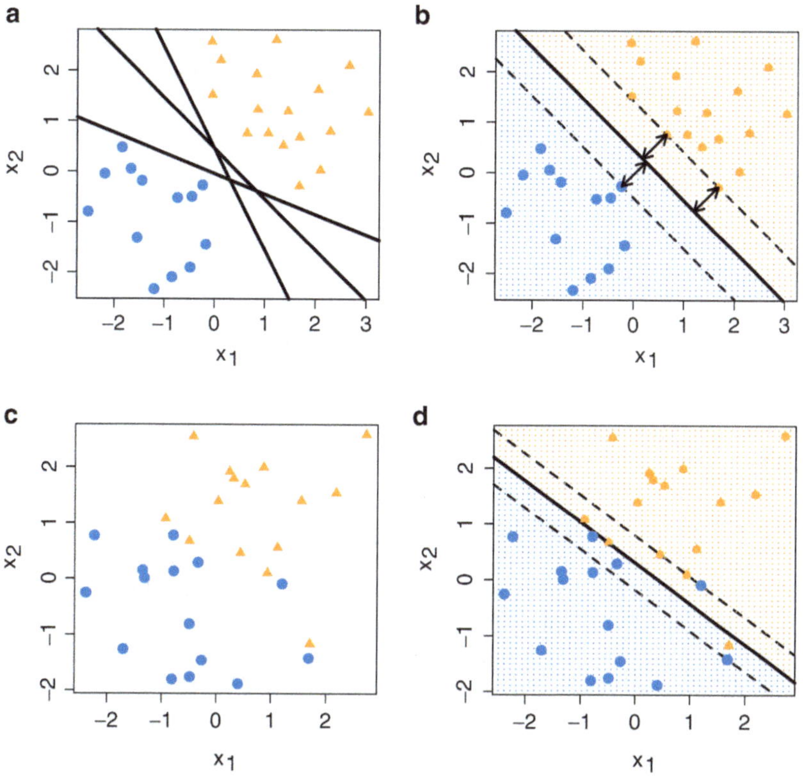

Fig. 11.2 An illustration of the maximal margin classifier and support vector classifier. (**a**) The observations are linearly separable, and three of many possible separating hyperplanes are shown (*solid line*). (**b**) The optimal separating hyperplane is displayed (*solid line*), along with the boundaries of the margin (*dashed line*). The background grid indicates the classification rule that results. (**c**) The observations are not linearly separable. (**d**) The hyperplane and margin of the support vector classifier are shown, for a given value of the tuning parameter. The background *grid* indicates the classification rule that results

around the hyperplane. This approach is often referred to as the *maximal margin classifier*, or the *optimal separating hyperplane*. An example of an optimal separating hyperplane, and the associated margin, is shown in Fig. 11.2b.

2. When $n > p$, a separating hyperplane often does not exist—that is, the training observations may not be linearly separable (Fig. 11.2c). The *support vector classifier* extends the maximal margin classifier to this setting, by choosing the hyperplane that correctly separates most of the training observations, while allowing some to be misclassified (Fig. 11.2d). Briefly, the support vector classifier finds a hyperplane that maximizes the margin, while allowing a user-specified number of observations to fall on the wrong side of the margin.

It turns out that even if a separating hyperplane *does* exist (as will generally be the case when $p > n$), one can often obtain better results by using the support vector classifier instead of using a separating hyperplane. In a sense, by allowing for some training observations to be misclassified, the support vector classifier avoids overfitting. It can be shown that the support vector classifier is the solution to the optimization problem

$$\underset{\beta_0 \in R,\, \beta \in R^p}{\text{minimize}} \quad \sum_{i=1}^{n} \max(1 - y_i f(\mathbf{x}_i), 0) + \lambda \|\beta\|^2, \tag{11.23}$$

where $f(\mathbf{x}) = \beta_0 + \mathbf{x}^T \beta$ and λ is a tuning parameter that controls the magnitude of β. Note that (11.23) involves minimizing the sum of a loss function and an ℓ_2 penalty, and is very similar to the optimization problem for penalized logistic regression (Sect. 11.2.2). The only difference is the loss function used. In fact, the form of (11.23) provides insight into the fact that the support vector classifier can be applied without modification (and can avoid overfitting) when $p > n$: the $\lambda \|\beta\|^2$ term is an ℓ_2 penalty that serves to regularize the solution when λ is sufficiently large.

In some data sets, observations cannot be well-separated with a linear decision boundary. SVM generalizes the support vector classifier to obtain a non-linear decision boundary by mapping the feature space into a higher dimension through the use of a *kernel*, a generalized notion of the distance between a pair of observations. In fact, the support vector classifier just described can be thought of as an SVM with a *linear* kernel. SVMs are often applied with non-linear kernels in order to achieve non-linear decision boundaries (radial and polynomial kernels are two popular choices). However, in high dimensions, using a non-linear kernel is often not warranted, as it leads to a classifier that is too complex given the very limited number of observations. We refer the reader to [42] for a more detailed discussion of SVMs and kernels. A comprehensive list of references for SVM can be found at `http://www.kernel-machines.org`.

The ideas described in this section lend themselves most naturally to the task of binary classification, where $K = 2$. However, many authors have extended SVM to the case of $K > 2$ [44, 52, 91]. We briefly describe one such approach, referred to as *one-versus-one* classification. We construct $\binom{K}{2}$ SVM classifiers, one for each pair of classes. A new observation $\mathbf{x}_*$ is then classified by assigning it to the class to which it is most frequently assigned by the $\binom{K}{2}$ classifiers.

11.7 Normalization of RNA-seq Data

RNA-seq data suffer from some systematic and non-systematic biases that can pose challenges for data analysis.

First of all, within a single RNA-seq experiment, each observation can have a vastly different number of total reads [83]. These differences in *sequencing depth* are due to technical artifacts, and are non-systematic, in the sense that technical replicates of the same observation might have vastly different sequencing depths. Thus x_{ij}, the count of the ith observation in the jth feature, depends not only on the expression level of the jth feature, but also on the sequencing depth that resulted from processing the ith observation. This can cause major problems for downstream analyses, since unless the data are properly normalized, the variation among the observations due to differences in sequencing depth can be far greater than the variation associated with the phenotype of interest.

Second, *transcript length bias* can have a major impact on the number of reads obtained per feature [12, 70]. For instance, suppose that the features correspond to genes. Then a longer gene will tend to have many more reads than a shorter gene. It has been well-documented that this bias leads to higher power to detect differential expression for long genes than for short genes [70]. A similar phenomenon can occur in the context of classification, in which long genes can play an outsize role in the classifier obtained unless the data are properly normalized before analysis.

The problem of how best to normalize RNA-seq data is far from trivial, and a number of approaches have been considered in the literature [4, 12, 23, 55, 66, 77]. There is no consensus as to which approach is best. We now briefly summarize several commonly-used approaches to normalize the observations in order to account for differences in sequencing depth. Each of these approaches involves dividing the p-vector of feature measurements for the ith observation by a scaling factor, s_i.

1. *Total count* [60, 66]: s_i is computed as the total count for the ith observation divided by the average total count across all observations.
2. *Upper quartile* [12]: let q_i be the 75th percentile, across all features, of the counts for the ith observation. The scaling factor is computed as $s_i = q_i / \frac{1}{n}\sum_{i=1}^{n} q_i$.
3. *Median ratio* [4]: let $m_i = \text{median}_j \left\{ \frac{x_{ij}}{(\prod_{i'=1}^{n} x_{i'j})^{1/n}} \right\}$; this is the median (across all features) of the ratio between the jth feature's count in the ith observation and the geometric mean of the jth feature's counts across all n observations. The scaling factor is then given by $s_i = m_i / \frac{1}{n}\sum_{i=1}^{n} m_i$.

In order to address differences in the total number of reads per feature due to transcript length bias or other issues, a simple option is to scale each feature to have standard deviation one. This scaling can be performed after carrying out the normalization described above.

We explore the normalization approaches described above in Sect. 11.8. We refer the reader to [23] for a comprehensive review of normalization approaches for RNA-seq data.

11.8 Evaluation of Methods for Classification

Here, we compare the performances of some of the classifiers described in the context of two RNA-seq data sets:

1. *Prostate cancer data*[2] [49]. This data set consists of expression levels of 62,706 gene transcripts from 20 patients with prostate cancer, as well as 10 benign matched controls.
2. *Cervical cancer data* [92]. This data set contains the expression levels of 714 microRNAs from 56 cervical tissue samples. Among the 56 observations, 27 are cancerous and 29 are non-cancerous.

We compare the performances of the following classifiers: ℓ_1-penalized logistic regression, NSC, Poisson LDA, PCC, sparse PCC, supervised PCC, sparse PLS, and SVM. Before discussing the specific implementation of each of the methods, we first discuss the general approach that we use to evaluate the methods.

11.8.1 Evaluation Criteria

In evaluating the classifiers, we consider two attributes: (1) *sparsity* and (2) *accuracy*. The sparsity of a classifier can be measured by the number of features involved in the decision rule. To assess the accuracy of a classifier, more care must be taken. Here we use *classification error*, the percentage of observations that are incorrectly classified, to measure accuracy. However, if classification error is evaluated on the original training data, the estimate will be overly optimistic [42,45]. Indeed, in high dimensions, it typically is possible to construct a classifier that perfectly classifies the training data. However, this classifier will perform poorly in classifying future observations—this phenomenon is known as *overfitting*. In practice, we are interested in developing a classifier that is accurate in classifying observations *not* used in training. In order to mimic the scenario of having a training set for model training and a test set for model evaluation, we split the available observations into a training set and a test set. The model can then be fit on the training set, and its performance can be evaluated on the test set.

Though we use classification error here as a measure of model accuracy, alternative measures, such as sensitivity and specificity, may be of interest depending on the application. In general, regardless of how accuracy is measured, it is very important to report accuracy on a test set rather than on a training set.

11.8.2 Evaluation Process

In order to obtain the results described in Sect. 11.8.4, we perform the following steps:

[2] We thank Liguo Wang for providing us the raw counts for the prostate cancer data set used in [49].

1. *Split data*: We randomly split the observations into a training set and a test set, with 70 % of observations allocated to the training set and 30 % to the test set. We denote the split data as $\mathbf{X}_{train}$, $\mathbf{y}_{train}$ and $\mathbf{X}_{test}$, $\mathbf{y}_{test}$.
2. *Screen data*: We filter the prostate cancer data by retaining only the 2000 features with the largest variance in $\mathbf{X}_{train}$. Additionally, we discard any features for which 10 % or fewer of the observations in $\mathbf{X}_{train}$ have non-zero counts.
3. *Normalize data*: The observations are normalized as discussed in Sect. 11.7.[3] We also consider simply transforming the data via $\log(x_{ij} + \varepsilon)$, where ε is a small nonnegative constant; this approach has been used in the context of microarray data [74]. Then, for all methods except Poisson LDA, each feature is standardized to have mean zero and variance one in the training set. Note that normalization and standardization are performed on the training set; then each test observation is normalized and standardized relative to the training set. For instance, for a test observation $\mathbf{x}_*$, the upper quartile scaling factor described in Sect. 11.7 is calculated as $q_*/\bar{q}_{train}$, where q_* is the 75th percentile of the counts for $\mathbf{x}_*$, and $\bar{q}_{train}$ is the mean of this quantity across the training observations. Similarly, the features of $\mathbf{x}_*$ are standardized by subtracting the means of the features in $\mathbf{X}_{train}$, and dividing by the standard deviations of the features in $\mathbf{X}_{train}$.
4. *Select tuning parameters*: The classification methods for high-dimensional data introduced in Sects. 11.2–11.6 involve at least one tuning parameter. Choosing an appropriate tuning parameter value is important, in order to avoid either overfitting (caused by fitting a model that is too complex given the data) or underfitting (caused by fitting a model that is not sufficiently complex). We use *R-fold cross-validation*[4] on the training set to select the tuning parameter [42,45]. If multiple tuning parameter values achieve the minimum cross-validation error, we choose the tuning parameter value corresponding to the most sparse classifier. For methods with multiple tuning parameters, cross-validation is performed over a multi-dimensional grid of tuning parameter values.
5. *Fit classifier*: We fit the classifier to the data $(\mathbf{X}_{train}, \mathbf{y}_{train})$ with the selected value of the tuning parameter.
6. *Assess classifier*: We apply the classifier to $\mathbf{X}_{test}$ in order to obtain predictions for $\mathbf{y}_{test}$, and calculate the test error, i.e., the proportion of observations in the test set that are incorrectly classified. We also report the number of features involved in the decision rule.

---

[3]In greater detail, for all methods except for Poisson LDA, we divided each observation by the scaling factors discussed in Sect. 11.7. In contrast, in applying Poisson LDA, observations were not divided by the scaling factor—instead, the scaling factor is directly incorporated into (11.14).

[4]Briefly, R-fold cross-validation involves splitting the observations in the training set into R sets. Then for $r = 1, \ldots, R$, we build classifiers for a range of tuning parameters using all observations except those in the rth fold. We then calculate the error e_r of each of these classifiers on the observations in the rth fold. Finally, we calculate the cross-validation error as $\frac{1}{R}\sum_{r=1}^{R} e_r$. The tuning parameter value corresponding to the minimum cross-validation error is selected.

Since both the prostate and cervical cancer data sets have a relatively small number of observations, we expect the test error to be highly variable across different splits of the observations into training and test sets. For this reason, we repeat the entire process (Steps 1–6 above) a total of ten times. In Sect. 11.8.4, we report a summary of the results across the ten repetitions.

11.8.3 Implementation of Specific Methods

Here we provide implementation details of the classifiers used in cases where further elaboration beyond the discussion in Sects. 11.2–11.6 is required.

Recall from Sects. 11.4 and 11.5 that PCC, sparse PCC, supervised PCC, and sparse PLS are two-step procedures—a lower-dimension feature set is obtained, and then a classifier is built using this new feature set. For all of these methods, logistic regression is used to perform classification using the new feature set. For supervised PCC, features are first selected based on differential expression, as computed using the `edgeR` method of [76]; then PCA is performed on the reduced set of features in order to obtain a lower-dimensional feature set. SVM is implemented with a linear kernel; this is a support vector classifier.

All analyses are performed using `R-CRAN`, a freely available language and environment for statistical computing, available at `www.r-project.org` [75]. Table 11.1 contains a list of `R` packages that implement the classification techniques described in Sects. 11.2–11.6. Additional guidance on the implementation of some of these methods can be found in [45].

11.8.4 Results

We now report the performances of the classifiers described earlier on the prostate and cervical cancer data sets.

11.8.4.1 Prostate Cancer

On the prostate cancer data, all methods average a test error of around 30 % (Fig. 11.3), and the type of normalization performed seems to have little effect. In these data, the test set contains only nine observations, six of which have prostate cancer. Hence a test error of 33 % could be achieved by simply assigning each observation to the cancer class, regardless of its feature measurements! In light of this fact, it is clear that all methods are performing quite poorly on these data.

To provide insight into this poor performance, we plot the first three principal component score vectors of the data, after performing total count normalization and standardizing each feature to have mean zero and variance one (Fig. 11.4).

Table 11.1 A list of R packages for implementing the classifiers described in Sects. 11.2–11.6

Method [citation]	R package
ℓ_1 penalized logistic/multinomial regression [30]	glmnet
Diagonal linear discriminant analysis [26]	sfsmisc
Sparse linear discriminant analysis	
Sparse discriminant analysis [15]	SparseLDA
Regularized linear discriminant analysis [36]	rda
Nearest shrunken centroids [86]	pamr
Penalized Fisher's linear discriminant [94]	penalizedLDA
Poisson linear discriminant analysis [93]	PoiClaClu
Principal component analysis	stats
Sparse principal component analysis	
Sparse PCA, [102]	elasticnet
Penalized matrix decomposition [95]	PMD
Supervised principal components [5]	superPC
Partial least squares	pls
Generalized partial least squares [24]	gpls
Partial least squares with ℓ_1 penalized logistic regression [28]	plsgenomics
Sparse partial least squares [13, 14]	spls
Support vector machine	e1071

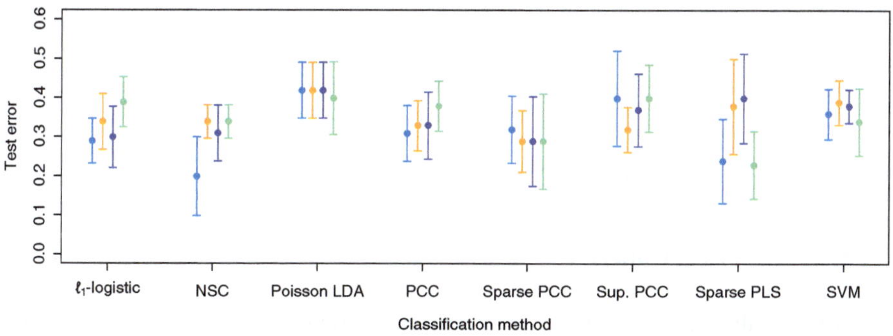

Fig. 11.3 Results for the prostate cancer data set [49]. The test error (averaged over ten splits of the data into training and test sets) is shown, along with 95 % confidence intervals. Four normalization methods are displayed: total count (*blue line*), median ratio (*orange line*), upper quartile (*purple line*), and log transformation (*green line*)

The figure reveals that in the first three principal component score vectors, the non-cancerous observations are intermixed with the cancerous ones. Therefore, it is not surprising that classification is very challenging on the basis of these data—there simply is not much difference between the two classes in terms of the feature measurements. Results similar to Fig. 11.4 are obtained using the other normalization techniques.

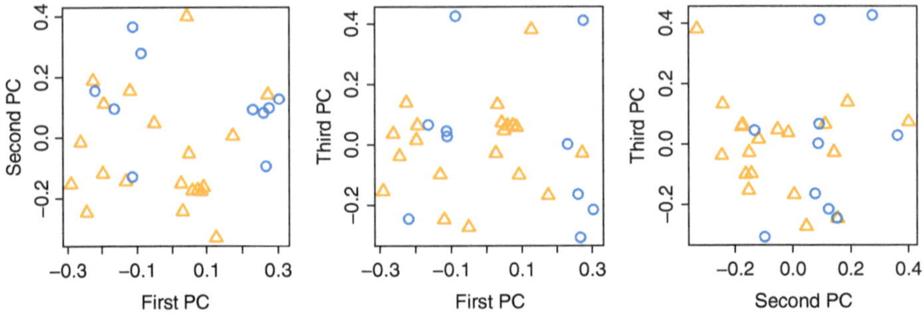

Fig. 11.4 The first three principal component score vectors (PCs) of the prostate cancer data, after normalization using total counts and then standardization of each feature to have mean zero and standard deviation one. Similar results are obtained if features are not standardized to have standard deviation one. The symbols indicate the class membership of the observations: prostate cancer (*orange triangle*) and non-cancer (*blue open circle*)

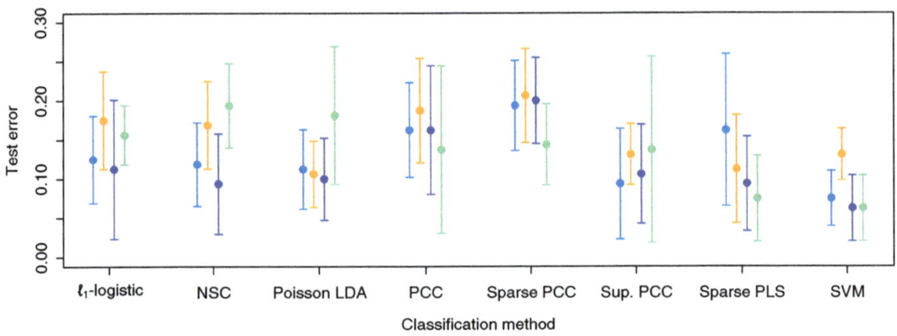

Fig. 11.5 Results for the cervical cancer data set [92]. Details are as in Fig. 11.3

11.8.4.2 Cervical Cancer

Results on the cervical cancer data set are presented in Fig. 11.5. Among the classification techniques that we consider, PCC and sparse PCC have the highest average test errors, and SVM has the lowest average test error.

The number of features used by each method is presented in Table 11.2. Recall that PCC and SVM[5] do not induce sparsity, and so all features are used by those classifiers. In contrast, ℓ_1-penalized logistic regression only uses 13 features on average. For this particular data set, we recommend choosing ℓ_1-penalized logistic regression, since its average test error is comparable to that of the other classification methods, and it uses substantially fewer features.

[5]Proposals have been made for an ℓ_1-penalized SVM that results in a sparse decision rule, but the standard SVM decision rule involves all of the features [100].

Table 11.2 The mean (and standard error) of the number of features used in each classifier across the ten splits of data

Method	Total count	Median ratio	Upper quartile	Log transformation
ℓ_1-logistic	13.2 (1.7)	15.2 (2.3)	17.8 (2.8)	14.5 (2.0)
NSC	121.6 (48.4)	179.8 (55.0)	201.3 (36.2)	34.7 (12.8)
Poisson LDA	96.4 (39.7)	101.8 (41.4)	97.9 (40.4)	38.1 (12.6)
Sparse PCC	217.6 (82.4)	193.8 (70.2)	150.0 (64.0)	267.5 (73.0)
Supervised PCC	198.6 (6.7)	209.7 (14.4)	218.1 (16.4)	202.3 (16.8)
Sparse PLS	383.0 (63.5)	517.9 (24.0)	294.2 (65.0)	421.1 (56.9)

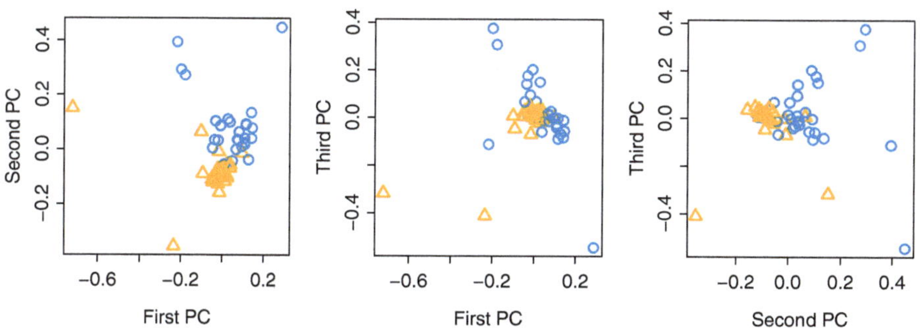

Fig. 11.6 The first three principal component score vectors (PCs) of the cervical cancer data. Details are as in Fig. 11.4

In this data set, the test set has 16 observations, of which seven belong to the cancer class. Thus a test error of 44 % can be achieved by always assigning observations in the test set to the non-cancer class. However, in contrast to the prostate cancer data, all of the classification methods provide a significant reduction in test error compared to not taking any expression data into account. This is consistent with the fact that the first three principal component score vectors show clear separation into the two classes (Fig. 11.6).

11.9 Discussion

We briefly discussed normalization of RNA-seq data in Sect. 11.7. Due to sequencing depth bias, transcript length bias, and other issues, RNA-seq data need to be normalized before any statistical analysis is performed. We note that normalization of RNA-seq data is currently an ongoing research area [4, 12, 55, 66, 70, 76, 77].

Batch effects are another important technical issue that comes into play in the analysis of RNA-seq data. Batch effects are defined by [53] as "sub-groups of measurements that have qualitatively different behaviour across conditions and are unrelated to the biological or scientific variables in a study". For instance, consider an extreme example: suppose that all of the observations in class 1 are processed at location A, and all of the observations in class 2 are processed at location B. Then some features may have differential expression between the two classes due not to inherent biological differences between the classes, but due to artifacts induced by the different locations in which the observations were processed. Failure to consider the presence of such batch effects can lead to a classifier that exploits this artificial signal; unfortunately, this classifier will not perform well on future observations.

In this chapter, we have focused on using RNA-seq data to build a classifier. A related problem, which we have mentioned in passing, involves testing individual features for differential expression across classes. While standard approaches such as the two-sample t-test can be applied to RNA-seq data, some authors have considered specialized approaches that exploit the fact that RNA-seq data are made up counts, which can perhaps be better modeled using a Poisson or negative binomial distribution [4, 12, 55, 60, 76, 77, 92].

In addition, we have presented classification methods in the context of RNA-seq data. However, with the exception of Poisson LDA (which is intended for count data), these methods are applicable to any type of high-dimensional data. All of the methods discussed here could be applied to other types of high-dimensional sequencing data, such as *chromatin immunoprecipitation sequencing* [73] and *methylation sequencing*, provided that classification is of interest.

In this chapter, we have presented some popular approaches for classification of high-dimensional data. However, we could barely scratch the surface of available methods, and a more detailed overview is outside the scope of this work. For instance, an interested reader might wish to also investigate ensemble-based approaches [10, 20, 22, 69], which are known to perform well in a variety of settings.

Acknowledgements D.W. received support for this work from NIH Grant DP5OD009145, NSF CAREER Award DMS-1252624, and a Sloan Foundation Research Fellowship.

References

[1] Agresti, A.: Categorical Data Analysis. Wiley, New York (2002)

[2] Aguilera, A.M., Escabias, M., Valderrama, M.J.: Using principal components for estimating logistic regression with high-dimensional multicollinear data. Comput. Stat. Data Anal. **50**(8), 1905–1924 (2006)

[3] Allen, D.M.: The relationship between variable selection and data augmentation and a method for prediction. Technometrics **16**(1), 125–127 (1974)

[4] Anders, S., Huber, W.: Differential expression analysis for sequence count data. Genome Biol. **11**, R106 (2010)

[5] Bair, E., Hastie, T., Paul, D., Tibshirani, R.: Prediction by supervised principal components. J. Am. Stat. Assoc. **101**(473), 119–137 (2006)

[6] Bair, E., Tibshirani, R.: Semi-supervised methods to predict patient survival from gene expression data. PLoS Biol. **2**(4), e108 (2004)

[7] Barshan, E., Ghodsi, A., Azimifar, Z., Zolghadri Jahromi, M.: Supervised principal component analysis: Visualization, classification and regression on subspaces and submanifolds. Pattern Recogn. **44**(7), 1357–1371 (2011)

[8] Bickel, P.J., Levina, E.: Some theory for Fisher's linear discriminant function, naive Bayes', and some alternatives when there are many more variables than observations. Bernoulli **10**(6), 989–1010 (2004)
[9] Boulesteix, A.L.: PLS dimension reduction for classification with microarray data. Stat. Appl. Genet. Mol. Biol. **3**(1), 1–33 (2004)
[10] Breiman, L.: Random forests. Mach. Learn. **45**(1), 5–32 (2001)
[11] Brown, M.P., Grundy, W.N., Lin, D., Cristianini, N., Sugnet, C.W., Furey, T.S., Ares, M., Haussler, D.: Knowledge-based analysis of microarray gene expression data by using support vector machines. Proc. Natl. Acad. Sci. **97**(1), 262–267 (2000)
[12] Bullard, J., Purdom, E., Hansen, K., Dudoit, S.: Evaluation of statistical methods for normalization and differential expression in mRNA-seq experiments. BMC Bioinform. **11**, 94 (2010)
[13] Chun, H., Keleş, S.: Sparse partial least squares regression for simultaneous dimension reduction and variable selection. J. Roy. Stat. Soc. Ser. B (Stat. Meth.) **72**(1), 3–25 (2010)
[14] Chung, D., Keles, S.: Sparse partial least squares classification for high dimensional data. Stat. Appl. Genet. Mol. Biol. **9**(1), Article 17 (2010)
[15] Clemmensen, L., Hastie, T., Witten, D., Ersbøll, B.: Sparse discriminant analysis. Technometrics **53**(4), 406–413 (2011)
[16] Collins, M., Dasgupta, S., Schapire, R.E.: A generalization of principal components analysis to the exponential family. In Advances in Neural Information Processing Systems, pp. 617–624 (2001)
[17] Cortes, C., Vapnik, V.: Support-vector networks. Mach. Learn. **20**(3), 273–297 (1995)
[18] d'Aspremont, A., Bach, F., Ghaoui, L.E.: Optimal solutions for sparse principal component analysis. J. Mach. Learn. Res. **9**, 1269–1294 (2008)
[19] d'Aspremont, A., El Ghaoui, L., Jordan, M.I., Lanckriet, G.R.: A direct formulation for sparse PCA using semidefinite programming. SIAM Rev. **49**(3), 434–448 (2007)
[20] Datta, S., Pihur, V., Datta, S.: An adaptive optimal ensemble classifier via bagging and rank aggregation with applications to high dimensional data. BMC Bioinformatics **11**(1), 427 (2010)
[21] De Leeuw, J.: Principal component analysis of binary data by iterated singular value decomposition. Comput. Stat. Data Anal. **50**(1), 21–39 (2006)
[22] Dietterich, T.G.: Ensemble methods in machine learning. In: Multiple Classifier Systems, pp. 1–15. Springer, Berlin (2000)
[23] Dillies, M.A., Rau, A., Aubert, J., Hennequet-Antier, C., Jeanmougin, M., Servant, N., Keime, C., Marot, G., Castel, D., Estelle, J., et al.: A comprehensive evaluation of normalization methods for Illumina high-throughput RNA sequencing data analysis. Brief. Bioinform. **14**(6), 671–683 (2013)
[24] Ding, B., Gentleman, R.: Classification using generalized partial least squares. J. Comput. Graph. Stat. **14**(2), 280–298 (2005)
[25] Donoho, D.L., Johnstone, I.M.: Adapting to unknown smoothness via wavelet shrinkage. J. Am. Stat. Assoc. **90**(432), 1200–1224 (1995)
[26] Dudoit, S., Fridlyand, J., Speed, T.P.: Comparison of discrimination methods for the classification of tumors using gene expression data. J. Am. Stat. Assoc. **97**(457), 77–87 (2002)
[27] Efron, B.: Estimating the error rate of a prediction rule: improvement on cross-validation. J. Am. Stat. Assoc. **78**(382), 316–331 (1983)
[28] Fort, G., Lambert-Lacroix, S.: Classification using partial least squares with penalized logistic regression. Bioinformatics **21**(7), 1104–1111 (2005)
[29] Frank, L.E., Friedman, J.H.: A statistical view of some chemometrics regression tools. Technometrics **35**(2), 109–135 (1993)
[30] Friedman, J., Hastie, T., Tibshirani, R.: Regularization paths for generalized linear models via coordinate descent. J. Stat. Software **33**(1), 1–22 (2010)
[31] Friedman, J.H.: Regularized discriminant analysis. J. Am. Stat. Assoc. **84**(405), 165–175 (1989)

[32] Fu, X., Fu, N., Guo, S., Yan, Z., Xu, Y., Hu, H., Menzel, C., Chen, W., Li, Y., Zeng, R., et al.: Estimating accuracy of RNA-seq and microarrays with proteomics. BMC Genom. **10**, 161 (2009)
[33] Furey, T.S., Cristianini, N., Duffy, N., Bednarski, D.W., Schummer, M., Haussler, D.: Support vector machine classification and validation of cancer tissue samples using microarray expression data. Bioinformatics **16**(10), 906–914 (2000)
[34] Geisser, S.: The predictive sample reuse method with applications. J. Am. Stat. Assoc. **70**(350), 320–328 (1975)
[35] Grosenick, L., Greer, S., Knutson, B.: Interpretable classifiers for FMRI improve prediction of purchases. IEEE Trans. Neural Syst. Rehabil. Eng. **16**(6), 539–548 (2008)
[36] Guo, Y., Hastie, T., Tibshirani, R.: Regularized linear discriminant analysis and its application in microarrays. Biostatistics **8**(1), 86–100 (2007)
[37] Guyon, I., Weston, J., Barnhill, S., Vapnik, V.: Gene selection for cancer classification using support vector machines. Mach. Learn. **46**(1–3), 389–422 (2002)
[38] Haas, B.J., Zody, M.C., et al.: Advancing RNA-seq analysis. Nat. Biotech. **28**(5), 421–423 (2010)
[39] Hastie, T., Buja, A., Tibshirani, R.: Penalized discriminant analysis. Ann. Stat. **23**(1), 73–102 (1995)
[40] Hastie, T., Tibshirani, R.: Discriminant analysis by Gaussian mixtures. J. Roy. Stat. Soc. Ser. B (Methodological) **58**(1), 155–176 (1996)
[41] Hastie, T., Tibshirani, R., Buja, A.: Flexible discriminant analysis by optimal scoring. J. Am. Stat. Assoc. **89**, 1255–1270 (1994)
[42] Hastie, T., Tibshirani, R., Friedman, J.J.H.: The Elements of Statistical Learning: Data Mining, Inference, and Prediction. Springer, New York (2009)
[43] Hoerl, A.E., Kennard, R.W.: Ridge regression: biased estimation for nonorthogonal problems. Technometrics **12**(1), 55–67 (1970)
[44] Hsu, C.W., Lin, C.J.: A comparison of methods for multiclass support vector machines. IEEE Trans. Neural Netw. **13**(2), 415–425 (2002)
[45] James, G., Witten, D., Hastie, T., Tibshirani, R.: An Introduction to Statistical Learning. Springer, New York (2013)
[46] Jolliffe, I.: Principal Component Analysis. Wiley, New York (2005)
[47] Jolliffe, I.T., Trendafilov, N.T., Uddin, M.: A modified principal component technique based on the LASSO. J. Comput. Graph. Stat. **12**(3), 531–547 (2003)
[48] Journée, M., Nesterov, Y., Richtárik, P., Sepulchre, R.: Generalized power method for sparse principal component analysis. J. Mach. Learn. Res. **11**, 517–553 (2010)
[49] Kannan, K., Wang, L., Wang, J., Ittmann, M.M., Li, W., Yen, L.: Recurrent chimeric RNAs enriched in human prostate cancer identified by deep sequencing. Proc. Natl. Acad. Sci. **108**(22), 9172–9177 (2011)
[50] Lee, S., Huang, J.Z., Hu, J.: Sparse logistic principal components analysis for binary data. Ann. Appl. Stat. **4**(3), 1579–1601 (2010)
[51] Lee, S.I., Lee, H., Abbeel, P., Ng, A.Y.: Efficient L1 regularized logistic regression. In: Proceedings of the National Conference on Artificial Intelligence, vol. 21, pp. 401–408. AAAI Press, Menlo Park (1999); MIT Press, Cambridge, London (2006)
[52] Lee, Y., Lin, Y., Wahba, G.: Multicategory support vector machines: theory and application to the classification of microarray data and satellite radiance data. J. Am. Stat. Assoc. **99**(465), 67–81 (2004)
[53] Leek, J.T., Scharpf, R.B., Bravo, H.C., Simcha, D., Langmead, B., Johnson, W.E., Geman, D., Baggerly, K., Irizarry, R.A.: Tackling the widespread and critical impact of batch effects in high-throughput data. Nat. Rev. Genet. **11**(10), 733–739 (2010)
[54] Leng, C.: Sparse optimal scoring for multiclass cancer diagnosis and biomarker detection using microarray data. Comput. Biol. Chem. **32**(6), 417–425 (2008)
[55] Li, J., Witten, D.M., Johnstone, I.M., Tibshirani, R.: Normalization, testing, and false discovery rate estimation for RNA-sequencing data. Biostatistics **13**(3), 523–538 (2012)

[56] Ma, Z.: Sparse principal component analysis and iterative thresholding. Ann. Stat. **41**(2), 772–801 (2013)
[57] Mai, Q., Zou, H., Yuan, M.: A direct approach to sparse discriminant analysis in ultra-high dimensions. Biometrika **99**(1), 29–42 (2012)
[58] Malone, J.H., Oliver, B.: Microarrays, deep sequencing and the true measure of the transcriptome. BMC Biol. **9**, 34 (2011)
[59] Mardia, K.V., Kent, J.T., Bibby, J.M.: Multivariate Analysis. Academic, New York (1980)
[60] Marioni, J.C., Mason, C.E., Mane, S.M., Stephens, M., Gilad, Y.: RNA-seq: an assessment of technical reproducibility and comparison with gene expression arrays. Genome Res. **18**(9), 1509–1517 (2008)
[61] Marx, B.D.: Iteratively reweighted partial least squares estimation for generalized linear regression. Technometrics **38**(4), 374–381 (1996)
[62] Marx, B.D., Smith, E.P.: Principal component estimation for generalized linear regression. Biometrika **77**(1), 23–31 (1990)
[63] McCarthy, D.J., Chen, Y., Smyth, G.K.: Differential expression analysis of multifactor RNA-seq experiments with respect to biological variation. Nucleic Acids Res. **40**(10), 4288–4297 (2012)
[64] McCullagh, P., Nelder, J.A.: Generalized Linear Models. Chapman and Hall, Boca Raton (1989)
[65] Meier, L., Van De Geer, S., Bühlmann, P.: The group lasso for logistic regression. J. Roy. Stat. Soc. Ser. B (Stat. Meth.) **70**(1), 53–71 (2008)
[66] Mortazavi, A., Williams, B.A., McCue, K., Schaeffer, L., Wold, B.: Mapping and quantifying mammalian transcriptomes by RNA-seq. Nat. Meth. **5**(7), 621–628 (2008)
[67] Nguyen, D.V., Rocke, D.M.: Multi-class cancer classification via partial least squares with gene expression profiles. Bioinformatics **18**(9), 1216–1226 (2002)
[68] Nguyen, D.V., Rocke, D.M.: Tumor classification by partial least squares using microarray gene expression data. Bioinformatics **18**(1), 39–50 (2002)
[69] Opitz, D., Maclin, R.: Popular ensemble methods: An empirical study. J. Artif. Intell. Res. **11**, 169–198 (1999)
[70] Oshlack, A., Wakefield, M.J.: Transcript length bias in RNA-seq data confounds systems biology. Biol. Direct **4**(14) (2009)
[71] Ozsolak, F., Milos, P.M.: RNA sequencing: advances, challenges and opportunities. Nat. Rev. Genet. **12**(2), 87–98 (2010)
[72] Park, M.Y., Hastie, T.: L1-regularization path algorithm for generalized linear models. J. Roy. Stat. Soc. Ser. B (Stat. Meth.) **69**(4), 659–677 (2007)
[73] Park, P.J.: ChIP–seq: advantages and challenges of a maturing technology. Nat. Rev. Genet. **10**, 669–680 (2009)
[74] Quackenbush, J.: Microarray data normalization and transformation. Nat. Genet. **32**, 496–501 (2002)
[75] R Core Team: R: A Language and Environment for Statistical Computing. R Foundation for Statistical Computing, Vienna, Austria (2013). http://www.R-project.org/
[76] Robinson, M.D., McCarthy, D.J., Smyth, G.K.: edger: a bioconductor package for differential expression analysis of digital gene expression data. Bioinformatics **26**(1), 139–140 (2010)
[77] Robinson, M.D., Oshlack, A.: A scaling normalization method for differential expression analysis of RNA-seq data. Genome Biol. **11**, R25 (2010)
[78] Schein, A.I., Saul, L.K., Ungar, L.H.: A generalized linear model for principal component analysis of binary data. In: Proceedings of the 9th International Workshop on Artificial Intelligence and Statistics, pp. 14–21 (2003)
[79] Shao, J.: Linear model selection by cross-validation. J. Am. Stat. Assoc. **88**(422), 486–494 (1993)
[80] Shen, H., Huang, J.Z.: Sparse principal component analysis via regularized low rank matrix approximation. J. Multivariate Anal. **99**(6), 1015–1034 (2008)
[81] Shendure, J.: The beginning of the end for microarrays? Nat. Meth. **5**(7), 585–587 (2008)

[82] Stone, M.: Cross-validatory choice and assessment of statistical predictions. J. Roy. Stat. Soc. Ser. B (Methodological) **36**, 111–147 (1974)

[83] Tarazona, S., García-Alcalde, F., Dopazo, J., Ferrer, A., Conesa, A.: Differential expression in RNA-seq: a matter of depth. Genome Res. **21**(12), 2213–2223 (2011)

[84] Tibshirani, R.: Regression shrinkage and selection via the lasso. J. Roy. Stat. Soc. Ser. B (Methodological) **58**, 267–288 (1996)

[85] Tibshirani, R., Hastie, T., Narasimhan, B., Chu, G.: Diagnosis of multiple cancer types by shrunken centroids of gene expression. Proc. Natl. Acad. Sci. **99**(10), 6567–6572 (2002)

[86] Tibshirani, R., Hastie, T., Narasimhan, B., Chu, G.: Class prediction by nearest shrunken centroids, with applications to DNA microarrays. Stat. Sci. **18**(1), 104–117 (2003)

[87] Trendafilov, N.T., Jolliffe, I.T.: Projected gradient approach to the numerical solution of the SCoTLASS. Comput. Stat. Data Anal. **50**(1), 242–253 (2006)

[88] Trendafilov, N.T., Jolliffe, I.T.: DALASS: variable selection in discriminant analysis via the LASSO. Comput. Stat. Data Anal. **51**(8), 3718–3736 (2007)

[89] Vapnik, V.: The Nature of Statistical Learning Theory. Springer, New York (2000)

[90] Wang, Z., Gerstein, M., Snyder, M.: RNA-seq: a revolutionary tool for transcriptomics. Nat. Rev. Genet. **10**(1), 57–63 (2009)

[91] Weston, J., Watkins, C.: Multi-class support vector machines. Technical report, Citeseer (1998)

[92] Witten, D., Tibshirani, R., Gu, S.G., Fire, A., Lui, W.O.: Ultra-high throughput sequencing-based small RNA discovery and discrete statistical biomarker analysis in a collection of cervical tumours and matched controls. BMC Biol. **8**(58) (2010)

[93] Witten, D.M.: Classification and clustering of sequencing data using a Poisson model. Ann. Appl. Stat. **5**(4), 2493–2518 (2011)

[94] Witten, D.M., Tibshirani, R.: Penalized classification using Fisher's linear discriminant. J. Roy. Stat. Soc. Ser. B (Stat. Meth.) **73**(5), 753–772 (2011)

[95] Witten, D.M., Tibshirani, R., Hastie, T.: A penalized matrix decomposition, with applications to sparse principal components and canonical correlation analysis. Biostatistics **10**(3), 515–534 (2009)

[96] Wold, H., et al.: Estimation of principal components and related models by iterative least squares. Multivariate Anal. **1**, 391–420 (1966)

[97] Wold, S.: Cross-validatory estimation of the number of components in factor and principal components models. Technometrics **20**(4), 397–405 (1978)

[98] Yuan, M., Lin, Y.: Model selection and estimation in regression with grouped variables. J. Roy. Stat. Soc. Ser. B (Stat. Meth.) **68**(1), 49–67 (2006)

[99] Zhu, J., Hastie, T.: Classification of gene microarrays by penalized logistic regression. Biostatistics **5**(3), 427–443 (2004)

[100] Zhu, J., Rosset, S., Hastie, T., Tibshirani, R.: 1-norm support vector machines. Adv. Neural Inform. Process. Syst. **16**(1), 49–56 (2004)

[101] Zou, H., Hastie, T.: Regularization and variable selection via the elastic net. J. Roy. Stat. Soc. Ser. B (Stat. Meth.) **67**(2), 301–320 (2005)

[102] Zou, H., Hastie, T., Tibshirani, R.: Sparse principal component analysis. J. Comput. Graph. Stat. **15**(2), 265–286 (2006)

Chapter 12
Isoform Expression Analysis Based on RNA-seq Data

Hongzhe Li

Abstract The development of novel high-throughput DNA sequencing methods has provided a powerful method for both mapping and quantifying transcriptomes. This method, termed RNA-seq (RNA sequencing), has advantages over microarray-based approaches in terms of wide dynamic range of expressions, less reliance on existing knowledge about genome sequence, and low background noise. After aligning the reads to the reference genomes, the first step of RNA-seq analysis is to infer relative transcript abundances. This can be done at the whole transcript level, at the isoform-specific relative abundance level assuming a known set of isoforms, and at the level where transcripts are identified and their abundances are quantified. We review these methods briefly and add some recent developments in dealing with non-uniform read distribution within a transcript. We focus on methods for simultaneous transcript discovery and quantification.

12.1 Introduction

The transcriptome is the complete set of transcripts and their respective quantities in a cell of a specific tissue for a specific developmental stage, including all expressed transcripts or isoforms. An important aspect of the transcriptome complexity is the generation of multiple transcript isoforms from a single gene in a genomic locus, due to the use of alternative initiation and/or termination of transcription and alternative splicing of pre-mRNAs. Understanding the transcriptome is important for revealing the molecular signatures of cells and tissues, and also for understanding development and disease. The primary aims of transcriptome analysis are to

H. Li (✉)
Department of Biostatistics and Epidemiology, Perelman School of Medicine,
University of Pennsylvania, Philadelphia, PA, USA
e-mail: hongzhe@upenn.edu

S. Datta and D. Nettleton (eds.), *Statistical Analysis of Next Generation Sequencing Data*, Frontiers in Probability and the Statistical Sciences, DOI 10.1007/978-3-319-07212-8_12,

catalogue all transcript isoforms (including mRNAs and non-coding RNAs) and to determine the transcriptional structure of genes in terms of their start sites, 5′ and 3′ ends, and splicing patterns [30]. Major applications of transcriptome analysis include quantification of the changing expression levels of each transcript during development, under different conditions and disease states.

The development of novel high-throughput DNA sequencing methods has provided a powerful method for both mapping and quantifying transcriptomes. This method, termed RNA-seq (RNA sequencing), has advantages over microarray-based approaches in terms of wide dynamic range of expressions, less reliance on existing knowledge about genome sequence, and low background noise. Large-scale RNA sequencing has been explored to understand mechanisms underlying human gene expression variation [20], to study the transcriptome genetics in a Caucasian population [17], and to uncover functional variations in humans [10].

Briefly RNA-seq involves the following steps. Long RNAs are first converted into a library of cDNA fragments through RNA fragmentation, followed by first-strand synthesis priming, which selects the 3′ fragment end (in transcript orientation), to make single stranded cDNA. Double stranded cDNA created during second-strand synthesis, which selects the 5′ fragment end, is then size selected, resulting in fragments suitable for sequencing. Sequencing adaptors are subsequently added to each cDNA fragment, and a short sequence is obtained from each cDNA fragment using high-throughput sequencing technology. The resulting sequence reads are then aligned to the reference genome or transcriptome, and classified as one of three types: exonic reads, junction reads, or poly(A) end-reads. These three types are used to generate a base-resolution expression profile for each gene [30]. The most advanced and commonly used tool for mapping the RNA-seq reads is *TopHat* [26], a fast splice junction mapper for RNA-seq reads. It aligns RNA-seq reads to the reference genome using the short read aligner *Bowtie* [9], and then analyzes the mapping results to identify splice junctions between exons.

After aligning the reads to reference genomes, the first step of RNA-seq analysis is to infer relative transcript abundances. This can be done at the whole gene level, at the isoform-specific relative abundance level assuming a known set of isoforms, and at the level where all transcripts are identified/assembled and their abundances are quantified. Pachter [19] gave a detailed review of methods for transcript quantification using RNA-seq assuming that the isoforms are known and pre-specified. We first review these methods for quantifying isoform specific expressions in Sect. 12.2. Special attention is given to some recent developments in dealing with non-uniformity of the read distribution within a transcript in Sect. 12.2.2. We then review methods for simultaneous transcript discovery and quantification in Sect. 12.3 and methods for allelic-specific expression analysis in Sect. 12.4. Finally, we present areas that need further methodological developments.

12.2 Transcript Quantification Assuming the Isoforms Are Known

A simple quantification of transcript abundance assumes that the transcriptome consist of a set isoforms with different abundances. In addition, the starting position of a read is assumed to be generated by choosing a site of an isoform uniformly at random among all the positions in the transcriptome. Then the number of reads per kilobase per millions reads (RPKM) [18] mapped to a gene can be used to quantify the gene's transcript abundance adjusting for the transcript length and total number of mapped reads. Alternatively, for paired-end RNA-seq experiments, fragments per kilobase per million mapped reads (FPRM) can be used to approximates the relative abundance of transcripts in terms of fragments observed from an RNA-Seq experiment.

12.2.1 Isoform-Level Transcript Quantification

Because most reads that are mapped to the gene are shared by more than one isoform, it is difficult to compute isoform-specific RPKM directly. Statistical methods are required to assign these reads to isoforms probabilistically. Typical alternative splicing (AS) events include alternative 5′ (or 3′) splice sites, exon skipping, intron retention, and mutually exclusive exons. Information about these events can be gained by partitioning a gene into a sequence of expressed segments (or simply segments) based on exon-intron boundaries [2, 14]. Figure 12.1 shows examples of six-exon genes with two and three isoforms.

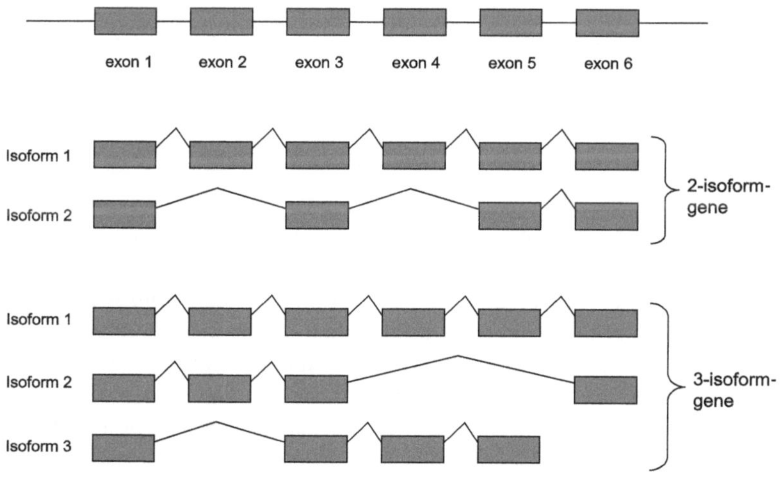

Fig. 12.1 Illustration of isoforms, where the top gene have two isoforms and the bottom gene has three isoforms. Both genes have six exons

For a given gene g, assuming the set of n isoforms is known (see Fig. 12.1), one can then define a set of expressed segments (e.g. all exons, and exon-exon junctions) denoted by S that define these n isoforms. With these ideas in mind, Jiang and Wong [8] developed a Poisson mixture model for estimating isoform-specific expression levels. Specifically, let $\mathbf{X} = \{X_s, s \in S\}$ be the set of observed read counts and S is an index set of the expressed segments. Each observation $X \in \mathbf{X}$ is a random variable representing the number of reads falling into some expressed segment in g, including exon or exon-exon junction. Assuming a uniform-sampling of the reads, we can assume that each $X \in \mathbf{X}$ follows a Poisson distribution with some mean parameter λ. Let θ_i be the relative abundance of the ith isoform, which is simply the proportion of transcripts. The mean number of reads falling into exon s is $l_s w \sum_{i=1}^n b_{is}\theta_i$, where l_s is the length (in bps) of exon s, w is the total number of reads, and b_{is} is 1 if isoform i contains exon s and 0 otherwise. For exon-exon junctions, the λ_s is $l_s w \sum_{i=1}^n b_{ij} b_{ik}\theta_i$, where l is the length of the junction region, and j and k are indices of the two exons involved in the junction. In general, for $s \in S$, λ_s is a linear function of $\theta_1, \theta_2, \cdots, \theta_n$, i.e., $\lambda_s = \sum_{i=1}^n a_{is}\theta_i$ for some known coefficients a_{is}. The likelihood function for the observed read counts $\mathbf{X}$ is then

$$l(\Theta|\mathbf{X}) = \sum_{s\in S}\left\{-\sum_{i=1}^{n}\theta_i a_{is} + X_s \log\left(\sum_{i=1}^{n}\theta_i a_{is}\right)\right\}. \tag{12.1}$$

This model can be interpreted as the read count X_s being sampled from a mixture of n Poisson distributions with different rate parameters determined by the isoform abundances and the lengths of the isoforms. An efficient EM algorithm can be developed to estimate the relative isoform abundances. When the number of isoforms specified in the model is large, there is potentially a problem of identifiability since different parameter settings in isoform abundances can explain the data equally well. Salzman et al. [22] discussed this identifiability problem in detail. In addition, the design matrix (a_{is}) can be further modified to account for pair-end reads mapping results and potential non-uniform sampling of the reads.

12.2.2 Accounting for Non-uniform Sampling

Non-uniform distributions of the sequenced fragments or reads over different positions across different isoforms have been observed in RNA-seq data [13]. Figure 12.2 shows the coverage plots of three genes, indicating the non-uniform distribution of the read coverage within the exons. Such non-uniformity can be due to sequencing and positional biases or due to mappability of the reads. Statistical methods are needed to correct for such biases in isoform abundance estimation. Hu et al. [6] presented a method that uses the empirical distribution of the read distribution along the transcripts to adjust for such biases in the EM iterations. Noting that uniform sampling of reads from a given isoform implies

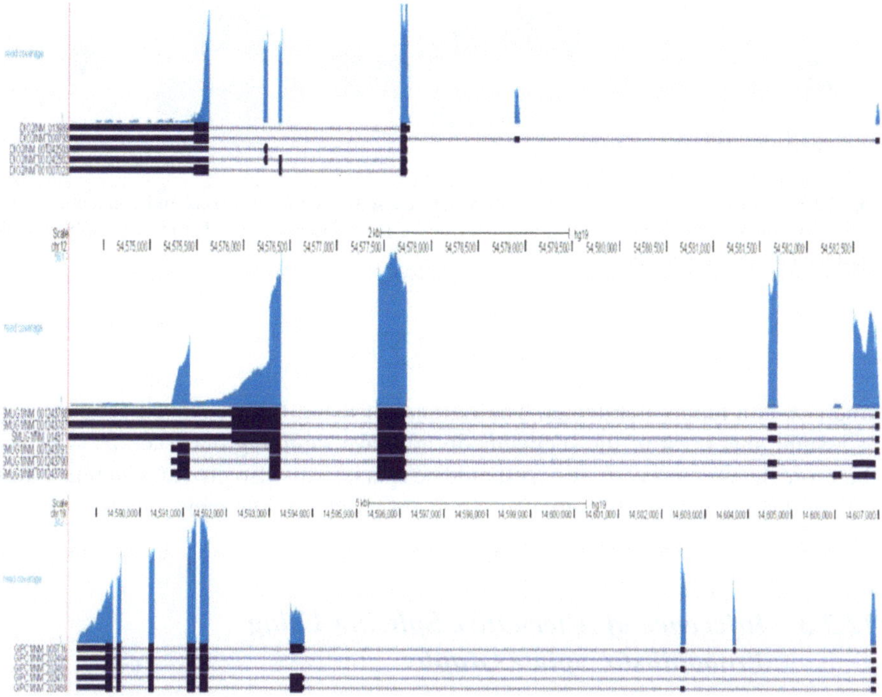

Fig. 12.2 Coverage plots of three genes in an RNA-seq dataset, showing non-uniform distributions of read coverage. This plot was reproduced from [6]

Poisson-distributed read counts, Li and Jiang [12] proposed to use a generalized Poisson distribution to model the read counts and proposed a component elimination EM algorithm to estimate the isoform abundances.

Alternatively, one can use a penalized likelihood approach to select the isoforms and estimate their abundances. Jiang and Salzman [7] proposed to assign a bias parameter β_s to each read type s and to reparameterize β_s as $\beta_s = \exp(b_s)$ for some $b_s \in \mathbb{R}$ to constrain $\beta_s \geq 0$. The actual effective sampling rate for read type s from isoform i now becomes $a'_{is} = a_{is}\beta_s$, and the corresponding log-likelihood function is

$$l(\Theta|\mathbf{X}) = \sum_{s\in S}\{-\lambda_s + X_s\log(\lambda_s)\}$$

$$= \sum_{s\in S}\left\{-\sum_{i=1}^{n}\theta_i a_{is}\exp(b_s) + X_s\log\left(\sum_{i=1}^{n}\theta_i a_{is}\exp(b_s)\right)\right\}. \qquad (12.2)$$

To estimate the read-type specific non-uniformity parameter b_s and the relative abundances θ_i, Jiang and Salzman [7] proposed an ℓ_1 penalized likelihood estimation method based on the likelihood function (12.2),

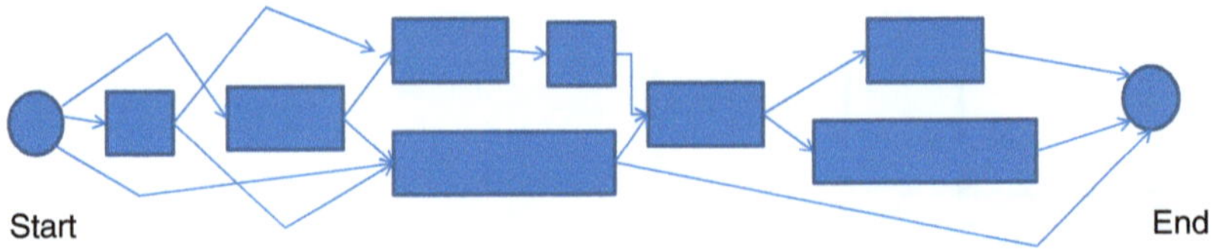

Fig. 12.3 An illustration of a splice graph, where cassettes represent exons and connecting lines represent splice paths. The splice graph provides a representation of all transcript splice paths represented in the transcript models

$$l(\Theta|\mathbf{X}) - \lambda \sum |b_s|,$$

where λ is a tuning parameter. The rational of using this penalized estimation is to assume that most of the segments do not have sampling bias (i.e., $b_s = 0$) and therefore the vector $b = (b_s, s \in S)$ should be sparse. This also provides an interesting way of dealing with outliers that are often observed in sequencing counts.

12.2.3 Inference of Alternative Splicing Using Probabilistic Splice Graphs

LeGault and Dewey [11] developed a method for inference of alternative splicing using an interesting idea of a probabilistic splice graph model. The method is still based on a set of annotated isoforms and provides a probability estimate of each alternative processing event. The key of the method is the use of splice graphs [5], a data structure that can represent all isoforms of a gene and show the structural relationships among these isoforms. The splice graph of a gene was originally defined by [5] as a directed acyclic graph, $G = (V,E)$, with a vertex for each exonic genomic position of the gene and an edge from vertex v to vertex u if the corresponding genomic position of v immediately precedes that of u in some isoform of the gene. One merges vertices v and u if $(v,u) \in E$ and outdegree(v) = indegree$(u) = 1$. In general, the vertices of a splice graph represent exonic segments of a gene, and an edge represents that one segment precedes another segment. The key property of a gene's splice graph is that every isoform of the gene corresponds to a path through the graph. Figure 12.3 illustrates one such splice graph, which provides a representation of all transcript splice paths represented in the transcript model.

The problem of identifying isoforms is equivalent to identifying the most probable pathways that explain the observed count data. The parameters associated with a given probability splice graph are the edge weights $a_{ij} \in [0,1]$ for each edge (u_i, v_j), where $\sum_j a_{ij} = 1$. These edge weight probabilities can then be used to quantify the probability of a transcript as the product of the edge weights of its corresponding path on the splice graph. LeGault and Dewey [11] developed an EM algorithm to estimate these edge weights treating the observed read sequences as the data for a pre-specified splice graph. Different from the penalized approaches given in Sect. 12.3.1, the method provides probability estimates of the different alternative processing events, which themselves can be biologically interesting. Since for a given set of gene annotations there can be different types of splicing graphs, it would be interesting to develop a statistical method to select probability splice graph that best fits the data. Based on this model, one can also test for differential alternative processing events between two RNA-seq samples using a likelihood ratio test.

12.3 Approaches for Simultaneous Isoform Discovery and Quantification

12.3.1 Penalized Regression Approaches for Simultaneous Isoform Discovery and Quantification Based on Known Annotations

Instead of pre-specifying the set of isoforms as in [8], Li et al. [15] considered all the possible isoforms by enumerating exons of every gene. For example, a gene of J nonoverlapping exons has $n = 2^J - 1$ possible isoforms, each composed of a subset of the J exons. However, because of the possible occurrence of alternative splicing within exons, isoforms of the same gene may have partially overlapping but different exons. Li et al. [15] further defined a subexon as a transcribed region between adjacent splicing sites in any annotated mRNA isoforms. Figure 12.4 illustrate how these subexons are defined based on the annotated isoforms. This uses the same idea of treating gene as being partitioned into a sequence of expressed segments based on exon-intron boundaries [2, 14]. With this definition, every gene has a set of nonoverlapping subexons, from which we can enumerate all the possible isoforms including the annotated ones. For a J-subexon gene, possible paired-end bins are $\{(i, j, k, l), 1 \leq i \leq j \leq k \leq l = J\}$. Then RNA-seq data are transformed into bin counts (i.e., number of reads in each bin), which are further normalized into a vector of proportions b. Let θ_i be the relative abundance of the ith possible isoform. Li et al. [15] relates the unknown $\theta = (\theta_1, \cdots, \theta_n)$ to observed b by a design matrix F, where

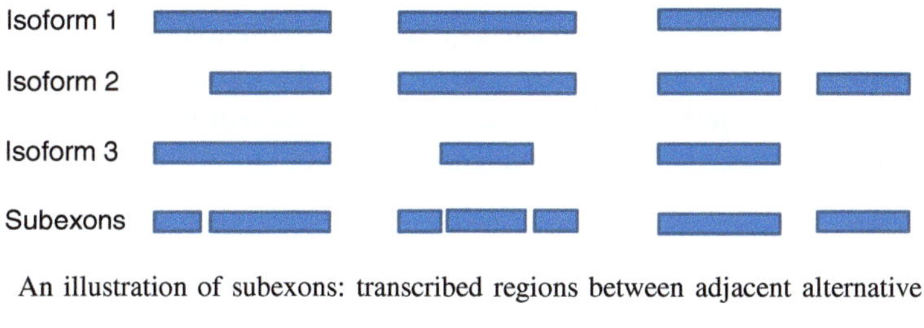

Fig. 12.4 An illustration of subexons: transcribed regions between adjacent alternative splicing sites

$F_{jk} = \Pr(j\text{th bin}|k\text{th isoform})$ (i.e., the conditional probability of observing a read in the jth bin given that the read is from the kth isoform). They assume the following linear model:

$$b_j = \sum_{i=1}^{2^J-1} F_{ji}\theta_i + \varepsilon_j, j = 1, \cdots, J,$$

where ε_j is the random noise whose components are independent and have mean 0. Li et al. [15] proposed to apply an ℓ_1 penalized least squares procedure for isoform identification and quantification,

$$\hat{\theta} = \text{argmin}_{\theta_1,\cdots,\theta_n \geq 0} \sum_{j=1}^{J} \left(b_j - \sum_{i=1}^{2^J-1} F_{ji}\theta_i \right)^2 + \lambda \sum_{i=1}^{n} \frac{|\theta_i|}{n_i},$$

where n_i is the number of subexons in the ith isoform. With n_i in the penalty term, θ_i would thus be favored if the length of the isoform n_i is large. Li et al. [15] implemented this method in a software package *SLIDE*.

Li et al. [14] developed a similar approach, *IsoLasso*, which also applies a Lasso regression approach to RNA-seq based transcriptome assembly. They use a simple 0/1 value for F_{jk}, indicating whether the kth isoform contributes to bin j. Using the same idea of regularized regression, Mezlini et al. [16] developed another penalized approach (named *iReckon*) that is different from *IsoLasso* in several ways. First, *iReckon* accepts a set of annotations but also aligns all of the reads to the genome using *TopHat* [26]. The alignments and the known isoforms are used to generate the set of all observed and known splice junctions, which in turn are used to construct splice graphs [5]. A splice graph can represent all the isoforms possibly present within the sample. This also includes pre-mRNA as a possible isoform. Second, instead of using an ℓ_1 penalty, *iReckon* uses a penalty of this form

$$\exp\left\{ \sum_{i=1}^{n} \sqrt[4]{\theta_i} \right\}.$$

The rationale of using this penalty function is that the isoform abundances (in RPKM) are similar to normalized frequencies; they have positivity constraints as well as a fixed sum. This penalty function is the exponential function of the bridge penalty, which can lead to sparse solutions.

The penalized regression approaches in Sect. 12.3 still assume that the all exons and sub-exons of a given gene are known from the current annotations of the transcriptomes. The method of [16] extends the possible isoform set to include those identified from spliced reads based on reads assembly.

12.3.2 *Ab initio Reconstruction of Cell-Type Specific Transcriptomes and lincRNA Quantification*

Methods based on *Ab initio* reconstruction of cell-type specific transcriptomes have also been developed, among which the most important and popular are *Cufflinks* [27] and *Scripture* [4]. These methods both align observed reads to a reference genome using *TopHat*, assemble transcripts, and estimate their abundances. *Cufflinks* constructs a parsimonious set of transcripts that are compatible with the reads observed by reducing the comparative assembly problem to a problem of maximum matching in a weighted bipartite graph.*Cufflinks* tends to choose a minimal set of isoforms. *Cufflinks* can estimate the abundances of the isoforms present in the sample, either using a known reference annotation, or after an ab-initio assembly of the transcripts using only the reference genome. The model used by *Cufflinks* is similar to the Poisson mixture model of Jiang and Wong [8].

Scripture constructs a connectivity graph of individual bases and then applies a statistical segmentation approach to identify paths in the graph that are enriched by reads compared to the background noise. In the path-scoring step, a threshold for genome-wide significance for each region is computed using a sliding window and a test statistic. *Scripture* then constructs a transcript graph that merges all significant windows to generate a set of transcripts in which each node represents an exon and each edge a splice junction. *Scripture* reports all possible isoforms that are comparable with the observed read data, which makes estimating abundances difficult.

Besides quantifying the transcripts that correspond to protein-coding genes, both *Cufflinks* and *Scripture* can also quantify the large intergenic noncoding RNAs (lincRNAs), which are non-protein coding transcripts longer than 200 nucleotides. These lincRNAs can serve as key regulators of diverse cellular processes [3]. Cabili et al. [1] presented an integrative approach to define a reference catalog of more than 8,000 human lincRNAs and found that lincRNA expression is tissue-specific compared with coding genes, and that lincRNAs are typically coexpressed with their neighboring genes. However, the ability of identifying and quantifying these lincRNAs is often limited by lower expression levels of non-coding transcripts relative to that of many protein-coding genes.

12.4 Allele-Specific Transcripts Quantification

RNA-seq can also be used to measure allele-specific expression (ASE) and to identify allelic expression imbalance (AEI) by assigning sequence reads to individual alleles (see Fig. 12.5 for an illustration). In order to do so, we need to have differential sites that are polymorphic. However, relative ASE can be systematically biased when sequence reads are aligned to a single reference genome [25] since allele-specific reads map preferentially to the reference allele when using a single

Fig. 12.5 Allele-specific expression represented as sequencing coverage per allele. Allelic expression imbalance is observed at site with alleles (A, G), but not at site with alleles (T, C)

reference genome to quantify ASE. The inability of mapping reads with more differences from the reference genome than mismatches allowed can lead to underestimation of the abundance of the alternative alleles and therefore cause measures of ASE to be biased toward the reference allele [25]. This has been clearly demonstrated in [25] using both simulated and real data sets. One approach to overcome this potential bias is to map the reads to the reference genome using the program *GSNAP* [31] which allows SNP-tolerant alignment. This was the approach taken by Skelly et al. [23]. Alternatively, one can construct a diploid personal genome sequence using genomic sequence variants (SNPs, indels, and structural variants), and then identify allele-specific events with significant differences in the number of mapped reads between maternal and paternal alleles [21].

For a given differentiating site, we can measure the relative ASE for individual variable sites. A simple binomial exact test can be performed to test the null hypothesis that each allele is equally expressed. It is often more interesting to estimate the relative ASE for individual exons or genes. Skelly et el. [23] proposed a hierarchical Bayesian model for testing the hypothesis of allele-specific gene expression, allowing different differential sites to have different degrees of ASEs. Turro et al. [28] presented another Bayesian approach for haplotype and isoform specific expression estimation using multi-mapping RNA-seq reads.

12.5 Discussion and Future Directions

We have reviewed some statistical methods for identifying all transcripts from RNA-seq data and quantifying the isoform-specific gene expression levels. Our reviews were focused mainly on isoform identification and quantification for one RNA-seq sample. Other very important areas related to RNA-seq data analysis include methods for isoform-specific differential expression analysis and differential exon usage analysis. Steijger et al. [24] have recently presented a comprehensive comparison of different methods for transcript reconstruction from RNA-seq and observed that expression-level estimates varied widely across methods. In addition, assembly of complete isoform structures also poses a major challenge even when all constituent elements are identified. Penalized regression using Lasso provides one approach for isoform structure discovery, however, its results can be sensitive to tuning parameter selection. Statistical methods are needed to quantify the uncertainty of the abundance estimates after the isoforms are identified through variable selection.

Interesting future directions include isoform-specific expression analysis based on multiple RNA-seq samples. Such multi-sample RNA-seq data analyses are especially important in controlling biases that show consistent patterns across multiple RNA-seq samples. An attempt for such a multi-sample RNA-seq data analysis was reported in [29], in which they proposed a Bayesian hierarchical model for multi-sample RNA-seq data analysis and showed a clear reduction of variance the isoform expression estimates. It is also interesting to extend the penalized estimation methods we reviewed to multi-sample RNA-seq data in order to increase the power of identifying the isoforms that are observed in most of the samples. Multiple-sample RNA-seq data can takes into account the population-level isoform expression abundances and allele frequency to improve quantification of allele-specific expression at the level of individual sample. Finally, large-scale RNA-seq data provide a unique opportunity to study how lincRNAs regulate expressions of protein-coding genes.

Acknowledgements This research is supported by NIH grants CA127334 and GM097505.

References

[1] Cabili, M.N., Trapnell, C., Goff, L., Koziol, M., Tazon-Vega, B., Regev, A., Rinn, J.L.: Integrative annotation of human large intergenic noncoding RNAs reveals global properties and specific subclasses. Genes Dev. **25(18)**, 1915–1927 (2011)

[2] Feng, J., Li, W., Jiang, T.: Inference of isoforms from short sequence reads. J. Comput. Biol. **8**(3), 305–321 (2011)

[3] Guttman, M., Rinn, J.: Modular regulatory principles of large non-coding RNAs. Nature **482**, 339–346 (2012)

[4] Guttman, M., Garber, M., Levin, J.Z., Donaghey, J., Robinson, J., Adiconis, X., Fan, L., Koziol, M.J., Gnirke, A., Nusbaum, C., Rinn, J.L., Lander, E.S., Regev, A.: Ab initio reconstruction of transcriptomes of pluripotent and lineage committed cells reveals gene structures of thousands of lincRNAs. Nat. Biotech. **28**(5), 503–510 (2010)

[5] Heber, S., Alekseyev, M., Sze, S., Tang, H., Pevzner, P.A.: Splicing graphs and EST assembly problem. Bioinformatics **18**, S181–S188 (2002)

[6] Hu, Y., Liu, Y., Mao, X., Jia, C., Ferguson, J., Xue, C., Reilly, M., Li, H., Li, M.: PennSeq: accurate isoform-specific gene expression quantification in RNA-seq by modeling non-uniform read distribution. Nucleic Acids Res. **42**(3), e20 (2014)

[7] Jiang, H., Salzman, J.: A penalized likelihood approach for robust estimation of isoform expression arXiv:1310.0379 (2013, preprint)

[8] Jiang, H., Wong, W.H.: Statistical inferences for isoform expression in RNA-seq. Bioinformatics **25**, 1026–1032 (2009)
[9] Langmead, B., Trapnell, C., Pop, M., Salzberg, S.L.: Ultrafast and memory-efficient alignment of short DNA sequences to the human genome. Genome Biol. **10**, R25 (2009)
[10] Lappalainen, T., Sammeth, A., Friedlander, M.R., et al.: Transcriptome and genome sequencing uncovers functional variation in humans. Nature **501**, 506–511 (2013)
[11] LeGault, L.H., Dewey, C.N.: Inference of alternative splicing from RNA-seq data with probabilistic splice graphs. Bioinformatics **29**(18), 2300–2310 (2013)
[12] Li, W., Jiang, T.: Transcriptome assembly and isoform expression level estimation from biased RNA-seq reads. Bioinformatics **28**(22), 2914–2921 (2012)
[13] Li, J., Jiang, H., Wong, W.H.: Modeling non-uniformity in short-read rates in RNA-seq data. Genome Biol. **11**, R50 (2010)
[14] Li, W., Feng, J., Jiang, T.: IsoLasso: a LASSO regression approach to RNA-seq based transcriptome assembly. J. Comput. Biol. **88**(11), 1693–1707 (2011)
[15] Li, J.J., Jiang, C.R., Brown, J.B., Huang, H., Bickel, P.J.: Sparse linear modeling of next-generation mRNA sequencing (RNA-seq) data for isoform discovery and abundance estimation. Proc. Natl. Acad. Sci. **108**(50), 19867–19872 (2012)
[16] Mezlini, A.M., Smith, E.J., Fiume, M., Buske, O., Savich, G.L., Shah, S., Aparicio, S., Chiang, D.Y., Goldenberg, A., Brudno, M.: iReckon: simultaneous isoform discovery and abundance estimation from RNA-seq data. Genome Res. **23**(3), 519–529 (2013)
[17] Montgomery, S.B., Sammeth, M., Gutierrez-Arcelus, M., Lach, R.P., Ingle, C., Nisbett, J., Guigo, R., Dermitzakis, E.T.: Transcriptome genetics using second generation sequencing in a Caucasian population. Nature **464**, 773–777 (2010)
[18] Mortazavi, A., Williams, B. A., McCue, K., Schaeffer, L., Wold, B.: Mapping and quantifying mammalian transcriptomes by RNA-seq. Nat. Meth. **5**, 621–628 (2008)
[19] Pachter, L.: Models for transcript quantification from RNA-seq. Technical Report. University of California, Berkeley (2013)
[20] Pickrell, J.K., Marioni, J.C., Pai, A.A., Degner, J.F., Engelhardt, B.E., Nkadori, E., Veyrieras, J.B., Stephens, M., Gilad, Y., Pritchard, J.: Understanding mechanisms underlying human gene expression variation with RNA sequencing. Nature **464**, 768–772 (2010)
[21] Rozowsky, J., Abyzov, A., Wang, J., Alves, P., Raha, D., Harmanci, A., Leng, J., Bjornson, R., Kong, Y., Kitabayashi, N., Bhardwaj, N., Rubin, M., Snyder, M., Gerstein, M.: AlleleSeq: analysis of allele-specific expression and binding in a network framework. Mol. Syst. Biol. **7**, 522 (2011)
[22] Salzman, J., Jiang, H., Wong, W.H.: Statistical modeling of RNA-seq data. Stat. Sci. **26** (1), 62–83 (2011)
[23] Skelly, D.A., Johansson, M., Madeoy, J., Wakefield, J., Akey, J.M.: A powerful and flexible statistical framework for testing hypotheses of allele-specific gene expression from RNA-seq data. Genome Res. **21**, 1728–1738 (2011)
[24] Steijger, T., Abril, J.F., Engström, P.G., Kokocinski, F., The RGASP Consortium, Hubbard, T.J., Guigó, R., Harrow, J., Berton, P.: Assessment of transcript reconstruction methods for RNRNA-seq. Nat. Meth. **10**, 1177–1184 (2013)
[25] Stevenson, K.R., Coolon, J.D., Wittkopp, P.J.: Sources of bias in measures of allele-specific expression derived from RNA-seq data aligned to a single reference genome. BMC Genom. **14**, 536 (2013)
[26] Trapnell, C., Pachter, L., Salzberg, S.L.: TopHat: discovering splice junctions with RNA-seq. Bioinformatics **25**(9), 1105–1111 (2009)
[27] Trapnell, C., Williams, B.A., Pertea, G., Mortazavi, A.M., Kwan, G., van Baren, M.J., Salzberg, S.L., Wold, B., Pachter, L.: Transcript assembly and quantification by RNA-seq reveals unannotated transcripts and isoform switching during cell differentiation. Nat. Biotech. **28**(5), 511–515 (2010)
[28] Turro, E., Su, S.Y., Gonçalves, Â., Coin, L.J., Richardson, S., Lewin, A.: Haplotype and isoform specific expression estimation using multi-mapping RNA-seq reads. Genome Biol. **12**(2), R13 (2011)

[29] Vardhanabhuti, S., Li, M., Li, H.: A hierarchical Bayesian model for estimating and inferring differential isoform expression for multi-sample RNA-seq data. Stat. Biosci. **5**(1), 244–258 (2013)
[30] Wang, Z., Gerstein, M., Snyder, M.: RNA-seq: a revolutionary tool for transcriptomics. Nat. Rev. Genet. **10**(1), 57–63 (2009)
[31] Wu, T.W., Nacu, S.: Fast and SNP-tolerant detection of complex variants and splicing in short reads. Bioinformatics **26**, 873–881 (2010)

Chapter 13
RNA Isoform Discovery Through Goodness of Fit Diagnostics

Julia Salzman

Abstract There is great interest from the biological community—basic scientists to clinicians—in determining the expressed RNA isoforms in cells. Determining the extent of RNA expression has potential implications for basic scientific models in biology and for diagnosing and treating diseases such as cancer. Next generation sequencing provides an opportunity to discover expressed RNA isoforms that have previously not been detected. Algorithms for detecting these isoforms from RNA-seq data have attracted great interest and have been quite successful. However, even the most widely used algorithms generally do not assess goodness of fit statistics, even when they are based on statistical models. This leads to high rates of false positives in algorithm output and makes real biological signal more difficult to detect. The goal of this chapter is to present a simple statistical method for isoform discovery based on assessing goodness of fit of a statistical model for mismatches of aligned reads to putative isoforms in RNA-seq data.

13.1 Introduction

DNA is essentially a simple quaternary code (A/C/G/T) that is highly stable and encodes almost all the information required for cellular function. Its sequence is generally highly conserved from generation to generation. In ways that are only partially understood, the same DNA quaternary code exists in every cell of organisms like flies, mice and humans, yet, different cells within an organisms have dramatically different shapes and behaviors called "phenotypes".

J. Salzman (✉)
Stanford University, Stanford, CA, USA
e-mail: julia.salzman@stanford.edu

S. Datta and D. Nettleton (eds.), *Statistical Analysis of Next Generation Sequencing Data*, Frontiers in Probability and the Statistical Sciences, DOI 10.1007/978-3-319-07212-8_13,

Much is known about some of the diverse set of phenotypes that can be generated by the same DNA: a major function of DNA is to serve as a template for the production of RNA, which is copied, with important and regulated modifications, from underlying DNA in a process called transcription. RNA performs a huge number of only partially understood functions in the cell: it codes for and regulates the production of proteins, thought to be 'work-horses' in biological systems, and also has functional roles, for example, performing chemical reactions, interacting with DNA to control transcription, performing immune-like roles, and other functions that are currently being brought to light. See [9, 14, 15, 23, 26] for examples of RNA regulation across the tree of life. While almost all human cells have two copies of each chromosome and the same DNA, no matter what cell type is being studied, RNA is much more complex and variable in its expression across cell types, and this complexity and variation is thought to control the diversification of an organism's cells.

There are two major characteristics of the complexity of the expressed RNA in a cell (the "transcriptome"). The first is its copy number or abundance: while each human gene exists in two copies at the level of DNA in a cell, the number of molecules of RNA copied from a gene (the expression level) can vary widely—from zero to tens of thousands of copies per cell. Expression of RNA can be studied using a variety of technologies: from very classical biochemical approaches (called "Northern blotting") to more advanced ones such as quantitative PCR and more recently, microarray analysis. Next generation sequencing (NGS) also provides an opportunity to study RNA expression levels, posing many interesting biological and statistical issues (see, e.g., [7,11,17,20,25], and Chapters 1 through 10 of this book).

The second characteristic of the complexity of RNA expression is the variation in expressed RNA sequences derived from a "gene", which can be thought of as a DNA sequence of n A/C/G/Ts ("nucleotides") that is subsequently processed into RNA. A very simplified model of how DNA is copied to RNA is that this copying is a 1-1 mapping: each DNA nucleotide is either present or absent in any RNA copied from the gene. This is an oversimplified model for most human genes, but can still be potentially useful in some contexts, although we will not discuss them here. Instead, we will focus on studying the only partially understood and very complex process by which DNA is copied into RNA.

The copying is extremely complex because a single DNA sequence can give rise to huge numbers of RNA sequences: essentially any substring of the DNA in a gene can be present in the RNA transcribed from it. In almost all eukaryotes, such as humans, worms, flies and even yeast, the copying mechanism that transcribes DNA into RNA is not 1-1. Instead, expressed RNA sequences are actually substrings of the DNA templates from which they are derived (see Fig. 13.1). This process is called splicing. Since in principle, any substring of the DNA can be processed into RNA, at least $2^n - 1$ linear RNA molecules could be generated from a gene with n bases, and more if reverse splicing is considered [21, 22]. In reality, there are biological mechanisms that restrict the number of RNA molecules that can be spliced from n DNA nucleotides to be much smaller than $2^n - 1$, but still, there is a huge diversity of potential RNA expression.

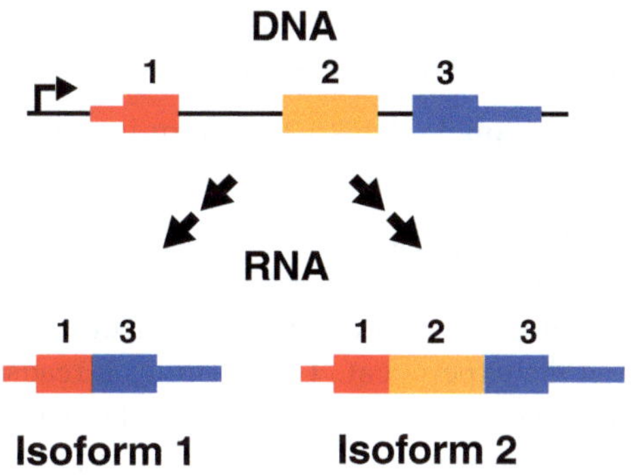

Fig. 13.1 DNA is represented at the top with introns as *thin lines* and exons as *thick colored boxes*. Two RNA isoforms are depicted, differing only in whether they include exon 2

This diversity is fascinating from a biological and statistical perspective, and much remains to be learned about it. Splicing is very important from a biological perspective because RNAs of even slightly different sequence are likely to perform a very different role in the cell. For example, expression of different substrings of RNA from the same gene may contribute to cancer metabolism [27]. The biological community is greatly interested in characterizing the expressed RNA sequences.

In fact, despite decades of study, models for which RNAs are expressed are still incomplete: new transcripts being discovered every month [4]. This is significant both because it means scientists have not fully characterized which RNAs are expressed in human cells—diseased or normal, and because incomplete annotation of isoforms cause the estimates of isoform abundance to be unreliable [1]. Thus, it is of great interest to determine the set of expressed RNAs, called RNA isoforms, and discover new annotated transcripts. In fact, in the past four years, more than 1,500 papers have been published that aim to make such discoveries.

Despite this interest and success, isoform discovery algorithms generally do not assess goodness of fit statistics, even when they are based on statistical models. This leads to high rates of false positives in the algorithm output and makes real biological signal more difficult to detect. The goal of this chapter is to present a simple statistical method for isoform discovery based on assessing goodness of fit of a statistical model for mismatches of aligned reads to putative isoforms in RNA-seq data that can reduce false positive rates in isoform discovery.

13.2 Biological and Statistical Background

To refresh the definition of DNA sequences that may or may not be retained in RNA, consider the following:

Definition 1. An intron is a continuous substring of DNA that is never represented in processed RNA.

Definition 2. An exon is a continuous substring of DNA that may or may not be represented in processed RNA.

Definition 3. An RNA isoform is a specific subset of exons contained in a single RNA molecule and processed from a single DNA sequence.

For simplicity, in this chapter, we assume each RNA string of length L (the "read length") is sampled independently in proportion to its abundance of the RNA in the population. Each letter (nucleotide) is observed with an error dependent only on the experiment. The resulting n reads of length L, $\{s_j\}_{j=1}^n$, are the data. This assumed sampling scheme is used to simplify and state our approach and methods. As is known, this assumption is overly simplifying (see, e.g., [18] and [16]), but all methods developed here can be extended to deal with more intricate parametric error models.

The biological problem motivating the statistical development below is to determine which RNA isoforms are expressed in a sample. The dataset that can be used to address this problem is the set of the nucleotide strings generated in a sequencing run. The following is simple overview of many statistical and bioinformatic methods that estimate the expression of RNA isoforms. It is conceptually outlined for the case of two isoforms (labeled 1 and 2) as follows:

1. The data are a sequence of strings $(A/C/G/T)$ of length L, $\{s_j\}_{j=1}^n$, e.g. $s_1 = AGGA \ldots TAA$.
2. For each $j = 1, \ldots, n$, an aligner (black box) takes the data and outputs the Hamming distance of s_j to a set of all RNA isoforms determined by exon combinations (in this case, just isoforms 1 and 2).
3. If there is an isoform with unique minimal Hamming distance to s_j, s_j is assumed to have been generated by this isoform. For simplicity, we do not consider the case where such a unique isoform does not exist.
4. The read alignments are considered counts and used to estimate the underlying RNA isoform expression through a Poisson model specified in the next section.

The above algorithm is simplified, but it is the basis of many popular alignment and isoform specific gene expression tools (see [17] for a review). Note that the effect of steps 2 and 3 in the above procedure can have significant impact on the resulting analysis but is not typically statistically modeled. This chapter addresses this key point.

For intuition, consider the classical case of sampling from a Gaussian mixture distribution with two means μ_1 and μ_2, which are known, where α, the parameter of interest, satisfies $0 \leq \alpha \leq 1$,

$$\alpha N(\mu_1, 1) + (1 - \alpha) N(\mu_2, 1).$$

If x is sampled from this distribution, one well studied algorithm for assigning the generating mean is to assign x to the distribution with mean μ_i where i satisfies

$$argmin_i ||x - \mu_i||_2^2.$$

In analogy to the problem considered in this chapter, estimating the set of all expressed isoforms of a gene, an observed sequence s_j would be assigned to the isoform i with minimum Hamming distance between s_j and i. In both scenarios, this assignment can also take important features of the underlying statistical model into account. In the case of the Gaussian mixture, clearly, and as is well known, the smaller the difference

$$||x-\mu_1||_2^2 - ||x-\mu_2||_2^2$$

and the larger the value of

$$||x-\mu_i||_2^2$$

the lesser the confidence that x is drawn from $N(\mu_i, 1)$. This is typically quantified using Bayesian posterior probabilities.

In analogy, in the RNA-seq framework, there may be interest in assigning s_j to an isoform. Some isoform assignment algorithms similarly provide a probabilistic estimate of the posterior probability that a particular read is generated from isoform i, and this is called an "alignment score" (see, e.g., [10]).

But importantly, the biological problem of interest is actually analogous to estimating α, not estimating the posterior probability that s_j is drawn from $N(\mu_i, 1)$. In other words, the biological problem of interest is in assessing the probability (analogous to estimating α) that isoform i is expressed and will generate a read, not primarily whether s_j is sampled from isoform i. Therefore, the above algorithm (steps 1–4) is formally a solution to estimating isoform expression levels. But, its use of an assignment of s_j to a particular isoform is not necessary, and moreover, even alignment scores don't properly estimate the underlying statistical quantity of interest. In fact, under mild assumptions, commonly used estimators of isoform abundance will be biased, as we will see below.

To continue the analogy with Gaussian mixtures, in many biological applications, the statistical problem of determining which isoforms are expressed can be formulated as sampling from a mixture of an unknown number of Gaussians (the number being possibly large or small) each with a mixing coefficients $\alpha_0, \ldots, \alpha_N$. The goal (and biological problem) is usually to estimate a potentially sparse solution to α_i. As a note, other authors have suggested a solution to this problem by introducing an L_1 penalty on the number of isoforms in the model (see [13]). The assumption of sparsity may be a useful assumption in many cases, but is not generally true. For example, some well known genes have thousands of RNA isoforms (see [24]).

Below, we develop a statistical framework and provide a disciplined statistical procedure that is analogous to finding a potentially sparse estimate of the α_i. The estimate is obtained by residual analysis from a model that takes into account the mismatches found in alignments and represents an estimate of the isoforms that are actually present in a pool of sequenced RNA.

13.3 Poisson Modeling

13.3.1 Review of Poisson Model Framework

To formulate the typical statistical analysis performed with RNA-seq, we will first introduce a particular case of the model in [20]. See this reference for further discussion of the terminology and background for this model. This model is actually quite general, and will be used subsequently to explore model selection for isoform expression using residual analysis. The notation $Po(\lambda)$ will denote the Poisson distribution with mean λ.

The model specifies that for a gene g with I annotated distinct transcript isoforms, the sequencing reads from g are comprised of J possible distinct read types. The simplest case is to consider a read type to be each unique sequence, although we will consider other statistical properties and definitions of read types in later in the chapter.

We will adopt the following terminology. Let θ be the $I \times 1$ isoform abundance vector whose i-th component represents the expression level of the RNA isoform i. Let A be the $I \times J$ sampling rate matrix with its (i, j)-th element a_{ij} denoting the rate that read type j is sampled from isoform i. For the purposes of this chapter, we will sometimes ignore sequencing depth, although this issue is discussed in [20] and can be modeled with A. Given θ and A, we assume that the $J \times 1$ read count vector n, where n_j denotes the number of reads of type j mapped to any of the I isoforms, follows a Poisson distribution

$$n_j|\theta,A \sim Poisson\left(\sum_{i=1}^{I} \theta_i a_{ij}\right).$$

The log-likelihood function is therefore

$$l(\theta;n,A) = \sum_{j=1}^{J}\left\{n_j \ln\left(\sum_{i=1}^{I} \theta_i a_{ij}\right) - \sum_{i=1}^{I} \theta_i a_{ij}\right\},$$

where the term $-\ln(n_j!)$ was ignored because it does not contain θ. Given this model, the primary goal is to estimate θ.

This is one of many model that have been proposed for isoform-specific RNA-seq (see [17] for a review); the simple most commonly applied special case of this model (and most others) has the following characteristics:

1. It ignores the effect of sequencing errors, with some notable exceptions in work on allele specific expression—see [2].
2. The estimated abundance of an RNA species does not depend on the average number of mismatches (Hamming distance) in the reads aligned to the RNA species.

The sampling rate matrix A is a set of parameters, assumed to be a known function of the sequencing library and gene. For single-end RNA-seq data, the simplest model is to assume the uniform sampling model which assigns a_{ij} as c where c is the sequencing depth (proportional to total number of mapped reads and the length of the transcript) of the experiment if isoform i can generate read type j and is 0 otherwise. For paired-end RNA-seq data, an insert length model can be used such that $a_{ij} = q(l_{ij})c$ if read type j can be mapped to isoform i with insert length (fragment length) l_{ij}, where $q()$ is the empirical probability mass based on all the mapped read pairs.

Because J (the number of distinct read types) is usually very large, especially for paired-end RNA-seq data, we introduced the following collapsing technique. This approach merges read types of proportional sampling rate vectors into read categories which we proved to be minimal sufficient statistics of the model. This does not change the model except that j now represents a read category rather than a read type. Another data reduction technique is to ignore all the read categories with zero read counts by introducing an additional term with the total sampling rates for each isoform $w_i = \sum_{j=1}^{J} a_{ij}$. In this case, the log-likelihood function becomes

$$l(\theta;n,A,w) = \sum_{n_j>0} \left\{ n_j \ln \left(\sum_{i=1}^{I} \theta_i a_{ij} \right) \right\} - \sum_{i=1}^{I} \theta_i w_i. \tag{13.1}$$

The sampling rate matrix A is supposed to be known in the model. Other work [6] has focused on methods for estimating A and assessing statistical properties of these models. Typically, the sampling rate matrix is used to improve statistical models by incorporating inference using paired end RNA-seq [20, 25] and sequence specific biases [16, 18]. However, the sampling rate matrix is quite general. In this chapter we use it to assess the goodness of fit of a model which specifies the expected rate of mismatches between a read and the isoform to which it best aligns. This will be discussed in Sect. 13.4.

13.3.2 Isoform Detection Is Confounded by Mismatches

This section contains the major point of this chapter: false discovery for isoforms expression can be reduced by the following simple observation. The probability of mismatches in alignments to isoforms which appear to be expressed, but are actually artifacts, is greater than the probability of mismatches in alignments to isoforms that are truly expressed. This idea is intuitive, but its statistical formulation is lacking in popular bioinformatic tools for isoform discovery. The concepts are formalized below.

Example 1 (Key example). Suppose an RNA molecule *GGG* is sequenced with an independent error rate per base that misreads G to T with probability p. Suppose n observations (reads) of this molecule are observed, each with length $L = 3$, and an aligner assigns read j (s_j) to the string $I_1 := GGG$ or $I_2 := TTT$ based on the minimum Hamming distance of s_j to I_1 or I_2. Thus, s_j is aligned to I_1 iff and only if its Hamming distance to I_1 is less than or equal to 1. Simple binomial modeling shows that if the true RNA molecule from which read j is derived has the sequence GGG, and if x_j is the number of mismatches between read j and the isoform to which it best aligns, the distribution of x_i is given in the following table:

Observed sequence	x_i	Probability
GGG	0	$(1-p)^3$
GGT, GTG, or TGG	1	$3p(1-p)^2$
GTT, TGT, or TTG	1	$3p^2(1-p)$
TTT	0	p^3

The above table of probabilities leads us to the following simple observation. Suppose for $i = 1$ or $i = 2$, μ_i is expected mismatch rate (Hamming distance) of $\{s_j\}_{j=1}^n$ assigned to I_i. Then, if $p < .5$,

$$\mu_1 < \mu_2, \tag{13.2}$$

i.e., the average number of mismatches in all aligning reads, conditional on being assigned to I_1, is smaller than the average number of mismatches of reads, conditional on being assigned to I_2.

Note that this example easily generalizes to the quaternary code (rather than the binary code (G,T) in the simplified model) as follows. The aligner determines that a read aligns to $I_1 = GGG$ or $I_2 = TTT$ if and only if it has Hamming distance to I_1 or I_2 less than or equal to 1. In this case, let p_T be the probability G is read as T and p_{AC} the probability G is read as A or C, and let $p = p_T + p_{AC}$. The above distribution of x_i conditional on reads aligning to I_1 remains the same; for reads aligning to I_2, $x_i = 1$ with probability $3p_T^2(1-p)$ and $x_i = 0$ with probability p_T^3. Again, if p is small, (13.2) still holds.

This example illustrates that under mild assumptions, if isoform 1 is the only expressed isoform, but is observed with some rate of error, most alignment algorithms, and consequently, most isoform-specific RNA-seq estimation algorithms will find evidence that isoform 2 is expressed, even when all reads aligning to it can be explained by observing isoform 1 with errors, which result in the reads with sequence different than GGG. Thus, while Example 1 seems trivial, it has real consequences: the concept illustrated this example is ignored by almost all RNA-seq algorithms, including those that attempt to discover structural sequence variants; the output of these algorithms will be noisier than necessary, and, as the preceding example illustrates, provably biased.

The following section introduces a statistical model that uses mismatches between reads and the isoform to which they best align (minimal Hamming distance) to detect isoforms whose expression could be attributed to noise. It is

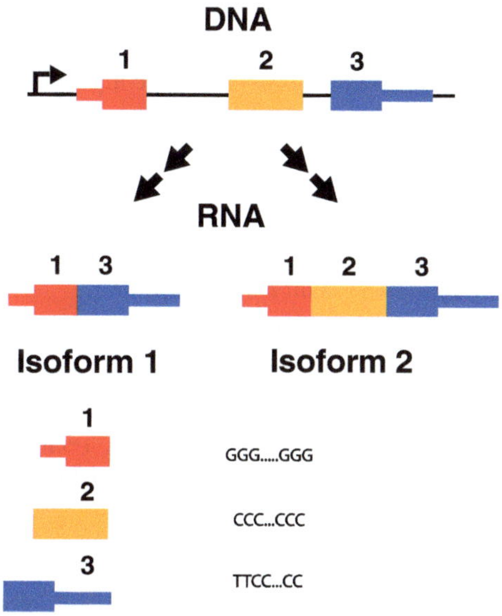

Fig. 13.2 DNA is represented at the top with introns as *thin lines* and exons as *thick colored boxes*. Two RNA isoforms are depicted, differing only in whether they include exon 2. The sequence of each exon is given schematically to represent the statistical challenge: the sequences of exons 2 and 3 are nearly identical, only differing by two bases at their start. If mismatches in the sequencing reads are present, a read that was derived from an exon 1 - exon 3 junction (boundary) may assigned by an alignment algorithm to the second isoform because the error-ridden sequence is the same as the sequence spanning the junction between exon 1 - exon 2

worth noting that some alignment algorithms do assign alignment quality scores to alignments (i.e. posterior probabilities that read j was generated from isoform i). However, this is conceptually and practically very different from Example 1.

We will first restate the intuitive motivation in the real biological problem with a commonly used bioinformatic approach and then give one very simple statistical approach to model selection for the number of expressed isoforms.

Example 2 (Biological example). Consider a case of two isoforms expressed from one gene, as in Fig. 13.2. Suppose exons 2 and 3 are highly homologous, differing only by 2 nucleotides. Suppose that only isoform 1 is expressed, but sequencing is used to test whether isoform 1 or isoform 2 is expressed. Assume that the error rate, that is, the rate a base is misread by the sequencer, is non-zero. If a read is said to align to isoform i ($i = 1$ or $i = 2$) where i is defined so that isoform i and the read have minimal Hamming distance, some reads will align to isoform 2 and its abundance will be estimated as strictly larger than zero as the sampling depth tends to infinity. Therefore, using the log likelihood in (13.1), the estimate of θ will be biased and is inconsistent.

By generalizing the example, with simple statistics, we can achieve much better performance of algorithms used to detect specific isoform expression. The examples provide a statistical basis for the following heuristic algorithm:

Algorithm 1.

1. *Use an empirical null distribution or simulation to model the distribution of* $\hat{\mu}_i$ *(an estimate of* μ_i*) when isoform i is actually not expressed (i.e.* $\theta_i = 0$*), but instead, reads from this isoform can be attributed to sequencing artifacts. See [22] for an example of modeling an empirical null.*
2. *Impose the following model selection procedure:*
   - *(a) Align all reads (e.g. assign a read to the closest isoform in the Hamming distance metric); compute* $\hat{\mu}_i$ *for isoform i.*
   - *(b) For a predefined quantile q where* $0 \leq q \leq 1$*, remove isoform i if* $\hat{\mu}_i$ *falls below the* qth *quantile of the empirical null distribution for* $\hat{\mu}_i$*.*

An implementation of this procedure is presented in [22] where it was empirically successful. This approach will not be further discussed here but has a fuller statistical development. Instead we will explore a residual-based analysis of model selection.

13.4 Model Selection Via the Sampling Rate Matrix

This section develops a basic theory of model selection in a simplified case of isoform-specific expression where we assume there is no multi-mapping, that is, each read can be uniquely assigned to a single isoform by the minimal Hamming distance. The theory can be extended in a straightforward way to the case of multi-mapping.

13.4.1 *Using the Sampling Rate Matrix A to Model Alignment Quality*

As in [20], a_{ij} is interpreted as the probability that isoform i generates a read of type j. Here "type" is quite general and can be used to model the effects of sequence biases, insert lengths or more. In [20], the probabilities corresponding to a_{ij} were introduced and modeled for deconvolving isoform specific expression, ignoring mismatches in alignment. Here we will show how the generality of the model allows it to be used to assess fit of RNA-seq models even in the case of estimating expression of a single isoform. Also, although sampling depth and RNA-seq normalization factors are incorporated into the $\{a_{ij}\}$ when differential expression is of interest, we ignore the issue of normalization in this model, since including a normalizing constant is a straightforward extension and tangential to the purpose of the exposition.

Suppose we are interested in the expression of only one isoform at a time and assume each read has a unique isoform to which it best aligns. In this case, i is repressed, and a_j is the probability that a read of type j is generated from the isoform to which it best aligns. In principle, a_j can be computed for each read as a function of the read start and any feature of the read. Here, we will use a simple model where a_j is a function of the Hamming distance between a read of type j and the isoform to which it is assigned. This leads to a simple minimal statistic for the model and a closed form maximum likelihood estimator.

Definition 4 (Simple mismatch model). Suppose p is the per base probability of an error, errors occur independently at each base, and sequencing reads are of length L. We assume p is known, a common assumption for NGS platforms. A read of type j will correspond to any read that aligns with j mismatches to its best alignment (to isoform i). We assume $a_j = \binom{L}{j} p^j (1-p)^{L-j}$. This assumption for the form of a_j is a reasonable simplifying assumption, but the true a_j parameters depend on the set of other isoforms and their sequence similarity to the isoform whose abundance we wish to estimate. The statistician can choose to model a_j in a way that takes this into account. Our choice of a_j assumes that all reads aligning to an isoform originate from this isoform and that all reads with j mismatches originating from the isoform of interest have a best alignment to that isoform.

Proposition 1. *For a fixed isoform i, let n_k denote the number of reads (of length L where isoform i is the closest alignment in Hamming distance) with k mismatches to isoform i, that is,*

$$n_k = \sum_{1 \le j \le n} 1_{d(s_j,i)=k},$$

where $d(s_j,i)$ is the Hamming distance between read s_j and isoform i. Suppose up to K mismatches are allowed in an alignment. Then the minimal sufficient statistic in the model (13.1) is $(n_0, n_1, \dots, n_K)$. Furthermore, with θ representing the abundance of isoform i,

$$n_k \sim Po(a_k \theta) = Po\left(\binom{L}{k} p^k (1-p)^{(L-k)}\, \theta\right),$$

and the MLE for θ is

$$\hat{\theta} = \frac{\sum_{0 \le k \le K} n_k}{\sum_{0 \le k \le K} \binom{L}{k} p^k (1-p)^{L-k}}.$$

Note that in this formulation, as in Example 1, for small pL, the more mismatches the read has, the less likely it is to have been generated from isoform i. Also, note that empirically, reads with greater than K mismatches are ignored; however, if

$$\sum_{K<k} \binom{L}{k} p^k (1-p)^{L-k}$$

is very small, the resulting bias in $\hat{\theta}$ will be small. Also, note that the above model will give an very biased estimate of θ if all reads aligning to isoform i have mismatches, because of situations such as in Example 1. The latter case can be easily identified with standard statistical techniques to assess the fit of the model, for example, residual analysis.

13.4.2 Residual Analysis

Under Proposition 1, we can assess the fit of each component of the minimal sufficient statistics and perform a standard goodness of fit test. For each k, let

$$\hat{\theta}_k = a_k\hat{\theta} = \binom{L}{k} p^k (1-p)^{L-k}\hat{\theta}.$$

Pearson residuals from this model are

$$\frac{n_k - \hat{\theta}_k}{\sqrt{\hat{\theta}_k}} \qquad \text{for } 0 \leq k \leq K.$$

For the case of isoform TTT in Example 1, n_0 will be much smaller than its expectation causing the residual to large and negative, whereas n_1 will be much larger than its expectation, causing the residual to be large and positive. This lack of fit can be formally detected with the Pearson chi-squared test.

An advantage of residual analysis, and especially χ^2, is that decompositions of χ^2 can used to produce interpretable summaries and ways of diagnosing lack of fit. This idea will just be sketched below; for more details, see [8] or [19]. In the context of sequencing data, this means that it is possible to obtain interpretable summaries to detect lack of fit of the model that is due to specific biases, such as read mismatches or bias in read distribution. An example of a decomposition of χ^2 and how it can be applied to Example 1 follows.

13.4.3 Detecting Lack of Fit

In our situation, the Pearson chi-square statistic is:

$$\chi^2 = \sum_{k=0}^{K} \frac{(n_k - \hat{\theta}_k)^2}{\hat{\theta}_k}.$$

A large value of χ^2 implies a lack of fit of the data to the model. In this case, the model is primarily an error model, and lack of fit is consistent with artifactual expression of the isoform due to sequencing error. If K is small, each individual residual can be examined. A real data example is presented below:

Example 3 (Leukocyte data [21]). In this example, we illustrate how residual analysis can be applied by considering alignments to the gene encoding a hemoglobin protein (HBG1, NM 000559) from the dataset [21]. As described in [21], evidence of exon rearrangement in the sample was identified by testing all reads that failed alignment to the human transcriptome as currently annotated. Using this method, a particular isoform of HBG1 was apparently expressed: 80 reads mapped to a junction between exon 3 and exon 1. Read lengths were 80 nucleotides.

Examining alignments of these reads reveals that 79 reads mapping to this putative junction have three mismatches and one read has two mismatches. This rate is much higher than what is expected, and intuitively means that this gene model (an exon 3 - exon 1 junction) will be identified as an artifact.

Fitting a Poisson model where $p = .01$ and $K = 3$ yields the following estimate of θ:

$$\hat{\theta} = \frac{0+0+1+79}{\left(\binom{80}{0}p^0(1-p)^{80} + \binom{80}{1}p^1(1-p)^{79} + \binom{80}{2}p^2(1-p)^{78} + \binom{80}{3}p^3(1-p)^{77}\right)}$$

Therefore, $\widehat{\theta_k} = \binom{80}{k}p^k(1-p)^{(80-k)}80.7$. The residuals r_k for $0 \le k \le 3$ are

$$r_0 = -6.01,\ r_1 = -5.40,\ r_2 = -3.12,\ \text{and}\ r_3 = 43.43.$$

It is clear that the Pearson χ^2 statistic rejects the fit of this model, and we conclude that the putative expression of a novel isoform of HBG1 is highly likely to be an artifact because of the size of r_3 which reflects the excess of reads with 3 mismatches to the gene.

13.4.4 Extensions

We have considered a simple method to detect lack of fit for residuals using a simple error model that models the probability of a particular read being generated by an isoform as a binomial random variable. This assumes each position in each read is equally likely to produce an error. However, this error model is overly simplified: it has been shown that real RNA-seq data has positional and sequence-dependent error rates.

Several approaches have been developed to model and correct such biases [3, 12, 18]. However, completely removing sampling biases is almost impossible because the technical procedure of sequencing and read mapping is often too complex.

More involved models of the mismatch rate, which take into account the position of the mismatch or its sequence specific context, can also be adapted for residual analysis by estimating the expression of an isoform and subsequently testing whether the residuals, through χ^2 are consistent with this model. In applied statistics, the decision to reject a particular model (isoform) can be made based on an empirical null distribution of χ^2 or a simulation. In addition, projections of χ^2 (see [19]) could instead be used to determine lack of fit and possibly to improve genome-wide error models.

13.5 Conclusion

This chapter presents two simple methods for detecting lack of fit between a simple one-isoform Poisson model and the raw sequencing alignments. We illustrated the method in this special case, and it is easy to see how it extends to more complex models of isoform specific expression (such as in [21]), non-Poisson models, such as the negative binomial, or more complex models of error rates (such as in [16]). Using these simple methods may dramatically reduce the false positive rates of isoform-discovery algorithms. Indeed, we have applied a more basic version of this algorithm [21] to successfully significantly reduce false positives.

Because we suggest model selection to be based on χ^2, the method as outlined here can be adapted to statistical measures of lack of fit of a model such as projections of χ^2, a simple case of which is considering the norm of individual residuals. Other examples include decomposing χ^2 to reflect many other features of interest in NGS data, such as position-specific distribution of mismatches in a read's alignment and lack of fit of a paired end insert length model. Pursuing these directions may yield useful statistical tools.

In summary, we have applied a simple statistical procedure to diagnose lack of fit, such as was suggested in [5], to NGS data. This is not a new statistical concept, but it rarely used by applied statisticians and practicing bioinformaticians. These simple statistical procedures are likely to significantly improve the performance of many commonly used bioinformatic algorithms. These algorithms are put into practice to address scientific questions that are investigated at a cost millions of dollars each year. Thus simple statistical ideas may lead to huge savings in costs required for validation and follow-up.

Acknowledgements I thank the editors for helpful comments that improved the exposition of this chapter.

References

[1] Black Pyrkosz, A., Cheng, H., Titus Brown, C.: RNA-Seq Mapping Errors When Using Incomplete Reference Transcriptomes of Vertebrates. ArXiv e-prints (2013)

[2] Degner, J.F., Marioni, J.C., Pai, A.A., Pickrell, J.K., Nkadori, E., Gilad, Y., Pritchard, J.K.: Effect of read-mapping biases on detecting allele-specific expression from RNA-sequencing data. Bioinformatics **25**(24), 3207–3212 (2009). doi:10.1093/bioinformatics/btp579. http://bioinformatics.oxfordjournals.org/content/25/24/3207.abstract

[3] Hansen, K.D., Brenner, S.E., Dudoit, S.: Biases in Illumina transcriptome sequencing caused by random hexamer priming. Nucleic Acids Res. **38**(12), e131 (2010)

[4] Harrow, J., Frankish, A., Gonzalez, J.M., Tapanari, E., Diekhans, M., Kokocinski, F., Aken, B.L., Barrell, D., Zadissa, A., Searle, S., Barnes, I., Bignell, A., Boychenko, V., Hunt, T., Kay, M., Mukherjee, G., Rajan, J., Despacio-Reyes, G., Saunders, G., Steward, C., Harte, R., Lin, M., Howald, C., Tanzer, A., Derrien, T., Chrast, J., Walters, N., Balasubramanian, S., Pei, B., Tress, M., Rodriguez, J.M., Ezkurdia, I., van Baren, J., Brent, M., Haussler, D., Kellis, M., Valencia, A., Reymond, A., Gerstein, M., Guigio, R., Hubbard, T.J.: Gencode: the reference human genome annotation for the encode project. Genome Res. **22**(9), 1760–1774 (2012)

[5] Hoaglin, D.: A poissonness plot. Am. Stat. **34**(3), 146–149 (1980)

[6] Jiang, H., Salzman, J.: A penalized likelihood approach for robust estimation of isoform expression. arXiv:1310.0379 (2013, preprint)

[7] Jiang, H., Wong, W.H.: Statistical inferences for isoform expression in RNA-seq. Bioinformatics **25**(8), 1026–1032 (2009)

[8] Kemp, A., Kemp, D.: Weldon's dice data revisted. Am. Stat. **45**(3), 216–222 (1991)

[9] Keren, H., Lev-Maor, G., Ast, G.: Alternative splicing and evolution: diversification, exon definition and function. Nat. Rev. Genet. **11**(5), 345–355 (2010). doi:10.1038/nrg2776. http://www.ncbi.nlm.nih.gov/pubmed/20376054

[10] Langmead, B.: Aligning short sequencing reads with Bowtie. In: Baxevanis, A.D., et al. (eds.) Current Protocols in Bioinformatics/Editoral Board, Chapter 11, Unit 11 7 (2010). doi:10.1002/0471250953.bi1107s32. http://www.ncbi.nlm.nih.gov/pubmed/21154709

[11] Li, B., Dewey, C.N.: Rsem: accurate transcript quantification from rna-seq data with or without a reference genome. BMC Bioinform. **12**, 323 (2011)

[12] Li, J., Jiang, H., Wong, W.H.: Modeling non-uniformity in short-read rates in rna-seq data. Genome Biol. **11**(5), R50 (2010)

[13] Li, J.J., Jiang, C.R., Brown, J.B., Huang, H., Bickel, P.J.: Sparse linear modeling of next-generation mRNA sequencing (RNA-seq) data for isoform discovery and abundance estimation. Proc. Natl. Acad. Sci. **108**(50), 19,867–19,872 (2011). doi:10.1073/pnas.1113972108. http://www.pnas.org/content/108/50/19867.abstract

[14] Lopez-Bigas, N., Audit, B., Ouzounis, C., Parra, G., Guigo, R.: Are splicing mutations the most frequent cause of hereditary disease? FEBS Lett. **579**(9), 1900–1903 (2005)

[15] Marquez, Y., Brown, J.W., Simpson, C., Barta, A., Kalyna, M.: Transcriptome survey reveals increased complexity of the alternative splicing landscape in Arabidopsis. Genome Res. **22**(6), 1184–1195 (2012). doi:10.1101/gr.134106.111. http://www.ncbi.nlm.nih.gov/pubmed/22391557

[16] Meacham, F., Boffelli, D., Dhahbi, J., Martin, D.I., Singer, M., Pachter, L.: Identification and correction of systematic error in high-throughput sequence data. BMC Bioinform. **12**, 451 (2011). doi:10.1186/1471-2105-12-451. http://www.ncbi.nlm.nih.gov/pubmed/22099972

[17] Pachter, L.: Models for transcript quantification from RNA-Seq. ArXiv e-prints (2011)

[18] Roberts, A., Trapnell, C., Donaghey, J., Rinn, J.L., Pachter, L.: Improving rna-seq expression estimates by correcting for fragment bias. Genome Biol. **12**(3), R22 (2011)

[19] Salzman, J.: Spectral analysis with markov chains. Ph.D. thesis, Stanford (2007)

[20] Salzman, J., Jiang, H., Wong, W.H.: Statistical modeling of RNA-Seq data. Stat. Sci. **26**(1), 62–83 (2011)

[21] Salzman, J., Gawad, C., Wang, P.L., Lacayo, N., Brown, P.O.: Circular rnas are the predominant transcript isoform from hundreds of human genes in diverse cell types. PLoS ONE **7**(2), e30,733 (2012)
[22] Salzman, J., Chen, R.E., Olsen, M.N., Wang, P.L., Brown, P.O.: Cell-type specific features of circular RNA expression. PLoS Genet. **9**(9), e1003,777 (2013)
[23] Sorber, K., Dimon, M.T., DeRisi, J.L.: RNA-Seq analysis of splicing in Plasmodium falciparum uncovers new splice junctions, alternative splicing and splicing of antisense transcripts. Nucleic Acids Res. **39**(9), 3820–3835 (2011). doi:10.1093/nar/gkq1223. http://www.ncbi.nlm.nih.gov/pubmed/21245033
[24] Sun, W., You, X., Gogol-Doring, A., He, H., Kise, Y., Sohn, M., Chen, T., Klebes, A., Schmucker, D., Chen, W.: Ultra-deep profiling of alternatively spliced Drosophila Dscam isoforms by circularization-assisted multi-segment sequencing. EMBO J. **32**(14), 2029–2038 (2013). doi:10.1038/emboj.2013.144. http://www.ncbi.nlm.nih.gov/pubmed/23792425
[25] Trapnell, C., Williams, B.A., Pertea, G., Mortazavi, A., Kwan, G., van Baren, M.J., Salzberg, S.L., Wold, B.J., Pachter, L.: Transcript assembly and quantification by rna-seq reveals unannotated transcripts and isoform switching during cell differentiation. Nat. Biotech. **28**(5), 511–515 (2010)
[26] Wang, E.T., Sandberg, R., Luo, S., Khrebtukova, I., Zhang, L., Mayr, C., Kingsmore, S.F., Schroth, G.P., Burge, C.B.: Alternative isoform regulation in human tissue transcriptomes. Nature **456**(7221), 470–476 (2008)
[27] Yang, W., Lu, Z.: Nuclear PKM2 regulates the Warburg effect. Cell Cycle **12**(19), 3154–3158 (2013). doi:10.4161/cc.26182. http://www.ncbi.nlm.nih.gov/pubmed/24013426

Chapter 14
MOSAiCS-HMM: A Model-Based Approach for Detecting Regions of Histone Modifications from ChIP-Seq Data

Dongjun Chung, Qi Zhang, and Sündüz Keleş

Abstract Chromatin immunoprecipitation followed by high throughput sequencing (ChIP-seq) experiments are routinely utilized for studying epigenomics of transcriptional regulation. We review some of the important statistical issues in the analysis of these experiments and extend our previous model for the analysis of ChIP-seq data of transcription factors, named MOSAiCS, with a hidden Markov model architecture (MOSAiCS-HMM). MOSAiCS-HMM provides a model-based approach for modeling read counts in histone modification ChIP-seq experiments and accounts for the spatial dependence in their ChIP-seq profiles. In addition, its R package implementation provides many functionality for summarizing these data and generating files that can be directly uploaded to the UCSC genome browser.

D. Chung (✉)
Department of Biostatistics, Yale School of Public Health, Yale University, 60 College Street, New Haven, CT 06520, USA
e-mail: dongjun.chung@yale.edu

Q. Zhang
Department of Biostatistics and Medical Informations, School of Public Health and Medicine, University of Wisconsin, 2130C Genetics/Biotechnology Center, 425 Henry Mall, Madison, WI 53706, USA
e-mail: qizhang@stat.wisc.edu

S. Keleş
Departments of Statistics and of Biostatistics and Medical Informatics, University of Wisconsin, 2124 Genetics/Biotechnology Center, 425 Henry Mall, Madison, WI 53706, USA
e-mail: keles@stat.wisc.edu

S. Datta and D. Nettleton (eds.), *Statistical Analysis of Next Generation Sequencing Data*, Frontiers in Probability and the Statistical Sciences, DOI 10.1007/978-3-319-07212-8_14,

14.1 Introduction

Regulation of gene expression is a multi-faceted process. DNA binding proteins, i.e., transcription factors, and histone modifications are two of the critical mechanisms for regulating gene expression. Transcription factors (TFs) interact with the DNA in a sequence specific or non-specific manner and can act alone or in protein complexes with co-factors. They promote (activate) or block (repress) expression of their specific target genes. In contrast, histones are a specific class of proteins that package DNA. Every 146 base pairs of DNA winds around a histone complex consisting of two of each of the H2A, H2B, H3, and H4 histone proteins, and form the structural unit of DNA called nucleosomes. The H3 and H4 histones have long tails that can be covalently modified at several places. Methylation, acetylation, and phosphorylation are some of the most commonly studied histone modifications and they affect diverse biological processes including gene regulation [30].

Chromatin immunoprecipitation followed by high throughput sequencing (ChIP-seq) has become a versatile experimental technique for profiling TF-DNA interactions, histone modifications, chromatin remodeling enzymes, RNA polymerase, and nucleosomes [2,15]. A typical TF or histone modification ChIP-seq experiment involves isolating regions of the genome interacting with the protein of interest or undergoing the targeted modification. This is accomplished by first cross-linking proteins and associated chromatin in a cell lysate and then shearing DNA to an average of 500 base pair fragments. Then, the DNA fragments associated with the protein of interest are selectively captured by immunoprecipitation with an antibody specific to that protein. In the case of histone modifications, antibodies targeting specific histone proteins with a specific modification are utilized. The associated DNA fragments are then purified and one (single-end sequencing) or both ends (paired-end sequencing) of the captured fragments are sequenced by using a high throughput sequencing platform.

These high throughput in vivo biological assays are embraced by large consortia projects such as ENCODE [10] and RoadMap EpiGenomics [4] and have resulted in large volumes of publicly available data. ChIP-seq experiments for transcription factors enable identification of where a protein binds to in the genome in vivo, whereas experiments targeting histone modifications identify which regions of the genome are undergoing the targeted histone modifications. Because both binding of transcription factors and histone modifications play important roles in cell specific gene regulatory programs, their genome-wide mapping is crucial for understanding and diagnosing human diseases.

Characteristics of data from ChIP-seq experiments vary based on what is being profiled (e.g., transcription factor, modified histone, RNA polymerase) and which sequencing parameters (e.g., single-end, paired-end) are being utilized. Illumina platform is by far the most popular choice for ChIP-seq experiments [2,15,21,27]. As a result of sequencing, reads of size 36–100 base pairs (bps) representing one or both ends of immunoprecipitated DNA fragments with varying lengths are obtained. The lengths of DNA fragments are typically kept around 150–300 bps for

Fig. 14.1 *Typical work flow of statistical analysis of ChIP-seq experiments*

A. Sequencing of the ChIP and control samples:
Single-end, Paired-end, Sequencing depth.

B. Alignment to the reference genome:
Choice of aligner (bowtie, bwa), Handling of multi-mapping reads.

C. Quality control:
Duplicate removal, Cross-correlation analysis.

D. Enrichment analysis:
Identify and filter enriched regions (FDR control, IDR control).

E. Downstream analysis:
Visualize on the genome browser, Binding event deconvolution for TFs, Motif analysis for TFs, Boundary adjustment for histone modifications.

optimal sequencing by a size selection step in the experimental protocol. ChIP-seq experiments are typically coupled with control experiments which either skip the immunoprecipitation step (Input control) or use a non-specific antibody (IgG control) to measure non-specific protein DNA interactions and characterize background read distribution. Compared to their array-based analogues (ChIP-chip experiments [5, 16]), ChIP-seq provides higher resolution and genomic coverage [38].

Analysis of ChIP-seq data involves multiple steps from quality assessment to downstream analysis for biological interpretation (Fig. 14.1). The main statistical task is, however, identifying regions of the genome that exhibit significantly higher levels of ChIP read counts compared to background read counts. Figure 14.2 displays ChIP and Input control profiles for such a region from a H3K4me3 experiment in GM12878 cell lines which was generated as part of the ENCODE project [10]. There are a plethora of computational and statistical approaches for analyzing data from ChIP-seq experiments (reviewed in [1]). Most of the well-studied approaches [7, 17, 18, 26, 37] are geared towards ChIP-seq experiments of transcription factors which generate punctuated peaks. In such data, ChIP reads concentrate on the TF-DNA interaction sites and have a clear summit. In contrast, ChIP-seq experiments profiling modified histones can result in punctuated, broad (e.g., H3K27me3, H3K36me3, and H3K9me3), or a mixture of punctuated and broad peaks and show larger variations in the widths of the enriched regions compared to TF ChIP-seq. Methods for analyzing ChIP-seq data of histone modifications either require running methods for punctuated signals in a special "broad" model [17, 37] or primarily focus on identifying differential histone modifications [23, 28, 32, 34]. Recently, a stochastic Bayesian Change-Point method named BCP [33] has been proposed for the analysis of diffuse histone ChIP-seq data and has been shown to be also effective in analyzing punctuate transcription factor ChIP-seq data.

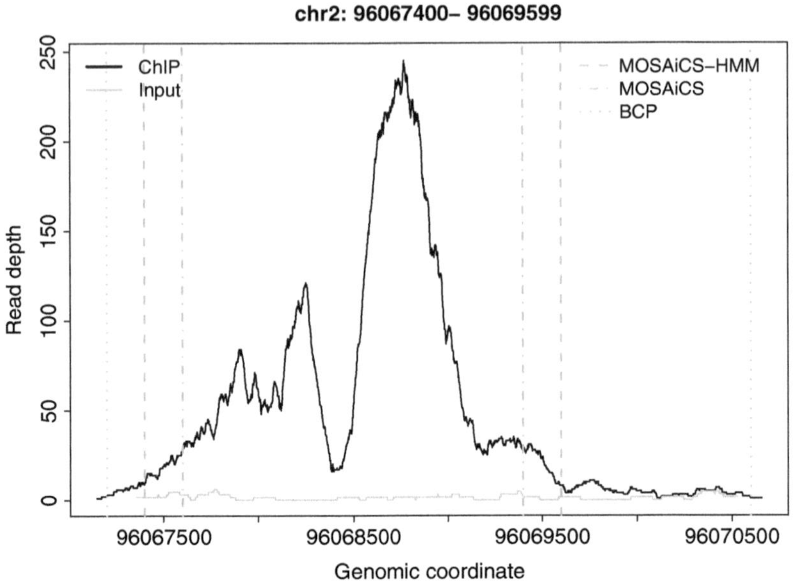

Fig. 14.2 *H3K4me3 ChIP-seq read profile generated by* R *package* dpeak *[7]. Black* and *gray curves* depict ChIP and sequencing depth normalized Input read counts for a peak identified by all of MOSAiCS-HMM, MOSAiCS, and BCP. Vertical lines depict the boundaries of the peak as determined by different peak callers

We have recently developed a model-based, versatile method, named MOSAiCS (Model-based One- and Two-Sample Analysis and Inference for ChIP-seq), for the analysis of ChIP-seq data [18]. MOSAiCS accommodates both one- (in the absence of a control sample) and two-sample analysis of ChIP-seq data. Unlike other popular ChIP-seq methods that consider explicit modeling of data only under the null hypothesis of no enrichment [26, 37], MOSAiCS provides biologically motivated statistical models for reads that arise under both non-enrichment (background) and enrichment (signal). Furthermore, MOSAiCS builds a parametric background model that takes into account biases such as GC content [8] and mappability [38] that are inherent to ChIP-seq data. MOSAiCS model does not assume punctuated or broad peak structures but instead quantifies whether the ChIP reads show enrichment compared to the background reads for every genomic interval (e.g., bin) of user defined size in the genome. Although such analysis captures most parts of the broad domains, large regions with low but consistent enrichment might be prone to misidentification. In this paper, we extend the MOSAiCS model with a hidden Markov model architecture to allow spatial dependence between adjacent bins and facilitate identification of broad enriched regions in ChIP-seq data. We conclude with a brief discussion of other issues concerning ChIP-seq data analysis (Fig. 14.1).

14.2 MOSAiCS-HMM Model

14.2.1 MOSAiCS

We first review the MOSAiCS model [18] that MOSAiCS-HMM builds on. Previous work by others and us have established that next generation sequencing datasets including naked DNA, Input DNA, and ChIP samples are prone to sequencing and other sources of biases [3, 8, 18, 26]. Specifically, observed read counts are affected by local sequence characteristics such as mappability and GC content. In order to correct these biases and obtain accurate measurements of enrichment signals, we developed MOSAiCS, a flexible mixture model that incorporates various sequence biases in modeling the background read distribution. We implemented the MOSAiCS model as `R` package `mosaics` which is available from *Bioconductor* (http://www.bioconductor.org/) [12]. In this `R` package, the MOSAiCS model is implemented in a computationally efficient way by using `Rcpp` and `parallel` `R` packages for *C++* implementation and parallel computing, respectively. `mosaics` package also provides various tools for exploratory analysis, model fitting, model selection, and diagnostics for ChIP-seq data analysis with MOSAiCS [31].

In the MOSAiCS model, reference genome is divided into non-overlapping intervals (e.g., bins) of typically 200 bps. We consider ChIP reads in each bin as arising from a mixture of non-enriched and enriched distributions. Let Y_j denote the ChIP read counts in j-th bin. Let M_j and GC_j be the bin-specific mappability and GC content scores. These quantities are defined as functions of base pair mappability and GC scores [6]. For a read length of k and library size of L, let $x_{(i):(i+k-1)}$ denote the $k$mer starting at position i and ending at position $i+k-1$ from $5'$ to $3'$. Let $x^c_{(i):(i-k+1)}$ denote the $k$mer starting at position i and ending in $i-k+1$ in the other strand. Then, the nucleotide-level mappability is defined as:

$$\delta_i = \begin{cases} 1 \text{ if} & x_{(i):(i+k-1)} \quad \text{is unique,} \\ 0 \text{ o.w.} \end{cases}$$

Mappability for a position in the reverse strand is similarly defined as:

$$\delta_i^c = \begin{cases} 1 \text{ if} & x^c_{(i):(i-k+1)} \quad \text{is unique,} \\ 0 \text{ o.w.,} \end{cases},$$

where $\delta_i^c = \delta_{i-k+1}$. The GC content at the nucleotide level is defined similarly by setting $\delta_i = I\{i\text{-th position is a G or C}\}$, where $I\{.\}$ is an indicator function. In the MOSAiCS model, bin-level versions of these quantities are utilized to account for the fact that the total number of observed counts at position i could be contributed by forward strand reads that originate between positions $i-L+1$ and i and get extended to L bps or the reverse strand reads that originate between positions i and $i+L-1$ and get extended to L bps. The bin-level mappability/GC content for single-end reads is defined as:

$$\delta_i^* = \frac{1}{2L}\left(\sum_{j=i-L+1}^{i} \delta_j + \sum_{j=i}^{i+L-1} \delta_j^c\right), \tag{14.1}$$

$$= \frac{1}{2L}\left(\sum_{j=i-L+1}^{i} \delta_j + \sum_{j=i-k+1}^{i+L-k} \delta_j\right). \tag{14.2}$$

Bin-level mappability and GC content scores for paired-end reads can be computed similarly by taking into account the actual lengths of the fragments that two end reads represent.

When a matching control sample, such as Input control, is available, we further denote the control read counts in j-th bin by X_j. Finally, we denote the indicator of enrichment status of j-th bin as Z_j, where $Z_j = 1$ if j-th bin is enriched, i.e., exhibiting TF binding or histone modification, and $Z_j = 0$ otherwise. We assume that enrichment status of individual bins are independent and is given as follows for $j = 1, 2, \cdots, M$,

$$\Pr(Z_j = 0) = \pi_0, \quad \Pr(Z_j = 1) = 1 - \pi_0. \tag{14.3}$$

Given these underlying enrichment states for j-th bin, we assume that

$$(Y_j|Z_j = 0) \sim N_j, \quad (Y_j|Z_j = 1) \sim N_j + S_j, \tag{14.4}$$

where N_j and S_j represent background and signal, respectively. MOSAiCS models reads from the background component with Negative Binomial regression:

$$N_j \sim NegBin(a, a/\mu_j), \tag{14.5}$$

where we model its mean, μ_j, slightly differently under three different scenarios. The specifications of these models emerged from exploratory analysis of a large collection of ENCODE datasets [18] and other datasets across multiple organisms [14, 22, 29]. The mappability scores contribute the mean model with a log transformation to account for the curvature that is apparent from the mappability versus ChIP read count relationship. Similarly, the piecewise linear B-spline model for the GC-content score enables a flexible way of capturing the GC content versus ChIP read count relationship observed in multiple ChIP-seq datasets. Next, we detail the three mean models and discuss the conditions under which they are appropriate.

- Case 1: In the absence of a control sample:

$$\log \mu_j = \beta_0 + \beta_M \log_2(M_j + 1) + \beta_{GC}' \mathbf{Sp}(GC_j),$$

where $\mathbf{Sp}(GC_j)$ is a vector of piecewise linear B-spline basis functions with knots at the first and third quantiles of the GC content. β_{GC} is vector-valued

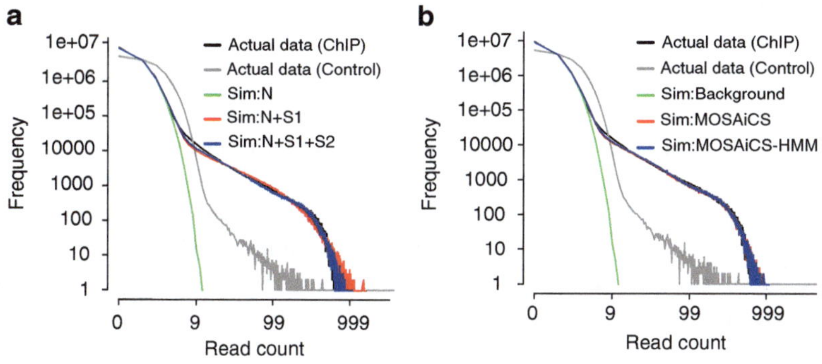

Fig. 14.3 *Goodness-of-fit (GOF) plots.* (**a**) MOSAiCS goodness-of-fit plot for replicate 1. Both axes are in the log10 scale. SIM:N: reads simulated from the estimated background read distribution. Sim:N+S1: reads simulated from the MOSAiCS model with one signal component for the ChIP reads. Sim:N+S1+S2: reads simulated from the MOSAiCS model with two signal components for the ChIP reads. Simulated data Sim:N+S1+S2 overlap the actual ChIP data well, indicating good overall fit. (**b**) MOSAiCS and MOSAiCS-HMM goodness-of-fit plot for replicate 1. Both axes are in the log10 scale. SIM:Background: reads simulated from the estimated background read distribution. Sim: MOSAiCS: reads simulated from the estimated MOSAiCS model with two signal components for the ChIP reads. Sim: MOSAiCS-HMM : reads simulated from the estimated MOSAiCS-HMM model with two signal components for the ChIP reads

and represents all the coefficients in the spline model. Current standard practice for ChIP-seq experiments is to couple each ChIP sample with a Input control sample. However, investigators occasionally generate ChIP samples without control samples especially when choosing among different antibodies for the same factor. This mean model facilitates the analysis of such samples without a control sample by approximating the background mean read counts using mappability and GC content scores.

- Case 2a: In the presence of a shallowly sequenced control sample:

$$\log \mu_j = \beta_0 + \left[\beta_M \log_2 (M_j + 1) + \beta'_{GC}\mathbf{Sp}(GC_j) + \beta_{X1}X_j^d\right] 1\left\{X_j \le s\right\}$$
$$+\beta_{X2}X_j^d 1\left\{X_j > s\right\},$$

where s and d are tuning parameters. This model is essentially performing a power transformation with exponent d on the control read counts and incorporating mappability and GC content values for bins with less than or equal to s control read counts. In our previous work [18], we have shown that even in the presence of a control sample, utilizing mappability and GC content values for estimating the background read distribution might improve detection power and eliminate false positives. From a practical point of view, inclusion of mappability and GC content values matters the most when the background read distribution cannot be estimated well just based on the control sample. MOSAiCS framework provides goodness-of-fit plots (Fig. 14.3a) which aid in this decision.

- Case 2b: In the presence of an adequately sequenced control sample:

$$\log \mu_j = \beta_0 + \beta_X X_j^d,$$

where d is again the exponent in the power transformation of the control read counts. This model is suitable for cases where the control sample is deeply sequenced and the fit can again be evaluated by the goodness-of-fit plots provided by MOSAiCS. Since its publications, we have applied MOSAiCS to tens to a few hundreds of datasets and observed that $s = 2$ and $d = 0.25$ work well in practice. Therefore, these values are currently the default values in the `mosaics` `R` package.

For the signal component, we consider both a single negative binomial distribution and a mixture of two negative binomial distributions, i.e.,

$$(1) \quad S_j \sim NegBin(b, c) + k,$$

$$(2) \quad S_j \sim p_1 NegBin(b_1, c_1) + (1 - p_1) NegBin(b_2, c_2) + k,$$

where k is a constant set to 3 and represents the minimum observable read count in an enriched region. The optimal model for signal component is determined based on Bayesian information criterion (BIC). The parameters in the MOSAiCS model are estimated using a computationally efficient Expectation-Maximization (EM) algorithm described in [18].

After we fit the MOSAiCS model, enriched regions are identified using a direct posterior probability approach [24] for false discovery rate (FDR) control based on the posterior probability that read counts for each bin are generated from the background component. Specifically, we first rank the bins according to increasing values of $\Pr\left(Z_j = 0|\mathbf{Y}; \hat{\Theta}\right)$, where $\hat{\Theta}$ denotes the final parameter estimates obtained from the EM algorithm. Let fdr_j denote the sorted $\Pr\left(Z_j = 0|\mathbf{Y}; \hat{\Theta}\right)$ values. Then, we increase the cutoff κ until the expected proportion of false discoveries given by

$$\frac{\sum_{j=1}^{M} fdr_j 1\left\{fdr_j \leq \kappa\right\}}{\sum_{j=1}^{M} 1\left\{fdr_j \leq \kappa\right\}},$$

reaches the pre-specified cutoff (α) for false discovery rate. Finally, using this determined cutoff $\hat{\kappa}$, bins satisfying the condition that $\Pr\left(Z_j = 0|\mathbf{Y}; \hat{\Theta}\right) \leq \hat{\kappa}$ are reported as enriched regions. This FDR control ensures that reported enriched regions achieve a certain level of statistical significance. However, in addition to statistical significance, investigators often would like to require each enriched region to have a minimum number of ChIP reads. Therefore, `R` package `mosaics` allows such a threshold as input. In practice, setting this threshold to a certain percentile (e.g., $0.90 - 0.99$) of the ChIP read count distribution works well if the control sample is shallowly sequenced (e.g., less than 20 million reads for human samples). In the presence of a deeply sequenced control sample, this threshold can also be set to a depth normalized percentile of the control read count distribution.

14.2.2 MOSAiCS-HMM

In the MOSAiCS model, enrichment states of adjacent bins are assumed to be independent. This assumption might be mildly violated in practice for the ChIP-seq data of TFs with narrow enrichment profiles that typically span 1–3 bins. However, it is more likely to be invalid for the ChIP-seq data of histone modifications which can easily cover a larger number of bins and might exhibit broad enrichment signals. In the case of broad signals, multiple adjacent bins constitute a wide block-shaped peak and a spatial correlation structure underlies the relation between enrichment status of adjacent bins. Hidden Markov Models (HMMs) provide a graceful way to handle these types of spatial correlations without losing spatial resolution (reviewed in [9] and [25] among many others). This observation motivates our development of the MOSAiCS-HMM framework to account for spatial correlations in ChIP-seq data.

In MOSAiCS-HMM, we assume that enrichment states constitute a Markov chain along each chromosome. Specifically, Eq. (14.3) of the MOSAiCS model is replaced by

$$\Pr\left(Z_{j+1}=b|Z_j=a\right)\equiv\pi_{ab},\quad a,b\in\{0,1\},\quad j=1,\cdots,M-1 \qquad (14.6)$$

and $\sum_{b=0}^{1}\pi_{ab}=1$ for $a=0,1$. Finally, conditional on these underlying enrichment states, ChIP read counts are assumed to follow the read count distributions of the MOSAiCS model, given in Eqs. (14.4), (14.5), and (14.6). This allows effective adjustment of sequence biases in the binding site identification, as shown in [18].

14.2.3 Parameter Estimation for the MOSAiCS-HMM Model

We estimate the parameters of MOSAiCS-HMM using the Baum-Welch algorithm, which is a special case of the EM algorithm. MOSAiCS-HMM inherits estimates of the emission distributions from the MOSAiCS fits to the data. Although this is in principle statistically inefficient, the MOSAiCS-HMM goodness-of-fit plots suggest that this procedure results in good fit to the data (Fig. 14.3b). More importantly, this approach accelerates fitting of MOSAiCS-HMM significantly because the Baum-Welch algorithm needs to only estimate the transition matrix and state probabilities for the starting bin. We fit the MOSAiCS-HMM model to each chromosome separately because a smooth transition between end of one chromosome and start of another chromosome is not expected. Furthermore, by analyzing each chromosome separately, the fitting of MOSAiCS-HMM can be easily parallelized to decrease computational cost.

Since MOSAiCS-HMM fit utilizes the background and signal distribution estimates of the MOSAiCS fit, the only parameters that need to be estimated for each chromosome with the Baum-Welch algorithm are $\Theta=(\pi_{00},\pi_{10},\pi_{01},\pi_{11},\pi_{*0},\pi_{*1})$, where π_{00}, π_{10}, π_{01}, and π_{11} are transition probabilities defined in Eq. (14.6), and π_{*0}

and π_{*1} are probabilities of enrichment states in the first bin of each chromosome, i.e., $\pi_{*0} \equiv \Pr(Z_1 = 0)$, $\pi_{*1} \equiv \Pr(Z_1 = 1)$, and $\pi_{*0} + \pi_{*1} = 1$. Then, the complete data likelihood function is given by

$$L_c = \prod_{k=0}^{1} \pi_{*k}^{Z_{0k}} \prod_{j=1}^{M-1} \prod_{k=0}^{1} \prod_{l=0}^{1} \pi_{kl}^{Z_{jk}Z_{(j+1)l}} \prod_{j=1}^{M} \prod_{l=0}^{1} \{\Pr(Y_j|Z_j = l)\}^{Z_{jl}}.$$

Because $\Pr(Y_j|Z_j = l)$ are obtained from the MOSAiCS fit, the MOSAiCS-HMM EM algorithm iterates between the following E- and M-steps until the likelihood or the parameter estimates converge or a fixed number of iterations specified by the user is reached. For the m-th iteration, we have the following and E- and M-steps.

E-step:

We first update the conditional probabilities of the enrichment states $k = 0, 1$ in the first bin of each chromosome as

$$z_{*k}^{(m)} = \Pr\left(Z_1 = k|\mathbf{Y};\Theta^{(m)}\right) = \frac{\pi_{*k}^{(m)} \Pr(Y_1|Z_1 = k)}{P(Y_1;\Theta^{(m)})}.$$

The conditional expectation of transition between the states can be computed efficiently using the forward and backward algorithms as follows. In the forward algorithm, we have

$$\begin{aligned} f_{1l}^{(m)} &= \pi_{*l}^{(m)} \Pr(Y_1|Z_1 = l), \quad l = 0, 1, \\ f_{jl}^{(m)} &= Pr\left(Y_1, Y_2, \cdots, Y_j, Z_j = l;\Theta^{(m)}\right) \\ &= \Pr(Y_j|Z_j = l) \sum_{k=0}^{1} f_{(j-1)k}^{(m)} \pi_{kl}^{(m)}, \quad j =, 2, 3 \cdots, M, \quad l = 0, 1. \end{aligned}$$

In the backward algorithm, we have

$$\begin{aligned} b_{Mk}^{(m)} &= 1, \quad , k = 0, 1, \\ b_{jk}^{(m)} &= Pr\left(Y_{j+1}, Y_{j+2}, \cdots, Y_M|Z_j = k;\Theta^{(m)}\right) \\ &= \sum_{l=0}^{1} \pi_{kl}^{(m)} \Pr(Y_{j+1}|Z_{j+1} = l) b_{(j+1)l}^{(m)}, \quad j = (M-1), (M-2), \cdots, 1, \quad k = 0, 1. \end{aligned}$$

Finally, we calculate the conditional probabilities of transition from state $k = 0, 1$ to $l = 0, 1$ based on the quantities from the forward and backward algorithms as

$$
\begin{aligned}
z_{jkl}^{(m)} &= \Pr\left(Z_j = k, Z_{j+1} = l|\mathbf{Y};\Theta^{(m)}\right) \\
&= \frac{f_{jk}^{(m)}\pi_{kl}^{(m)}\Pr(Y_{j+1}|Z_{j+1} = l)b_{(j+1)l}^{(m)}}{P(\mathbf{Y};\Theta^{(m)})}, \quad j = 1,2,\cdots,(M-1).
\end{aligned}
$$

M step:

For states $k = 0,1$ and $l = 0,1$, we update the transition probabilities as

$$
\pi_{kl}^{(m+1)} = \frac{\sum_{j=1}^{M-1} z_{jkl}^{(m)}}{\sum_{l'=0}^{1}\sum_{j=1}^{M-1} z_{jkl'}^{(m)}}
$$

and the probabilities of states $k = 0,1$ in the first bin of each chromosome as

$$
\pi_{*k}^{(m+1)} = \frac{z_{*k}^{(m)}}{\sum_{k'=0}^{1} z_{*k'}^{(m)}}.
$$

We use the scaling procedures provided in [9] to avoid numerical problems in the forward and backward algorithms.

With MOSAiCS-HMM, users can finalize the set of enriched regions by either the Viterbi algorithm or the posterior decoding. If the Viterbi algorithm is used, the most likely sequences of enrichment states are determined across each chromosome. With the posterior decoding approach, enrichment state of each bin is determined using the direct posterior probability approach [24] for FDR control. We next discuss the details of the decoding procedures.

14.2.3.1 Viterbi Algorithm for the MOSAiCS-HMM Model

The Viterbi algorithm for MOSAiCS-HMM identifies the most likely sequences of enrichment states across each chromosome, i.e.,

$$
\hat{\mathbf{Z}} = \arg\max_{\mathbf{Z}} \Pr\left(\mathbf{Y},\mathbf{Z};\hat{\Theta}\right),
$$

where $\hat{\Theta}$ is the final parameter estimates obtained from the EM algorithm. Specifically, the Viterbi algorithm is implemented in the following four steps. First, in the initialization step, for states $l = 0,1$, we set

$$
\begin{aligned}
v_{1l} &= \hat{\pi}_{*l}\Pr(Y_1|Z_1 = l), \\
ptr_{1l} &= 0,
\end{aligned}
$$

where $\hat{\pi}_{*l}$ is the final estimate for π_{*l}. Second, in the recursion step, from bin $j = 2, 3, \cdots$ to bin M,

$$v_{jl} = \Pr(Y_j|Z_j = l) \max_k \left\{ v_{(j-1)k} \hat{\pi}_{kl} \right\},$$

$$ptr_{jl} = \arg\max_k \left\{ v_{(j-1)k} \hat{\pi}_{kl} \right\},$$

where $\hat{\pi}_{kl}$ is the final estimate for π_{kl}. Third, in the termination step, we set

$$\hat{z}_M = \arg\max_k \left\{ v_{Mk} \right\}.$$

Finally, in the trace back step from bin $j = (M-1), (M-2), \cdots$ to bin 1,

$$\hat{z}_j = ptr_{(j+1)\hat{z}_{j+1}},$$

where $\hat{z}_j$ is the estimated state for j-th bin.

14.2.3.2 Posterior Decoding for MOSAiCS-HMM Model

In the posterior decoding approach, the enrichment state for j-th bin is determined using the direct posterior probability approach of [24] for false discovery rate control based on the following posterior probabilities:

$$\Pr\left(Z_j = k|\mathbf{Y}; \hat{\Theta}\right) = \frac{\hat{f}_{jk} \hat{b}_{jk}}{P(\mathbf{Y}; \hat{\Theta})},$$

where $\hat{\Theta}$ denotes the final parameter estimates obtained from the EM algorithm, and $\hat{f}_{jk}$ and $\hat{b}_{jk}$ are the values from the forward and backward algorithms based on the final parameter estimates. Specifically, we first rank the bins according to increasing values of $\Pr\left(Z_j = 0|\mathbf{Y}; \hat{\Theta}\right)$ and denote these sorted values with fdr_j. Then, we increase the cutoff κ until the expected proportion of false discoveries given by

$$\frac{\sum_{j=1}^{M} fdr_j 1\left\{fdr_j \le \kappa\right\}}{\sum_{j=1}^{M} 1\left\{fdr_j \le \kappa\right\}},$$

reaches the pre-specified false discovery rate α. Finally, using this determined cutoff $\hat{\kappa}$, we report the bins satisfying the condition that $\Pr\left(Z_j = 0|\mathbf{Y}; \hat{\Theta}\right) \le \hat{\kappa}$ as enriched regions.

The MOSAiCS-HMM model is now part of the `R` package `mosaics` ($\geq$ 1.6.0).

14.3 Case Study: H3K4me3 Profiling in GM12878 Cells

We used ChIP-seq data of H3K4me3 in GM12878 cells from the ENCODE project to evaluate performances of MOSAiCS, MOSAiCS-HMM, and BCP. The dataset contained two ChIP replicates with 21.3 and 18.1 million aligned reads each. Each ChIP sample was analyzed with respect to a common Input control sample of 13.4 million aligned reads. We used the default parameter values for BCP and a false discovery rate of 0.05 and a threshold value equal to the 99-th percentile of the bin-level ChIP read counts for MOSAiCS and MOSAiCS-HMM with the posterior decoding approach. Overall, MOSAiCS-HMM fits had better BIC values than the MOSAiCS fits for both replicates (36,273,306 (MOSAiCS) versus 33,169,356 (MOSAiCS-HMM) for replicate 1; 32,568,329 (MOSAiCS) versus 29,652,020 (MOSAiCS-HMM) for replicate 2). The goodness-of-fit plots indicate that both models fit the data adequately (Fig. 14.3b).

BCP identified 17664 and 16964 peaks for the two replicates whereas MOSAiCS and MOSAiCS-HMM identified 16438 and 20079 peaks for replicate 1 and 16730 and 20294 peaks for replicate 2, respectively. We allowed MOSAiCS to merge enriched bins that are within 200 bps of each other to facilitate identification of wide enriched regions. We then evaluated the replicate consistency of the methods by ranking and comparing the peaks from the two replicates of each method. For BCP, peak-specific posterior means, which are the only statistical measurements of enrichment reported in the BCP output, were used for ranking whereas for MOSAiCS and MOSAiCS-HMM, maximum signal which denotes the maximum bin-level ChIP read count within the peak region was used. Similar results were obtained when the MOSAiCS and MOSAiCS-HMM peaks were ranked with respect to their maximum posterior probability of enrichment over the bins within the enriched regions. Figure 14.4a depicts the percentage overlap between the peak

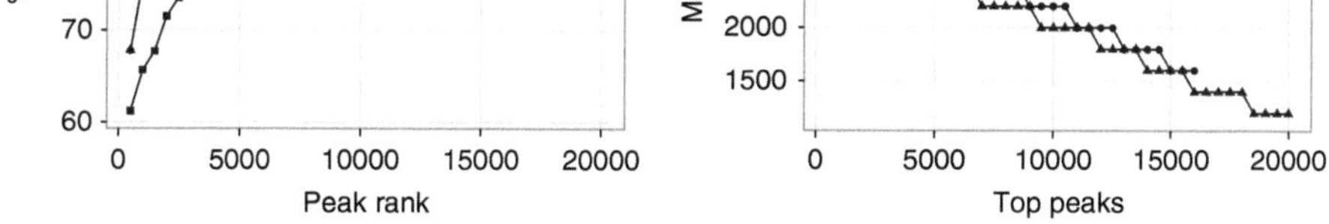

Fig. 14.4 *Comparison of MOSAiCS, MOSAiCS-HMM, and BCP on H3K4me3 ChIP-seq data from GM12878.* (**a**) Overlap percentages of enriched regions identified by two independent replicates as a function of the peak rank. (**b**) Median widths across top (500, 1000, 1500, ··· , 20000) ranked peaks

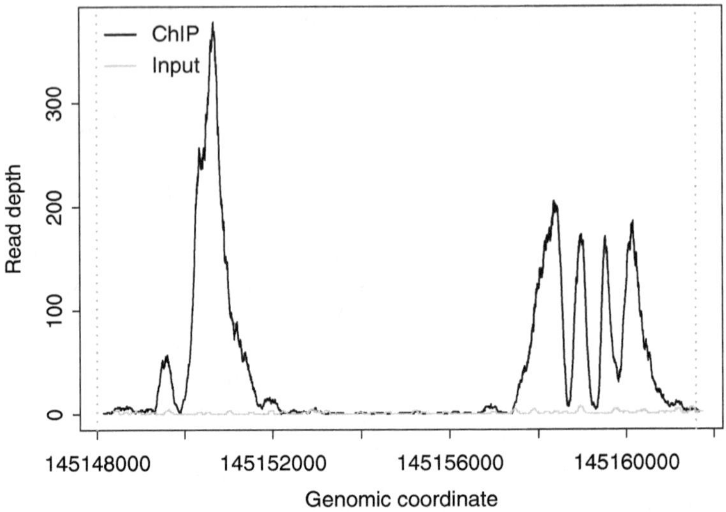

Fig. 14.5 *Comparison of MOSAiCS, MOSAiCS-HMM, and BCP on H3K4me3 ChIP-seq data from GM12878.* H3K4me3 ChIP-seq read profile generated by the R package dpeak [7] for a wide BCP peak. *Dashed, vertical gray lines* mark the boundaries of the BCP peak. Both MOSAiCS and MOSAiCS-HMM identify two peaks within this enriched region

sets of the two replicates for each method as a function of peak rank. We note that both MOSAiCS and MOSAiCS-HMM provide better ranking of the peaks than BCP. The final overlap percentages of the peak sets of the replicates are comparable between the methods, indicating that MOSAiCS-HMM is identifying more peaks with the same overlap consistency rate. This overlap analysis is based on the original widths of the peaks reported by each method. Figure 14.4b displays the median widths across top $500, 1000, 1500, \cdots, 20000$ peaks for the peak sets of replicate 2 from each method. Similar results are obtained with replicate 1 (data not shown). We observe that BCP peaks are the widest. Despite this, overlap percentages of the ranked BCP peaks are the smallest as illustrated in Fig. 14.4a. BCP peaks often have long flanking regions lacking enrichment (Fig. 14.2) or a single BCP peak harbours multiple enriched regions separated by long regions lacking enrichment. An example of the latter case is provided in Fig. 14.5, where two enriched regions separated by about 5000 bps are reported as a single peak. We also note that MOSAiCS-HMM actually provides slightly narrower peaks than MOSAiCS. This indicates that the gain from the HMM architecture cannot simply be attained by merging of enriched bins within close proximity of each other in the MOSAiCS output.

H3K4me3 is a promoter-specific histone modification associated with active transcription; therefore H3K4me3 enrichment is expected at the promoter regions of genes that are transcribed in GM12878. To evaluate biological relevance of peaks identified by each method, we overlapped promoter regions of the expressed genes in GM12878 with each of the peak sets. Expressed genes are defined based

Table 14.1 *H3K4me3 peak coverage of the promoters of the 5979 expressed genes in GM12878*

	BCP	MOSAiCS	MOSAiCS-HMM
# of overlapping promoters	2782 (5484)	4514 (5360)	4745 (5363)
# of completely covered promoters	546	656	704

The numbers of overlapping promoters are based on the intersection of promoters overlapping with peaks of both replicates. Numbers in parentheses denote the numbers of promoters overlapping with the peaks when the original peak widths are used. The numbers of completely covered promoters are based on the minimum of the number of promoters completely covered by the peaks of each of the replicates

on ENCODE2 RNA-seq data from GM12878 by subsetting genes with transcripts per million larger than 20. For each gene, we defined the promoter region as the [-1000, +500] bps interval anchored at the transcription start site. Since wider peaks are expected to provide higher overlap by definition, we resized the peaks of each method to 2000 bps by using the midpoint of the peak as the anchoring point. MOSAiCS pipeline reports a summit. Ideally, a summit denoting the location of the highest signal would be a better anchoring point for all the methods; however since BCP only reports intervals of enrichment, using the midpoint as the anchor minimized the summit selection bias between the methods. Table 14.1 summarizes the total number of promoters that overlap with peak lists of each method and also specifies how many of the promoters are completely within a H3K4me3 peak. We observe that MOSAiCS-HMM peaks overlap with a larger fraction of the active promoters and completely cover the largest number of promoters. When the promoter overlap of the peaks is calculated using the original widths (numbers reported in parentheses in Table 14.1), a slightly higher number of promoters are overlapping with the BCP peaks; however as depicted in Fig. 14.5, this gain comes at the price of many base pairs that lack any enrichment within the peak regions.

14.4 Discussion

We presented an extension of MOSAiCS, named MOSAiCS-HMM, for analyzing ChIP-seq data of histone modifications. MOSAiCS-HMM can analyze ChIP-seq experiments with or without a Input control experiment and provides FDR control.

We conclude by discussing some other key issues related to histone ChIP-seq data, and more generally ChIP-seq data (Fig. 14.1). The commonly used read lengths in ChIP-seq protocols are 50 to 100 bps. This results in about 10–25 % of the reads aligning to multiple locations on the reference genome for human and mouse samples. These reads are commonly referred to as *multi-reads* and are typically discarded from the analysis. This leads to missing read data for highly repetitive regions of the genome and such reads are important to recover for studying TFs or histone modifications that interact with repetitive DNA. To address this issue, we developed a ChIP-seq-specific read mapper [6] named CSEM. This mapper utilizes Bowtie [20] alignments of the reads, where multi-reads are retained, and fractionally allocates multi-mapping reads by considering the local read contents of the mapping positions. As a result, it can generate both an alignment file with all the mapping reads and their allocation probabilities and a pseudo alignment file in bed file format where each multi-read is allocated to its most probable mapping location. The latter alignment file is accepted as input by multiple peak callers.

There are multiple quality control procedures developed for ChIP-seq data. Most notable of these is the cross-correlation analysis which is built on calculating cross-correlation between strand-specific genome-wide ChIP read profiles [19]. In ChIP-seq experiments with high signal-to-noise ratios, cross-correlation between the base-pair level forward and reverse strand read counts is expected to attain its maximum value around the average fragment length. A maximum cross-correlation value at a length vastly different from that of the average fragment size indicates potential problems with the ChIP-seq data and requires further attention. ChIP-seq experiments are prone to a wide range of amplification biases. A commonly encountered bias is the extreme amplification of local regions. For such abnormally amplified regions, the same set of nucleotides covering the region appears in the data set hundreds and even thousands of times. The common practice to alleviate problems due to abnormal amplification effects is the removal of multiple copies of a given read. More specifically, only a single read is allowed to start at each distinct genomic position. This feature is also part of the `mosaics` R package. Many ChIP-seq analysis methods provide some level of FDR control. However, the reliability of the FDR control typically depends on how well the assumed model fits the data. An alternative approach, which relies on the consistency between two independent replicates of the ChIP-seq data, is control of irreproducible discovery rate (IDR). This approach has been widely adapted by the ENCODE project [19] and is shown to stabilize the number of peaks obtained from the same data set by different methods. When the MOSAiCS-HMM GOF plots indicate a lack of fit, IDR provides a robust alternative for choosing the number of peaks in MOSAiCS-HMM.

Once the enriched regions are identified in a ChIP-seq experiment, downstream analysis depends on the specific application. For TFs, especially in compact genomes, an important issue is the deconvolution of closely located binding events. Most of the commonly used ChIP-seq analysis methods [17, 18, 26, 37] are not specifically designed to deconvolve closely located binding; however, the number of methods which can perform such a task is on the increase [7, 13, 36]. In TF ChIP-seq experiments, summits of the peaks (predicted binding locations) are the main parameters of interest. In contrast, for histone ChIP-seq experiments, the boundaries of the enriched regions constitute one of the most important features. Most of the commonly used histone ChIP-seq analysis methods operate by binning the genome into small non-overlapping intervals. As a result, the resulting enriched regions might have inaccurate boundaries and require post trimming and extension procedures. It is often of interest to study multiple histone modifications simultaneously and divide genome into regions exhibiting different combinations of histone modifications [11]. To this end, we developed jMOSAiCS [35], which efficiently analyses multiple TF or histone modification datasets simultaneously and identifies regions showing combinatorial enrichment of the studied factors.

Acknowledgements We thank Professor Colin Dewey of University of Wisconsin, Madison, for providing us with a RSEM-processed version of ENCODE GM12878 RNA-seq data. This research was supported by National Institutes of Health Grants HG007019 and HG003747 to S.K.

References

[1] Bailey, T., Krajewski, P., Ladunga, I., Lefebvre, C., Li, Q., Liu, T., Madrigal, P., Taslim, C., Zhang, J.: Practical guidelines for the comprehensive analysis of ChIP-seq data. PLoS Computat. Biol. **9**(11), e1003,326 (2013)
[2] Barski, A., Cuddapah, S., Cui, K., Roh, T.Y., Schones, D.E., Wang, Z., Wei, G., Chepelev, I., Zhao, K.: High-resolution profiling of histone methylations in the human genome. Cell **129**(4), 823–837 (2007)
[3] Benjamini, Y., Speed, T.P.: Summarizing and correcting the GC content bias in high-throughput sequencing. Nucleic Acids Res. **40**, e72 (2012)
[4] Bernstein, B.E., Stamatoyannopoulos, J.A., Costello, J.F., Ren, B., Milosavljevic, A., Meissner, A., Kellis, M., Marra, M.A., Beaudet, A.L., Ecker, J.R., Farnham, P.J., Hirst, M., Lander, E.S., Mikkelsen, T.S., Thomson, J.A.: The NIH Roadmap Epigenomics Mapping Consortium. Nat. Biotechnol. **28**(10), 1045–1048 (2010)
[5] Buck, M.J., Lieb, J.D.: ChIP-chip: considerations for the design, analysis, and application of genome-wide chromatin immunoprecipitation experiments. Genomics **84**, 349–360 (2004)
[6] Chung, D., Kuan, P.F., Li, B., Sanalkumar, R., Liang, K., Bresnick, E.H., Dewey, C., Keleş, S.: Discovering transcription factor binding sites in highly repetitive regions of genomes with multi-read analysis of ChIP-seq data. PLoS Computat. Biol. **7**, e1002,111 (2011)
[7] Chung, D., Park, D., Myers, K., Grass, J., Kiley, P., Landick, R., Keleş, S.: dPeak: High resolution identification of transcription factor binding sites from PET and SET ChIP-seq data. PLoS Computat. Biol. **9**(10), e1003,246 (2013)
[8] Dohm, J., Lottaz, C., Borodina, T., Himmelbauer, H.: Substantial biases in ultra-short read data sets from high-throughput DNA sequencing. Nucleic Acids Res. **36**(16), e105 (2008)
[9] Durbin, R., Eddy, S., Krogh, A., Mitchison, G.: Biological sequence analysis: Probabilistic models of proteins and nucleic acids. Cambridge University press, Cambridge (1998)
[10] ENCODE Project Consortium, Bernstein, B.E., Birney, E., Dunham, I., Green, E.D., Gunter, C., Snyder, M.: An integrated encyclopedia of DNA elements in the human genome. Nature **489**(7414), 57–74 (2012)
[11] Ernst, J., Kellis, M.: Discovery and characterization of chromatin states for systematic annotation of the human genome. Nat. Biotechnol. **28**, 817–25 (2010)
[12] Gentleman, R.C., Carey, V.J., Bates, D.M., others: Bioconductor: Open software development for computational biology and bioinformatics. Genome Biol. **5**, R80 (2004)
[13] Guo, Y., Mahony, S., Gifford, D.K.: High resolution genome wide binding event finding and motif discovery reveals transcription factor spatial binding constraints. PLoS Computat. Biol. **8**, e1002,638 (2012)
[14] Jang, S.W., Srinivasan, R., Jones, E.A., Sun, G., Keles, S., Krueger, C., Chang, L.W., Nagarajan, R., Svaren, J.: Locus-wide identification of egr2/krox20 regulatory targets in myelin genes. J. Neurochem. **115**(6), 1409–1420 (2010)
[15] Johnson, D.S., Mortazavi, A., Myers, R.M., Wold, B.: Genome-wide mapping of in vivo protein-DNA interactions. Science **316**(5830), 1497–1502 (2007)

[16] Keleş, S.: Mixture modeling for genome-wide localization of transcription factors. Biometrics **63**, 10–21 (2007)

[17] Kharchenko, P.V., Tolstorukov, M., Park, P.J.: Design and analysis of ChIP-seq experiments for DNA-binding proteins. Nat. Biotechnol. **6**, 1351–1359 (2008)

[18] Kuan, P., Chung, D., Pan, G., Thomson, J., Stewart, R., Keleş, S.: A Statistical Framework for the Analysis of ChIP-seq data. J Am. Stat. Assoc. **106**(459), 891–903 (2011)

[19] Landt, S.G., Marinov, G.K., Kundaje, A., Kheradpour, P., Pauli, F., Batzoglou, S., Bernstein, B.E., Bickel, P., Brown, J.B., Cayting, P., Chen, Y., DeSalvo, G., Epstein, C., Fisher-Aylor, K.I., Euskirchen, G., Gerstein, M., Gertz, J., Hartemink, A.J., Hoffman, M.M., Iyer, V.R., Jung, Y.L., Karmakar, S., Kellis, M., Kharchenko, P.V., Li, Q., Liu, T., Liu, X.S., Ma, L., Milosavljevic, A., Myers, R.M., Park, P.J., Pazin, M.J., Perry, M.D., Raha, D., Reddy, T.E., Rozowsky, J., Shoresh, N., Sidow, A., Slattery, M., Stamatoyannopoulos, J.A., Tolstorukov, M.Y., White, K.P., Xi, S., Farnham, P.J., Lieb, J.D., Wold, B.J., Snyder, M.: ChIP-seq guidelines and practices of the ENCODE and modENCODE consortia. Genome Res. **22**(9), 1813–1831 (2012)

[20] Langmead, B., Trapnell, C., Pop, M., Salzberg, S.L.: Ultrafast and memory-efficient alignment of short DNA sequences to the human genome. Genome Biol.**10**, R25 (2009)

[21] Mikkelsen, T.S., Ku, M., Jaffe, D.B., Issac, B., Lieberman, E., Giannoukos, G., Alvarez, P., Brockman, W., Kim, T.K., Koche, R.P., Lee, W., Mendenhall, E., O'Donovan, A., Presser, A., Russ, C., Xie, X., Meissner, A., Wernig, M., Jaenisch, R., Nusbaum, C., Lander, E.S., Bernstein, B.E.: Genome-wide maps of chromatin state in pluripotent and lineage-committed cells. Nature **448**, 553–560 (2007)

[22] Myers, K.S., Yan, H., Ong, I.M., Chung, D., Liang, K., Tran, F., Kele, S., Landick, R., Kiley, P.J.: Genome-scale analysis of *escherichia coli* fnr reveals complex features of transcription factor binding. PLoS Genetics **9**(6), e1003,565 (2013)

[23] Nair, N.U., Sahu, A.D., Bucher, P., Moret, B.M.E.: ChIPnorm: A statistical method for normalizing and identifying differential regions in histone modification ChIP-seq libraries. PLoS ONE **7**(8), e39,573 (2012)

[24] Newton, M.A., Noueiry, A., Sarkar, D., Ahlquist, P.: Detecting differential gene expression with a semiparametric hierarchical mixture method. Biostatistics **5**(2), 155–176 (2004)

[25] Rabiner, L.R.: A tutorial on hidden Markov models and selected applications in speech recognition. Proc. IEEE **77**(2), 257–286 (1989)

[26] Rozowsky, J., Euskirchen, G., Auerbach, R., Zhang, D., Gibson, T., Bjornson, R., Carriero, N., Snyder, M., Gerstein, M.: PeakSeq enables systematic scoring of ChIP-Seq experiments relative to controls. Nat. Biotechnol. **27**(1), 66–75 (2009)

[27] Seo, Y.K., Chong, H.K., Infante, A.M., In, S.S., Xie, X., Osborne, T.F.: Genome-wide analysis of SREBP-1 binding in mouse liver chromatin reveals a preference for promoter proximal binding to a new motif. PNAS **106**(33), 13,765–13,769 (2009)

[28] Song, Q., Smith, A.D.: Identifying dispersed epigenomic domains from ChIP-Seq data. Bioinformatics **27**(6), 870–871 (2011)

[29] Srinivasan, R., Sun, G., Keles, S., Jones, E.A., Jang, S.W., Krueger, C., Moran, J.J., Svaren, J.: Genome-wide analysis of egr2/sox10 binding in myelinating peripheral nerve. Nucleic Acids Res. **40**(14), 6449–6460 (2012)

[30] Strahl, B.D., Allis, C.D.: The language of covalent histone modifications. Nature **403**(6765), 41–45 (2000)

[31] Sun, G., Chung, D., Liang, K., Keleş, S.: Statistical analysis of ChIP-seq data with MOSAiCS. In: Shomron, N. (ed.) Deep Sequencing Data Analysis. Methods in Molecular Biology, vol. 1038, pp. 193–212. Humana Press, NewYork (2013)

[32] Taslim, C., Huang, T., Lin, S.: DIME: R-package for identifying differential ChIP-seq based on an ensemble of mixture models. Bioinformatics **27**(11), 1569–70 (2011)

[33] Xing, H., Mo, Y., Liao, W., Zhang, M.Q.: Genome-wide localization of protein-DNA binding and histone modification by a Bayesian change-point method with ChIP-seq data. PLoS Computat. Biol. **8**(7), e1002613 (2012)

[34] Xu, H., Wei, C.L., Lin, F., Sung, W.K.: An HMM approach to genome-wide identification of differential histone modification sites form chip-seq data. Bioinformatics **24**(20), 2344–2349 (2008)

[35] Zeng, X., Sanalkumar, R., Bresnick, E.H., Li, H., Chang, Q., Keleş, S.: jMOSAiCS: Joint analysis of multiple ChIP-seq datasets. Genome Biol. **14**, R38 (2013)

[36] Zhang, X., Robertson, G., Krzywinski, M., Ning, K., Droit, A., Jones, S., Gottardo, R.: PICS: probabilistic inference for ChIP-seq. Biometrics **67**(1), 151–163 (2011)

[37] Zhang, Y., Liu, T., Meyer, C.A., Eeckhoute, J., Johnson, D.S., Bernstein, B.E., Nussbaum, C., Myers, R.M., Brown, M., Li, W., Liu, X.S.: Model-based analysis of ChIP-Seq (MACS). Genome Biol. **9**(9), R137 (2008)

[38] Zhang, Z.D., Rozowsky, J., Snyder, M., Chang, J., Gerstein, M.: Modeling ChIP sequencing in silico with applications. PLoS Computat. Biol. **4**(8), e1000,158 (2008)

Chapter 15
Hierarchical Bayesian Models for ChIP-seq Data

Riten Mitra and Peter Müller

Abstract Histone modifications (HMs) are post-translational modifications of the nucleosome. Studying the presence or absence of these modifications in genomic regions is a central topic in modern epigenetics. HMs regulate various biological processes by overwriting the DNA-inscribed code. Experimental evidence suggests that they perform this task through a complex biological network. In other words, HMs combinatorially influence gene expression. We present two model-based approaches to decode this mechanism using ChIP-seq data. Both approaches are based on hierarchical Bayesian models. The first model derives a conditional independence structure among the HMs through a graphical model. The challenge here is to model the unobserved binary (presence/absence) status of HMs on the basis of read counts. The other critical aspect is to model the dependence between these latent binaries in a way that allows tractable posterior inference. The second model relates HMs and functional genomics through a local bi-clustering approach. Here HMs are clustered and each HM cluster gives rise to a (nested) partition of genomic loactions, with respect to that subset of HMs. These models are, to the best of our knowledge, the first model-based fully Bayesian approaches to discovering epigenetic associations. Validation with known experimental findings suggests the importance and usefulness of these approaches in our understanding of gene regulation.

R. Mitra (✉)
University of Louisville, 2301 S 3rd St, Louisville, KY 40292, USA
e-mail: riten82@gmail.com

P. Müller
University of Texas at Austin, Austin, TX, USA
e-mail: pmueller@math.utexas.edu

S. Datta and D. Nettleton (eds.), *Statistical Analysis of Next Generation Sequencing Data*, Frontiers in Probability and the Statistical Sciences, DOI 10.1007/978-3-319-07212-8_15,

15.1 Introduction: Histone Codes

The advent of high-throughput technologies such as ChIP-seq [25, 26] has greatly facilitated the study of histone modifications (HMs). This in turn has led to a proliferation of computational tools specially designed for ChIP-seq data. Most of these tools are useful aids for data preprocessing and visualization. However, to extract significant insights about the epigenome, probabilistic model-based approaches beyond pre-processing of the raw data are required. Such algorithms require to incorporate expert knowledge on the experimental setup and a mathematically rigorous, yet interpretable, statistical framework.

Understanding the dynamics of HMs and their occurrences are important scientific goals in themselves. However, HMs are critical to epigenetic research primarily because of their roles in the transcriptional process [13, 18, 31]. For example, consider a specific modification, say histone acetylation, which is generally associated with increased transcriptional activity. This modification occurs when HAT (histone acetyl transferase) enzymes adds an acetyl group on the lysine residue of a H3 tail. The positive charge on the residue is neutralized by the negative charge on the acetyl ions. This reduces electrostatic attraction between the histone and the charged DNA backbone, loosening the chromatin structure. A loosened up chromatin is more accessible to binding by transcription factors. In contrast, removal of the acetyl group make the wrapping of DNA around nucleosome tighter. Similarly histone methylation (caused by addition of methyl groups by methyltransferases) is mostly associated with repressive activity [15, 30]

Current scientific knowledge about HMs precludes the possibility for a single HM to cause transcription activation or repression. On the contrary, accumulating evidence suggests that the regulation of gene expression by HMs is not as simple as an on-off switch, but involves convoluted combinatorial effects, some times referred to as the 'histone code' [5]. The combinatorial nature of the effects suggests a complex cross-talk mechanism among the different HMs. Such interactions are the subject of substantial scientific interest and can shed considerable light on the epistatic relationships among the related nearby genes. In general, HMs could compete antagonistically with each other if there are multiple modification pathways targeting the same site. While one modification may be totally dependent upon another, the binding of a protein to a particular modification can be disrupted by an adjacent modification. There may also be cooperation between modifications in order to efficiently recruit specific factors and enzymes. For a comprehensive review, we refer the readers to [1] which discusses different types of crosstalk, describes their level of complexity and provides interesting examples. These examples suggest that HM occurrences are very likely to be correlated, but we still lack a unified picture of the interactions between a group of HMs.

In Sect. 15.3 we build a comprehensive model for inferring statistical relationships among various HMs. However, network inference itself is not sufficient for unraveling the histone code. To go a step further, we need to integrate information about genome functionality with HMs in a co-association model. Mounting experimental evidence of recruitment by modified histones of their respective enzymes hint towards a clustering of HM data across genomic locations. We implemented related inference by means of a recently developed nonparametric Bayesian biclustering approach. Related work is summarized in Sect. 15.4. The biclustering yields two partitions – a genomic partition nested within another partition of HMs. In other words, we allow for a separate partition of genomic loci with respect to each HM cluster. Both models offer coherent posterior inference.

The aim of this chapter, therefore, is twofold. First, we describe a network model to capture the HM interactions. The model facilitates inference on the dependence structure across HMs. We refer the interested readers to [23] for a detailed discussion of the inference strategies from this model. Second, we describe a method to capture co-clustering patterns of HMs. Applying both of these models to human CD4+ T cell data, we recovered many existing relationships and hypothesized many more interesting ones Before describing the models, we give a brief overview of ChIP-seq technology and the associated statistical issues in Sect. 15.2.

15.2 ChIP-seq Technology

ChIP-seq [27] is a new high throughput technology developed to directly sequence DNA fragments at low cost. It combines chromatin immunoprecipitation (ChIP) of DNA fragments with next generation sequencing. The ChIP part of the workflow is similar to the older microarray based ChIP-ChIP protocols. First, the HM specific antibodies are crosslinked to DNA in vivo by treating cells with formaldehyde. An enzyme called micrococcal nuclease (MNase) is then used to extract the nucleosomal DNA. Next, an antibody specific to the HM of interest is used to immunoprecipitate the DNA-HM complex. Finally, the crosslinks are reversed and the released DNA fragments are prepared for direct sequencing. That is, the fragments of interest are not hybridized on an array. Millions of short sequence reads are generated from regions bound to the antibody target (the signal).

ChIP-seq offers single base-pair resolution and greater coverage than ChIP-chip, and thus significantly improves data quality. The first comprehensive genome-wide maps using ChIP-seq were created in 2007 [4]. Twenty histone methylation marks, as well as the histone variant H2A.Z, RNA Polymerase II, and the DNA-binding protein CTCF, were profiled in human T cells, with an average of 8 million tags per sample using Solexa 1G. This was followed by a map of 18 histone acetylation marks in the same cell type [33]. However, as with the development of microarray data platforms a decade ago, this new technology raises important statistical problems and issues. A major challenge lies in modeling the experimental noise. This noise originates from multiple sources including non-specific binding to the antibody, errors in amplification (local PCR), and sequencing errors. The quality of the antibody is usually the most critical factor. Even high-quality ChIP-seq libraries are likely to consist mainly of noise rather than signal, with 80–90% background signal possible. As we shall see next, an important contribution of our proposed models is an automated model-based procedure to filter away such noise.

15.3 Inference on HM Networks: A Hierarchical Graphical Model

15.3.1 A Graphical Model for Conditional Independence Structure

A hierarchical model traces the model construction from the unknown variables to the observed data. It thus provides a roadmap for hypothetical data generation. In Fig. 15.1 we illustrate this process with a flow chart.

At the root lies a graph G that encodes conditional independence structure among the HMs. For the moment, we only note that G represents the association among HMs. It is the main target of inference. The bottommost layer is the data matrix y. The intermediate layers include a matrix of latent indicators e. Its entries e_{it} are indicators for the presence of HM i at location t. The use of e_{it} formalizes the biologically meaningful notion of presence versus absence of HMs. The raw data y can be viewed as a noise-corrupted version of e. For reasons specified in the introduction, it makes more biological sense to model the dependence structure among the latter rather than the former. We find a description of similar data reduction approaches in the microarray literature. For example, in [24] the probability of expression (POE) model was proposed to account for the latent categories of gene expression. The latent indicator matrix e has the same dimension as y.

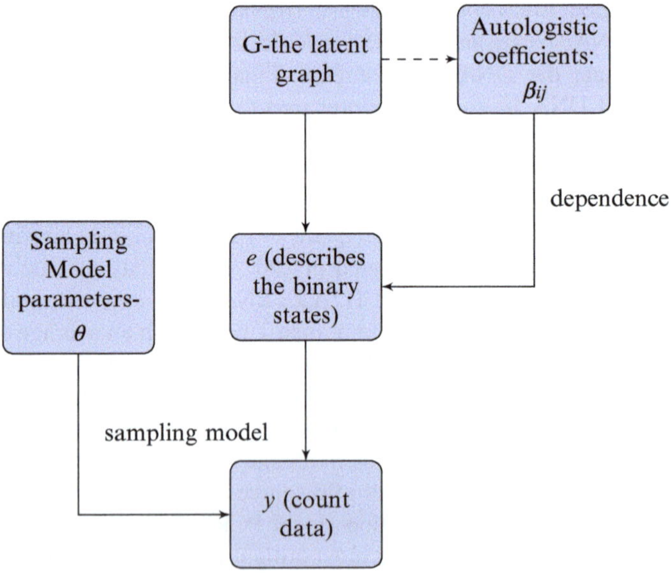

Fig. 15.1 Data generating mechanism

Mathematically, a probabilistic network (or graphical model) G represents the dependence structure that is implied by a joint distribution of a set of random variables. It can be pictorially depicted by a graph. The graph consists of a set of vertices (or nodes) and a set of edges (or connections) between pairs of nodes. The vertices correspond to the variables. The presence of a connection between two vertices implies interaction between the nodes. Formally, the absence of an edge (i,j) between nodes i and j indicates independence of variables i and j, conditional on all other variables. Graphical models have been widely applied to reconstruct gene regulatory networks and protein interaction networks. However, applications to analyzing HM data are still limited. In [23] we report an approach of using graphical models for inference about the dependence structure of HMs. Our approach is based on modeling the dependence at the level of the latent binary indicators e_{it}. That is, we model dependence at the level of a meaningful biological signal of absence versus presence of HMs, rather than at the level of noisy measurements. This approach requires us to define a joint prior probability model $p(e_t \mid G)$ for $e_t = (e_{it}; i = 1, \ldots, m)$. The model is defined in such a way that the implied conditional independence structure for e_t matches the structure that is encoded in G, as described earlier. It is always possible to define such a joint probability model $p(e_t \mid G)$. This is guaranteed by the Hammersley Clifford Theorem [7]. The latter posits necessary and sufficient conditions for a joint probability model to be identified from the set of conditional distributions. This result allows us to write the joint probability model $p(e_t)$ indexed by a set of parameters β as follows:

$$p(e_t \mid \beta, G) = p(0 \mid \beta, G) \cdot \exp\left\{ \sum_i \beta_i e_{it} + \sum_{i<j} \beta_{ij} (e_{it} - \nu_i)(e_{jt} - \nu_j) \right\} \tag{15.1}$$

where $\nu_i = \exp(\beta_i)/\{1 + \exp(\beta_i)\}$ is a deterministic function of β_i that centers the model. The centering of e_{it} with ν_i improves the convergence of posterior simulation [8]. The assignment of edges in G is a deterministic function of the restrictions on β. A pair of vertices (i,j) is not connected by an edge if and only if $\beta_{ij} = 0$.

Model (15.1) is known as the autologistic model. In our statement, we restricted the model to cliques, that is, interaction terms in the exponential, of size at most 2. In an extension with arbitrary size interaction terms, the family of autologistic models includes all possible joint probability models $p(e_t)$.

We continue the prior model construction with a prior model $p(G)$ for the conditional independence graph G. We define the prior $p(G)$ as a uniform distribution over all possible subgraphs with vertex set V. Next, we complete the model construction with a sampling model

$$p(y \mid e, \theta) = \prod_{t,i} p(y_{it} \mid e_{it}, \theta),$$

Fig. 15.2 Fit of a Poisson/lognormal mixture model to the count data of an HM. The *red (peaked) curve* is the density of $0.5 \times Pois(1)$ $I(y_{it} < 2) + 0.3 \times LN(1, 0.4) +$ $0.2 \times LN(2, 0.6)$. The histogram shows the empirical distribution of the data

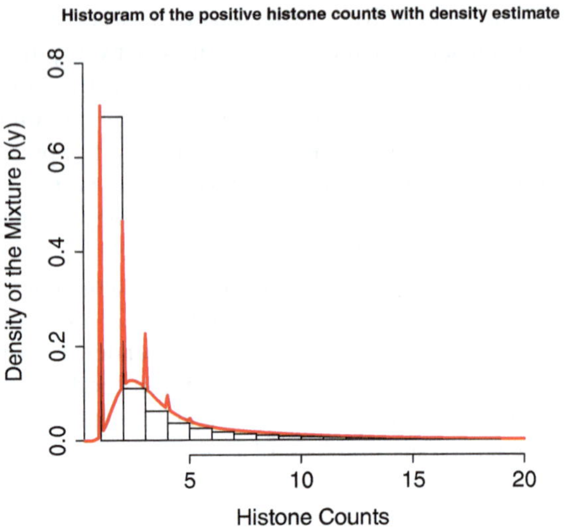

for the observed counts y_{it} given the latent e_{it}. Here θ are additional parameters that index the sampling model. In particular, we propose a mixture model with a Poisson distribution (Poi) for the low counts, say $y_{it} < c_i$, and a mixture of two log-normal (LN) distributions for moderate to high counts. In the specification of the sampling model we give meaning to the binary indicators e_{it} by defining $p(y_{it} \mid e_{it} = 0)$ as the Poisson distribution, and $p(y_{it} \mid e_{it} = 1)$ as the mixture of two log normal distributions. This particular specification is motivated by an inspection of the empirical distribution, that is, the histograms of observed HM counts as shown below. The red curve in Fig. 15.2 is an example of a fit using the proposed mixture model. As stated before, the model hierarchy implies the conditional independence of y given e across both HMs i and loci t. This simplifying assumption is motivated by a preference for parsimony. As an additional empirical verification, we implemented a simple diagnostic check. Specifically, we investigated the variance-covariance matrix of residuals obtained after a model fit. The low correlations reaffirmed the validity of our conditional independence assumptions. A discussion of related issues and conditional independence appears in [23].

Posterior inference is carried out as posterior Markov chain Monte Carlo (MCMC) simulation, using a combination of Gibbs and Metropolis-Hastings transition probabilities. Specifically, the MCMC iterates over the following transition probabilities:

$$[e \mid G, \beta, \theta, y], [\theta \mid y, [\beta \mid e, G], [G \mid \beta, e].$$

In implementing the above transition probabilities, we exploit the fact that the autologistic model 15.1 implies easily interpreted and understood complete conditional models.

Conditional on the other variables, the distribution of e_{it} at node i is a logistic regression with two-way interaction coefficients. This is a desirable property of the joint distribution and simplifies the Gibbs sampling of e in the following way. Suppose $e_{-it} = (e_{ht},\ h \neq i)$ denote the indicators for all HMs other than i at genomic location t and let $i \sim j$ indicate that i and j are neighbors in the graph G. We update e_{it}, $i = 1,\dots,m$, using

$$p(e_{it} \mid e_{-it}, \beta, \theta, y) \propto \exp\left\{ \beta_i e_{it} + \sum_{j:\, j\sim i} \beta_{ij}(e_{it} - \nu_i)(e_{jt} - \nu_j) \right\} p(y \mid e, \theta),$$

$e_{it} \in \{0,1\}$, and repeat the same loop for each $t = 1,\dots,n$. Note that e_t, $t = 1,\dots,n$, are conditionally independent given all other parameters and y.

15.3.2 Results: A ChIP-Seq Graph

We analyzed ChIP-seq counts from randomly sampled 50,000 genomic locations in a ChIP-seq experiment for CD4+ T lymphocytes [4, 33]. We considered inference for a selected subset of 17 HMs. Using the previously described model, we estimated the conditional independence structure G for these 17 HMs. We summarize $p(G \mid y)$ by recording for each edge (i,j) a posterior inclusion probability defined by $P_{ij} = p(G_{ij} = 1 \mid y)$. We summarize these estimates into a single posterior graph by thresholding of the P_{ij}. The thresholds are chosen to achieve a posterior expected false discovery rate (FDR) close to 0.05. That is, we obtain a graph G by including all edges with $P_{ij} > c$. The posterior expected FDR corresponding to any given threshold c is calculated by

$$FDR_c = \frac{\sum_{i\,j}[(1 - P_{ij}) I(P_{ij} > c)]}{\sum_{i,j} I(P_{ij} > c)}.$$

Figure 15.3 shows the estimated HM network. In [23] we discuss the implications of the estimated dependence structure and relate it to known results. Many of the reported edges corroborate the existing literature on HMs.

Building on the inference for (global) conditional independence structure, next we compare dependence structure across different genomic locations. We address this problem in [20] and [21]. We briefly describe the extension below. Mainly, we exploit the flexible nature of the hierarchical setup to consider inference on network G^k across different regions, $k = 1,\dots,K$.

First it is reasonable to expect that HM networks in different regulatory regions share an underlying global mechanism but differ in small but important aspects. The formal representation of such prior information requires a joint model over multiple regions. We describe an approach based on a hierarchical model across graphs G^k, $k = 1,\dots,K$. Here we assume that we have K genomic regions, each characterized by its unique network structure G^k. For example, we can consider $K = 3$ different types of genomic regions: *promoters*, *insulators* and *enhancers*.

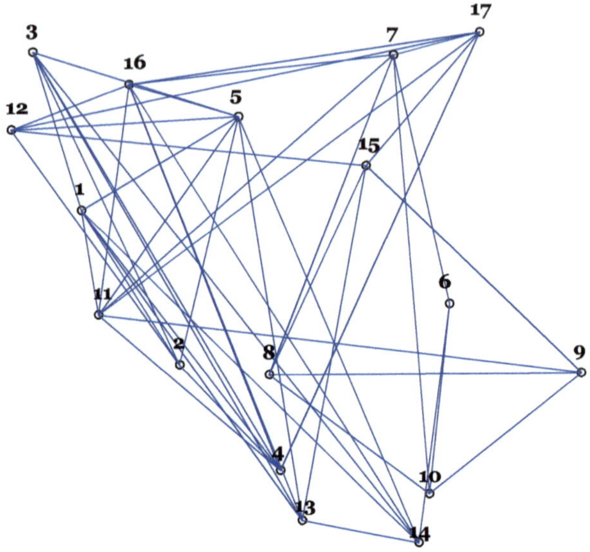

Fig. 15.3 Estimated graph G for the conditional independence structure of the selected 17 HMs. n the figure we indicate the different HMs with running indices 1 through 17, using H2BK120ac (1), H2BK12ac (2), H2BK20ac (3), H2BK5ac (4), H3K4AC (5), H3K4ME1 (6), H3K4ME2 (7), H3K4ME3 (8), H3K9AC (9), H3K9ME1 (10), H3K27AC (11), H3K36AC (12), H3K18AC (13), H4K91AC (14), H2A.Z (15), H4K5AC (16) and H4K8AC (17)

We now add a latent common graph G^0 to the model and construct a hierarchical prior for G^k. Let $G^k_{ij} = I(\{i,j\} \in E^k)$ denote the event that an edge connects nodes i and j in graph G^k. We assume

$$p(G^k_{ij} = 1 \mid G^0_{ij}) = \begin{cases} \rho_1 & \text{for } G^0_{ij} = 1 \\ \rho_2 & \text{for } G^0_{ij} = 0, \end{cases} \quad k = 1,2,3,$$

$$p(G^0_{ij} = 1) = \rho_0. \tag{15.2}$$

Here ρ_j, $j = 0,1,2$ are unknown hyperparameters. These equations represent a hierarchical prior for G^k, with a hyperprior on the common latent graph G^0. The model is completed with uniform hyperpriors for the probabilities ρ_0, ρ_1 and ρ_2. The proposed hierarchical graphical models borrow strength across regions and allow inference on the differences between the regions.

The model for the observed data y and the latent indicators e_{it} remains almost unchanged, except that $\{G, \beta, e, y, \theta\}$ in the single graph model are now replaced by $\{G^k, \beta^k, e^k, y^k, \theta^k\}$. Here y^k refers to the data restricted to the k-th region. Similarly, β^k denotes parameters that index the prior for the binary indicators in the k-th submodel, etc. Let θ^k denote the parameters that index the sampling model for y^k. The joint model is summarized as

$$p(G^0)\, p(G^1 \ldots G^K \mid G^0)\, p(\beta^1 \ldots \beta^K \mid G^1 \ldots G^K)\, p(e^1 \ldots e^k \mid \beta^1 \ldots \beta^K)$$
$$\times\, p(y^1 \ldots y^K \mid e^1 \ldots e^K, \theta^1 \ldots \theta^K)\, p(\theta^1 \ldots \theta^K). \tag{15.3}$$

The first factor is the hyperprior on G^0. The next term denotes the prior on the individual graphs conditional on the common network, as in (15.2). The dependence is induced at the level of the graphical models. Each of factors beyond the first one can be written as product over K factors, corresponding to the K submodels. For details of the posterior inference in multiple graphs and the results from a larger ChiP-seq data, we refer the readers to [22].

In summary, network inference highlights some interesting features of the human epigenome. The high degree of connectivity in the posterior global network and in the subsequent domain specific networks is striking evidence of a thriving cross-talk within HMs. Moreover, the differential edges among the domain-specific networks suggest that the regulatory processes of HM vary with transcriptional activity. Thus the network inference is a significant step towards elucidating the 'histone code'. However, cracking the code would demand a model that explicitly incorporates a variety range of genomic domains. In fact, it should rely less on segments a-priori known to be functionally important, and instead be able to discover associations *de novo*. Since the direction of causality between HMs and transcription is still a matter of ongoing debate, the desired model should depict associations between HMs and genomic features, rather than be interpreted as conditional regressions. In the next section, we discuss a novel method for biclustering that accounts for all these features in a single hierarchical framework.

15.4 NoB-LCP: A Bi-Clustering Method to Crack the 'Histone Code'

15.4.1 Clustering of HMs

A model was developed ([19, 35]) that allows formal inference about clustering HMs, in particular allowing for the tendency of HMs to cluster along certain genetic domains. Many HMs are known to cluster into broad groups [4, 33]. Moreover, such domains are reported to propagate through cell division [6]. There are several notable findings on individual HM domains, e.g., the trimethylation of histone H3 at lysine residue 9. H3K9me3 recruits HP1 which in turn recruits a specific H3K9 methyltransferase enzyme. The latter modifies H3K9 on other histones in the neighborhood, thereby self-propagating the heterochromatin state [2, 17]. Another notable example is the trimethylation of histone H3 lysine 27 (H3K27me3). In Drosophila, the spread of H3K27ME3 is associated with the looping action of two polycomb complexes PRC1 and PRC2 [28]. PRC2 generates the HM which in turn recruits PRC1. Histone acetylation marks also show individual clustering pattern [10, 16].

However, little is known about the joint co-clustering patterns of these HMs. The latter is absolutely essential to test the hypothesized histone code which suggests that some degree of HM co-localization determines the functionality of genomic regions. One dimensional clustering would fail to capture this complex interdependence between HMs and the genome. This motivated us to formulate a local clustering approach. As the name suggests, this model is designed to cluster the rows (genomic locations) in a way that is nested within the clusters of HMs (columns). In other words, the model co-clusters any two HMs which correspond to the same genomic partition. Methodologically it charts a definite departure from traditional Bayesian clustering models that rely on unique and distinct parameters to distinguish clusters. We describe the model below. The model builds on a nested clustering model developed in [19]. It generalizes the model to accommodate the data formats that are required for the application to HM data.

15.4.2 A Nested Clustering Model

Let y denote the observed data matrix with N rows and G columns. In the application to HM data G will be the number of HMs and N will be the number of genomic locations. Let $C = \{C_0, C_1, C_2 \ldots C_Q\}$ be a family of mutually exclusive subsets that define a latent random partition of $\{1, \ldots, G\}$ (columns). For each $q \in 1 \ldots Q$, $R_q = \{R_{q1}, R_{q2} \ldots R_{qN}\}$ defines a local clustership indicator of genomic locations (rows). By this formulation, we allow a different partition of genomic locations for each subset of HMs. We construct a prior $p(C, R_1, \ldots, R_Q)$ in a hierarchical fashion by first formulating the prior $p(C)$ along HMs and then, conditional on C, extending the model along the genome. To do the former, we first distinguish between two HM types. In the context of epigenetic pathways, it is natural that only a selected subset of the HMs would contribute to a nested partition of genomic locations, but not all HMs necessarily do. This is similar to the concept of variable selection in regression where we filter out the insignificant variables to improve power in detecting a signal for the remaining significant variables. In our specific model, we incorporate this assumption by creating a latent category (indexed by C_0) of HMs that do not give rise to a nested partition of genomic locations. We refer to C_0 as idle HMs and to the remainder as active HMs.

The prior $p(C)$ for the random partition is given by a zero-enriched Polya Urn model [29] as follows. Let $G' < G$ denote the number of active HMs. Also, let $p_q = |C_q|$ denote the cardinality of the q-th set of HMs. We define,

$$p(C) = \pi_0^{G'} (1-\pi_0)^{G-G'} \frac{\alpha^Q \prod_{q=1}^{Q} \Gamma(p_q)}{\prod_{g=1}^{G'} (\alpha + g - 1)}. \tag{15.4}$$

Here α represents the total mass parameter of the underlying Polya Urn prior. The mass parameter in Polya Urn priors represents how the prior probability mass is distributed among clusters. For example, with values of α much lesser than 1, the mass will be highly concentrated in a handful of clusters. In addition, the zero-enriched prior assigns each HM with probability $(1-\pi_0)$ to the idle set. Among those HMs that are not in the idle set, the ratio in the third factor in (15.4) determines the probability of building clusters $C_1, \ldots, C_Q$, including the actual number of clusters. In fact, this third factor defines a random partition that is well known as the Polya Urn prior. A Polya Urn prior is a well known non-parameteric prior for clustering of experimental units. The specific form of this prior can be obtained by integrating out the random probability measure in a model with iid sampling, where the random probability measure itself has a Dirichlet Process (DP) prior. Conditional on the random measure, the experimental units are assumed to be generated from an iid (independent and identically distributed) model. In this way, a prior on clustering is induced marginally. Thus, while the support of DP is on random distributions of the experimental units, the Polya Urn acts on the space of clusters. As a result of this relationship, it is a common practice to refer to the underlying DP process while describing such clustering models.

Having defined the partitioning for HMs, we now impose a prior for the random genomic partitions, nested within HM clusters. Let D_q be the number of genomic clusters defined by $R_q = \{R_{q1}, \ldots, R_{qD_q}\}$, that is, the partition of genomic locations with respect to the HMs in the qth HM set. We shall use d as the running index for the active clusters in the qth HM set and let $n_{qd} = |R_{qd}|$ be the number of genomic locations in the dth active cluster. Here again, we allow for an idle cluster R_{q0}. This is the set of inactive genomic locations that do not meaningfully co-cluster with other loci with respect to the qth HM set. The probability of belonging to this set is denoted by $1 - \pi_1$. Let $m_q = \sum_{d>=1} n_{qd}$. Now assuming independent zero-enriched Polya Urn priors for each of the R_qs, we get

$$P(R \mid C) = \prod_{q=1}^{Q} P(R_q) \text{ with } P(R_q) = \pi_1^{m_q}(1-\pi_1)^{N-m_q} \frac{\beta^{D_q} \prod_{d=1}^{D_q} \Gamma(n_{qd})}{\prod_{i=1}^{m_q} (\beta + i - 1)}. \quad (15.5)$$

Here β is the mass parameter of the underlying Polya Urn prior. We observe that the functional form of the $P(R_q)s$ is very similar to that in 15.4. However, these terms are now indexed by the number of genomic locations specific to each active cluster. This is an important feature of our model that underlines the local clustering paradigm. Again, as in the case of α defined before, the mass parameter is a hyper parameter indexing the Polya Urn prior. The nonparametric prior relaxes the requirement to fix the number of clusters. Recall that Q is random and depends on the HM clustering C.

We finally complete the hierarchy with a sampling model,

$$y_{ig} \sim \text{Poi}(\theta_{ig}).$$

Here Poi represents the Poisson likelihood, i indexes a genomic location and g represents a HM.

Note that for any particular HM, we force all locations in the same cluster to share the same rate i.e., $\theta_{ig} = \theta^{\star}_{dg}$ for all $i \in R_{qd}$. Here $\theta^{\star}_{dg}$ denotes the common unique value of θ_{ig} for all loci in R_{qd}. However we allow HMs within the same

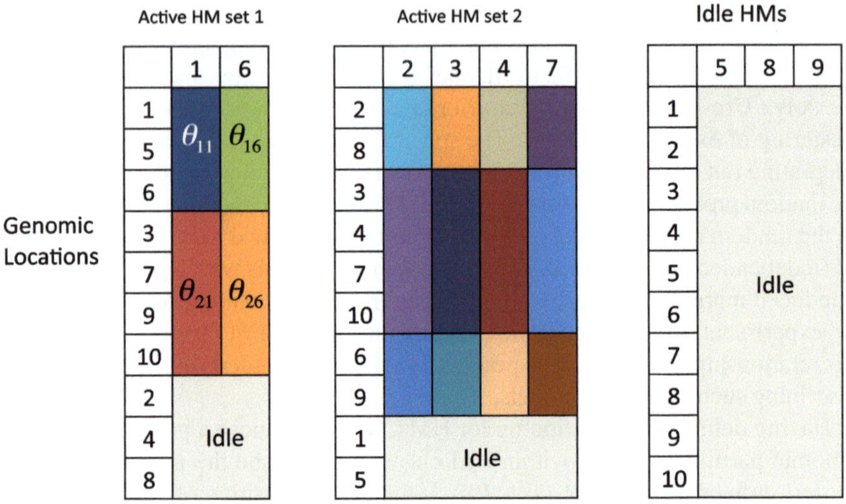

Fig. 15.4 A sample realization of the NoB-LCP model with 9 HMs and 10 genomic locations. There are two active HM sets and an idle HM set, including HMs 5, 8, 9. This figure is adapted from [35]

partition to have different rates for the same location segment, i.e., HMs in the same HM cluster share the same partition of locations only, but not the same rate.

The remaining step is to specify the priors for the rate parameters. Here, we allow them to have their own independent gamma priors. Specifically, we use the same gamma prior for all active genomic locations within active HM sets. For an idle HM set and for the inactive locations of active HM sets, we assign gamma distributions with different hyperparameters. We index the set of all gamma hyper parameter pairs as k and λ.

These prior specifications for the Poisson rates imply a negative binomial distribution for the HM counts marginally. In this way, we implicitly accounted for the over dispersion in the sequencing data. Beta priors are allocated for π_0 and π_1 – the probabilities for the idle set. The hyper-parameters for the Beta priors are carefully chosen to control multiplicity and induce sparsity. When strong prior knowledge are not available for real data, they could be easily set to reflect our lack of information. In summary, we state the joint model as

$$\begin{aligned} P(y,C,R_1,\ldots,R_Q,\theta,k,\lambda,\pi_0,\pi_1) = \\ P(y \mid \theta)P(\theta \mid C,R_1,\ldots,R_Q,k,\lambda) \\ \times p(k,\lambda)P(R_1,\ldots,R_Q \mid C)P(C)P(\pi_0)P(\pi_1). \quad (15.6) \end{aligned}$$

In Fig. 15.4, we show the structure of the proposed non-parametric Bayesian local clustering for Poisson (NoB-LCP) model. The vertical gray block indicates the idle

HMs. Cells marked with the same color share the same Poisson rates θ_{dg}. The different colors across the HM columns in the same active HM set show that the rates are allowed to vary, keeping the partitions fixed. There are gray horizontal blocks within each HM set indicating the inactive genomic locations with respect to that HM set. Note that there is no partition along the columns (HMs) in those regions.

To sample from this joint model, we construct a Gibbs sampler that iterates over the following transition steps

$$[R_1,\ldots,R_Q \mid y,C,\pi_1],[C \mid y,R_1,\ldots,R_Q,\pi_0],[\theta \mid y,C,R_1,\ldots,R_Q]$$
$$[\pi_0 \mid C],[\pi_1 \mid C,R_1,\ldots,R_Q] \quad (15.7)$$

The NoB-LCP model considers two axes of variation– the genome and the HMs. The primary objectives of inference are the partitions along the two axes. Posterior distributions of the partitions are characterized by the MCMC posterior samples. However, summarizing them is a major challenge. We use the algorithm defined by [9] to propose a summary of partitions. Briefly, the algorithm is as follows. We first define a matrix H of co-clustering probabilities. The $(i,j)^{th}$ entry of H is the posterior probability that HM i shares the same cluster with HM j. The entries of H can be easily estimated from the posterior distributions of clusters as the proportion of sampled clusters where HM i coclusters with HM j.

Now for any imputed partition C in the MCMC output, we define a matrix S^C. The $(i,j)^{th}$ entry of S^C is the indicator that HM i co-clusters with HM j in the partition C. Thus S^C can be thought of as a discretized version of H. We now define the optimum clustering as one which minimizes the Frobenius distances between S^C and H. For details, we refer the readers to [9].

15.4.3 Results: Co-Clustering ChIP-seq Data

We selected two important regulatory regions for illustrating our approach, namely, promoters and insulators. Promoters are regions of DNA upstream of a gene that initiate transcription of the particular gene. Promoters are sometimes regulated by enhancers, which are short genomic regions that bind to transcription factors. The enhancers can influence a set of genes and unlike promoters, need not be in close proximity to those genes. An insulator, on the other hand, is a boundary region that blocks the interaction between enhancers and promoters. The promoter data was obtained from the UCSC Genome Browser [12]. The insulator information was obtained from the CTCFBSDB [3], a CTCF binding site database to identify insulators. The rows of the data matrix are a sample of 50 genomic locations in promoter regions and 50 genomic locations in insulator regions.

We employed a total of 10,000 MCMC iterations with the initial 5,000 samples discarded as burn-in. The initial clustering is determined from a hierarchical

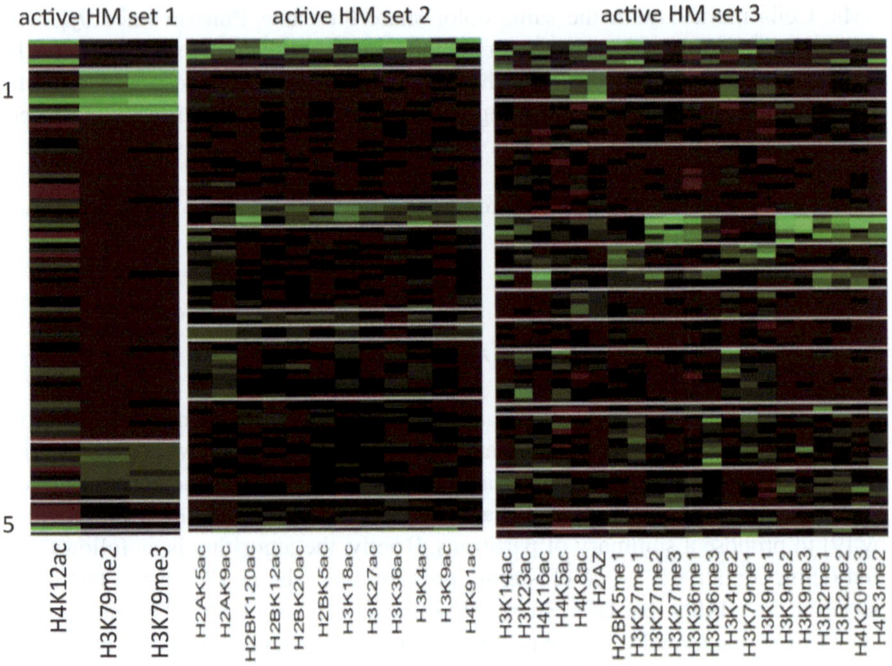

Fig. 15.5 Heatmaps of three active HM sets for ChIP-seq data. White horizontal lines indicate division of location clusters. This figure is adapted from [35]

clustering algorithm. Summarizing the posterior distribution by Dahl's method as described above, we finally obtained 3 active HM sets. Each of them is characterized by its unique pattern of clustering genomic locations. The results confirm that posterior inference distinguishes different types of regulatory elements and clusters similar elements together reasonably well. Figure 15.5 shows the heatmaps of all estimated active HM sets. These three sets are candidates of co-localized HMs that relate to gene transcription. In addition, the heatmap shows genomic location clusters nested within each active HM set. The local clustering patterns are clearly noticeable. In the active HM sets, there is hardly any significant pattern in the idle genomic locations based on the observation that colors are more or less randomly scattered. In contrast, the active genomic locations nested within active HM sets show more homogeneity. The single idle HM set combines all genomic regions and hence show greater variability.

Evidence from epigenetic literature [4, 13, 14, 30, 31] has broadly categorized HMs into either activating or repressing modifications for transcription of protein coding genes. It is well known that the transcription activating HMs generally correspond to acetylations while the transcription repressing ones map to methylations. A remarkable result is that the clusters obtained after inference are clearly distinguished by methylations and acetylation sets. Out of the 12 HMs in active HM set 2, all of them turned out to be acetylations; out of the 21 HMs in the active HM set 3 only two of them are methylations. Another notable finding is the list of HMs known to be highly correlated in previous studies that were identified by the NoB-LCP model. This list includes H2BK120ac, H2BK12ac, H2BK20ac, H2BK5ac, H3K18ac, H3K27ac, H3K36ac, H3K4ac, H3K9ac, and H4K91ac. Here we find a considerable overlap with the HM pairs reported to have strong edges from our network analysis. Similarly, the clustering pattern along the genomic axis validates known properties of the regulatory regions. The active HM set 1 include only promoter regions, in which H4K12ac, H3K79me2 andH3K79me3 clearly show relatively high expression. This is consistent with previous findings which show that H4K12ac counts are elevated in the promoter and transcribed regions of active genes [33]. It is also well known that H3K79me2 and H3K79me3 are important histone markers for the prediction of promoter regions [32, 34].

15.5 Conclusion

We describe a class of hierarchical Bayesian models which hold a special relevance to an important topic in next generation genomics and proteomics, namely histone modifications. Compared to traditional graphical models, the proposed models induce greater flexibility by means of latent variables at different levels of the model hierarchy. This is critical, for example, in modeling interactions among latent indicators of presence of HMs, thus discovering biologically more meaningful associations than raw correlations. The bottommost layers of the hierarchy allow widely different choices of sampling distributions. This is a highly desirable property for sequencing data whose distributions deviate markedly from standard models.

One of these approaches is a graphical model that represents the global dependencies among HMs. The latent graphs characterize the conditional independence structure among HMs. Though initially built for single networks, the associated graphical priors allow easy extension to models for dependent families of graphs. The results of the single network model point towards the existence of a strong cross-talk mechanism among the HMs. To complete the picture, we also consider inference related to the association of these HMs with gene expression. Again, previous studies had confirmed that ChIP signals from many HMs form diffuse, broad domains. Some of the relationships have been tested previously in mammalian embryonic stem (ES) cells and fruitflies. Based on this, we introduce a non-parametric clustering model to provide a formal structure of co-localization between HMs and regulatory elements. We emphasize the role of Bayesian paradigm all through our inference. The 'unknown' model parameters are random variables having prior distributions. This provides us with much more than a point estimate, viz, the entire posterior distribution of partitions and networks. Posterior inference mainly requires Gibbs sampling, which relies on the full conditionals from the data generating model.

Not all relationships inferred from our models have been tested experimentally and represent potentially new causal and/or combinatorial relationships. Such relationships provide a blueprint for mapping the histone code. The objectives attained by our models could also be approached through multistep procedures. However, such ad-hoc measures would propagate biased uncertainty estimates. In contrast, our models offer the advantages of full model-based inference. Flexibility is an important feature of the two proposed models. For example, covariates (e.g., transcription binding sites) could be included to provide a more realistic picture of regulatory processes. Some other possible extensions include the sampling models. We note that these parameters represent the leaves of the model hierarchy and could be specified independently of the graphical priors. Here, they are used to model count data obtained from ChIP-seq experiments. These could be replaced by essentially arbitrary other models without any change in the underlying probability models on the random partitions. Also, the auto logistic model for networks can be easily extended to accommodate inference for cliques of size 3 or more. The coefficients $\beta_{i_1 \cdots i_k}$ will have interpretations as many-way interaction coefficients. Cliques of size 3, for example, would model three-way interactions between triplets of connected HMs that are connected to each other. This would imply that statistical relationships between two HMs would depend on the value of the third HM.

Finally, we acknowledge that formal statistical approaches to histone modifications are still in their infancy. The biology of HMs and their implications in medical research is attaining higher relevance with new experimental evidence pouring in at a fast rate. These findings are attracting novel and insightful computational approaches. Recently [11] made a genome wide map of HMs based on multivariate HMMs. We find ample scope as Bayesian statisticians to follow that direction and formally extend their approach. Currently we incorporate dependence among genomic regions through clustering. This assumes an exchangeable prior across locations. However, the underlying biology suggests a more complex relationship. The length of regions over which histone marks remains stable are known to vary across different domains. Genomic regions contiguous to one another are more likely to have the same histone signatures. Non-exchangeable segmentation priors like Hidden Markov Models can potentially exploit these features. However, we are uncertain if a simple multi-dimensional HMM would be optimal for this purpose. There are two concerns. First, there could be several HMs which would not participate in the segmentation. Second, the assumption of modeling individual HMs independently is itself questionable. It would be best, both for dimension-reduction and biological interpretability, if we could capture this varying effect of HMs through non-overlapping clusters. Briefly, this would demand a combined approach of genome segmentation and HM clustering. Flexible length-based priors may be considered for the former. Such priors should be able to accommodate prior information on the functionality of genomic regions. We hope to pursue these ideas in the future.

References

[1] Bannister, A.J., Kouzarides, T.: Regulation of chromatin by histone modifications. Cell Res. **21**(3), 381–395 (2011)

[2] Bannister, A.J., Zegerman, P., Partridge, J.F., Miska, E.A., Thomas, J.O., Allshire, R.C., Kouzarides, T.: Selective recognition of methylated lysine 9 on histone h3 by the hp1 chromo domain. Nature **410**(6824), 120–124 (2001)

[3] Bao, L., Zhou, M., Cui, Y.: Ctcfbsdb: a ctcf-binding site database for characterization of vertebrate genomic insulators. Nucleic Acids Res. **36**(suppl 1), D83–D87 (2008)

[4] Barski, A., Cuddapah, S., Cui, K., Roh, T.Y., Schones, D.E., Wang, Z., Wei, G., Chepelev, I., Zhao, K.: High-resolution profiling of histone methylations in the human genome. Cell **129**, 823–837 (2007)

[5] Berger, S.: The complex language of chromatin regulation during transcription. Nature **447**, 407–412 (2007)

[6] Bernstein, B.E., Meissner, A., Lander, E.S.: The mammalian epigenome. Cell **128**(4), 669–681 (2007)

[7] Besag, J.: Spatial interaction and the statistical analysis of lattice systems. J. Roy. Stat. Soc. Ser. B **135**, 192–236 (1974)

[8] Caragea, C., Kaiser, S.: Autologistic models with interpretable parameters. J. Agric. Biol. Environ. Stat. **14**, 281–300 (2009)

[9] Dahl, D.: Model-based clustering for expression data via a Dirichlet process mixture model. In: Do, K.A., Müller, P., Vannucci, M. (eds.) Cambridge University Press, Cambridge (2006)

[10] Dodd, I.B., Micheelsen, M.A., Sneppen, K., Thon, G.: Theoretical analysis of epigenetic cell memory by nucleosome modification. Cell **129**(4), 813–822 (2007)

[11] Ernst, J., Kellis, M.: Discovery and characterization of chromatin states for systematic annotation of the human genome. Nat. Biotechnol. **28**(8), 817–825 (2010)

[12] Fujita, P.A., Rhead, B., Zweig, A.S., Hinrichs, A.S., Karolchik, D., Cline, M.S., Goldman, M., Barber, G.P., Clawson, H., Coelho, A., et al.: The ucsc genome browser database: update 2011. Nucleic Acids Res. **39**(suppl 1), D876–D882 (2011)

[13] Grunstein, M.: Histone acetylation in chromatin structure and transcription. Nature **389**(6649), 349–352 (1997)

[14] Heintzman, N.D., Stuart, R.K., Hon, G., Fu, Y., Ching, C.W., Hawkins, R.D., Barrera, L.O., Van Calcar, S., Qu, C., Ching, K.A., Wang, W., Weng, Z., Green, R.D., Crawford, G.E., Ren, B.: Distinct and predictive chromatin signatures of transcriptional promoters and enhancers in the human genome. Nat. Genet. **39**, 311–318 (2007)

[15] Hublitz, P., Albert, M., Peters, A.H.: Mechanisms of transcriptional repression by histone lysine methylation. Int. J. Dev. Biol. **53**(2), 335 (2009)

[16] Jacobson, R.H., Ladurner, A.G., King, D.S., Tjian, R.: Structure and function of a human tafii250 double bromodomain module. Science **288**(5470), 1422–1425 (2000)

[17] Lachner, M., O'Carroll, D., Rea, S., Mechtler, K., Jenuwein, T.: Methylation of histone h3 lysine 9 creates a binding site for hp1 proteins. Nature **410**(6824), 116–120 (2001)

[18] Lee, D.Y., Hayes, J.J., Pruss, D., Wolffe, A.P.: A positive role for histone acetylation in transcription factor access to nucleosomal dna. Cell **72**(1), 73–84 (1993)

[19] Lee, J., Mueller, P., Zhu, Y., Ji, Y.: A nonparametric Bayesian model for local clustering with application to proteomics. J. Am. Stat. Assoc. **108**, 775–778 (2013)

[20] Mitra, R., Müller, P., Ji, Y.: Bayesian graphical models for differential pathways. Tech. Representative, ICES, University of Texas at Austin (2012)

[21] Mitra, R., Müller, P., Ji, Y.: Bayesian multiplicity control for graphs. Tech. Representative, University of Texas at Austin (2012)

[22] Mitra, R., Müller, P., Liang, S., Xu, Y., Ji, Y.: Towards breaking the histone code-bayesian graphical models for histone modifications. Circulation: Cardiovasc. Genetics. **6**(4), 419–426 (2013)

[23] Mitra, R., Müller, P., Liang, S., Yue, L., Ji, Y.: A bayesian graphical model for chip-seq data on histone modifications. J. Am. Stat. Assoc. **108**(501), 69–80 (2013)
[24] Parmigiani, G., Garrett, S., Anbazhagan, R., Gabrielson, E.: A statistical framework for expression-based molecular classification in cancer. J. Roy. Stat. Soc. Ser. B **64**, 717–736 (2002)
[25] Schmid, C.D., Bucher, P.: Chip-seq data reveal nucleosome architecture of human promoters. Cell **131**(5), 831–832 (2007)
[26] Schmidt, D., Wilson, M.D., Spyrou, C., Brown, G.D., Hadfield, J., Odom, D.T.: Chip-seq: Using high-throughput sequencing to discover protein–dna interactions. Methods **48**(3), 240–248 (2009)
[27] Schwartz, Y.B., Pirrotta, V.: Polycomb silencing mechanisms and the management of genomic programmes. Nat. Rev. Genet. **8**, 9–22 (2007)
[28] Schwartz, Y.B., Pirrotta, V.: Polycomb complexes and epigenetic states. Curr. Opin. Cell Biol. **20**(3), 266–273 (2008)
[29] Sivaganesan, S., Laud, P.W., Müller, P.: A bayesian subgroup analysis with a zero-enriched polya urn scheme. Stat. Med. **30**(4), 312–323 (2011)
[30] Stewart, M.D., Li, J., Wong, J.: Relationship between histone h3 lysine 9 methylation, transcription repression, and heterochromatin protein 1 recruitment. Mol. Cell. Biol. **25**(7), 2525–2538 (2005)
[31] Struhl, K.: Histone acetylation and transcriptional regulatory mechanisms. Genes Dev. **12**(5), 599–606 (1998)
[32] Wang, X., Xuan, Z., Zhao, X., Li, Y., Zhang, M.Q.: High-resolution human core-promoter prediction with coreboost_hm. Genome Res. **19**(2), 266–275 (2009)
[33] Wang, Z., Zang, C., Rosenfeld, J.A., Schones, D.E., Barski, A., Cuddapah, S., Cui, K., Roh, T.Y., Peng, W., Zhang, M.Q., Zhao, K.: Combinatorial patterns of histone acetylations and methylations in the human genome. Nat. Genet. **40**, 897–903 (2008)
[34] Weishaupt, H., Sigvardsson, M., Attema, J.L.: Epigenetic chromatin states uniquely define the developmental plasticity of murine hematopoietic stem cells. Blood **115**(2), 247–256 (2010)
[35] Xu, Y., Lee, J., Yuan, Y., Mitra, R., Liang, S., Müller, P., Ji, Y., et al.: Nonparametric bayesian bi-clustering for next generation sequencing count data. Bayesian Anal. **8**(4), 759–780 (2013)

Chapter 16
Genotype Calling and Haplotype Phasing from Next Generation Sequencing Data

Degui Zhi and Kui Zhang

Abstract In this chapter, we will review current statistical and computational approaches for genotype calling and haplotype phasing from next generation data. We will focus on statistical ideas and ignore many practical bioinformatics issues such as image processing for base calling, read mapping, sequencing error rate recalibration, etc, each of which is a topic in its own right. We will give derivations of commonly used approaches, emphasize their assumptions, and aim to unify them in an all-encompassing Bayesian framework. We will point out limitations of single-site genotype likelihood methods that dominate current practice and discuss strategies to use haplotype informative reads.

16.1 Introduction and Overall Pipeline of Analysis of NGS Data

One of the primary applications of next-generation sequencing (NGS) technologies is to sequence human individuals and derive their genotype information. Traditionally, after a reference genome sequence was assembled and a set of genetic variants was discovered, microarrays can be used to profile genotypes of samples. However, microarrays can only identify the genotype information over a predefined set of variants, typically of high minor allele frequencies. NGS technologies can identify all variants including single nucleotide polymorphisms (SNPs), short insertion and deletions (INDELs), and structural variations (SVs) across all frequency spectrum.

D. Zhi (✉) • K. Zhang
Department of Biostatistics, University of Alabama at Birmingham, Birmingham, AL 35294, USA
e-mail: dzhi@uab.edu; kzhang@uab.edu

S. Datta and D. Nettleton (eds.), *Statistical Analysis of Next Generation Sequencing Data*, Frontiers in Probability and the Statistical Sciences, DOI 10.1007/978-3-319-07212-8_16,

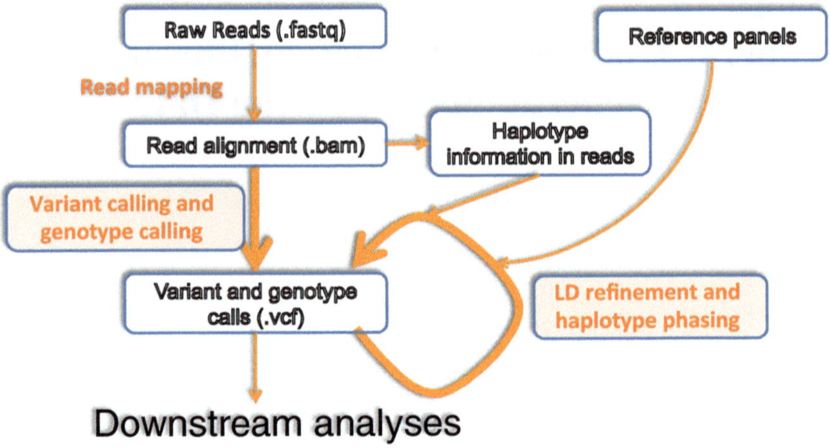

Fig. 16.1 Diagram of NGS analytical pipeline

In addition, exome sequencing or candidate gene sequencing can efficiently profile all genotype information in given regions with much lower cost. Therefore, NGS is becoming the method of choice of many human genetic studies.

However, data generated from NGS technologies are far more complex than data from microarray technologies. Due to the inherent random sampling nature of sequencing technologies, flexible experimental designs, and various practices in bioinformatics processing, many advanced methods are required to handle such analytic challenges of NGS data.

In a typical human population sequencing study, the DNAs of a number of samples are broken into fragments and tips of a set of fragments are read out from sequencing machines. Continuous strings of nucleotides of A, C, G, and T, or *reads*, are generated from sequencing machines containing the base calls. Typically, for each called nucleotide (also called a base call), there is an associated quality score indicating the confidence of that base call. Base calls and quality scores are routinely encoded in files of the FASTQ format. These reads are subject to the following analyses (Fig. 16.1).

First, these files are mapped to the reference genome sequence, forming a pile-up alignment of reads over the genome, typically in the compressed BAM format. These alignments are subject to estimated alignment quality [8] or re-alignment around potential short insertion/deletion sites [5]. Methods have also been developed to adjust the base quality scores from read alignments. Once the alignment files are generated, genotype likelihoods, summary statistics for the marginal probability of genotypes at individual positions, are calculated. Afterwards, with a certain specification of the prior distribution for genotypes, the posterior probability of a position being a potential genetic variant is calculated. Sites with posterior probability above a certain predefined threshold are elected as potential polymorphic sites (PPSs), and the likelihoods of genotypes of each individual over these PPSs are calculated. The results from variant calling and genotype calling are typically packaged in a VCF format file. Various additional information can be used to refine these preliminary variant and genotype calls and to phase haplotypes. While the current practice is to use the linkage disequilibrium (LD) information for refinement, methods that can also leverage direct haplotype information from sequencing reads are available.

In this chapter, we will focus on reviewing the current statistical and computational approaches for genotype calling and haplotype phasing from next generation data but ignore many practical bioinformatics issues such as image processing for base calling, read mapping, sequencing error rate recalibration, etc, each of which is a topic of its own right. We will give the derivations of commonly used approaches, emphasize their assumptions, and aim to unify them in an all-encompassing framework. For simplicity, we will mainly focus on biallelic variants (e.g., single nucleotide polymorphisms (SNPs)) from unrelated, diploid samples. We acknowledge the existence but will not discuss the following topics: short insertion/deletion (indel) calling, structural variation and copy number variation (CNV) calling, multiploidy (such as in cancer) sample calling, and family-based methods.

16.2 Basic Notations

In a population sequencing project, N unrelated diploid individuals (or samples, we use these two terms interchangeably) are sequenced, covering a number of base pairs. After the preliminary variant call, L sites are identified as potential polymorphic sites (PPSs). For each individual n, K_n reads are generated. For simplicity, two paired-end reads are considered as a single read. The read data for nth sample are a pair of K_n-by-L matrices, $R^n = \{r^n_{kl}\}$ and $E^n = \{e^n_{kl}\}$, where

$$r^n_{kl} = \begin{cases} 0, & \text{read } k \text{ of individual } n \text{ has the alternative allele at site } l; \\ 1, & \text{read } k \text{ of individual } n \text{ has the reference allele at site } l; \\ \text{NA}, & \text{otherwise.} \end{cases} \tag{16.1}$$

and

$$e^n_{kl} = \begin{cases} \Pr(\text{sequencing error}), & \text{read } k \text{ of individual } n \text{ covers site } l \\ \text{NA}, & \text{otherwise.} \end{cases} \tag{16.2}$$

Please refer to Fig. 16.2 for an example of the r-matrix for an individual.

In this chapter, we only consider biallelic variants, while the equations for multi-allelic variants can be derived similarly. Throughout this chapter, we denote N as the number of samples, $D^n = (R^n, E^n) = (\{r^n_{kl}\}, \{e^n_{kl}\})$ as sequencing read data and error from nth sample, G^n as the genotype (# of ref-alleles) of nth sample,

Fig. 16.2 Example for a read data matrix for an individual, converted from three reads covering three variant sites

```
Ref: ACCGATCTA
Alt: ATCGAACGA        r-matrix
--------------
R1 : .T..             0  NA NA
R2 :   C...T.G        1  1  0
R3 :      ..T.G.      NA 1  0
```

and H^n as the haplotype pair of nth sample. When it is clear from the context, we will omit subscripts and/or superscripts for the sake of simplicity. For example, for a fixed individual and a fixed variant site, we will omit the sample index and the site index and use $D = (R,E)$ to represent the sequencing read data of that specific sample at that specific variant site. For a fixed individual, we will only omit the sample index and use $D_l = (R_l, E_l)$ to represent the sequencing read data of that specific sample at site l.

The goal of deriving genotype information from NGS data can be divided into three related tasks: variant detection, genotype calling, and haplotype phasing. *Variant detection*, also called *site promotion*, is to identify all potential polymorphic sites among a set of samples. *Genotype calling* is the task of determining the genotype of each sample at all variant sties. *Haplotype phasing* is the task of determining the haplotypes at heterozygous sites for each sample.

16.3 Basic Statistical Approaches for Genotype Calling and Haplotype Phasing

Modern methods for genotype calling generally follow a Bayesian framework. Specifically, the posterior probability of a genotype given the sequencing reads is

$$\Pr(G|D) \propto \Pr(D|G) * \Pr(G), \tag{16.3}$$

where D represents the sequencing reads and G represents the underlying genotypes at all variant sites across all samples. The likelihood $\Pr(D|G)$ describes the conditional probability of the observed sequencing reads given genotypes. The prior term $\Pr(G)$ captures our prior knowledge about the genotypes of these samples. Often $\Pr(G)$ can be written as a parametric form $\Pr(G) = \Pr(G;\theta)$. The genotypes that maximize the posterior probability are selected as true genotypes.

The benefit of such approaches is that the likelihood of sequencing reads and the prior term of genotypes can be computed separately. The likelihood term summarizes only the read data and nothing else. Also, it implicitly assumes that all information in sequencing reads that are relevant to genotype calling is summarized in the likelihood, and thus raw read alignments can be discarded and only the likelihood needs to be stored. However, as will be discussed below, the genotype

Table 16.1 Classes of genotype calling and haplotype phasing methods and their representatives

		Genotype likelihood	
Prior		Single-site	Multi-site
Single site	Single-sample	Samtools-pileup [10]	HASH [2]
	Multi-sample	Sampletool-mpileup [7], GATK [5]	
Multi-site, multi-sample		MVNcall [14]	HapSeq [19], HapSeq2 [18]
		Thunder [12]	HARSH [17]

likelihood may not be a sufficient summary of information in sequencing reads, resulting in loss of power in the subsequent analysis.

The prior term encodes the conditional distributions of the genotype of one sample given the population information, which includes both population parameters and the data from other samples. For single-sample calling, it is often straightforward to find the maximum posterior probability $\Pr(G|D)$ over genotypes considered. For multi-sample calling, genotypes of all samples are jointly modeled. Such approaches are more powerful than single sample calling approaches. Unfortunately, it is often difficult to find the genotypes of a set of samples that maximize the posterior probability due to the complicated formula of $\Pr(G|D)$ and the large number of possible genotypes. Two general strategies are commonly used: expectation-maximization (EM) based approaches and Gibbs-sampling based approaches.

The EM based approaches are suited when the parametric form $\Pr(G) = \Pr(G;\theta)$ is relatively simple. In such approaches, in the E-step, the genotype G of each sample will be treated as "missing data" and $E(G)$ is estimated given the sequencing reads and the current estimate of the population parameter θ. Then the θ that maximizes $\Pr(G;\theta)$ given $E(G)$ from the E-step will be used as updated estimates in the M-step.

If the form $\Pr(G)$ is too complicated, it is often not possible to estimate it directly. In many situations, it is often possible to write out the conditional probability of the genotype of one sample given the other samples. In such case, a Gibbs-sampling like approach can be used to iteratively sample the genotypes of each sample in turn.

Depending on the forms of the likelihood term and the prior term in the calculation of $\Pr(G|D)$, current approaches of genotype calling and haplotype phasing can be largely classified into 6 categories (Table 16.1).

16.4 Single-Site Genotype Likelihood

In the Bayesian framework of genotype calling, the genotype likelihood, $\Pr(D|G)$, summarizes the information of sequencing reads in individual alignment columns for a sample. Assuming unrelated samples, the genotype likelihood for each sample can be calculated independently.

In current practice, most existing methods assume that read data at different sites are independent. As a result, the term genotype likelihood is often a synonym of single-site genotype likelihood. However, even though NGS reads are often considered "short", they may still cover multiple sites and provide important genotype and haplotype information. Below, we follow this convention but we will also discuss multi-site genotype likelihood or haplotype-likelihood. This independent site assumption is obviously with a heritage from the practice of genotyping microarray analyses. The major benefit of this assumption is that read data for an individual at a site can be represented succinctly by just three numbers (of 2 degree of freedom): $LG(g) = \Pr(D|G = g), g = 0,1,2$, and the information on individual reads is no longer needed for downstream analyses.

Here is how the genotype likelihood is calculated. Suppose for an individual at a site there are K reads. The likelihood of a pair of haploid alleles of an individual $h = (h^{(1)}, h^{(2)})$ is

$$\begin{aligned} LH(h) &= \Pr((D,E)|h = (h^{(1)}, h^{(2)})) \\ &= \prod_{k=1}^{K} \left(\frac{1}{2}\Pr((r_k, e_k)|h^{(1)}) + \frac{1}{2}\Pr((r_k, e_k)|h^{(2)}) \right), \end{aligned} \tag{16.4}$$

where the sequencing error term is used to calculate

$$\Pr((r_k, e_k)|h) = \begin{cases} 1 - e_k, & r_k = h; \\ e_k, & r_k \neq h. \end{cases} \tag{16.5}$$

Suppose for an individual at a site, $\mathbf{K}_1$ is the set of reads having the reference allele and $\mathbf{K}_2$ is the set of reads having the alternative allele, and $|\mathbf{K}_1| + |\mathbf{K}_2| = K$, we have [7]:

$$\begin{aligned} LG(g) = \Pr(D|G = g) = & \frac{1}{2^K} \prod_{j_1 \in \mathbf{K}_1} ((2-g)e_{j_1} + g(1 - e_{j_1})) \\ & * \prod_{j_2 \in \mathbf{K}_2} ((2-g)(1 - e_{j_2}) + g e_{j_2}). \end{aligned} \tag{16.6}$$

This information can be extracted from the read alignments, and once calculated, the raw alignment files are no longer needed. Li [7] even advocated that genotype likelihood may be preferred over the actual genotype calls as it captures the uncertainties in genotype calling. The common practice, however, is still to select the genotype that maximize the posterior likelihood as the true underlying genotype.

In practice, the genotype likelihood becomes the central interface between the low-level read alignment and quality recalibration and the high-level genotype probability calculation and downstream analyses. Informatically, the genotype likelihood is typically encoded by the GL or PL fields in the Variant Call Format (VCF) files. The GL field is three real numbers ($L(0)$, $L(1)$, and $L(2)$) for a biallelic variant. PL field contains three integers corresponding to "[N]ormalized,

Phred-scaled likelihoods for genotypes" as defined in the VCF specification, i.e., $round[-10 * \log_{10}(L(0))]$, $round[-10 * \log_{10}(L(1))]$, and $round[-10 * \log_{10}(L(2))]$.

While the genotype likelihood is convenient and has been widely used, one should be aware of the limitation of the site independence assumption. After all, a sequencing read is a continuous "readout" of the genetic information along a single physical chromosome. Therefore, sequencing reads indeed carry haplotype information. In the Sects. 16.6–16.7, we will show how the haplotype information in reads can be used to improve genotype calling and haplotype phasing.

16.5 Single-Sample Genotype Calling

Within the Bayesian framework, once the genotype likelihood is calculated, one only needs to assign a proper prior and a cutoff of the posterior probability for genotype calling. The prior can be from a public database of genetic variants with allele frequency information. But importantly, the prior for any site to be a variant should be non-zero.

There is a danger for doing single-sample calling when multiple samples are sequenced and analyzed together in the downstream, e.g., for genetic association studies after genotype calling. For single-sample calling, variant detection and genotype calling are essentially the same task. The result of single-sample calling is only a list of discovered variants, i.e., sites where the sample has a different allele from the reference genome. However, different samples may have different set of variants discovered when they are individually processed. There are two possible reasons why a variant at a site is not called by one sample but is called in others: (i) that sample is homozygote reference allele with sufficient sequencing coverage, or (ii) that sample does not have sufficient sequencing coverage. Single sample calling does not have a way to distinguish between these two scenarios. Therefore, the standard practice is to use multi-sample calling. Admittedly, single-sample calling can be straightforwardly implemented, and can be appropriate if a sufficient sequencing depth is obtained. However, when multiple samples are sequenced together, it is always more powerful to consider all samples together [15]. Even if a single sample is sequenced, say, in a clinical lab setting, it is still desired to use information from other samples, either from the same lab, or from a public database, for variant-calling.

16.6 Multi-Sample Calling

16.6.1 *Joint Likelihood*

Jointly calling genotypes from multiple samples is a standard practice for NGS data analysis. The main additional information that multi-sample calling uses over single-sample calling is the genotype frequency: $p_{11} = \Pr(G = 0)$, $p_{12} = \Pr(G = 1)$, and $p_{22} = \Pr(G = 2)$. There are 2 free parameters as $p_{11} + p_{12} + p_{22} = 1$. Often in practice, the Hardy-Weinberg equilibrium (HWE) is assumed. This assumption will reduce the number of parameters by one, but it is not required. This classical result of population genetics states that for a site with reference allele frequency ϕ, the genotype frequencies are ϕ^2, $2\phi(1-\phi)$, and $(1-\phi)^2$, for genotypes 2, 1, and 0 (number of reference alleles), respectively, i.e.,

$$\Pr_{HWE}(G = g|\phi) = \binom{2}{g} \phi^g (1-\phi)^{2-g}. \tag{16.7}$$

$$\begin{aligned}\Pr(G^n = g, (R^n, E^n)|\phi) &= \Pr((R^n, E^n)|G^n = g) * \Pr(G^n = g|\phi) \\ &= LG(G^n = g)\Pr_{HWE}(G^n = g|\phi),\end{aligned} \tag{16.8}$$

where $LG(G^n = g)$ is the genotype likelihood defined in Eq. 16.6. Based on that and assuming sample independence, we can write out the joint genotype likelihood of a collection of N samples from a population with the HWE assumption:

$$LP(\phi) = \prod_{n=1}^{N} \Pr((R^n, E^n)|\phi) = \prod_{n=1}^{N} \sum_{g=0}^{2} LG(G^n = g)\Pr_{HWE}(G^n = g|\phi). \tag{16.9}$$

16.6.2 Maximum Likelihood Estimation

From this joint likelihood, the analytical solution for maximization is difficulty but we can maximize the likelihood according to ϕ by numerical methods. Martin et al. [13] developed an Expectation Maximization algorithm (EM) to estimate ϕ.

In the E-step, we estimate the parameter $E(G^n|(R^n, E^n), \phi^{(t)})$ given the sequencing read data and current estimate $\phi^{(t)}$:

$$\begin{aligned}E(G^n|(R^n, E^n), \phi^{(t)}) &= \sum_{g=0}^{2} g * \Pr(G^n = g|(R^n, E^n), \phi^{(t)}) \\ &= \sum_{g=0}^{2} g * \frac{\Pr(G^n = g, (R^n, E^n)|\phi^{(t)})}{\Pr((R^n, E^n)|\phi^{(t)})} \\ &= \frac{\sum_{g=0}^{2} g * \Pr(G^n = g, (R^n, E^n)|\phi^{(t)})}{\Pr((R^n, E^n)|\phi^{(t)})} \\ &= \frac{\sum_{g=0}^{2} g * \Pr(G^n = g, (R^n, E^n)|\phi^{(t)})}{\sum_{g=0}^{2} \Pr(G^n = g, (R^n, E^n)|\phi^{(t)})},\end{aligned} \tag{16.10}$$

where $\Pr(G^n = g, (R^n, E^n)|\phi^{(t)}) = LG(G^n = g)\Pr_{HWE}(G^n = g|\phi^{(t)})$ (Eq. 16.8).

In the M-step, the multi-sample likelihood of ϕ is maximized at

$$\begin{aligned}\phi^{(t+1)} &= \frac{\sum_{n=1}^{N} E(G^n|(R^n,E^n),\phi^{(t)})}{2N} \\ &= \frac{1}{2N}\sum_{n=1}^{N} \frac{\sum_{g=0}^{2} g * \Pr(G^n = g,(R^n,E^n)|\phi^{(t)})}{\sum_{g=0}^{2} \Pr(G^n = g,(R^n,E^n)|\phi^{(t)})}. \end{aligned} \tag{16.11}$$

The EM algorithm will run to convergence when the update $\Delta\phi = \phi^{(t+1)} - \phi^{(t)}$ is small. In practice, the EM algorithm may converge very slowly when the sample size is small or read depth is not high, i.e., the signal from the data is weak [7]. In such case, numerical methods such as Brent's method [4] is recommended [7].

Once the maximum likelihood (ML) estimate of ϕ is obtained, one can obtain the estimate of $G^n, n = 1,\cdots,N$, through the maximization of the posterior probability $\Pr(G|D)$. Based on the maximum likelihood genotype call, $\hat{G}^n$, one can obtain a variant call: claiming any position with $\sum_{n=1}^{N} \hat{G}^n < 2N$ as a variant. However, this approach only gives a point estimate of genotypes. Since it does not consider the distribution of G^n's, it cannot offer a posterior probability assessment of a given position to be a true variant site. As discussed below, a more appropriate method for variant calling is to estimate the posterior probability $\Pr(\sum_{n=1}^{N} G^n < 2N)$.

16.6.3 Estimating the Number of Non-Reference Alleles

For a potential polymorphic site, we define the random variable $X = \sum_{n=1}^{N} G^n$, the number of reference alleles in all samples. In order to call a variant (by calculating the posterior probability that the site has non-reference alleles from all samples), we need to derive the distribution of X.

Assuming HWE, we have

$$\Pr(G^1 = g^1,\cdots,G^N = g^N|X = q) = \begin{cases} \sum_{n=1}^{N} \binom{2}{g^n}/\binom{2N}{q}, & \sum_{n=1}^{N} g^n = q; \\ 0, & \text{otherwise.} \end{cases} \tag{16.12}$$

Then the overall likelihood of allele count is

$$\begin{aligned} LQ(q) &= \Pr(D|X = q) \\ &= \sum_{\sum_{n=1}^{N} g^n = q} \Pr(G^1 = g^1,\cdots,G^N = g^N|X = q) \prod_{n=1}^{N} LG(G^n = g^n) \\ &= \frac{1}{\binom{2N}{q}} \sum_{\sum_{n=1}^{N} g^n = q} \prod_{n=1}^{N} \binom{2}{g^n} LG(G^n = g^n). \end{aligned} \tag{16.13}$$

However, this likelihood function is difficult to compute as it involves summing over all elements in the set $\{(g^1,\cdots,g^N)|\sum_{n=1}^{N} g^n = q\}$. Li [7] proposed the following efficient dynamic programming algorithm to calculate $L(q), q = 0, 1, \cdots, 2N$. First define the partial sum:

$$z_{n,q} = \sum_{\sum_{n'=1}^{n} g^{n'} = q} \prod_{n'=1}^{n} \binom{2}{g^{n'}} LG(G^{n'} = g^{n'}), \tag{16.14}$$

for $0 \le q \le 2N$. Setting $z_{0,0} = 0$, $z_{n,q}$ can be calculated iteratively:

$$z_{n,q} = \sum_{g^n=0}^{2} z_{n-1,q-g^n} \binom{2}{g^n} LG(G^n = g^n), \tag{16.15}$$

and finally

$$LQ(q) = \frac{z_{n,q}}{\binom{2N}{q}}. \tag{16.16}$$

16.6.4 *Variant Detection*

For calling potential polymorphic site or site promotion, we are interested in the posterior probability $\Pr(X < 2N|D)$. Commonly, assuming an infinite-sites neutral model, the probability distribution of the allele count in $2N$ chromosomes is:

$$\Pr(X = q) = \begin{cases} \theta/(2N-q), & q < 2N; \\ 1 - \theta \sum_{n=1}^{2N} \frac{1}{n}, & q = 2N. \end{cases} \tag{16.17}$$

where θ is the population-specific heterozygosity parameter. In practice, θ is set to a small number reflecting the fact that most sites are not a variant. For European population θ is usually set to about $0.8 * 10^{-3}$.

Using Bayes' rule, we have the probably that the site is not a variant is:

$$\Pr(X = 2N|D) = \frac{\Pr(X = 2N)\Pr(D|X = 2N)}{\sum_{q=0}^{2N} \Pr(X = q)\Pr(D|X = q)}. \tag{16.18}$$

With this, we a can assign a Phred-scale variant quality score as $Q = -10 * \log_{10}[\Pr(X = 2N|D)]$, and call the site as a variant site if Q is greater than a threshold.

16.7 Multi-Site Multi-Sample Methods

The above multi-sample variant and genotype calling methods only work on one site at a time. It is well-known that genotypes at adjacent variant sites are correlated due to the linkage disequilibrium (LD). Moreover, the assumption that genotype likelihoods at different variant sites are independent is invalid when reads are long enough to cover more than one site, i.e., we need both multi-site priors and multi-site genotype likelihoods. In this section we will describe statistical models that can capture the correlation structures among multiple sites across multiple individuals, i.e., the LD information and/or the haplotype information in reads of each individual.

With the added complexity in these models, advanced algorithms have been developed to find the maximum likelihood estimates. Most of the popular algorithms are based on Markov Chain Monte Carlo (MCMC). Because these methods are usually much slower than single-site methods, they are typically run for a set of potential polymorphic sites (PPSs) that are preliminarily called by single-site methods, in order to refine the genotype calls made by single-site methods.

When the LD information is used, these methods are often called "LD-based refinement" methods or "LD-refinement" for short. Often these methods infer the genotypes through inference on the underlying haplotypes, therefore LD-refinement methods are also the mainstream methods for haplotype phasing. The resulted haplotypes out of LD-refinement methods from NGS projects, such as the 1,000 Genomes Project [1], become standard resources that serve as reference haplotypes for high-resolution genotype imputation of existing genome-wide association studies (GWAS) data [6].

16.7.1 Multi-Site Prior: Models for Chromosome Sharing Among Multiple Individuals

Following the literature of haplotype phasing and genotype imputation, Li and Stephens' PAC model [9] becomes the standard method for LD-based refinement of genotype calling and haplotype phasing from NGS data. The essence of the Li and Stephens' PAC model is that the haplotype pair of an individual is the recombinant of a set of reference haplotypes, which can be described by the Hidden Markov Model (HMM): $\Pr(D,S) = \Pr(D|S) * \Pr(S)$, where D is the observed data and $S = \{S_1, \cdots, S_L\}$ are the "hidden state" variables that indicate from which of the reference haplotypes that the current haplotype pair of that will be generated. In this formula, $\Pr(S) = \Pr(S_1)\prod_{l=2}^{L}\Pr(S_l|S_{l-1})$ forms a Markov chain and is considered as the prior information of the genotypes across all L PPSs of an individual.

Although the calculation of the above likelihood function is straightforward given the hidden state, it is extremely difficult to find the hidden state that maximize the likelihood function due to the huge number of possible hidden states, which is $O(W^{2L})$ if we have W reference haplotypes across L sites. In practice, the number of reference haplotypes is set to be at least a few hundred to achieve high accuracy. Therefore, the parameters (and hidden states) are estimated through MCMC sampling based on the Baum's forward algorithm [3] and backward selection.

Notably, there are alternative multi-site prior distributions proposed. Wen and Stephens [16] used multi-variate normal distribution over a pre-defined number of sites among a set of reference haplotypes. This prior can be efficiently calculated as only the first and the second moments are needed to estimate a normal distribution. However, Menalaou and Marcini [14] implemented a genotype calling algorithm in a Gibbs sampler fashion and found that it offers suboptimal performance compared to methods based on PAC models. This is likely due to that multi-variate normal distribution is only an approximation of the underlying chromosomal sharing process and it is limited to normal distributions over a fixed number of sites, which may not captures long-range LD information among sites.

16.7.2 *The HMM for Genotype Calling and Haplotype Phasing*

In this section, we give the detailed description of the Hidden Markov Model (HMM) for genotype calling and haplotype phasing from NGS data implemented in Thunder [11]. Specifically, the model can be described as follows:

$$\Pr(D,S) = \Pr(S_1)\prod_{l=2}^{L}\Pr(S_l|S_{l-1})\prod_{l=1}^{L}\Pr(D_l|S_l). \tag{16.19}$$

Again, D is the observed data and $S = \{S_1,\cdots,S_L\}$ are the "hidden state" variables. At each site l, we use a pair of indicator variables, $S_l = (x_l, y_l)$ to indicate from which of the reference haplotypes the current haplotypes will be generated.

To make inference from this HMM, we need to specify the prior probability $\Pr(S_1)$, the emission probability $\Pr(D_l|S_l)$, and the transition probability $\Pr(S_l|S_{l-1})$.

The prior probability, $\Pr(S_1)$, is often assumed to be equal for all possible compatible haplotype configurations of each individual.

To specify the emission probability $\Pr(D_l|S_l)$, we further denote $T_l(i)(l = 1,\cdots,L$ and $i = 1,\cdots,H)$ as the number of reference allele observed at the site l in the reference haplotype i, so $T_l(S_l^n) = T_l(x_l^n) + T_l(y_l^n)$ is the observed genotype at the site l for the nth individual given the underlying hidden state S_l^n. When this model is applied to genotype imputation where D is the observed genotype data and $G = 0,1$, or 2 derived from microarrays, the emission probability is typically of the form:

$$\Pr(G_l|S_l) = \begin{cases} (1-\varepsilon_l)^2 + \varepsilon_l^2, & T_l(S_l) = 1 \;\&\; G_l = 1; \\ 2(1-\varepsilon_l)\varepsilon_l, & T_l(S_l) = 0 \text{ or } T_l(S_l) = 2 \;\&\; G_l = 1; \\ (1-\varepsilon_l)^2, & T_l(S_l) = G_l = 0 \text{ or } T_l(S_l) - G_l = 2; \\ (1-\varepsilon_l)\varepsilon_l, & T_l(S_l) = 1 \;\&\; G_l = 0 \text{ or } G_l = 2; \\ \varepsilon_l^2, & T_l(S_l) = 0 \;\&\; G_l = 2 \text{ or } T_l(S_l) = 2 \;\&\; G_l = 0. \end{cases} \tag{16.20}$$

where ε_l is an error parameter that reflects the combined effect of gene conversion, mutation, and genotyping error. For next generation sequencing data, the emission probability includes the genotype likelihood [11] and is the summation over all possible genotypes:

$$\Pr(D_l|S_l) = \sum_{G_l=0}^{2} \Pr(D_l|G_l)\Pr(G_l|S_l). \tag{16.21}$$

The transition probability, $\Pr(S_l|S_{l-1})$, following the Li and Stephens' paper [9], is defined as a function of the crossover parameter θ_l. The probability of no recombinant staying at the same haplotype (no recombinant) at site l is $1-\theta_l$, and the probability of jumping to any of the W reference haplotypes (including itself) is θ_l/W. We will see below that this parameterization allows efficient computation of forward probabilities. The reference haplotypes can either be internal (i.e., haplotypes from other samples—in the situation, the set of haplotypes is changed in each HMM iteration) or external (i.e., haplotypes from other projects such as the 1,000 Genomes Project - in the situation, the set of haplotypes remains same in each HMM iteration) or a mixture of both. Therefore for the pair of haplotypes, we have the following transition probability:

$$\Pr(S_l|S_{l-1}) = \begin{cases} \frac{\theta_l^2}{W^2} & \text{if } x_l \neq x_{l-1} \text{ and } y_l \neq y_{l-1}; \\ \frac{(1-\theta_l)\theta_l}{W} + \frac{\theta_l^2}{W^2}, & \text{if } x_l \neq x_{l-1} \text{ or } y_l \neq y_{l-1} \text{ but not both;} \\ (1-\theta_l)^2 + \frac{2(1-\theta_l)\theta_l}{W} + \frac{\theta_l^2}{W^2}, & \text{if } x_l = x_{l-1} \text{ and } y_l = y_{l-1}. \end{cases} \tag{16.22}$$

After these terms are defined, the genotype calling and haplotype phasing is based on a Gibbs sampler: a random pair of haplotypes of each individual is assigned according to the observed sequencing reads data. That is, $S_1,\cdots,S_L$ for each individual n are sampled separately according to the likelihood function $LS(S|D) \propto \Pr(D,S)$. Specifically, S_L is first sampled according to $\Pr(D,S)$, then $S_l(l=L-1,\cdots,1)$ are sampled according to the following conditional probability:

$$\Pr(S_l = (x_l,y_l)|D = (D_1,\cdots,D_l), S_{l+1} = (x_{l+1},y_{l+1})) \propto$$
$$\Pr(S_{l+1} = (x_{l+1},y_{l+1})|S_l = (x_l,y_l))\Pr(S_l = (x_l,y_l), D_1,\cdots,D_l), \tag{16.23}$$

where $\Pr(S_l = (x_l,y_l), D_1,\cdots,D_l)$ is the forward probability and can be efficiently calculated through Baum's forward algorithm [3]. Then $S_1,\cdots,S_L$ are used to impute genotype $G_1,\cdots,G_L$ of that individual according to $\Pr(D_l|S_l)$ and determine the new pair of haplotypes of that individual. Then new pair of haplotypes replaces the old pair of haplotypes and is used as the reference haplotypes for other individuals. The sampling procedure is performed over all individuals and repeated for a number of rounds (e.g., 50–100). The consensus genotype and pair of haplotypes of each individual can then be determined by averaging results over replicates.

It is worth mentioning the calculation of the forward probability here. The forward probability is the summation over all W^2 states and the overall complexity of calculation can be $O(W^4)$ without the simplification. As we have mentioned that the HMM often uses $W = 2(N-1)$ internal reference haplotypes, the direct calculation can be time consuming. However, from the design of transition probability

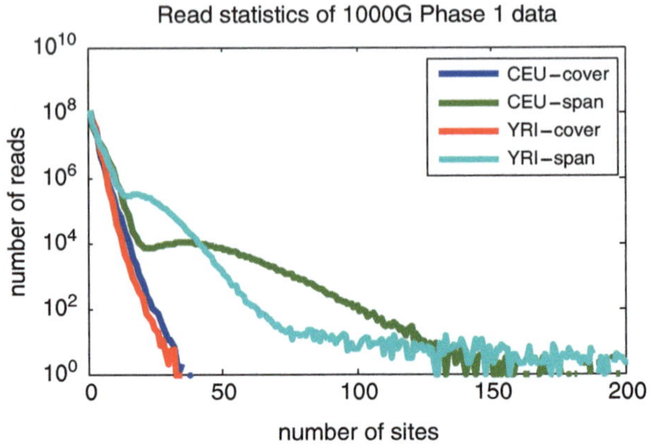

Fig. 16.3 Read cover and span distributions in the 1,000 Genomes Project data sets [1]. Only chromosome 20 for Utah residents with ancestry from northern and western Europe (CEU) and Yoruba in Ibadan of Nigeria(YRI) data are shown. The following measures are relevant to LD-refinement algorithms based on haplotype-informative reads: (i) read-cover: the numbers of sites that are covered by a read; (ii) read-span: the number of sites between the first site and the last site covered by a read. The only difference between these two measures is that sites skipped by the gap between the paired-end reads are counted in read-span, but not in read-cover. For non-paired-end reads, the two statistics are same. In the 1,000G Phase 1 data we see read cover tapers off quickly and few reads cover more than 40 sites. The patterns of read-span are more complex. There are pairs of reads that span over 1,000 sites. This is likely due to structural variations or technical artifacts. Still, bimodal patterns of readspan are evident in both CEU and YRI samples. Note that CEU data have a fraction of 454 reads that are longer, are thus CEU data have longer read-cover and read-span. Adapted from Fig. 4 of [18]

in Li and Stephens' model, all hidden states are symmetric, and the transition probability, $\Pr(S_l = (x_l, y_l)|S_{l-1} = (x_{l-1}, y_{l-1}))$, only depends on weather there is a recombination between S_l and S_{l-1}. Therefore the calculation can be simplified so that the complexity of the computation is $O(W^2)$ rather than $O(W^4)$ [11].

16.7.3 Incorporating Haplotype Information in Reads for Genotype Calling and Phasing Haplotype Information in Reads

After preliminary variant calling, reads that span more than one potential polymorphic sites (PPSs) contain important haplotype information. Such reads are sometimes called "haplotype informative reads" [18]. Fig. 16.3 provides an different variant sites that is used in the HMM of Thunder is invalid. New models are needed to capture the haplotype information from reads.

Based on the haplotype information contained, we identify three sets of read information: $\mathbf{R}_1$, $\mathbf{R}_2$, and $\mathbf{R}_3$: $\mathbf{R}_1$ captures single site information, $\mathbf{R}_2$ captures haplotype information across two adjacent sites, and $\mathbf{R}_3$ captures long-range haplotype information (see HapSeq2 paper for more details [18]). Different methods use different aspects of the read data. Traditional methods like Thunder [11] break all information in $\mathbf{R}_2$ and $\mathbf{R}_3$ into $\mathbf{R}_1$. HapSeq [19] uses $\mathbf{R}_1$ and $\mathbf{R}_2$, but breaks $\mathbf{R}_3$. HapSeq2 is the first method that efficiently uses all information in $\mathbf{R}_1$, $\mathbf{R}_2$, and $\mathbf{R}_3$. It is worth noting that the assignment of a read to $\mathbf{R}_1$, $\mathbf{R}_2$, and $\mathbf{R}_3$ is based on the number of sites not the number of heterozygote sites that it covers. Such construction is only performed once, when the set of potential variant sites are fixed at the beginning of algorithm, and not changed according to genotypes or haplotypes of samples.

16.7.4 Multi-Site Haplotype Likelihood

For the reads spanning multiple variant sites, we can define their corresponding multi-site haplotype likelihood. Following the notations, we define the haplotype likelihood (HL) for a multi-site haplotype pair $H = \{h,\overline{h}\} = \{\{h_m\},\{\overline{h}_m\},m = l_1,\cdots,l_M\}$ for a read that covers M sites $l_1,\cdots,l_M$ with the corresponding bases $r_{l_1},\cdots,r_{l_M}$ and the error rates $e_{l_1},\cdots,e_{l_M}$, as the joint probability:

$$\begin{aligned} HL(H) &= \Pr(r_{l_1},\cdots,r_{l_M},e_{l_1},\cdots,e_{l_M}|(h,\overline{h})) \\ &= 0.5*\Pr(r_{l_1},\cdots,r_{l_M},e_{l_1},\cdots,e_{l_M}|h) \\ &\quad +0.5*\Pr(r_{l_1},\cdots,r_{l_M},e_{l_1},\cdots,e_{l_M}|\overline{h}), \end{aligned} \tag{16.24}$$

where

$$\Pr(r_{l_1},\cdots,r_{l_M},e_{l_1},\cdots,e_{l_M}|(h_{l_1},\cdots,h_{l_M})) = \prod_{m=1}^{M} \Pr(r_{l_m},e_{l_m}|h_{l_m}) \tag{16.25}$$

and $\Pr(r_{l_m},e_{l_m}|h_{l_m}) = 1 - e_{l_m}$ if $r_{l_m} = h_{l_m}$ and $\Pr(r_{l_m},e_{l_m}|h_{l_m}) = e_{l_m}$ otherwise. Here, $\Pr(r_{l_m},e_{l_m}|h_{l_m})$ is similar with that is defined in genotype likelihood (GL) as specified in VCF files. The site-specific error rate, e_{l_m}, can be readily obtained from BAM files.

16.7.5 The HMM Model in HapSeq

Zhi et al. [19] extended the HMM in Thunder by incorporating the haplotype information in reads, $\mathbf{R}_2$, and implemented it in the HapSeq program. Essentially, the HapSeq's HMM is defined as follows:

$$\begin{aligned} \Pr(D = (\mathbf{R}_1,\mathbf{R}_2),S) &= \Pr(S)\Pr(\mathbf{R}_1|S)\Pr(\mathbf{R}_2|S) \\ &= \Pr(S_1)\textstyle\prod_{l=2}^{L}\Pr(S_l|S_{l-1})\prod_{l=2}^{L}\Pr(R_{2,l}|S_{l-1},S_l)\prod_{l=1}^{L}\Pr(R_{1,l}|S_l) \end{aligned} \tag{16.26}$$

Here $\mathbf{R}_1$ and $\mathbf{R}_2$ are set of reads that are non-overlapping and independent as described above. $\mathbf{R}_3$ reads are broken to $\mathbf{R}_1$ and $\mathbf{R}_2$ reads. Here we introduce some more notations. We denote $R_{1,l}^n$ as the set of reads that cover the single site l for the individual n, $R_{2,l}^n$ as the set of reads that cover two adjacent site l and $l+1$ for the individual n.

The major difference between the HapSeq's HMM and the Thunder's HMM is that HapSeq's HMM introduces a new emission probability term, $\Pr(R_{2,l}|S_{l-1},S_l)$, to model the 2 site read information. $\Pr(R_{2,l}|S_{l-1},S_l)$ is the probability of observed jumping reads conditioning on the underlying state at the PPSs $l-1$ and l. In addition, HapSeq' HMM still has the computational complexity of $O(W^2)$ rather than $O(W^4)$ when there are W reference haplotypes [18].

16.7.6 The HMM Model and Metropolis-Hastings Procedure in HapSeq2

Theoretically, the $\mathbf{R}_3$ reads can also be incorporated into the HMM:

$$\begin{aligned}\Pr(\mathbf{R}_1,\mathbf{R}_2,\mathbf{R}_3,S) &= \Pr(S)\Pr(\mathbf{R}_1|S)\Pr(\mathbf{R}_2|S)\Pr(\mathbf{R}_3|S)\\ &= \Pr(S_1)\textstyle\prod_{l=2}^{L}\Pr(S_l|S_{l-1})\prod_{l=1}^{L}\Pr(R_{1,l}|S_l)\\ &\textstyle\prod_{l=2}^{L}\Pr(R_{2,l}|S_{l-1},S_l)\displaystyle\prod_{l=2}^{L}\Pr(R_{3,l}|S_l,S_{l-1},\cdots,S_1)\end{aligned} \quad (16.27)$$

It can be seen that the emission probability, $\Pr(R_{3,l}|S_l,S_{l-1},\cdots,S_1)$ depends on not only S_l and S_{l-1} if $R_{3,l}$ covers sites l and $l-1$ and some sites from 1 to $l-2$. This greatly increases the computational complexity when we perform the forward probability calculation in Monte-Carlo sampling. The computation increases rapidly with the inclusion of reads that cover more sites. Note when we only consider the reads that cover a single site and two adjacent sites, the complexity of calculation of the forward probability is still $O(W^2)$ with W reference haplotypes. Therefore, the above pure HMM is not practical to handle R_3 type of reads due to the highly computational complexity.

To incorporate the haplotype information of reads spanning two or more adjacent and nonadjacent sites, Zhang and Zhi [18] developed a Metropolis-Hastings procedure to sample the haplotype pair of each individual according given the sequencing reads, the reference haplotypes, and the genotypes obtained from the HMM in HapSeq.

16.8 Results

Multi-sample joint calling and LD-refinement are standard practice in large population sequencing projects such as the 1,000 Genomes Project [1]. In the phase 1 release of the project, 1,092 individuals from multiple populations around the world are sequenced using exome (50–100×) sequencing and low-coverage (2–6×) whole sequencing. As shown in Fig. 16.4, multi-sample variant calling for low-coverage whole-genome sequencing reaches the level of accuracy to the high-coverage exome sequencing when the non-reference alleles are present over 10 times out of the 2,184 chromosomes. Moreover, LD-refinement greatly improved the genotype calling across the entire spectrum of variants except for very rare (singletons and doubletons).

Zhang and Zhi [18] compared the performance of Thunder, HapSeq, and HapSeq2 in terms of accuracy of genotype calling and haplotype phasing using simulated data as well as data from the 1,000 Genomes Project Phase 1 Chromosome 20.

Fig. 16.5 show the results with the data from the 1,000 Genome Project. Again, haplotype phasing by interlaced MH-flipping produced longer SEF-blocks, while genotype calling accuracy is also improved (Fig. 16.5). For all pairs of methods one with and one without interlaced HM flipping, interlaced HM flipping increase the average length of SEF-block by 70 KB to 86 KB. This represents 23.6 % to 44.6 % improvement. Between Thunder to HapSeq2, the improvement of the length of the SEF-block is 148 KB (77.6 %) for CEU and 148 KB (66.7 %) for YRI. All these improvements coincide with improvement of genotype calling accuracy.

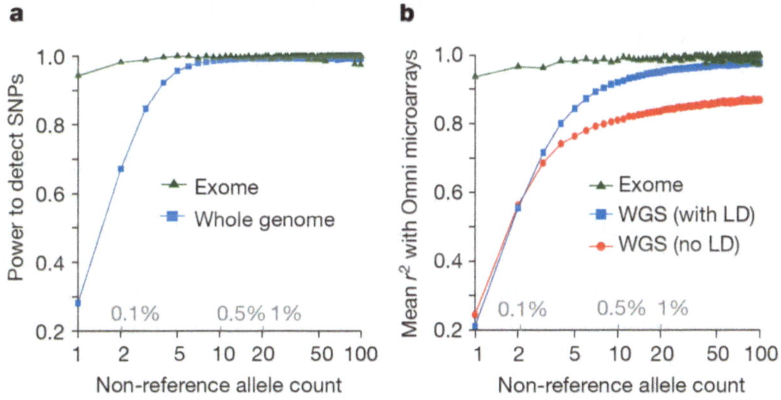

Fig. 16.4 Power to discover and call rare variants. Adapted from Fig. 1 of the 1,000 Genomes Project phase 1 paper [1]

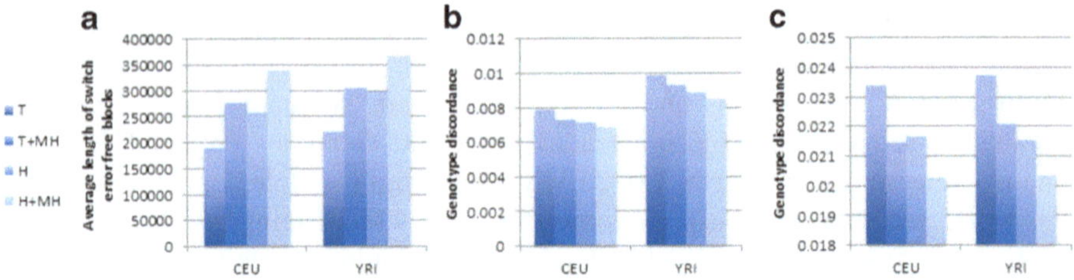

Fig. 16.5 Haplotype phasing (**a**) and genotype calling (**b**) in real data. Results are obtained from the 1,000 Genomes Phase 1 data for CEU and YRI individuals. Methods labels: T = Thunder, H = HapSeq, MH = interlaced MH-flipping. Adapted from Fig. 5 of [18]

16.9 Conclusions

In this chapter, we reviewed statistical approaches for variant and genotype calling as well as haplotype phasing from NGS data. We focused on the settings of low-coverage sequencing from unrelated individuals. Under an overall Bayesian framework, various models and algorithms for representing the information of sequencing reads (likelihoods) and for representing population knowledge (priors) of variants and their corresponding genotypes were discussed. We specifically made the distinction between single-site and multi-site models. While previous methods focused on single-site likelihoods, we emphasized the importance of using multi-site likelihoods. We described efficient algorithms that can make genotype calling and haplotype phasing more accurate under multi-site likelihoods and multi-site prior models. With these methodological advancements, genotype calling and haplotype phasing are quite accurate. According to the 1,000 Genomes Project [21], the power of detecting variants with frequency of 1 % is 99.3 % and the genotype calling accuracy for heterozygous sites is more than 99 % for common SNPs and 95 % for SNPs at frequency of 0.5 % (Fig. 16.4). It is also estimated that switch error for haplotype phasing over common variants occurs at about every 300–400 kb.

While these are exciting achievements, there is still room for improvement. In particular, the power for detecting rare variants, especially for singletons and doubletons, is still low in low-coverage sequencing. While this problem is largely due to the limited availability of sequencing data, more advanced methods taking advantage of haplotype informative reads may help. Moreover, with the advancements of sequencing technologies, longer reads are becoming available. Current practice of ignoring haplotype information in reads will not be optimal. However, efficient computational strategies are required to handle reads covering a large number of potential polymorphic sites. Finally, NGS genotype data are being generated and accumulated with a fast pace and there will be a need to jointly analyze one's in-house data together with the rich resources of publicly available data. Efficient statistical and informatics methods, along with well designed and implemented software packages, are indispensable for meeting the analytical challenges of this kind of big-data research.

Acknowledgements This work is partly supported by National Institute of Health (NIH) grant R00 RR024163. Computational portions of this research were supported by NIH S10RR026723.

References

[1] Abecasis, G.R., et al.: An integrated map of genetic variation from 1,092 human genomes. Nature **491**, 56–65 (2012)

[2] Bansal, V., et al.: An MCMC algorithm for haplotype assembly from wholegenome sequence data. Genome Res. **18**, 1336–1346 (2008)

[3] Baum, L.E.: An inequality and associated maximization technique in statistical estimation for probabilistic functions of Markov processes. Inequalities **3**, 1–8 (1972)

[4] Brent, R.P.: Algorithms for Minimization Without Derivatives. Courier Dover Publications, New York (1973)

[5] DePristo, M.A., et al.: A framework for variation discovery and genotyping using next-generation DNA sequencing data. Nat. Genet. **43**, 491–498 (2011)

[6] Howie, B., et al.: Fast and accurate genotype imputation in genome-wide association studies through pre-phasing. Nat. Genet. **44** 955–959 (2012)

[7] Li, H.: A statistical framework for SNP calling, mutation discovery, association mapping and population genetical parameter estimation from sequencing data. Bioinformatics. **27**, 2987–2993 (2011)

[8] Li, H.: Improving SNP discovery by base alignment quality. Bioinformatics **27**, 1157–1158 (2011)

[9] Li, N., Stephens, M.: Modeling linkage disequilibrium and identifying recombination hotspots using single-nucleotide polymorphism data. Genetics **165**, 2213–2233 (2003)

[10] Li, H., et al.: The sequence alignment/map format and SAMtools. Bioinformatics **25**, 2078–2079 (2009)

[11] Li, Y., et al.: MaCH: using sequence and genotype data to estimate haplotypes and unobserved genotypes. Genet. Epidemiol. **34**, 816–834 (2010)

[12] Li, Y., et al.: Low-coverage sequencing: Implications for design of complex trait association studies. Genome Res. **21**, 940–951 (2011)

[13] Martin, E.R., et al.: SeqEM: an adaptive genotype-calling approach for nextgeneration sequencing studies. Bioinformatics **26**, 2803–2810 (2010)

[14] Menelaou, A., Marchini, J.: Genotype calling and phasing using nextgeneration sequencing reads and a haplotype scaffold. Bioinformatics **29**, 84–91 (2013)

[15] Nielsen, R., et al.: Genotype and SNP calling from next-generation sequencing data. Nat. Rev. Genet. **12**, 443–451 (2011)

[16] Wen, X., Stephens, M.: Using linear predictors to impute allele frequencies from summary or pooled genotype data. Ann. Appl. Stat. **4**, 1158–1182 (2010)

[17] Yang, W.Y., et al.: Leveraging reads that span multiple single nucleotide polymorphisms for haplotype inference from sequencing data. Bioinformatics **29**, 2245–2252 (2013)

[18] Zhang, K., Zhi, D.: Joint haplotype phasing and genotype calling of multiple individuals using haplotype informative reads. Bioinformatics **29**, 2427–2434 (2013)

[19] Zhi, D., et al.: Genotype calling from next-generation sequencing data using haplotype information of reads. Bioinformatics **28**, 938–946 (2012)

Chapter 17
Analysis of Metagenomic Data

Ruofei Du and Zhide Fang

Abstract In this chapter, we first briefly introduce the background of next-generation sequencing metagenomics, including the special properties in this research field and the challenges for statistical analysis. A metagenomic study typically consists of sampling, filtering, DNA extraction, sequencing, binning, assembly, profiling and down-stream analysis. We describe the widely used statistical methods in determining the sufficiency of a metagenomic sample size, or classifying metagenomic sequencing reads into taxonomic bins, or assessing the accuracy of metagenomic assembly. In addition, we outline the steps and statistical methods for correcting the systematic errors in metagenomic profiling. Last, statistical methods for metagenomic comparison are discussed, including both parametric and non-parametric methods for comparison among different groups of multiple samples. The multiple comparison problem is also simply discussed.

17.1 Introduction

17.1.1 The Emergence of Metagenomics

As tiny as they are, microbes (or microorganisms) cannot be recognized by human naked eyes. There exist various types of microbes, including bacteria and archaea, and eukaryotes such as fungi and algae. These tiny creatures are everywhere in the ecosystem and play an important role in our daily life. Thus, deep and intensive explorations of the microbial world are appealing.

R. Du • Z. Fang (✉)
Louisiana State University Health Sciences Center, 433 Bolivar Street, New Orleans, LA 70112, USA
e-mail: rdu@lsuhsc.edu; zfang@lsuhsc.edu

S. Datta and D. Nettleton (eds.), *Statistical Analysis of Next Generation Sequencing Data*, Frontiers in Probability and the Statistical Sciences, DOI 10.1007/978-3-319-07212-8_17,

Traditional microbial experimentation emphasizes pure culture. That is a single microbial species is isolated and cultured in laboratory conditions for investigation. Pure culture was successful in identifying disease causation, and/or association between a bacterial species and particular infectious disease. In terms of power and precision, pure culture had become a gold standard for microbiological experimentation until the middle of the twentieth century, and most knowledge of the modern microbiology was obtained by using pure culture [32]. However, the majority of the microbial species are unculturable. It is estimated that less than 1 % microbial species have been identified through laboratory culture [20, 38]. Additionally, we cannot, by the pure culture approach, detect the diversity of species in a microbial community and their functional properties. New approaches are needed to systematically study microbial communities.

After the discovery of the DNA double helix structure, techniques based on DNA sequencing, such as the Sanger method [35], have helped scientists to make breakthroughs in biological research. With microbial DNA sequencing, a pioneering work by Carl Woese [44] revealed that rRNA genes carry phylogenetic information of microbes. Researchers in Pace's group studied the sequences of rRNA genes and successfully obtained the microbial phylogenetic structure of an environmental sample [34]. This sequencing approach avoids nurturing a microbial community for investigation under a microscope. Since then, the paradigm for researchers to study the microbial world has gradually shifted, not only on microscopic individuals, but also on microbial community which is accessed by its genetic material. Strictly speaking, analyzing rRNA sequences of a community to access its phylogenic structure is not a typical metagenomic study, because only the specific genes are included instead of the entire genomic information of the microbes. Nonetheless, it offers a new avenue to look into the microbial world wider (for new species) and deeper (for new functions).

Metagenomics, sometimes called environmental or community genomics, emphasizes that the object of the study is the collective set of genomes (compared to microscopic species); and it is a culture-independent approach (compared to a culture-based approach).

17.1.2 A Brief View of Metagenomic Studies

Metagenomic studies can be broadly split into function-based analysis and sequence-based analysis [14]. In function-based analysis, the cloned environmental genomic DNAs are screened for desired phenotypes, such as enzyme activity or antibiotic production [11]. Once effective clones are identified, their sequences can be determined. The key advantage of function-based analysis is the discovery of novel genes related to given functions. In sequence-based metagenomics, the entire genomic DNAs will go through the sequencing process. Since statistics is mainly involved in the sequence-based approach, in this Chapter, unless otherwise specified, by metagenomics or metagenomic study, we refer to the sequence-based metagenomics.

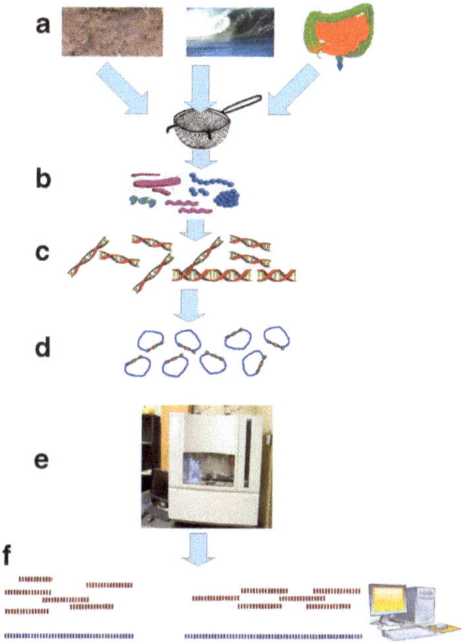

Fig. 17.1 The typical steps involved in a metagenomic study: (**a**) sampling; (**b**) filtering; (**c**) genomic material extraction; (**d**) preparation prior to sequencing; (**e**) DNA sequencing; (**f**) computational or statistical analysis (This figure has been reproduced from Wooley et al. (2010))

Fig. 17.1 exhibits the typical steps in a metagenomic study, including sampling, filtering, genomic material extraction, DNA sequencing, sequencing read binning, sequencing read assembly (optional), sequencing read functional annotation and the down-stream analysis [45]. Environmental samples contain genomic information which can be used to identify a population, regarding the diversity and relative abundance of microbiota. Filtering is necessary due to the fact that these samples may contain information on more than just microbial communities. Typically, a molecular particle of size outside the microbe scope, either smaller or larger, will be left out. Then, genomic DNA is extracted from the filtered samples. If the goal of a study is to assess the species diversity and abundance, 16S rRNA sequence calling may be performed based on the extracted genomic DNA [41]. On the other hand, to profile gene expression or expression dynamics, cDNA can be synthesized from environmental RNA samples [13]. The sequencing step is to determine the exact nucleotide base orders of the DNAs. In the binning step we put sequencing reads into the bins (e.g., phylogenetic categories) for future analysis. By assembly, the sequencing reads are concatenated and merged into a longer sequence, called a contig. Similar to the binning step, sequencing reads are associated to biological functional tags for future functional diversity analysis. The down-stream analysis includes profiling of the microbial communities and conducting comparisons between communities.

17.1.3 Next-Generation Sequencing Metagenomics

A metagenomic study is referred to as next generation sequencing (NGS) metagenomics if the metagenomic DNA is sequenced on a NGS platform from Roche 454 Life Sciences, or Illumina/Solexa, or Applied Biosystems, or others. NGS technologies adopt an array-based work flow by setting up millions of DNA fragments on a single chip and determining the nucleotide base compositions of all these fragments in parallel. This work flow is far more efficient than the traditional chain-termination approach used in the Sanger method, and thus makes NGS a high throughput cost-effective technology.

For currently available sequencing technologies, the first required step of the process is always to break up the intact metagenomic DNA into small fragments, due to the fact that we will become concerned about sequencing errors after the sequencing circles run for a certain number of times. This fragment process actually bestows a nickname on NGS as shotgun sequencing because it has the firing pattern of a shotgun. The another crucial step is to amplify the copies of DNA fragments in order to get enough copies for sequencing cycles. In the Sanger method, this is usually achieved by cloning the DNA into plasmid vectors and letting them self-replicate in *E. coli* cells. However, cloning bias exists because some DNA sequences do not adjust well or may even be lethal to *E. coli* so that they are cloned less often than the others. Cloning-free NGS technologies can eliminate the possibility of this bias, and thus, are more suitable to metagenomic studies [30, 31].

17.2 Statistical Analyses in a NGS Metagenomic Experiment

As in other NGS technologies such as RNA-seq, ChIP-seq, etc., variations may be introduced in different steps of a NGS metagenomic experiment. Statistical methods can be applied to model such variations for achieving validity and efficiency. In this section, we describe the statistical applications in NGS metagenomics.

17.2.1 Sufficiency of Sample Size

As a necessary step of the analysis for a metagenomic study, we need to know how well a collected sample represents the microbial population in the environment. That is, we want to know whether or not the sample is sufficient for the study. Ideally, a sample should contain all the species presented in the environment and reflect the relative abundance among them.

17.2.1.1 Species Richness Estimators

Assume there is a total of N individual microbes in a microbial community, each belonging to one of S different species. Let p_i be the probability that a randomly selected microbe is from the ith species, where $i = 1,2,\dots,S$, and $\sum p_i = 1$. A random sample of n microbes is collected from the community. Let Y_i be the number of microbes representing the ith species in the sample so that $\sum_{i=1}^{S} Y_i = n$. Define S_T as the total number of distinct species in the sample, and C_k as the count of species which occur exactly k times in the sample, $0 \le k \le n$, then $C_k = \sum_{i=1}^{S} I(Y_i = k)$, with $I(\cdot)$ being the indicator function, $S_T = \sum_{k=1}^{n} C_k$ and $\sum_{k=1}^{n} kC_k = n$. We will use corresponding lower-case letters for the observed values of these random variables in a sample of n microbes. Obviously,

$$S = S_T + C_0.$$

But C_0 is unobservable. Hence, we need to find an estimate for the expected value of C_0 in order to estimate S.

It is clear that we have the probability, for any $i = 1,2,\dots,n$,

$$P\{Y_i = k\} = \binom{n}{k} p_i^k (1-p_i)^{n-k},$$

and thus the expectation

$$E(C_k) = \sum_{i=1}^{S} \binom{n}{k} p_i^k (1-p_i)^{n-k}.$$

When k is very small compared to n, this expectation can be approximated as follows [16].

$$E(C_k) \approx \frac{1}{k!} \sum_{i=1}^{S} (np_i)^k e^{-np_i}. \tag{17.1}$$

Let X be a random variable with the cumulative distribution function

$$F(x) = \frac{\sum_{np_i \le x} np_i e^{-np_i}}{\sum_{i=1}^{S} np_i e^{-np_i}}. \tag{17.2}$$

Then, the rth moment of X can be derived as [16]:

$$\mu_r = \frac{\sum_{i=1}^{S} (np_i)^{r+1} e^{-np_i}}{\sum_{i=1}^{S} (np_i) e^{-np_i}} \approx (r+1)! \frac{E(C_{r+1})}{E(C_1)},$$

where $r \geq 1$, and the approximation comes from (17.1). Thus, given a sample of microbes from the community, we can obtain an estimate of μ_r by replacing the expectations with the corresponding observed values:

$$m_r = (r+1)!\frac{c_{r+1}}{c_1}. \tag{17.3}$$

Furthermore, from (17.1) and (17.2), we have

$$\begin{aligned} E(C_0) &\approx \sum_{i=1}^{S} e^{-np_i} = \sum_{i=1}^{S} np_i e^{-np_i} \frac{\sum_{i=1}^{S} e^{-np_i}}{\sum_{i=1}^{S} np_i e^{-np_i}} \\ &\approx E(C_1) \int_0^n x^{-1} dF(x). \end{aligned} \tag{17.4}$$

Now, denote $\mathscr{F}(\mu_1, \mu_2)$ as the class of all cumulative distribution functions in $[0, n]$ with fixed first two moments, μ_1, μ_2, and define $G(x)$ as

$$G(x) = \frac{(n-\mu_1)^2}{(n-\mu_1)^2 + (\mu_2 - \mu_1^2)} I\left(\frac{n\mu_1 - \mu_2}{n - \mu_1} \leq x < n\right) + I(x \geq n).$$

Then $G(x) \in \mathscr{F}(\mu_1, \mu_2)$. It can be shown that [5, 16], for $F \in \mathscr{F}(\mu_1, \mu_2)$,

$$\begin{aligned} \int_0^n x^{-1} dF(x) &\geq \int_0^n x^{-1} dG(x) \\ &= \frac{1}{(n-\mu_1)^2 + (\mu_2 - \mu_1^2)} \left(\frac{(n-\mu_1)^3}{n\mu_1 - \mu_2} + \frac{\mu_2 - \mu_1^2}{n}\right). \end{aligned}$$

Then, by (17.4) and replacing $E(C_1)$ with the observed value c_1, we have a lower bound estimate of $E(C_0)$ [5]:

$$\begin{aligned} \widehat{E(C_0)}_{min} &\approx \frac{c_1}{(n-\mu_1)^2 + (\mu_2 - \mu_1^2)} \left(\frac{(n-\mu_1)^3}{n\mu_1 - \mu_2} + \frac{\mu_2 - \mu_1^2}{n}\right) \\ &\to \frac{c_1}{\mu_1}, \quad as\ n \to \infty. \end{aligned}$$

Replacing μ_1 by its estimate in (17.3), we have an estimate $\widehat{E(C_0)} = c_1^2/(2c_2)$. Thus, the number of species in the community can be estimated by

$$\hat{S} = s_T + \widehat{E(C_0)} = s_T + \frac{c_1^2}{2c_2},$$

which is widely known as the Chao1 estimator if $c_2 > 0$ [5, 10, 21]. A modification can be used when $c_2 = 0$ [6]:

$$\tilde{S} = s_T + \frac{c_1(c_1 - 1)}{2(c_2 + 1)}.$$

For a community with many rare species, the abundance-based coverage estimate (*ACE*), which makes use of all c_k instead of only c_1 and c_2, appears to be more appropriate for the species richness [7,8]. Specifically, with a pre-selected frequency cutoff r (usually $r = 10$) and three estimated quantities: $s_{abund} = \sum_i I(y_i > r)$ as the observed number of abundant species, $s_{rare} = \sum_i I(0 < y_i \leq r)$ as the observed number of rare species and $y_{rare} = \sum_i y_i I(0 < y_i \leq r)$ as the observed number of microbes belonging to the rare species, the *ACE* of the species richness is defined as

$$\hat{S} = s_{abund} + \frac{s_{rare}}{\widehat{SC}_{ACE}} + \frac{c_1}{\widehat{SC}_{ACE}} \hat{\gamma}^2_{ACE},$$

where $\widehat{SC}_{ACE} = 1 - c_1/y_{rare}$ is the estimated sample coverage and

$$\hat{\gamma}^2_{ACE} = max\left\{ \frac{s_{rare} \sum_{k=1}^{r} k(k-1)c_k}{y_{rare}(y_{rare} - 1)\widehat{SC}_{ACE}} - 1,\ 0 \right\}$$

is the estimated coefficient of variation. Readers are referred to [7, 8] for detailed discussions about these estimators.

17.2.1.2 Using Species Richness Estimates to Evaluate Sufficiency of Sample Size

The 16S rRNA genes are usually used as the biomarkers for bacterial species analysis. Specifically, the 16S rRNA gene fragments are detected from metagenomic sequencing reads, and grouped into Operational Taxonomic Units (OTUs), so that the reads in an OTU are likely sampled from the same species [36, 47]. The species richness can then be estimated through the number of distinct OTUs (that is, s_T) and the numbers of the fragments grouped into each OTU (i.e., the y_i values).

Rarefaction curves are usually used to estimate the fraction of species being sequenced [17, 21, 25]. A rarefaction curve plots the estimated number of species in a community as a function of the number of microbes sampled. By subsampling within the detected 16S rRNA fragment pool, for a subsample size, the Chao1 estimator or *ACE* is computed accordingly. Starting from a small subsample size and ending at n, the rarefaction curve can be drawn. The curve usually begins with an increasing slope. If, after a certain point, the curve consistently flattens, it suggests that the richness estimates reach a stable asymptotic value, and we may determine that the sample size is sufficient. Otherwise, more observations should be collected.

17.2.2 Statistical Methods in Metagenomic Binning

Metagenomic binning refers to the process of classifying metagenomic sequencing reads into taxonomic groups (e.g., OTUs). Depending on the information taken for the classification, the widely used binning methods may be divided into two categories: similarity-based methods and composition-based methods [33]. In similarity-based methods, the measurements of the sequence similarities between the metagenomic reads and the genomic references are employed to determine the taxonomic groups from which the reads come. Composition-based methods usually take into account the patterns of conserved short oligonucleotides (also called K-mers, or K-tuples) appearing in the metagenomic reads. If a method first learns the patterns from the reference database and applies the learned result for classification, it is a supervised method. On the other hand, if the patterns of the short oligonucleotides are purely counted from the reads themselves, the approach is an unsupervised method. In the following two subsections, we describe the statistical aspects in both similarity-based and composition-based methods.

17.2.2.1 A Similarity-Based Statistical Framework

This method was proposed in [24]. Suppose that there are N metagenomic sequencing reads being aligned against a database of genomic references, and that S genomes (species) are detected being hit by at least one read. Denote A_{ij} as the alignment length, and M_{ij} as the number of matched base pairs when the ith read $(i = 1,2,\ldots,N)$ is aligned to the genome j $(j = 1,2,\ldots,S)$. Set M_{ij} equal to 0 if the read is not aligned to the genome. Define R_j as the probability of a read from the genome j in the metagenome, $\sum_{j=1}^{S} R_j = 1$. We assume that, for any aligned base pair within two aligned sequences, the probability of having a mismatch is a constant, denoted by p. Let $A_i = max\{A_{ij}, j = 1,2,\ldots,S\}$ be the maximum alignment length between the ith read and the S candidate genomes. We believe that if the ith read is generated from one of the S genomes, then A_i is the length of the alignment when it is compared to that genome. Under the assumption that the base pair matching is independent across sequence positions, the probability of the ith read coming from the genome j with $(A_i - M_{ij})$ aligned mismatches is $R_j p^{A_i - M_{ij}}(1-p)^{M_{ij}}$. Let Z_i be the genome that the ith read comes from. Then the probability of observing the ith read with similarity measurements A_i and M_{ij} is $\sum_{j=1}^{S} I(Z_i = j) R_j p^{A_i - M_{ij}}(1-p)^{M_{ij}}$.

Assume that the sequencing reads are independent of one another, then given the alignment results $\{A_i : i = 1,\ldots,N\}$ and $\{M_{ij} : i = 1,\ldots,N, j = 1,\ldots,S\}$, the likelihood function for the parameters p and R_i is

$$L(p, R_j, j = 1,\ldots,S) = \prod_{i=1}^{N} \sum_{j=1}^{S} I(Z_i = j) R_j p^{A_i - M_{ij}}(1-p)^{M_{ij}}.$$

Since a read comes from a single genome, the summation in the likelihood function actually includes only one addend, and thus it is easy to obtain the log likelihood function. However, the genome that a read is generated from is unobservable, that is, Z_i are latent variables. Hence the Expectation-Maximization (EM) algorithm can be applied to find the maximum likelihood estimates (MLEs) of p and R_j.

Let $\hat{p}$ and $\hat{R}_j$ be the MLEs obtained by the EM algorithm. Then the probability that the ith read is generated from genome g, given the similarity measurements, can be calculated as

$$P_{ig} = P(Z_i = g|A_i, M_{ij}, j = 1,2,\ldots,S) = \frac{\hat{R}_g \hat{p}^{A_i - M_{ig}}(1-\hat{p})^{M_{ig}}}{\sum_{j=1}^{S} \hat{R}_j \hat{p}^{A_i - M_{ij}}(1-\hat{p})^{M_{ij}}},$$

for $i = 1,2,\ldots,N$, and the ith read is classified as the genome which corresponds to the maximum of $\{P_{ig}, g = 1,2,\ldots,S\}$. Other similarity-based binning tools, such as MEGAN [22] and CARMA3 [15], are also available in literature. Based on simulated datasets, Jiang et al. [24] demonstrated that their procedure has the best performance.

17.2.2.2 Composition-Based Statistical Approaches

In this subsection, we further introduce two more metagenomic binning tools which are based on genome sequence composition.

17.2.2.3 Relative Abundance Index (RAI)

In this approach, the RAI profile of a genome consists of K-mer scores, each of which is defined to measure the relative abundance of the K-mer in the metagenomic reads [33]. The probability of observing any K-mer, $x_1,\ldots,x_K$, can be written as $P(x_1,\ldots,x_K) = P(x_K|x_1,\ldots,x_{K-1})P(x_1,\ldots,x_{K-1})$. The conditional probability can be reduced to $P(x_K)$ if x_K occurs independently to all previous bases. Under the assumption that the sequence $\{x_k\}$ follows a Markov chain of order r, we have

$$P(x_K|x_1,\ldots,x_{K-1}) = P(x_K|x_{K-r},\ldots,x_{K-1}) = \frac{P(x_{K-r},\ldots,x_K)}{P(x_{K-r},\ldots,x_{K-1})}.$$

Given a K-mer, $x_1,x_2,\ldots,x_K$, the relative abundance index of order r is defined as the log base 2 of the ratio between the probability of observing $(x_1,x_2,\ldots,x_K)$ and the expected probability under the assumption on specific r. That is,

$$RAI_r(x_1,\ldots,x_K) = log_2 \frac{P(x_1,\ldots,x_K)P(x_{K-r},\ldots,x_{K-1})}{P(x_{K-r},\ldots,x_K)P(x_1,\ldots,x_{K-1})},$$

where $r = 0, 1, \ldots, K-2$. Note that when $r = 0$, $P(x_{K-r}, \ldots, x_{K-1}) = 1$, and RAI_0 is for the independent case. Adding these RAI_r, over r, gives the RAI of the K-mer on a genome:

$$RAI(x_1, \ldots, x_K) = \sum_{r=0}^{K-2} RAI_r(x_1, \ldots, x_K).$$

For a sample metagenome the probabilities in RAI are estimated by relative frequencies of the K-mer in a genome reference. To classify a sequencing read, say R, calculate the frequency of one K-mer in R, $f_R(x_1, \ldots, x_K)$, and define the membership score of R belonging to the genome j as

$$M_{R(j)} = \sum_{\text{all } K\text{-mers in } R} f_R(x_1, \ldots, x_K) \widehat{RAI}^{(j)}(x_1, \ldots, x_K),$$

where $\widehat{RAI}^{(j)}$ is the estimated RAI on the genome j. Note that if the pattern of the K-mer matches between the sequencing read and the genome j, the positive $\widehat{RAI}$ values will be weighted by higher frequencies, and the negative $\widehat{RAI}$ will be weighted by lower frequencies. Thus, the membership score is larger for a matched pattern than a mismatched one. So the rule is to classify R to the genome with the largest score, $max\{M_{R(j)} : j = 1, 2, \ldots, S\}$. Interested readers are referred to [33] for more details and discussions.

17.2.2.4 AbundanceBin

This approach was proposed in [46] and is based on the Lander-Waterman model [26, 27]. It assumes that the number of occurrences of a K-tuple in the sequencing reads generated from the genomes of the same species follows a Poisson distribution. Let M be the total number of K-tuples. Let w_i be the ith K-tuple, $i = 1, 2, \ldots, M$. Let $n(w_i)$ be the count of the sequencing reads covering w_i, and let λ_j be the parameter of the Poisson distribution for the jth species s_j (λ_j depends on the abundance of the species, $j = 1, 2, \ldots, S$). Let l_j be the total genome size of the jth species and L be the total size of the metagenomes. Denote Z_i as the species that the ith K-tuple belongs to. Since Z_i are latent variables, the EM algorithm is applied to find the maximum likelihood estimates of λ_j and g_j, using the conditional probability of w_i being from the jth species given the observations $\{n(w_i)\}$ in the E-step:

$$P(Z_i = s_j | n(w_i),\ i = 1, 2, \ldots, M) = \frac{l_j}{\sum_{k=1}^{S} \left(\frac{\lambda_k}{\lambda_j}\right)^{n(w_i)} e^{-\lambda_k + \lambda_j} l_k},$$

which can be derived by Bayes' rule with the prior probabilities $P(Z_i = s_j) \propto l_j$.

For a metagenomic sample, the number of species, S, is generally unknown. A recursive procedure is suggested in [46] to search for this number. Once the EM algorithm converges, the proposed classification rule claims that a sequencing read, r, comes from the species corresponding to the maximum of the probabilities:

$$P(r \in s_j) = \frac{\prod_{w_i \in r} P(Z_i = s_j | n(w_i))}{\sum_{k=1}^{S} \prod_{w_i \in r} P(Z_i = s_k | n(w_i))}, \quad j = 1, 2, \ldots, S.$$

17.2.3 *Assessing the Accuracy of a Metagenomic Study*

Assembling is an important step in a metagenomic study, especially for the species without existing reference genomes. Thus, given an assembly result of the metagenomic reads, it is very crucial to check the accuracy before the assembly is used for further analysis. On the other hand, facing many assembly tools with different pros and cons, one may need to select the one with the best performance. Assembly Likelihood Evaluation (*ALE*) was proposed to fulfill this need [9].

Let R be the metagenomic reads utilized to assemble a genome. Let A be the event that the assembled sequence is equal to the true genome sequence. Then, by Bayes' rule, the conditional probability that the assembly is correct, given the reads, is $P(A|R) = P(R|A)P(A)/P(R)$. The *ALE* score is defined as the logarithm of this probability. For the comparison of two assemblies, the difference of the *ALE* scores does not depend on $P(R)$, which is difficult to calculate.

Under the assumption of independence, the conditional probability $P(R|A)$ can be decomposed as

$$P(R|A) = P_{placement}(R|A)P_{insert}(R|A)P_{depth}(R|A),$$

where "placement" represents the event that the reads are placed/mapped to the genome, "insert" represents the event that, for paired-end reads, the lengths of the inserts found in reads agree with the lengths of inserts used to generate the reads, and "depth" represents the event that the read depth/coverage at each location of the genome is equivalent to the expected depth. Note that for single-end reads, the insert probability is set to be one. Next we briefly explain each term and how the probability is computed. Readers are referred to [9] for technical details.

The probability, $P_{placement}(R|A)$, quantifies the accuracy of the reads being placed back to the assembled genome. If the bases in a read are independently generated by the sequencer, then the probability, $P_{match}(r_i|A)$ that a read r_i matches to the genome is the product of $P(base_j|A)$, for all $base_j$ in r_i. Here,

$$P(base_j|A) = Q_j I(base_j \text{ is correctly matched}) + \frac{1-Q_j}{4} I(base_j \text{ is not matched}) + \frac{1}{4} I(base_j \text{ is not determined}),$$

where Q_j is the probability of the $base_j$ being correctly called by the sequencer. Note that Q_j can be calculated from the quality score provided in the sequencing result. The orientation of a read can be either forward or backward, so the orientation probability, $P_{orientation}(r_i|A)$, is calculated as the empirical frequency of the observed orientation of the read r_i. Thus, we can calculate the placement probability as:

$$P_{placement}(R|A) = \prod_{r_i \in R} P_{placement}(r_i|A) = \prod_{r_i \in R} P_{match}(r_i|A) P_{orientation}(r_i|A).$$

For paired-end reads, the insert lengths from all the mappings of the reads are collected and the mean μ and the variance σ^2 of the lengths are then calculated. The computation is unnecessary if these two quantities are known from the sequencing procedure. The insert likelihood, based on the observed insert length L_i, is set to be $(\sqrt{2\pi}\sigma)^{-1} exp(-(L_i - \mu)^2/(2 * \sigma^2))$, under the normality assumption on L_i. This, along with the assumption of independently inserting, gives

$$P_{insert}(R|A) = \prod_{r_i \in R} \frac{1}{\sqrt{2\pi}\sigma} e^{-\frac{(L_i-\mu)^2}{2\sigma^2}}.$$

For the depth probability, by the Lander-Waterman theory, the read depth at a position ideally follows a Poisson rule. However, this distribution may be affected by the GC content at the position, because DNA fragments from different GC-content areas are unequally amplified [1]. Thus, Clark et al. [9] model the read depth at a position by a Poisson distribution, with Poisson mean following a Gamma distribution. That is, the read depth is modeled by a Negative Binomial distribution. Then, the depth likelihood is calculated as:

$$P_{depth}(R|A) = \prod_p P_{depth}(d_p|A) = \prod_p \binom{d_p + \lambda_p - 1}{d_p} \left(\frac{1}{2}\right)^{d_p + \lambda_p},$$

where d_p is the depth in position p, and λ_p is the larger of a pre-defined integer (say, 10) and the average of the observed depths over positions in the region, defined by GC content, at which the position p locates.

The prior probability, $P(A)$, of an assembly A is determined based on the authors' belief that a single genome has its own specific K-mer profile (for any user-defined integer K). Let M be the number of all possible K-mers, and n_j be the number of times the jth K-mer appears in the assembly A. For $m = 1, 2, \ldots, M$, define $f_m = n_m / \sum_{j=1}^{M} n_j$ as the observed frequency of the mth K-mer. Then the probability $P(A)$ is proportional to

$$P_{K\text{-}mer}(S) = \prod_{m=1}^{M} f_m^{n_m}.$$

After obtaining these three probabilities for each assembly, the difference in *ALEs* for any two assemblies, A_1 and A_2, can then be determined by

$$ALE_1 - ALE_2 = \log \frac{P_{placement}(R|A_1)P_{insert}(R|A_1)P_{depth}(R|A_1)}{P_{placement}(R|A_2)P_{insert}(R|A_2)P_{depth}(R|A_2)},$$

and a positive difference indicates that the assembly A_1 has higher likelihood of being correct than the assembly A_2. The example of applying *ALE* on real data, and the study on $ALE's$ sensitivity and specificity can be found in [9].

17.3 Statistical Methods in Metagenomic Profiling and Comparison

17.3.1 Statistical Methods to Adjust Metagenomic Profiling

With metagenomic sequencing reads, researchers are interested in obtaining the taxonomic structure and the functional properties of a microbial community. These have been vividly presented as questions like "Who is there?" and "What are they doing?" [32]. By sequence homology searching, a count dataset representing the relative abundances of the features (i.e, OTUs or functional families) associated to the microbial sample can be generated. This is called metagenomic profiling. However, when using NGS metagenomic reads for profiling, systematic errors can be introduced during sequence homology searching and quantifying the abundance of a feature with the number of reads being aligned. Here we use the functional profiling of a metagenomic sample as an example to explain how the errors are produced.

The metagenomic reads are compared against a protein sequence database. A read is usually assigned a function according to its best-hit alignment. The list of all the detected functional families and the read counts to these families present the functional profile of the metagenomic sample. Due to the short length of reads and local sequence similarity among different functions in the database, a short read originating from a non-coding sequence may be assigned a function; or a short read originating from a coding sequence associated to a specific function may be erroneously assigned a different function. This introduces artificial functions to the profile, and incorrect read counts are assigned to the functions as well. In [12], a proposed method suggests taking a justified BLAST similarity score cut-off strategy instead of empirical BLAST E-value cut-offs, which may be only good for long reads, to filter out the artificial functions in alignments. Then Fisher's Quadratic Discriminate function is applied on the values of a ratio, computed from the outputs by different alignment tools (e.g., BLAST and RPS-BLAST), to remove the artificial functions that cannot be filtered out by the cut-off and have large counts. Interested readers are referred to [12] for technical details of the methods.

Developing statistical methods to adjust the read counts for the functions they are truly associated with is our current research topic. In addition, Sharon et al. [37] pointed out that it is not proper to use the raw read counts to quantify the relative abundances among different functional families, since more short reads are usually generated from a longer coding sequence than from a shorter one.

We comment that the errors for metagenomic profiling lie in the constructed list of features (e.g., artifacts), the read counts to the features (e.g., reads being assigned to wrong functions), and the relative abundance assessed by the read count (e.g., read count bias). They are systematic errors, similar to those in microarray experiments (corrected through the normalization process), but more complex. It is an ongoing active research area. In order to make the downstream comparisons meaningful, these errors need to be corrected with sophisticated statistical methods/models.

17.3.2 Statistical Methods for Metagenomic Comparison

For many metagenomic studies, the goal is to carry out the comparison among the metagenomic profiles, to detect the features with significantly different abundances. The result will help us to correlate the enrichment or the scarcity of specific features (i.e., species or functions) with the environmental/clinical characteristics of microbial communities. The comparison is among different groups of samples and follows the setting in a general clinical comparison–two or more treatment populations with each comprising of multiple samples. Note that multiple testing corrections should be considered because many features are being compared simultaneously.

For a metagenomic profile, it is a common observation that there is a large between-sample variation in the abundance counts to a feature. Different statistical methods have been proposed to address this problem. We describe four approaches in the following.

17.3.2.1 Beta-Binomial Approach

This method is for a two-group comparison. The goal is to test whether the proportions, p_{g1} and p_{g2}, of the abundance for each feature are significantly different. More details about the Beta-binomial approach can be found in [2].

Denote n_j as the total abundance count in the jth sample ($j = 1, \cdots, J$) of a group, and p_j as the true proportion of the abundance of the feature in this sample. Let Y_j be the random variable representing the abundance count of the feature. Assume that Y_j follows a $Binomial(n_j, p_j)$ distribution with p_j having a $Beta(\alpha, \beta)$ prior distribution. A weighted linear combination of single sample estimates, $\hat{p}_j = Y_j/n_j$, was proposed to construct the test statistic. That is, define $\hat{p} = \sum_{j=1}^{J} w_j \hat{p}_j$, with any positive weights w_j whose sum is one. Then we have the mean $E(\hat{p}) = \alpha/(\alpha+\beta)$, and the variance,

$$Var(\hat{p}) = \sum_{j=1}^{J} \frac{w_j^2 \alpha\beta}{(\alpha+\beta)(\alpha+\beta+1)} \left[\frac{1}{\alpha+\beta} + \frac{1}{n_j} \right].$$

Thus, we can choose the w_j's by minimizing this variance. By the method of Lagrange multipliers, it is easy to get $w_j \propto \left[\frac{1}{\alpha+\beta} + \frac{1}{n_j} \right]^{-1}$. Baggerly et al. [2] obtained an unbiased estimator of $Var(\hat{p})$: $\hat{V}_0 = (\sum w_j^2 \hat{p}_j^2 - (\sum w_j^2)\hat{p}^2)/(1 - \sum w_j^2)$. They claimed that this estimate might be too small in some cases and suggested a modified estimator, $\hat{V}$, as the larger of $\hat{V}_0$ and $(\sum Y_j/(\sum n_j)^2)(1 - \sum Y_j/\sum n_j)$.

In practice, since α and β are unknown parameters, an iteration process can be used to find w_j. Starting from the initial values of $w_j^{(0)} = n_j/\sum n_j$, and the corresponding $\hat{p}^{(0)}$, $\hat{V}^{(0)}$, we can find the method of moment estimators of α, β from the above expressions of the mean and variance of $\hat{p}$, and then update w_j. Repeat the process until the values of w_j stabilize.

Based on these estimates for each of two groups separately, the test statistic for the equality of the proportions of the abundance of the feature is,

$$t_w = \frac{\hat{p}_{g1} - \hat{p}_{g2}}{\sqrt{\hat{V}_{g1} + \hat{V}_{g2}}},$$

where $g1$ and $g2$ represent group 1 and group 2 correspondingly. The null distribution of t_w is approximately a t-distribution with degrees of freedom,

$$df = \frac{(\hat{V}_{g1} + \hat{V}_{g2})^2}{\frac{\hat{V}_{g1}^2}{n_{g1}-1} + \frac{\hat{V}_{g2}^2}{n_{g2}-1}}.$$

17.3.2.2 Overdispersed Logistic Regression Approach

Using the t statistic in the **Beta-binomial** approach confines its application to two-group comparison only. To address the overdispersion problem in the comparison of G ($G \geq 3$) groups, logistic regression can be used [3].

We adopt the notation in the previous section except that the observation index j now includes all G groups. Assume that $x_{jg} = 1$ if the jth observation belongs to the gth group, or 0 otherwise, $g = 1, \ldots, G-1$. The logistic model for proportions is $logit(p_j) = \beta_0 + \sum_{g=1}^{G-1} \beta_g x_{jg}$, with $Var(Y_j) = n_j p_j (1-p_j)[1 + (n_j - 1)\phi]$, where ϕ is the parameter to reflect the dispersion scale. Instead of computing the statistics separately for each group, the model is fitted by using all the observations. The parameter β_g reflects the group effect. In other words, the hypothesis of $\beta_1 = \cdots = \beta_{G-1} = 0$ is equivalent to that of the abundance proportions of the feature being the same among all groups. The model also provides opportunity to look at other contrasts. Note that more x variables can be added to the model should more covariates have an effect on the feature abundance. The estimates of the model parameters can be obtained using the iteratively reweighted least squares (IRLS) procedure [43], where the weight w_j is $1/[1+(n_j-1)\phi]$. The model can also be fitted by the quasi-likelihood method [29].

17.3.2.3 Overdispersed Log-Linear Regression Approach

Instead of the proportions, this approach models the expected abundance count, that is, $\log(\lambda_j) = \beta_0 + \Sigma_{g=1}^{G-1} \beta_g x_{jg}$ with $E(Y_j) = \lambda_j$ and $Var(Y_j) = \lambda_j(1+\lambda_j\phi)$. Note that Y_j can be viewed as the random sample from a Gamma-Poisson rather than a Beta-Binomial population. The estimates of the model parameters can be obtained by the IRLS procedure with the weight w_j equal to $1/(1+\lambda_j\phi)$ [23, 28].

17.3.2.4 Nonparametric T-Test

A nonparametric approach is used when there is no assumption on the distribution. A nonparametric t-test, proposed and discussed in [40, 42], is one of these approaches.

Let $y_{ijl}, i = 1,2; j = 1,\cdots J_i; l = 1,\cdots L$, be the abundance count of the sample j in group i for a feature l. A simple normalization converts the raw count to a fraction of the feature in the sample: $f_{ijl} = y_{ijl}/\Sigma_l y_{ijl}$. For the feature l, denote $\bar{f}_{il}$ and s_{il}^2 as the sample mean and variance of the fractions in group i. Then the t statistic is defined as $t_l = (\bar{f}_{1l} - \bar{f}_{2l})/\sqrt{\frac{s_{1l}^2}{J_1} + \frac{s_{2l}^2}{J_2}}$, and the p-value is determined by permutation. We randomly permute the f_{ijl}'s within feature l and keep J_1 samples in group 1 and J_2 samples in group 2. Repeat the permutation B times, and obtain a t statistic $t_l^{(b)}$ for the bth permutation. Then the p-value of the nonparametric t-test is calculated as the percentage of the t statistics from the permutations as extreme or more extreme than the observed t_l:

$$p_l = \frac{\#\left\{b : |t_l^{(b)}| \geq |t_l|,\ b = 1,\cdots,B\right\}}{B}.$$

Obviously, $1/B$ is the lowest attainable p-value, so B should be set large enough (e.g., 1000). In the case that J_1, J_2 are small (say, < 8), this is not a proper way to compute the p-value. We instead use the permuted t statistics from all the features to calculate the p-values. That is,

$$p_l = \frac{\Sigma_{b=1}^{B} \#\left\{m : |t_m^{(b)}| \geq |t_l|,\ m = 1,\cdots,L\right\}}{BL}.$$

17.3.3 Multiple Testing Correction

As in gene expression data analysis, statistical analysis of metagenomic data involves testing multiple hypotheses simultaneously since there are thousands of features in one metagenomic experiment. That is, we have to address the problem of an inflated false positive rate caused by multiple hypothesis testing.

There are plenty of methods in the literature to address this problem. They either adjust the p-values from tests for individual features, or calculate the corresponding q-values [39, 40]. Methods to adjust p-values include Bonferroni's single-step adjusted p-values, Holm's step-down adjusted p-values [19], Hochberg's step-up adjusted p-values [18], Benjamini-Hochberg's adjusted p-values [4], and many more. These methods have been implemented in most statistical software such as *R* and *SAS*.

Acknowledgements We thank the referees for their helpful comments. Zhide Fang was supported by 1 U54 GM104940 from the National Institute of General Medical Sciences of the National Institutes of Health which funds the Louisiana Clinical and Translational Science Center of Pennington Biomedical Research Center.

References

[1] Aird, D., Ross, M.G., Chen, W.S., Danielsson, M., Fennell, T., Russ, C., Jaffe, D.B., Nusbaum, C., Gnirke, A.: Analyzing and minimizing PCR amplification bias in Illumina sequencing libraries. Genome Biol. **12**(2), R18 (2011)

[2] Baggerly, K.A., Deng, L., Morris, J.S., Aldaz, C.M.: Differential expression in SAGE: accounting for normal between-library variation. Bioinformatics **19**(12), 1477–1483 (2003)

[3] Baggerly, K.A., Deng, L., Morris, J.S., Aldaz, C.M.: Overdispersed logistic regression for SAGE: modelling multiple groups and covariates. BMC Bioinform. **5**, 144 (2004)

[4] Benjamini, Y., Hochberg, Y.: Controlling the false discovery rate: a practical and powerful approach to multiple testing. J. Roy. Statist. Soc. Ser. B **57**(1), 289–300 (1995)

[5] Chao, A.: Nonparametric estimation of the number of classes in a population. Scand. J. Statist. **11**(4), 265–270 (1984)

[6] Chao, A.: Species richness estimation. Encyclo. Statist. Sci. **12**, 7907–7916 (2005)

[7] Chao, A., Lee, S.M.: Estimating the number of classes via sample coverage. J. Amer. Statist. Assoc. **87**(417), 210–217 (1992)

[8] Chao, A., Yang, M.C.: Stopping rules and estimation for recapture debugging with unequal failure rates. Biometrika **80**(1), 193–201 (1993)

[9] Clark, S.C., Egan, R., Frazier, P.I., Wang, Z.: ALE: a generic assembly likelihood evaluation framework for assessing the accuracy of genome and metagenome assemblies. Bioinformatics **29**(4), 435–443 (2013)

[10] Colwell, R.K., Chao, A., Gotelli, N.J., Lin, S.Y., Mao, C.X., Chazdon, R.L., Longino, J.T.: Models and estimators linking individual-based and sample-based rarefaction, extrapolation and comparison of assemblages. J. Plant Ecol. **5**(1), 3–21 (2012)

[11] Dinsdale, E.A., Edwards, R.A., Hall, D., Angly, F., Breitbart, M., Brulc, J.M., Furlan, M., Desnues, C. et al.: Functional metagenomic profiling of nine biomes. Nature **452**(7187), 629–632 (2008)

[12] Du, R., Mercante, D., Fang, Z.: An artificial functional family filter in homolog searching in next-generation sequencing metagenomics. PLoS ONE **8**(3), e58669 (2013)
[13] Frias-Lopez, J., Shi, Y., Tyson, G.W., Coleman, M.L., Schuster, S.C., Chisholm, S.W., Delong, E.F.: Microbial community gene expression in ocean surface waters. Proc. Natl. Acad. Sci. USA **105**(10), 3805–3810 (2008)
[14] Gabor, E., Liebeton, K., Niehaus, F., Eck, J., Lorenz, P.: Updating the metagenomics toolbox. Biotechnol. J. **2**(2), 201–206 (2007)
[15] Gerlach, W., Stoye, J.: Taxonomic classification of metagenomic shotgun sequences with CARMA3. Nucleic Acids Res. **39**(14), e91 (2011)
[16] Harris, B.: Determining bounds on integrals with applications to cataloging problems. Ann. Math. Statist. **30**(2), 521–548 (1959)
[17] Heck Jr, K.L., van Belle, G., Simberloff, D.: Explicit calculation of the rarefaction diversity measurement and the determination of sufficient sample size. Ecology **56**(6), 1459–1461 (1975)
[18] Hochberg, Y.: A sharper bonferroni procedure for multiple tests of significance. Biometrika **75**(4), 800–802 (1988).
[19] Holm, S.: A simple sequentially rejective multiple test procedure. Scand. J. Statist. **6**(2), 65–70 (1979)
[20] Hugenholtz, P., Goebel, B.M., Pace, N.R.: Impact of culture-independent studies on the emerging phylogenetic view of bacterial diversity. J. Bacteriol. **180**(18), 4765–4774 (1998)
[21] Hughes, J.B., Hellmann, J.J., Ricketts, T.H., Bohannan, B.J.: Counting the uncountable: statistical approaches to estimating microbial diversity. Appl. Environ. Microbiol. **67**(10), 4399–4406 (2001)
[22] Huson, D.H., Auch, A.F., Qi, J., Schuster, S.C.: MEGAN analysis of metagenomic data. Genome Res. **17**(3), 377–386 (2007)
[23] Ismail, N., Jemain, A.A.: Handling overdispersion with negative binomial and generalized Poisson regression models. In: Casualty Actuarial Society Forum, pp. 103–158. United Book Press, Baltimore (2007)
[24] Jiang, H., An, L., Lin, S.M., Feng, G., Qiu, Y.: A statistical framework for accurate taxonomic assignment of metagenomic sequencing reads. PLoS ONE, **7**(10), e46450 (2012)
[25] Kemp, P.F., Aller, J.Y.: Estimating prokaryotic diversity: when are 16S rDNA libraries large enough? Limnol. Oceanogr. Methods **2**(4), 114–125 (2004)
[26] Lander, E.S., Waterman, M.S.: Genomic mapping by fingerprinting random clones: a mathematical analysis. Genomics **2**(3), 231–239 (1988)
[27] Li, X., Waterman, M.S.: Estimating the repeat structure and length of DNA sequences using l-Tuples. Genome Res. **13**(8), 1916–1922 (2003)
[28] Lu, J., Tomfohr, J., Kepler, T.: Identifying differential expression in multiple SAGE libraries: an overdispersed log-linear model approach. BMC Bioinform. **6**, 165 (2005)
[29] MacCullagh, P., Nelder, J.A.: Generalized Linear Models, 2nd edn. Chapman and Hall, London (1989)
[30] Mardis, E.R.: Next-generation DNA sequencing methods. Annu. Rev. Genomics Hum. Genet. **9**, 387–402 (2008a)
[31] Mardis, E.R.: The impact of next-generation sequencing technology on genetics. Trends Genet. **24**(3), 133–141 (2008b).
[32] National Research Council: The new science of metagenomics: revealing the secrets of our microbial planet. National Academies Press (US), Washington (DC) (2007)
[33] Nalbantoglu, O.U., Way, S.F., Hinrichs, S.H., Sayood, K.: RAIphy: phylogenetic classification of metagenomics samples using iterative refinement of relative abundance index profiles. BMC Bioinform. **12**, 41 (2011)
[34] Pace, N.R.: A molecular view of microbial diversity and the biosphere. Science **276**(5313), 734–740 (1997)
[35] Sanger, F., Nicklen, S., Coulson, A.R.: 1977. DNA sequencing with chain-terminating inhibitors. Proc. Natl. Acad. Sci. USA, **74**(12), 5463–5467 (1977)

[36] Shah, N., Tang, H., Doak, T.G., Ye, Y.: Comparing bacterial communities inferred from 16s rRNA gene sequencing and shotgun metagenomics. Pac. Symp. Biocomput. **16**, 165–176 (2011).

[37] Sharon, I., Pati, A., Markowitz, V.M., Pinter, R.Y.: A statistical framework for the functional analysis of metagenomes. In: Research in Computational Molecular Biology, pp. 496–511. Springer, Berlin/Heidelberg (2009)

[38] Staley, J.T., Konopka, A.: Measurement of in situ activities of nonphotosynthetic microorganisms in aquatic and terrestrial habitats. Annu. Rev. Microbiol. **39**, 321–346 (1985).

[39] Storey, J.D.: A direct approach to false discovery rates. J. Roy. Statist. Soc. Ser. B **64**(3), 479–498 (2002).

[40] Storey, J.D., Tibshirani, R.: Statistical significance for genomewide studies. Proc. Natl. Acad. Sci. USA **100**(16), 9440–9445 (2003)

[41] Tringe, S.G., von Mering, C., Kobayashi, A., Salamov, A.A., Chen, K., Chang, H.W., Podar, M., Short, J.M., Mathur, E.J., Detter, J.C., et al.: Comparative metagenomics of microbial communities. Science **308**(5721), 554–557 (2005).

[42] White, J.R., Nagarajan, N., Pop, M.: Statistical methods for detecting differentially abundant features in clinical metagenomic samples. PLoS Comput. Biol. **5**(4), e1000352 (2009).

[43] Williams, D.A.: Extra-binomial variation in logistic linear models. J. Roy. Statist. Soc. Ser. C (Appl. statist.) **31**(2), 144–148 (1982)

[44] Woese, C.R.: Bacterial evolution. Microbiol. Rev. 51(2), 221–271 (1987).

[45] Wooley, J.C., Godzik, A., Friedberg, I.: A primer on metagenomics. PLoS Comput. Biol. **6**(2), e1000667 (2010)

[46] Wu, Y.W., Ye, Y.: A novel abundance-based algorithm for binning metagenomic sequences using l-tuples. In: Research in Computational Molecular Biology, pp. 535–549. Springer, Berlin/Heidelberg (2010)

[47] Ye, Y.: Identification and quantification of abundant species from pyrosequences of 16S rRNA by consensus alignment. In: Proceeding of IEEE International Conference on Bioinformatics and Biomedicine (BIBM), pp. 153–157 (2010)

Chapter 18
Detecting Copy Number Changes and Structural Rearrangements Using DNA Sequencing

Venkatraman E. Seshan

Abstract Chromosomal abnormalities in the form of deletions, duplications, inversions and translocations are common in cancer. These changes feed the oncogenic process by affecting genes that are involved in tumor growth. Next generation sequencing has aided our ability to study these changes at very high resolution. In this chapter we will describe the nature of these data and the information contained in them for the detection of the structural changes. We will present the circular binary segmentation algorithm for the segmentation of array based copy number data and adapt it to NGS data. We will also present a method for the detection of somatic structural rearrangement. We will illustrate these methods using data from breast cancer cell line (tumor) along with its blood (normal) counterpart generated by the cancer cell-line encyclopedia project.

18.1 Introduction

The flow of genetic information in cells [3, Chap. 5] occurs primarily through the transcription of DNA into RNA which is then translated into proteins that carry out the cellular functions. This is stated as *DNA makes RNA, RNA makes proteins, proteins make us* [18] and referred to as the central dogma of molecular biology [8]. This implies that changes to DNA can have an effect on the biological processes. These changes in DNA can be mutations as well as structural changes. In humans, autosomal chromosomes appear in pairs, one from each parent, and thus have two copies of every gene; the allosomes (sex chromosomes) are XX in females (two copies of X) and XY in males (one copy each of X and Y). Gains and losses of all or parts of chromosomes are known as copy number changes

V.E. Seshan (✉)
Department of Epidemiology and Biostatistics, Memorial Sloan Kettering Cancer Center, 307 East 63rd St., 3rd Floor, New York, NY 10065, USA
e-mail: seshanv@mskcc.org

S. Datta and D. Nettleton (eds.), *Statistical Analysis of Next Generation Sequencing Data*, Frontiers in Probability and the Statistical Sciences, DOI 10.1007/978-3-319-07212-8_18,

and are implicated in many diseases. These changes could either be germline (inherited) or somatic (acquired). Examples of germline changes are 3 copies of chromosome 21 (copy number gain) resulting in Down's syndrome [15, Chap. 5] or single X (copy number loss) resulting in Turner's syndrome [15, Chap. 5]. Somatic changes are very common in cancer, where a gene is gained and it promotes growth (oncogene, e.g., ERBB2 (HER2/Neu) in breast cancer [14]), or a gene is deleted and the ability to control growth is lost (tumor suppressor gene, e.g., PTEN in prostate cancer [38]). Other changes to DNA such as the Philadelphia chromosome [23], a reciprocal translocation between chromosomes 9 and 22, is another type of structural change implicated in cancer (chronic myelogenous leukemia or acute lymphocytic leukemia). Thus, studying copy number changes and other structural rearrangements is important for understanding the oncogenic process.

Karyotyping, which is the study of the number and appearance of chromosomes, was the earliest method used for detecting chromosomal aberrations and provides information at a low resolution. The development of comparative genomic hybridization (CGH) [13, 20] allowed measurement of copy number changes over the entire genome and enabled it to be localized to a chromosome at an improved resolution of 10 to 20 megabase. High throughput methods such as BAC (bacterial artificial chromosome), aCGH (array comparative genomic hybridization) and SNP (single nucleotide polymorphism) arrays, based on the microarray technology have systematically increased the resolution and thereby our ability to detect gains and losses of smaller chromosomal regions; see [27] for a review of array CGH technology. Whereas a karyotype assay can clearly show trisomy of chromosome 21, the loss of PTEN cannot be readily visualized in a Affymetrix SNP 6.0 array with over 1.8 million markers. Thus sound analytic methods are required for the large volume of noisy data generated by the high throughput methods.

The analysis of copy number data is composed of two parts—the identification of regions of gains and losses in each subject followed by combining this information across samples to identify recurrent events associated with cancer. Several methods have been proposed for the per sample analysis of copy number data which can be characterized as "smoothing and thresholding" or "segmentation" methods. A comprehensive comparison of the performance of several of these methods was done by [16]. Overall, segmentation methods were found to be most suitable for the per sample analysis of copy number data. The principal concept behind the segmentation methods is that since copy number for a cell is integer valued, gains and losses are discrete events and thus along a chromosome the gain or loss induces a jump discontinuity. Note that the tissue sample being assayed is a collection of cells all of which will not have the same changes. However, the distinct clones that make up the tissue sample is expected to be far fewer than the number of cells and hence the average copy number will have the form of a step function. We formulated this as a change point problem to develop the circular binary segmentation (CBS) algorithm [25, 36] which is one of the widely used methods. GISTIC [4], GRIN [28] and RAE [34] are frequently used algorithms to combine the copy number changes detected in the per sample analysis in order to identify recurrent events and implicated genes.

Next generation sequencing (NGS) of genomic DNA enables us to obtain information on somatic mutations and structural changes. The structural changes include copy number gains and losses as well as rearrangements such as inversions and translocations. [Note: Inversions and translocations are explained in Sect. 18.3.1] Several algorithms such as BreakDancer [6], CNVnator [1], CNVseq [40], CREST [37], SegSeq [7], seqCBS [33], SVminer [12] are currently used for obtaining structural change information from NGS data. In the following sections we will describe the CBS algorithm, adapt it to sequencing data, and demonstrate it using cell line data. We will finish the chapter by presenting a simplified summary of the procedure for identifying other structural variations.

18.2 Background

In this section we will describe the design and techniques used to generate the data that are to be analyzed. The first step in the process of obtaining the data is the generation of a library of genomic DNA composed of short DNA fragments, typically 100 to 500 nucleotides long, from the sample of interest. This library can encompass the entire genome (whole genome sequencing) or selected genomic regions (targeted sequencing). The creation of the library in either case begins with generating DNA fragments by randomly breaking the entire genome using a technique such as sonication. The fragments are then sorted by molecular weight to enable the selection of fragments of the desired length. In targeted sequencing an additional selection process is employed where the DNA is hybridized to arrays with probes that are designed to capture DNA fragments that cover the genomic regions of interest. A specific case of targeted sequencing is whole exome sequencing where the genomic regions selected are all the exons (coding regions) of all known genes (approximately 20,000). Custom gene panels [11] that cover a smaller collection of genes known to be most commonly associated with cancer are also currently in use. The regions in targeted sequencing span a small fraction of the whole genome, 1–2% in the case of whole exome and even less for custom panels, allowing for high coverage of the target.

The library that is generated is then sequenced to obtain reads, which are the strings of bases or nucleotides, that make up a part of the fragment. Sequencing can either be single-end or paired-end where the DNA fragments are sequenced (read) from either one end or both ends, respectively. Read length, which is the number of nucleotides sequenced, can be specified in the instrument for an experiment. The reads are then mapped to a reference genome to obtain positional information on where the reads, and hence the fragments, come from, *i.e.*, their locations. These locations follow a probability distribution that is influenced by factors such as the GC content and mappability. The data used for identifying structural changes are various characteristics of the reads such as their locations and fragment size.

In cancer research, the principal goal of DNA sequencing is to identify changes in DNA (mutations and structural) acquired by the tumor. Hence, typically, both tumor and normal cells are sequenced. The sequencing of normal cells will help identify any germline events, for example, BRCA1 mutation, that may be present. In paired tumor-normal sequencing, the comparison of the tumor to its matched normal will benefit from the canceling out of the influence of the technical factors that affect sequencing. Running them in the same batch would additionally ensure that batch effects are minimized. Although the use of paired tumor-normal samples is ideal for identifying changes that are specific to the tumor, it may not always be feasible. For instance, the analysis of archival tumors in which only tissue samples from the tumor are available will need an external pool of normal samples to identify tumor specific changes. However, large scale copy number polymorphisms have been seen in the germline [31] and Redon et al. (2006) [29] created a first-generation copy number variation (CNV) map from copy number profiling of the HapMap samples. Thus, the comparison of tumor data to an external normal needs to account for technical artifacts that may not cancel as well as inherited copy number events.

Unlike karyotying, both sequencing and array based measurement of copy numbers query the DNA fragments directly and do not contain information on individual cells. This introduces an identifiability problem as follows. Let us suppose that a global change in copy number has occurred in which every single chromosome in the cell is duplicated resulting in a total copy number of four. Whole genome duplication such as these and aneuploidy in general are common in cancer [9]. In terms of information contained in the DNA, a tissue with cells of this kind is indistinguishable from a tissue of normal cells. In general, both the array based and sequencing approaches to copy numbers can only provide relative copy number changes and require external information to resolve the relative numbers into absolute copy numbers. The ABSOLUTE algorithm [5] provides a method to use the ploidy (which is the average copy number) and tumor purity to obtain the absolute copy numbers.

In the next section, we will present a method for analyzing copy numbers from matched tumor-normal sequencing data. Furthermore, the changes identified will be based on the relative copy numbers and thus gains and losses will be relative to the average copy number of the tumor.

18.3 Methods

The read data generated from DNA sequencing contains not only information on the nucleotides that make up the subject's genome but also the relative abundance of a locus as well as distances between loci. These additional elements can be leveraged to detect structural changes to the DNA. In the following subsections we will develop a method to obtain the copy number profile from the relative abundance measure.

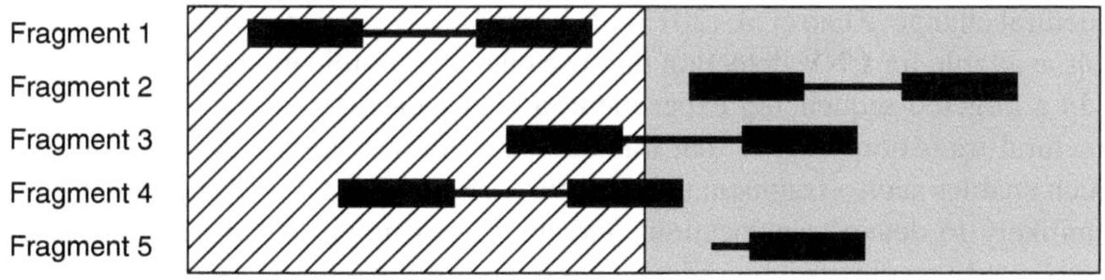

Fig. 18.1 Different types of fragments in a paired end DNA sequencing data

18.3.1 Structural Change Information in NGS Data

In the background section, we described how the reads data that are to be used for identifying structural changes are generated. We will now describe the information contained in these data using Fig. 18.1 which shows a portion of the tumor genome and five fragments from it.

The displayed portion of the tumor genome consists of a striped and a shaded part both of which are contiguous in the germline but the transition from the striped to the shaded does not occur in the germline and thus represents a structural change boundary. The fragments shown are from paired end reads where the thick rectangles are the reads and the thin one connecting them is the inferred intermediate region once the reads are mapped.

In the germline, both the striped and the shaded regions will appear exactly twice in a cell provided they are not polymorphic but in a tumor cell they appear more than twice if the region is gained and fewer than twice if it is lost. The transition corresponds to a translocation if both the regions have the same orientation as in the reference genome and an inversion if their orientations are opposite. The translocations can be intra- or inter-chromosomal depending on whether both regions come from the same chromosome and different chromosomes, respectively.

The fragments shown in the figure are read pairs for which at least one of the two ends is mapped to either the striped or the shaded genomic region. The top three fragments have both ends mapped and the bottom two have only one end mapped. Note that, although both reads of Fragment 4 are shown, only the read contained in the striped region will be mapped using a standard alignment procedure and the other end would require a partial read mapping algorithm such as CREST [37] to be mapped. Unmappable reads, such as the mate pair of Fragment 5, can occur if the read contains repetitive elements that are not uniquely identifiable.

A region that is gained in the tumor will contribute more fragments to the tumor reads and one that is lost will contribute fewer fragments. So the counts of the reads within a region, namely its abundance measure, is related to the copy number. Since the reads in Fragments 1, 2 and 5 are completely contained within the striped and shaded regions, they only contribute to the abundance measure. Since the two ends of Fragments 3 and 4 are mapped to the two regions, not only do they contribute to the abundance measure, they can also inform directly on the possible location of a structural change. Zhao et al. (2013) [41] provide a review of various computational tools available for CNV detection that use one or a combination of these features.

In a targeted sequencing experiment, a read pair will contain the location of a structural transition, only if that transition occurs near a targeted genomic region which enables such a fragment to be captured. So a targeted sequencing experiment is unlikely to detect translocations and inversions unless the regions where such events could occur are specifically targeted, for example, the translocation in Philadelphia chromosome. Hence de novo structural rearrangements are rarely identifiable in targeted sequencing. The abundance measure however, is available and effective for copy number profiling both in whole genome and targeted sequencing. We will describe our method based on abundance measure (read-depth) in detail.

18.3.2 Circular Binary Segmentation

Let $X_1, X_2, \ldots$ be a sequence of random variables. A change-point is an index ν such that the random variables $X_1, \ldots, X_\nu$ have a common distribution F_0 and $X_{\nu+1}, \ldots$ have a different distribution F_1 (until the next change-point or the end of the sequence). For the copy number problem using data from array CGH the X_is are the log-transformed normalized intensities (or log-ratios) of the markers which are ordered by the position along the chromosome and thus is a finite sequence of length m. Since the copy number of a cell is integer valued and the tumor consists of far fewer distinct clones than cells, it is appropriate to view the locations where the copy number changes to be the change-points that we wish to detect.

The test statistic introduced in the CBS algorithm [25] to detect the change-points is the maximal t-statistic given by:

$$T = \max_{1 \le i < j \le m} \left\{ \hat{\sigma}_{ij} \sqrt{\frac{1}{j-i} + \frac{1}{m-j+i}} \right\}^{-1} \left| \frac{S_j - S_i}{j-i} - \frac{S_m - S_j + S_i}{m-j+i} \right|$$

where $S_i = X_1 + \cdots + X_i$ is the partial sum and $\hat{\sigma}_{ij}^2$ is the mean-squared error given by

$$\hat{\sigma}_{ij}^2 = \frac{1}{m-2} \left[\sum_1^m X_i^2 - (S_j - S_i)^2/(j-i) - (S_m - S_j + S_i)^2/(m-j+i) \right].$$

The motivation for this test statistic is as follows. If the X_is are normally distributed with a common variance then the change-points correspond to a change in mean. Suppose the change-points are fixed at i and j then the optimal statistic to test the equality of the means of the two sets $\{X_{i+1}, \ldots, X_j\}$ and $\{X_1, \ldots, X_i, X_{j+1}, \ldots, X_m\}$ is the t-statistic. Because the change-points are unknown we obtain our test statistic by maximizing the t-statistic over all possible i and j. Note that $j = m$ corresponds to the case of a single change-point. The null hypothesis of no change-points is rejected if the p-value of the test statistic is below the significance threshold. Since the log-ratio data may not be normally distributed the CBS procedure was made robust by using a permutation reference distribution. The algorithm begins by testing for the presence of change-points in whole chromosomes. If the null hypothesis of no change-points is rejected, then the change-points that are detected will segment the chromosome into two (test detects one change-point) or three contiguous regions (test detects two change-points). The test procedure is applied recursively to each of the regions until no change-points are detected in any of them.

In comparative studies, the CBS algorithm was found to perform well consistently [39] and had the best operational characteristics [16] amongst several methods for analyzing copy number data. However, since the test statistic is maximizing over both i and j the computing time grew as the square of the number of markers which made the analysis burdensome as the resolution of arrays increased. To address this, [36] developed a faster CBS algorithm using tail probability approximations of Gaussian random fields as well as sequential testing. These and additional algorithmic improvements have made the use of this procedure routine for the analysis of array based copy number data.

18.3.3 Adapting CBS to NGS Data

In a sequencing experiment, the DNA fragments are sampled randomly and thus, a region that has a higher copy number contributes a larger number of fragments than a region with a lower copy number. The locations that the reads are mapped to is a function of several factors such as sequence composition and fragment size. Although the distribution of the locations of the mapped reads is non-uniform, the ratio of the read counts between tumor and normal will be proportional to the tumor to normal copy number ratio. Two additional scaling factors are needed for the read count ratio to reflect the true copy number ratio. The first is the ratio of total number of reads in the tumor and normal, which adjusts for the fact that tumors are often sequenced to a higher coverage than normal. The second factor depends on the purity and ploidy of the tumor. Thus the read count ratio data enables us to detect the regions of copy number change but will only give us a relative copy number. For instance, suppose we are interested in knowing whether the ERBB2 gene is gained (relative to the average copy number of the tumor) in a breast cancer sample; we can count the fragments that map to this gene in the tumor and normal samples and compare that ratio to the ratio of total number of fragments mapped in the two samples.

The independent elements in a sequencing experiment are the DNA fragments which are represented by a read pair, if both ends are mapped, and a single read, if only one end is mapped. If the abundance measure is calculated at the nucleotide level, then a DNA fragment contributes to the read count of all the positions within the read as well as all those in its mate pair. This induces a serial correlation in the abundance measure data indexed by genomic position. Read pairs that span a structural transition, such as Fragment 3, can induce a longer range correlation. In order to obtain copy number data that are independent, we need that each fragment be counted towards only one abundance measure data point.

A deterministic approach to ensure that each fragment is counted towards one copy number data point only, is to represent each fragment by its mid-point. This presents a problem for fragments where only one end is mapped as well as those fragments with both ends mapped that are not consistent with the lengths of the fragments selected for sequencing. Such fragments can be removed from the copy number calculations and since they typically represent a small fraction of the reads, it is expected to have minor effect on the copy number profile. Alternately, we can include them as follows: for fragments with only one end mapped, use the midpoint of the read; for fragments with both ends mapped, pick one of the reads at random and pick its midpoint. In targeted sequencing, we expect only one of the two reads in a read pair that needs such probabilistic assignment, to be near a target interval and can choose the midpoint of that read to represent it. We will calculate the abundance measure for copy number profiling from these positional data. Note that under this data representation, the average number of fragments per position will be the average coverage divided by the read length for single-end sequencing and average coverage divided by twice the read length for paired-end sequencing. For example, in an experiment with $50\times$ coverage using 2×75bp sequencing this translates to a read count of $1/3$ fragments per base. Since this number is small, we will require that the data be binned to aggregate information and provide reliable copy number profile. We recommend a bin size that gives an average bin count of 25 or higher which for this example will result in a bin size of 100 bases.

A final feature of the data that requires attention is specific to targeted sequencing where capture technique is used to enrich DNA fragments from genomic regions of interest. Although these capture technologies have high specificity, it is not perfect, i.e., the library being sequenced will consist of DNA fragments that are not on target. This will lead to a large number of bins, far exceeding the bins that cover the regions being targeted, with very low counts (typically singletons). Since these bins are spread over the entire genome, the data from them will have an undue influence on the copy number profile and should be discarded prior to analysis. We address this by using primarily bins that span the regions of interest with target intervals enlarged to allow for fragment overhang.

With these preliminaries in place, let N_1 and N_0 be the total number of mapped fragments for the tumor and normal samples, respectively. Let (n_{1i}, n_{0i}) be the tumor and normal fragment counts for the ith bin, and m be the number of bins. Similar to the log-ratios from copy number arrays we define the copy number data used for the segmentation as $X_i = \log_2[(1+n_{1i})/(1+n_{0i})] - \log_2(N_1/N_0)$, where the 1 is added to address bins with zero counts. The average fragment counts for bins within a region of constant copy number is proportionally increased or decreased and thus the log-ratio has a constant mean. However, the variability of fragment count is proportional to the average and thus we expect the variability of the log-ratio to be inversely proportional to the average fragment count. While the test statistic shown in Sect. 18.3.2 is adequate, a weighted version of the statistic will be more appropriate.

Suppose $\{Y_1,\ldots,Y_k\}$ and $\{Z_1,\ldots,Z_l\}$ are two sets of random variables where Y_is have mean μ and variance σ_i^2 and Z_js have mean θ and variance τ_j^2. Then the minimum variance estimate of the difference in means $\mu - \theta$ is the difference in weighted average with weights given by the inverse of the variances, i.e., $(\sum \sigma_i^{-2} Y_i / \sum \sigma_i^{-2}) - (\sum \tau_j^{-2} Z_j / \sum \tau_j^{-2})$. Thus the optimal statistic for testing the hypothesis $\mu = \theta$ is the weighted t-statistic based on this difference in weighted average. The maximal t-statistic we will use for change-point detection will be the maximum over all i and j of the weighted t-statistic suggested by the minimum variance estimate. Note that we need to know the parameters σ_i^2 and τ_j^2, at least up to a constant, to obtain the weighted t-statistic.

For the fragment count data, we expect the variance of the counts to be proportional to the mean. The proportionality constant is 1 if the counts have a Poisson distribution and the relationship holds for distributions with extra variation such as negative binomial. So the variance of the log of the counts will be inversely proportional to the mean counts and thus the weight will be proportional to counts. Note that the tumor counts in the log ratio is affected by gains and losses and which can influence the weights. Thus we recommend using only the normal counts for the weights which is consistent with the null hypothesis of no change. In order to dampen the effect of bins with very large counts we suggest that the weights grow as the logarithm of the counts. In the next section we will present an example of the copy number analysis of sequencing data to demonstrate all these.

An alternate approach to the analysis is to use a variance stabilizing transformation. Anscombe (1948) [2] showed that for a Poisson random variable X, the transformation $\sqrt{X+3/8}$ is variance stabilizing, if the rate parameter is large enough (≥ 5). However, in order to allow for extra variation if we posit that the count data are distributed as negative binomial, then the variance stabilizing transformation is either $\sinh^{-1}\left[\sqrt{(X+3/8)/(k-3/4)}\right]$ or $\log(X + k/2)$ where k is the dispersion parameter. Ignoring the transformation's dependence on the dispersion parameter k, one can define the copy number data as $\sqrt{n_{1i}+3/8} - \sqrt{n_{0i}+3/8}$ and segment them using the unweighted CBS algorithm. Note that these data will not be centered at zero and hence the sign of the segment mean does not indicate a gain or a loss from the average tumor copy number. However, the underlying true regions of constant copy number will be the same as in the log-ratio.

18.4 An Example

In this section we will illustrate in detail the steps involved in the analysis of DNA sequencing data for copy number changes using data from a cancer cell line. The data are from the breast cancer cell line HCC1143 and its blood (normal) counterpart

HCC1143BL which are part of the cancer cell line encyclopedia (CCLE) project (http://www.broadinstitute.org/ccle/home). Whole exome sequencing (paired-end 2×75bp) was done for these two samples and the read data, aligned to the HG19_Broad_variant (Human reference GRCh37) reference genome, are available at the Cancer Genomics Hub (https://cghub.ucsc.edu/datasets/data_sets.html). The size of these two data sets are 10.8 Gb and 8.3 Gb, respectively, and the analysis will require powerful computers. Software requirement for this analysis are: *Bioconductor* [10], specifically the *Rsamtools* [22] and *DNAcopy* [32] packages, *Integrative Genomics Viewer* [30], *Picard* [26] and *samtools* [19]. Note that all dependencies of these software should also be available.

We begin with using *samtools* to summarize the data file that was downloaded from CCLE. The summary data (with line numbers added) for the normal sample are:

```
1   68629600 + 6562518 in total (QC-passed reads +
                                     QC-failed reads)
2   10054468 + 1517557 duplicates
3   67842739 + 5593779 mapped (98.85%:85.24%)
4   68629600 + 6562518 paired in sequencing
5   34314800 + 3281259 read1
6   34314800 + 3281259 read2
7   66854380 + 5314442 properly paired (97.41%:80.98%)
8   67196156 + 5353240 with itself and mate mapped
9   646583 + 240539 singletons (0.94%:3.67%)
10  301196 + 35316 with mate mapped to a different chr
11  260127 + 29485 with mate mapped to a different chr
                                           (mapQ>=5)
```

The first line says that there are approximately 75 million reads in total in this sample which are decomposed into those that passed quality control (QC) and those that did not. This QC flag is platform and aligner specific. We will restrict the analysis to only those reads that passed QC (over 90% of the total). Lines 4 through 6 give the breakdown of the reads in Line 1, namely they are paired (Line 4) and that each end contributes half of the reads (Lines 5 and 6). Line 3 gives the number of reads that are mapped to the reference genome among the number listed in Line 1. The reasons the reads are unmapped are varied, such as structural rearrangement as seen in Fragment 4 of Fig. 18.1 or viral DNA mixed in with the sample. Line 7 gives the number of reads from fragments with both ends mapped and the two reads are consistent with the expected fragment sizes and the reads are in the proper direction (5′ to 3′ and vice versa, respectively). Line 8 gives the reads from fragments for which both ends are mapped. This number is larger than the one in Line 7 as it includes improperly paired reads as well. The counts of improperly paired reads with the two ends mapped to two different chromosomes is given in Line 10 and the subset that meets a mapping quality threshold is given in Line 11. Line 9 gives the number of fragments for which only one of the two reads is mapped. Finally, Line 2 gives the numbers of reads that are considered duplicates.

Duplicates are fragments for which the two ends, when mapped, give the same start and end locations and (nearly) identical read sequences. Since it is very unlikely that two identical DNA fragments are generated during the original DNA preparation, these are considered to have risen at the PCR amplification step where some fragments can get overamplified. Thus, duplicate reads do not provide independent information on the DNA of the sample and hence, only the read pair with the best read qualities is kept and the rest are removed from further analysis. We accomplish the deduplication step using the *Picard* software (MarkDuplicates option) which unlike *samtools* can also remove inter-chromosomal duplicates.

In summary, these data come from approximately 34 million fragments of which 5 million are potential PCR duplicates resulting in 29 million fragments of usable data. The pairs that are not proper (the excess of Line 8 over Line 7), especially, the ones with the mate mapped to a different chromosome (Lines 10 and 11), are the informative ones for non copy number structural changes (translocations and inversions). Additionally, information in the mate pair of the singletons in line 9 can potentially be extracted using partial read alignment for use in detecting structural variations. A similar breakdown of the summary data of the tumor file tells us that there are approximately 34 million usable fragments in that sample. The target enrichment intervals used for the whole exome sequencing is available in the CGHUB website (https://cghub.ucsc.edu/datasets/whole_exome_agilent_1.1_refseq_plus_3_boosters_plus_10bp_padding_minus_mito.Homo_sapiens_assembly19.targets.interval_list.tsv). There are a total of 36.6 million bases in these intervals (31.8 million if the targets labeled `new_exome_1.1_content` are excluded) which results in an expected count of 1 fragment per base in the target region.

Note that for variant (somatic mutation) detection, it is customary to conduct indel realignment and recalibration of quality score steps on the sequence data using *GATK* [21]. The copy number analysis can be performed after these steps and can benefit from them, particularly if read quality is accounted for in the analysis since the influence of poor quality reads can be eliminated. The quality recalibration step is valuable for identifying other structural variations reliably.

Once the data have been deduplicated, we extract the properly paired reads from both the tumor and normal samples. Since the data are from a cancer cell that originated in a female, we only extracted the reads that mapped to the 22 autosomes and the X chromosome which resulted in 28.5 and 33.5 million fragments, respectively, for tumor and normal. The number of reads mapped to the Y chromosome is approximately 16,000 for both the tumor and the normal which is reassuringly negligible. The densities of the fragment lengths for the tumor and normal samples are shown in Fig. 18.2. Fragments with lengths smaller that 76 or larger than 750 were not included in this figure for visual clarity. Although the fragments not included in the density plot can provide alternate information on structural changes, their contribution to the abundance measure is minimal as they represent 0.49% and 0.66% of normal and tumor fragments, respectively. The fragment lengths of the normal sample (median 163) are slightly larger than that for

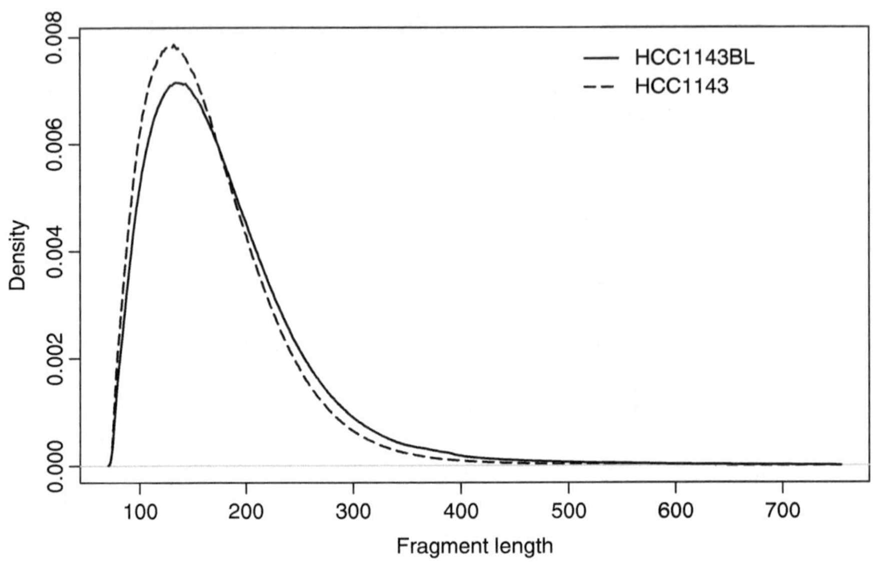

Fig. 18.2 The distribution of fragment lengths in tumor (*dashed line*) and normal (*solid line*) samples. The densities were generated using fragments whose lengths are between 76 and 750 bases

Table 18.1 The target intervals in the TP53 gene

Chr	Start	End	Width	Target
17	7572915	7573020	106	Target_128140
17	7573915	7574045	131	Target_128141
17	7576841	7576938	98	Target_128142
17	7577007	7577167	161	Target_128143
17	7577487	7577620	134	Target_128144
17	7578165	7578301	137	Target_128145
17	7578359	7578566	208	Target_128146
17	7579300	7579602	303	Target_128147
17	7579688	7579733	46	Target_128148
17	7579827	7579924	98	Target_128149

the tumor sample (median 154), and a vast majority of fragments (93.8% of normal and 96.6% of tumor) are fewer than 300 bases in length.

In order to provide further insight into the nature of targeted sequencing data, we will take an in depth look at the well known cancer gene TP53. This gene spans a 10 kilobase region on chromosome 17 with target intervals of different widths which are shown in Table 18.1. A figure of the data from this region for the normal sample generated using *Integrative Genomics Viewer* is in Fig 18.3. In the top part of the figure, the chromosome is shown with the region of interest in p13.1 highlighted in red and the genomic positions in bases. Below that are the genes in that region and the exons. The gene display is packed to show various forms of the gene present in RefSeq (http://www.ncbi.nlm.nih.gov/refseq/); the tall blue rectangles are the exons

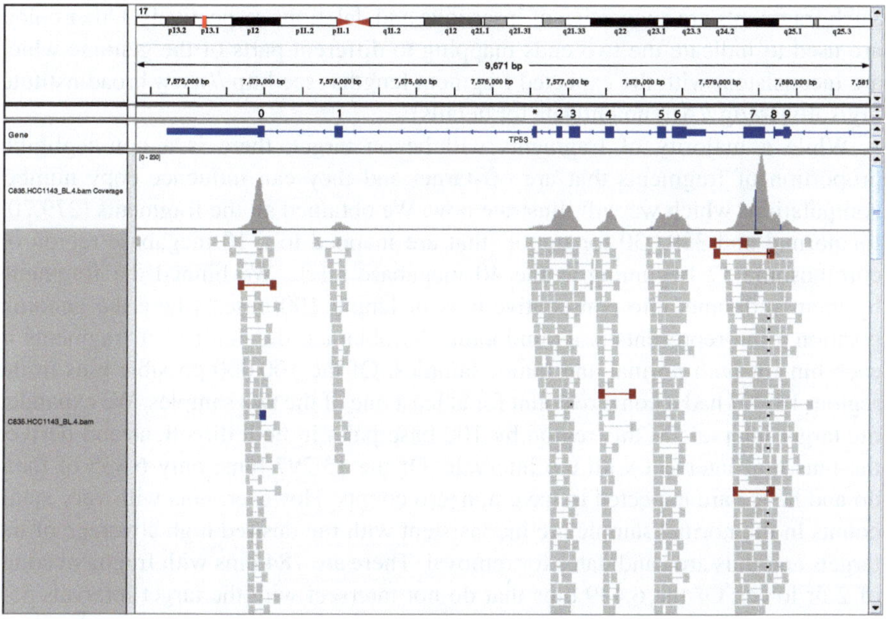

Fig. 18.3 The normal sample coverage plot for TP53 as obtained from the *Integrative Genomics Viewer*. The tall *blue rectangles* are the exons targeted in sequencing

and the shorter ones are start and end of untranslated regions (UTR). The labels for the target intervals in Table 18.1 were added to the figure generated by IGV. The labels are 0 to 9 for the 10 intervals in the table, and the third rectangle is labeled with a star as it does not appear to be a target interval in this sequencing experiment. In the bottom three-quarters of the figure, the coverage histogram is shown in the upper part and a stacked display of individual reads in the lower part.

Aspects of the data seen in the figure are:

- In order to achieve target coverage, the capture probes must be designed such that either end of the fragment falls on the target interval. This leads the coverage to extend beyond the target intervals (overhang on all target intervals).
- Overlap of fragments leads to non-uniform coverage within a target interval. This is attributable to varying widths of the fragments as well as tiling of capture probes. (Notice the bimodality of the coverage histogram for the eighth target.)
- Targets need not achieve the same average coverage as seen in the intervals labeled 0 and 7 having much higher coverage than the rest and the interval labeled 1 having a low coverage. Possible reasons for this are capture probe efficiency and interval characteristics such as size and GC content.

The figure provides several pieces of information on the individual reads. It color codes fragments in red to indicate that they are too wide compared to expected width and blue to indicate that these fragments are narrower than read length.

Such fragments are suggestive of insertions and deletions respectively. Other colors are used to indicate the two ends mapping to different parts of the genome which are inconsistent with the expected fragment lengths (see http://www.broadinstitute.org/software/igv/AlignmentData for details).

While a majority of fragments will be on target, there is a non-negligible proportion of fragments that are off-target and they can influence copy number computations which we will illustrate now. We obtained all the fragments (279,702 for normal and 339,030 for tumor) that are mapped to a 10 megabase region on chromosome 17 beginning at the 40 megabase mark. We binned the fragments by their midpoints into consecutive bins of length 100 bases where the genomic position a bin represents is its midpoint. We obtained the number of fragments in each bin for both normal and tumor samples. Of the 100,000 possible bins in the region, 13,297 had a nonzero count for at least one of the two samples. We expanded the target intervals in this region by 100 base pairs in both directions and derived the bins that intersect with the intervals. Of the 13,297 bins, only 6,835 of them do and hence are expected to have nonzero counts. However, bins with very small counts in the normal sample are inconsistent with the desired high coverage of the targets and thus are candidates for removal. There are 784 bins with fragment count of 2 or lower. Of the 6,439 bins that do not intersect with the target intervals 553 have fragment counts in the normal sample of at least 10, far more than the small number expected due to off-target fragments. Therefore, we included them in the copy number analysis. This results in 6,604 bins that are to be used in the copy number analysis and 6,693 bins to be discarded. The discarded bins account for just 5,807 fragments in the normal and 8,582 in the tumor (less than 3%). Fig. 18.4 shows the log-ratio computed as the ratio of scaled fragment counts where the grey and red points correspond to the included and discarded bins, respectively. Note that the red points form a band around zero with a significant presence near 1 and -1, which are the bins with one fragment in the tumor sample and zero in the normal, and vice versa. Despite the clear gain visible at the 46 megabase location, the loss in the 40–41.3 megabase region and focal loss around 42.7 megabase, the large number of red points in those regions will have an adverse effect on the copy number analysis, demonstrating the utility of pruning these bins. For the whole genome, binning the data results in 1,723,210 bins with nonzero counts in either sample of which 1,039,881 are to be discarded using the same consideration; they account for less than 4% of the total fragment count which is far lower than that expected from target efficiency.

The final piece of information needed for applying weighted CBS to the data are the weights assigned to the bins. The optimal weight for a bin is proportional to the variance of the fragment count for that bin which is a function of the unobserved rate parameter. The fragment counts which are the estimates of the rates are also very skewed thus using weights proportional to counts will make a handful of bins with large counts highly influential. Thus, we chose weights proportional to the logarithm of bin counts assigning greater weights to bins with large counts but protects against undue influence of bins with extreme counts. Although the optimal weights for the weighted t-statistic will depend on the mean counts of both the normal and the tumor

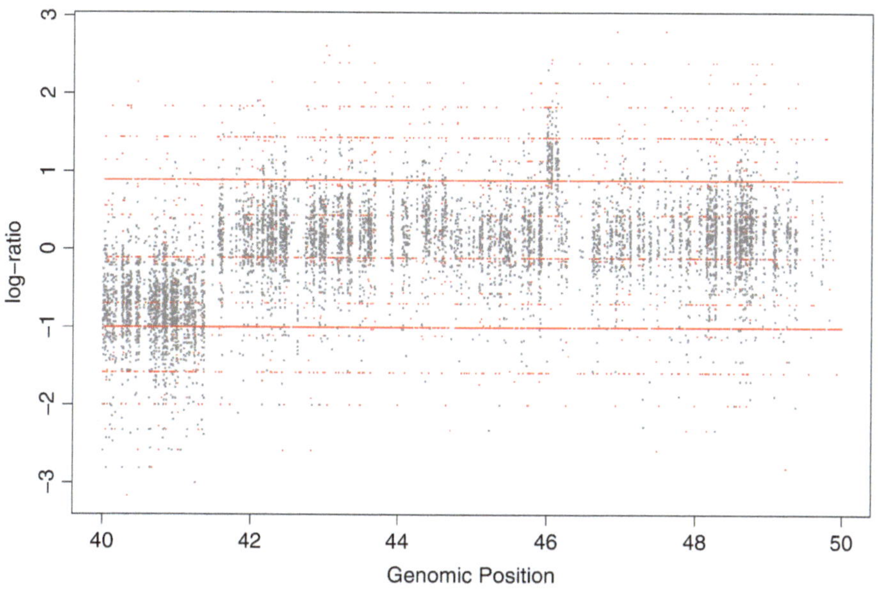

Fig. 18.4 The copy number log-ratio plot of the 40–50 megabase region on chromosome 17. The bins included in the analysis are in *grey* and the ones excluded are in *red*

samples, the tumor counts can change dramatically due to gains and losses. Thus, a more suitable choice of weights is to use the logarithm of just the normal counts (or median of several normal samples, if available).

Using the *DNAcopy* package, we segmented the logarithm of the ratio of scaled fragment counts for the bins to be used in the analysis. In Fig. 18.5, we show the whole genome copy number profile for this sample. The points are the log-ratio of the bins which are shown in alternate shades of grey to indicate different chromosomes. The algorithm segmented the genome into 419 regions with constant copy number which are shown as blue lines drawn at the level of the segment mean. The number of segments vary between chromosomes with the largest number (44) in chromosome 1 and the smallest number (7) in chromosome 22. The figure also shows the segmentation results from a SNP array analysis as red lines. Note that the SNP array data are not necessarily in the same scale and thus the red and blue lines may not overlap. Furthermore, since the SNP array probes cover the genome more uniformly than the targeted exome sequencing, there are far more red segments. However, the two sets of results show remarkable concordance except for chromosomes 2 and X, where the systematic large gap between the blue and red lines suggests that the cells used for exome sequencing have one fewer copy of these two chromosomes compared to the cells used for the SNP array.

In Fig. 18.6, we present a 25 megabase region on chromosome 17 to highlight the results. Note that while the exome segments (blue) and SNP segments (red) are similar, there are some locations where they differ. There is a small region

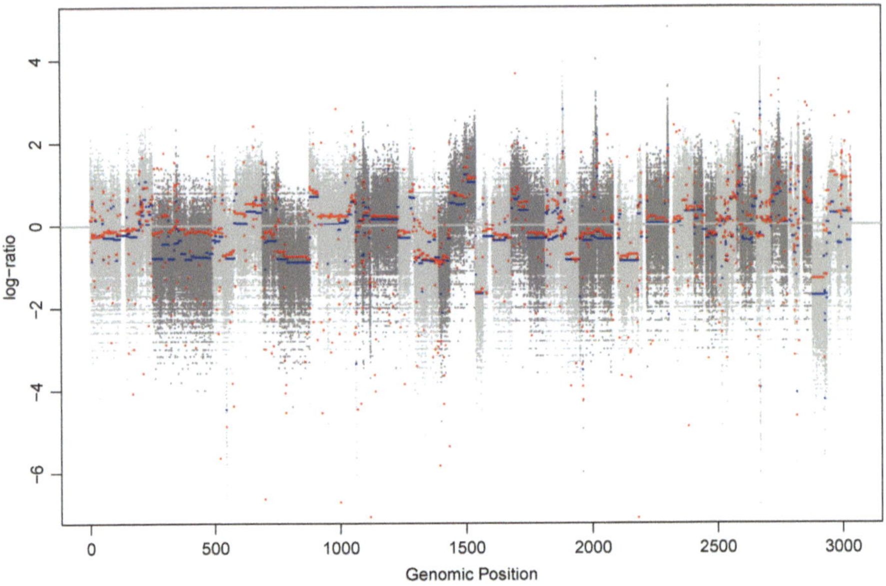

Fig. 18.5 The copy number profile of the whole genome. The chromosomes are colored in alternate shades of *grey*. The *blue lines* are regions of constant copy number identified from exome sequencing data. The *red lines* are the regions from SNP array data

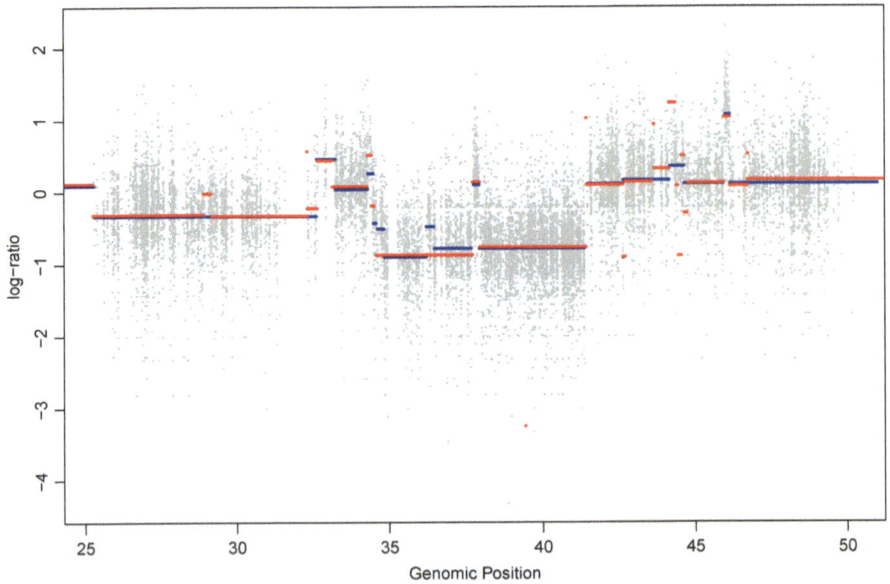

Fig. 18.6 The copy number profile of the 25–40 megabase region on chromosome 17. The *blue lines* are the segment means from exome sequencing data. The *red lines* from SNP array data

around 29 megabase and several small regions around 44 megabase the SNP array identifies that are not seen in the exome data. This may be attributable to the lack of data since the exome target intervals do not span the genome uniformly. In order to ascertain this, we reviewed the intervals identified by the SNP array and compared it to the target interval. The interval around 29 megabase spanned from positions 28,931,871 to 29,187,109 which is a 255 kilobase region. There are 17 target intervals in the whole exome sequencing in the last one third of this region starting from positions 29,111,193 and ending at 29,185,353 that covered just 4,676 bases. Likewise, the seven regions identified around 44 megabase in the SNP array covered an area that is 1.1 megabase long but were target poor for exome sequencing in that the target intervals only spanned 23 kilobases. The region between 35 and 38 megabase shows three segments for the exome where as just one for the SNP. This could be either due to higher resolution of exome data or the cell lines not being static.

It is common practice to undo small changes that do not meet a magnitude threshold. This occurs when a gentle wave in the data due to technical artifact looks like a change in mean. This step was not applied in the results presented as the goal was to present the full results. The overarching message from this analysis is that DNA sequencing, in particular targeted sequencing, can be successfully used to obtain copy number profiles.

18.5 Other Structural Variations

DNA sequencing can be used to identify other structural variants such as inversions and translocations. As seen in Sect. 18.2, the informative fragments for identifying these are those of Type 3 in Fig. 18.1. These are fragments that have high quality reads on both ends that are reliably mapped to the genome but are not properly paired. The improper pairing can occur due to an inter-chromosomal translocation, where the two reads are mapped to two different chromosomes, an intra-chromosomal translocation, where the reads from the two ends are mapped to the same chromosome but are directed away from each other rather than towards each other, or an inversion, where the reads from the two ends are mapped in the same direction. In all cases, the inferred fragment size is far larger than the expected fragment size. [Note: A proper pair can result in a large fragment size when there is a deletion in between the two reads; Fragments of type 4 in Fig. 18.1 can also be used for identifying these structural variations provided partial alignments can be done.]

The "bam" files used in this step have been deduplicated, realigned and their base quality scores recalibrated. The first step in identifying inversions and translocations is to extract all the improperly paired fragments where both reads are mapped to chromosome 1 through X and pass the instrument's quality control. There are 158,433 such fragments in the normal sample and 145,858 in the tumor. Note that these counts are just 0.5% of the total number of fragments in the sample. This is to be expected since the striped-shaded region junction (seen in Fig. 18.1) needed for these structural events are uncommon as most fragments are interior fragments (of Types 1 and 2). Additionally fragments of Types 4 and 5 will be unmapped using standard alignment software. Although a read may pass quality control as determined by the sequencing machine, the mapping quality of the read may not be high enough to provide valid information. Thus, we will use the mapping quality filter of 20 (possible error in alignment of 1%) to restrict the analysis to high quality reads. This reduced the number of fragments where both ends are mapped with a quality greater than 20 to 108,055 for the normal and 98,385 for the tumor. Note that there are more improperly paired fragments in the normal than the tumor. This might be due to the sequence similarity between different regions in the genome and hence mapping may not be unique and absolute.

A single fragment suggesting a structural variation is not a proof of it. The more the number of fragments indicating a structural variation the stronger the evidence. However, a somatic structural change acquired in the tumor will not be present in the germline. Hence one must verify that any structural variant identified in the tumor is present only in the tumor and not the germline. We begin this by counting the number of fragments from both tumor and normal samples that are in a neighborhood of every fragment. In the Example section earlier, we noted that most fragments are between 75 and 300 bases long. Thus, we define the neighborhood of a fragment to be within 1,000 bases of the starting location of the reads from both ends. Note that the neighborhood of each fragment will contain itself and hence the minimum fragment count is 1. Of the 206,440 combined fragments, only 3,042 have fragment counts greater than 1. Furthermore, if a fragment has several other fragments in its neighborhood, then all of them have this fragment in their neighborhoods as well. In fact, they cluster strongly and the 3,042 fragments with neighborhood count of more than one reduce to a far smaller number of clusters.

In Fig. 18.7, we display the fragment counts in the tumor plotted against the counts in the normal. The scatterplot shows that, in this data, there is a strong relationship between the tumor and normal counts suggesting that most of the suggested changes are present in both tumor and normal cells. In order to identify possible tumor specific changes we restricted ourselves to the fragments for which the normal count in the neighborhood is at most 3 and conducted a Binomial test for the hypothesis that the proportion of tumor counts out of the total is greater than 0.5. Table 18.2 lists the three clusters of fragments that are significant after adjusting for multiple comparison. The table gives the chromosome to which the fragments are mapped, the median start location of the first and second read, and the number of fragments in tumor and normal.

In Fig. 18.8, we show the copy number profiles, obtained using the abundance measure data, for these two regions. The top and bottom row of figures correspond to chromosomes 21 and 14, respectively. The first figure in each row shows the entire region where the start locations of the respective reads are marked by a vertical line. For chromosome 14, the two lines at position 105.412 megabase appear as one due to their closeness. The second and third figures in the top

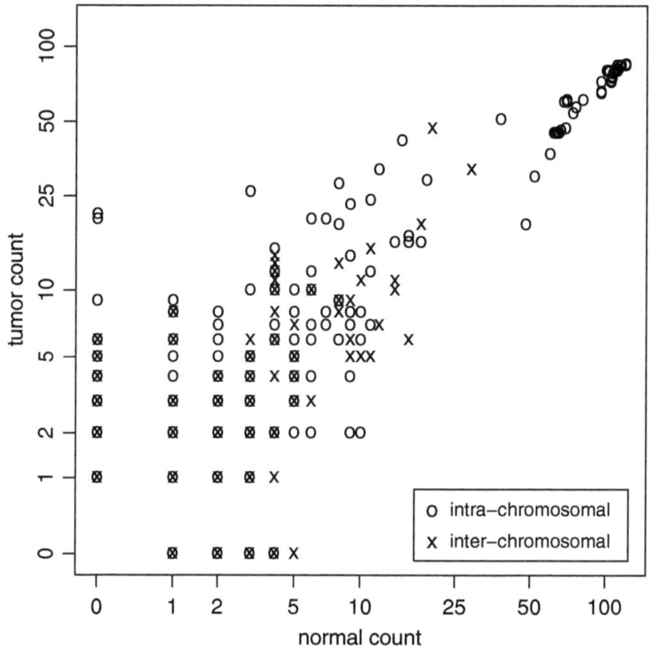

Fig. 18.7 The number of fragments in the neighborhood of an improperly paired fragment that belong to tumor and normal samples

Table 18.2 The details of the three clusters of fragments identified as present in tumor only

First read		Second read		Fragment count	
Chromosome	Location	Chromosome	Location	Tumor	Normal
21	43,246,325	21	47,347,121	21	0
14	106,471,416	14	107,282,893	20	0
14	105,412,008	14	105,412,453	26	3

row show the read locations of the first and second read are close to breakpoints identified in the copy number segmentation in the previous section and thus this rearrangement is consistent with copy number data. The second figure in the bottom row corresponds to the read location of 106.471 megabase in the second row of Table. 18.2. While this location is close to a break point, its companion is close to the end of chromosome with just two target intervals in its vicinity, and thus no additional information on the structural change is available. The third figure in the bottom row shows the two read locations in the third row and the two points in the interval between them that are seen in the figure are below the majority of the points in their vicinity. This suggests a small deletion since the two locations are just 445 bases apart. However the copy number segmentation does not pick them up as the magnitude of the change is within the noise of the copy number ratio. In all,

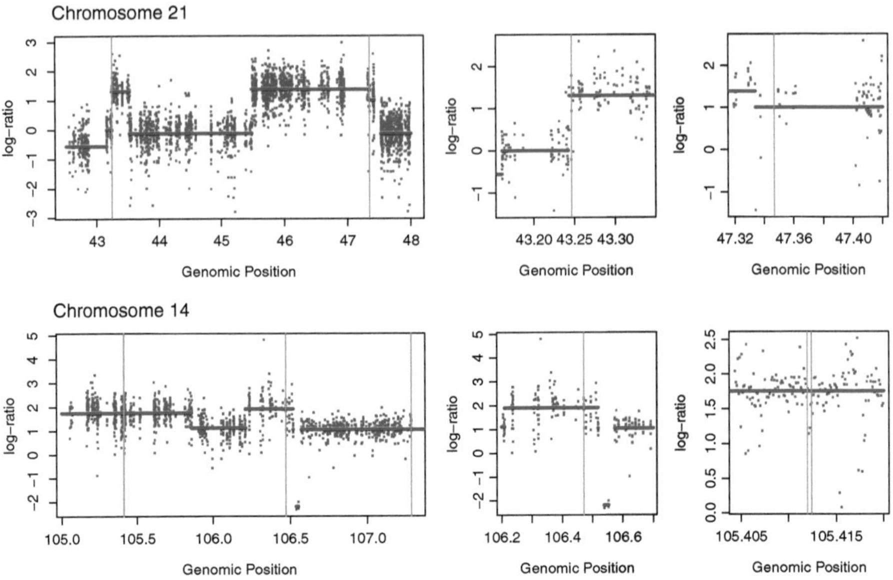

Fig. 18.8 The copy number profiles of the regions with plausible structural variants listed in Table 18.2

the presence of breakpoints in the copy number profile near the read locations of the plausible structural variations, lend support to their presence.

Although there are extensive structural rearrangements present in this cell line (the spectral karyotype of these cells is at http://www.path.cam.ac.uk/~pawefish/BreastCellLineDescriptions/HCC1143.html), we identified just 3 of them and none of the inter-chromosomal ones. Our inability to detect such an event is primarily due to fact that these data rose from a targeted sequencing and hence has large gaps in information. In order for targeted sequencing to be able to detect inversions and translocations, the junction (the striped-shaded region boundary in Fig. 18.1) should be close to a target interval and the capture probe should fully reside within the striped or the shaded region. This makes the likelihood of a fragment that contains an inversion or a translocation event being captured and sequenced very low. Thus whole genome sequencing is more apt for identifying structural rearrangements as it does not select for specific fragments to be sequenced and is thus far more likely to contain fragments with such events.

18.6 Summary

DNA sequencing, in particular targeted sequencing, is widely used in cancer research with the primary purpose of identifying somatic mutations. In this chapter, we adapted the Circular Binary Segmentation algorithm for the analysis of copy numbers using DNA sequencing data. We showed using a whole exome sequencing dataset that copy number profile can be obtained from it. Despite the target intervals covering less than 2% of the genome, this profile is highly concordant with profile obtained from SNP array with whole genome coverage. The high coverage used in exome sequencing has the potential to identify intragenic changes such as deletion of a few exons which may not be feasible with current whole genome arrays.

DNA sequencing also provides information on polymorphic sites (SNPs) within the target intervals which in turn provides allele specific copy number information. We adapted CBS to obtain parent specific copy number profile from SNP array data [24] which can in turn be adapted to sequencing data. Similarly, the ASCAT algorithm, developed by Van Loo et al. (2010) [35], for the analysis of allele specific copy numbers can also be applied in the sequencing context. Such an analysis can provide additional information such as copy neutral loss of heterozygosity or uniparental disomy which enhances our understanding of the oncogenic process.

DNA sequencing contains three types of information - copy number, genotype and structural rearrangement. The methods we described treat these separately. However, since these data elements are not orthogonal to each other, there is potential to borrow information from all three types of data to develop a unified method to detect these structural variations. Other considerations such as the optimal bin size and the choice of filtering parameters and their effect should be studied for existing methods as well as those being developed.

The purpose behind studying structural variations is their impact on gene expression and their consequences. There is a positive correlation between copy numbers and gene expression. Likewise, the *bcr-abl* fusion protein provides a powerful example for the consequences of translocations. However, a comprehensive catalog of all possible events will require several tens of thousands of samples [17]. Thus careful consideration of the design of these experiments is essential. As we noted, targeted sequencing may not provide information on structural rearrangements but the high coverage that they can achieve to detect somatic mutations will be prohibitively resource intense for whole genome sequencing. Additionally, fusion transcripts are best detected using RNA sequencing. These aspects present a vibrant area for future research on how best to combine different sequencing methodologies to extract the information in a sample. A related problem is how best data from multiple samples can be combined to identify the affected biological processes and pathways and how they can be prioritized for further study.

Finally, the volume of data from these experiments are immense and will require efficient software for processing them. This presents a venue for the development of efficient methods and algorithms. In summary, DNA sequencing provides a wealth of data which can add to our knowledge with further research and proper analytic tools.

Acknowledgements The author thanks Arshi Arora for her programming assistance in processing the bam files and Drs. Glenn Heller, Jennifer Levine and Ronglai Shen for their valuable comments and suggestions. Supported by grants from the National Cancer Institute (CA163251, CA008748) and the Susan G. Komen for the Cure Foundation (IIR12221291).

References

[1] Abyzov, A., Urban, A.E., Snyder, M., Gerstein, M.: Cnvnator: an approach to discover, genotype, and characterize typical and atypical cnvs from family and population genome sequencing. Genome Res. **21**(6), 974–984 (2011)

[2] Anscombe, F.J.: The transformation of poisson, binomial and negative-binomial data. Biometrika **35**(3–4), 246–254 (1948)

[3] Berg, J.M., Tymoczko, J.L., Stryer, L.: Biochemistry. Seventh Edition. W. H. Freeman & Company, New York (2011)

[4] Beroukhim, R., Getz, G., Nghiemphu, L., Barretina, J., Hsueh, T., Linhart, D., Vivanco, I., Lee, J.C., Huang, J.H., Alexander, S., et al.: Assessing the significance of chromosomal aberrations in cancer: methodology and application to glioma. Proc. Natl. Acad. Sci. **104**(50), 20,007–20,012 (2007)

[5] Carter, S.L., Cibulskis, K., Helman, E., McKenna, A., Shen, H., Zack, T., Laird, P.W., Onofrio, R.C., Winckler, W., Weir, B.A., et al.: Absolute quantification of somatic dna alterations in human cancer. Nat. Biotechnol. **30**(5), 413–421 (2012)

[6] Chen, K., Wallis, J.W., McLellan, M.D., Larson, D.E., Kalicki, J.M., Pohl, C.S., McGrath, S.D., Wendl, M.C., Zhang, Q., Locke, D.P., et al.: Breakdancer: an algorithm for high-resolution mapping of genomic structural variation. Nat. Meth. **6**(9), 677–681 (2009)

[7] Chiang, D.Y., Getz, G., Jaffe, D.B., O'Kelly, M.J., Zhao, X., Carter, S.L., Russ, C., Nusbaum, C., Meyerson, M., Lander, E.S.: High-resolution mapping of copy-number alterations with massively parallel sequencing. Nat. Meth. **6**(1), 99–103 (2009)

[8] Crick, F., et al.: Central dogma of molecular biology. Nature **227**(5258), 561–563 (1970)

[9] Ganem, N.J., Storchova, Z., Pellman, D.: Tetraploidy, aneuploidy and cancer. Curr. Opin. Genet. Dev. **17**(2), 157–162 (2007)

[10] Gentleman, R.C., Carey, V.J., Bates, D.M., Bolstad, B., Dettling, M., Dudoit, S., Ellis, B., Gautier, L., Ge, Y., Gentry, J., et al.: Bioconductor: open software development for computational biology and bioinformatics. Genome Biol. **5**(10), R80 (2004). URL http://www.bioconductor.org/

[11] Han, S.W., Kim, H.P., Shin, J.Y., Jeong, E.G., Lee, W.C., Lee, K.H., Won, J.K., Kim, T.Y., Oh, D.Y., Im, S.A., et al.: Targeted sequencing of cancer-related genes in colorectal cancer using next-generation sequencing. PloS One **8**(5), e64,271 (2013)

[12] Hayes, M., Pyon, Y.S., Li, J.: A model-based clustering method for genomic structural variant prediction and genotyping using paired-end sequencing data. PloS One **7**(12), e52,881 (2012)

[13] Kallioniemi, A., Kallioniemi, O.P., Sudar, D., Rutovitz, D., Gray, J.W., Waldman, F., Pinkel, D.: Comparative genomic hybridization for molecular cytogenetic analysis of solid tumors. Science **258**(5083), 818–821 (1992)

[14] King, C.R., Kraus, M.H., Aaronson, S.A.: Amplification of a novel v-erbb-related gene in a human mammary carcinoma. Science **229**(4717), 974–976 (1985)

[15] Kumar, V., Abbas, A.K., Fausto, N., Aster, J.C.: Robbins and cotran pathologic basis of disease, Professional Edition: Expert Consult-Online. Elsevier Health Sciences, New York (2009)

[16] Lai, W.R., Johnson, M.D., Kucherlapati, R., Park, P.J.: Comparative analysis of algorithms for identifying amplifications and deletions in array cgh data. Bioinformatics **21**(19), 3763–3770 (2005)

[17] Lawrence, M.S., Stojanov, P., Mermel, C.H., Robinson, J.T., Garraway, L.A., Golub, T.R., Meyerson, M., Gabriel, S.B., Lander, E.S., Getz, G.: Discovery and saturation analysis of cancer genes across 21 tumour types. Nature **505**, 495–501 (2014)

[18] Leavitt, S.A., et al.: Deciphering the genetic code: Marshall Nirenberg. http://history.nih.gov/exhibits/nirenberg/glossary.htm (2010)

[19] Li, H., Handsaker, B., Wysoker, A., Fennell, T., Ruan, J., Homer, N., Marth, G., Abecasis, G., Durbin, R., et al.: The sequence alignment/map format and samtools. Bioinformatics **25**(16), 2078–2079 (2009). URL http://samtools.sourceforge.net/

[20] du Manoir, S., Speicher, M.R., Joos, S., Schröck, E., Popp, S., Döhner, H., Kovacs, G., Robert-Nicoud, M., Lichter, P., Cremer, T.: Detection of complete and partial chromosome gains and losses by comparative genomic in situ hybridization. Human Genet. **90**(6), 590–610 (1993)
[21] McKenna, A., Hanna, M., Banks, E., Sivachenko, A., Cibulskis, K., Kernytsky, A., Garimella, K., Altshuler, D., Gabriel, S., Daly, M., et al.: The genome analysis toolkit: a mapreduce framework for analyzing next-generation dna sequencing data. Genome Res. **20**(9), 1297–1303 (2010). URL http://www.broadinstitute.org/gatk/
[22] Morgan, M., Pagès, H., Obenchain, V.: Rsamtools: Binary alignment (BAM), variant call (BCF), or tabix file import. URL http://bioconductor.org/packages/release/bioc/html/Rsamtools.html
[23] Nowell, P.C., Hungerford, D.A.: A minute chromosome in chronic granulocytic leukemia. Science **132**, 1497–1501 (1960)
[24] Olshen, A.B., Bengtsson, H., Neuvial, P., Spellman, P.T., Olshen, R.A., Seshan, V.E.: Parent-specific copy number in paired tumor–normal studies using circular binary segmentation. Bioinformatics **27**(15), 2038–2046 (2011)
[25] Olshen, A.B., Venkatraman, E., Lucito, R., Wigler, M.: Circular binary segmentation for the analysis of array-based dna copy number data. Biostatistics **5**(4), 557–572 (2004)
[26] Picard: Command-line utilities that manipulate SAM files. URL http://picard.sourceforge.net/index.shtml
[27] Pinkel, D., Albertson, D.G.: Array comparative genomic hybridization and its applications in cancer. Nat. Genet. **37**, S11–S17 (2005)
[28] Pounds, S., Cheng, C., Li, S., Liu, Z., Zhang, J., Mullighan, C.: A genomic random interval model for statistical analysis of genomic lesion data. Bioinformatics **29**(17), 2088–2095 (2013)
[29] Redon, R., Ishikawa, S., Fitch, K.R., Feuk, L., Perry, G.H., Andrews, T.D., Fiegler, H., Shapero, M.H., Carson, A.R., Chen, W., et al.: Global variation in copy number in the human genome. Nature **444**(7118), 444–454 (2006)
[30] Robinson, J.T., Thorvaldsdóttir, H., Winckler, W., Guttman, M., Lander, E.S., Getz, G., Mesirov, J.P.: Integrative genomics viewer. Nat. Biotechnol. **29**(1), 24–26 (2011). URL http://www.broadinstitute.org/igv/
[31] Sebat, J., Lakshmi, B., Troge, J., Alexander, J., Young, J., Lundin, P., Månér, S., Massa, H., Walker, M., Chi, M., et al.: Large-scale copy number polymorphism in the human genome. Science **305**(5683), 525–528 (2004)
[32] Seshan, V.E., Olshen, A.: DNAcopy: DNA copy number data analysis. URL http://bioconductor.org/packages/release/bioc/html/DNAcopy.html
[33] Shen, J.J., Zhang, N.R.: Change-point model on non-homogeneous poisson processes with application in copy number proling by next-generation dna sequencing. https://statistics.stanford.edu/sites/default/files/BIO%20257.pdf (2011)
[34] Taylor, B.S., Barretina, J., Socci, N.D., DeCarolis, P., Ladanyi, M., Meyerson, M., Singer, S., Sander, C.: Functional copy-number alterations in cancer. PLoS One **3**(9), e3179 (2008)
[35] Van Loo, P., Nordgard, S.H., Lingjærde, O.C., Russnes, H.G., Rye, I.H., Sun, W., Weigman, V.J., Marynen, P., Zetterberg, A., Naume, B., et al.: Allele-specific copy number analysis of tumors. Proc. Natl. Acad. Sci. **107**(39), 16,910–16,915 (2010)
[36] Venkatraman, E., Olshen, A.B.: A faster circular binary segmentation algorithm for the analysis of array cgh data. Bioinformatics **23**(6), 657–663 (2007)
[37] Wang, J., Mullighan, C.G., Easton, J., Roberts, S., Heatley, S.L., Ma, J., Rusch, M.C., Chen, K., Harris, C.C., Ding, L., et al.: Crest maps somatic structural variation in cancer genomes with base-pair resolution. Nat. Meth. **8**(8), 652–654 (2011)
[38] Wang, S., Gao, J., Lei, Q., Rozengurt, N., Pritchard, C., Jiao, J., Thomas, G.V., Li, G., Roy-Burman, P., Nelson, P.S., et al.: Prostate-specific deletion of the murine pten tumor suppressor gene leads to metastatic prostate cancer. Canc. cell **4**(3), 209–221 (2003)

[39] Willenbrock, H., Fridlyand, J.: A comparison study: applying segmentation to array cgh data for downstream analyses. Bioinformatics **21**(22), 4084–4091 (2005)

[40] Xie, C., Tammi, M.T.: Cnv-seq, a new method to detect copy number variation using high-throughput sequencing. BMC Bioinform. **10**(1), 80 (2009)

[41] Zhao, M., Wang, Q., Wang, Q., Jia, P., Zhao, Z.: Computational tools for copy number variation (cnv) detection using next-generation sequencing data: features and perspectives. BMC Bioinform. **14**(Suppl 11), S1 (2013)

Chapter 19
Statistical Methods for the Analysis of Next Generation Sequencing Data from Paired Tumor-Normal Samples

Mengjie Chen*, **Lin Hou***, **and Hongyu Zhao**

Abstract The emergence of next generation sequencing technologies has brought cancer studies into a genomic era. With reasonable cost, cancer genomes can be scrutinized with unprecedented resolution and sensitivity. In this chapter, we discuss statistical methods that have been proposed to detect somatic variations at the DNA level using paired tumor and normal sequencing data, including single nucleotide alterations (SNAs) and copy number alterations (CNAs). We describe selected statistical methods, their strengths and limitations, and discuss future directions.

19.1 Introduction

Starting from 2006, large-scale profiling of mutational landscapes of cancer genomes has brought remarkable advances in our understanding of tumorigenesis. Represented by large community efforts such as The Cancer Genome Atlas (TCGA) [25, 42], a great number of discoveries of cancer-specific alterations and cancer drivers have been enabled by next generation sequencing technology. With sequencing cost continuously decreasing, cancer genome profiling will continue to shift from whole genome array/Sanger sequencing to whole exome/genome sequencing. Next generation platforms can produce data to characterize DNA

---

* joint first authors

M. Chen
Program of Computational Biology and Bioformatics,
Yale University, New Haven, CT 06520, USA,
e-mail: mengjie.chen@yale.edu

L. Hou • H. Zhao (✉)
Department of Biostatistics,
Yale School of Public Health, New Haven, CT 06520, USA
e-mail: lin.hou@yale.edu; hongyu.zhao@yale.edu

S. Datta and D. Nettleton (eds.), *Statistical Analysis of Next Generation Sequencing Data*,
Frontiers in Probability and the Statistical Sciences, DOI 10.1007/978-3-319-07212-8_19,

alteration, transcriptome, methylation, nucleosome positioning and many other features. With a pair of tumor and normal samples, this strategy can potentially yield a comprehensive characterization of genomic alterations for the tumor. In this chapter, we focus on the analysis of DNA alteration, including single nucleotide alterations (SNAs), representing changes at the single base pair level, and copy number alterations (CNAs), representing changes at the structural level. We describe strengths and weaknesses of selected statistical methods and future methodology development directions.

19.2 Single Nucleotide Aberration Detection

Somatic mutations refer to genetic abnormalities in a cell that can be passed on to the daughter cells during the process of cell division. In this section we are interested in a specific type of somatic mutations, namely, the SNAs which are defined as single nucleotide variants that are present in tumor tissues, but not in the matched normal tissue. At first, researchers used a "subtraction" approach to identify SNAs [45]. Genotypes of tumor/normal tissues are called separately, and SNAs are then inferred at genomic positions of inconsistent results between the two calling results. In other words, SNAs are inferred by subtracting the variants identified in normal tissues from those identified in tumor tissues. As intuitive as it sounds, the "subtraction" method fails to take into account the uncertainty in the genotyping results in tumor/normal tissue and the dependencies between tumor/normal genotypes. Later on, instead of comparing the genotypes in tumor/normal tissues, the alignments of two genomes are directly compared to call SNAs [15, 17, 18, 33, 35].

Before describing the statistical models of each SNA detection method, we first introduce some notations. For a genomic position, we denote the reference allele as A and the non-reference allele as B. For simplicity, we assume, as in all somatic mutation calling algorithms discussed here, each position is bi-allelic with neutral ploidy (diploid). In the normal tissue, the numbers of reads that are mapped to this position with allele A and allele B, named "allelic counts", are denoted by n_A and n_B respectively; and the corresponding numbers in the tumor tissue are denoted by t_A and t_B. Let G_t and G_n denote the genotypes of the tumor and normal samples, respectively. Let **T** and **N** denote all the information collected from sequencing data of tumor and normal samples respectively, including allelic counts, read depths, base qualities, and other information. Let **D** represent the union of tumor and normal information, $\mathbf{D} = \{\mathbf{T}, \mathbf{N}\}$. Lastly, let S_i denote the somatic state of variant i, where $S_i = 1$ if the genotype of variant i in the tumor sample (G_t^i) differs from the corresponding genotype in the normal sample (G_n^i), and $S_i = 0$ otherwise. We will omit the position indicator i hereafter.

Many SNA calling methods have been proposed to compare the sequencing data from tumor and normal tissues to identify SNAs. Based on the underlying methodology, we can classify these algorithms into three categories: heuristic methods, statistical methods, and machine learning methods. The remainder of

Table 19.1 Contingency Table of Fisher's Exact Test in VarScan2

	Reference Allele	Non-reference Allele
Tumor	t_A	t_B
Normal	n_A	n_B

this section is organized as follows: first, we will introduce various types of SNA detection algorithms; second, we will discuss the empirical filters and compare the performance of different algorithms.

19.2.1 Heuristic Methods

In heuristic approaches, tumor and normal samples are analyzed separately, and the results are then combined and compared to infer somatic mutations. The "subtraction" method falls into this category. VarScan2 [17] first compares the genotypes between normal and tumor samples. When the two genotypes do not match, the one-tailed Fisher's exact test is utilized to test whether the non-reference allele is more abundant in tumor reads (see Table 19.1 for the contingency table). If the p-value of the test meets the significance level (set to 0.1 by default), the variant is called somatic. Shimmer [15] also uses Fisher's exact test, but the p-values are corrected for multiple testing. A variant is called somatic if the corresponding false discovery rate is below a user-specified threshold.

Besides the inferred genotype, single sample genotyping algorithms usually calculate the genotype likelihood, $P(\mathbf{D}|Genotype)$. There are ten possible genotypes in a diploid genome (AA, AC, AG, AT, CC, CG, CT, GG, GT, TT). Thus *Genotype* is coded as 0 to 9. SomaticSniper [18] combines the genotype likelihood of tumor samples and normal samples in a Bayesian framework (Equations (19.1) and (19.2)), to assign a phred-scale somatic score to each candidate variant (Equation (19.3)):

$$P(S=0|\mathbf{T},\mathbf{N}) = P(G_t = G_n|\mathbf{T},\mathbf{N}) = \sum_{j=0}^{9} P(G_t = G_n = j|\mathbf{T},\mathbf{N}), \tag{19.1}$$

$$\begin{aligned} &P(G_t = G_n = i|\mathbf{T},\mathbf{N}) \\ &= \frac{P(\mathbf{T},\mathbf{N}|G_t = G_n = i)P(G_t = G_n = i)}{\left(\sum_{j=0}^{9} P(\mathbf{T}|G_t = j)P(G_t = j)\right)\left(\sum_{j=0}^{9} P(\mathbf{N}|G_n = j)P(G_n = j)\right)} \\ &= \frac{P(\mathbf{T}|G_t = i)P(\mathbf{N}|G_n = i)P(G_t = i)P(G_n = i)}{\left(\sum_{j=0}^{9} P(\mathbf{T}|G_t = j)P(G_t = j)\right)\left(\sum_{j=0}^{9} P(\mathbf{N}|G_n = j)P(G_n = j)\right)}, \end{aligned} \tag{19.2}$$

$$\text{SomaticScore} = -10\log_{10} P(S=0|\mathbf{D}). \tag{19.3}$$

In the initial derivation of SomaticSniper, the genotypes of tumor sample and normal sample are assumed to be independent. However, the independence assumption,

$P(G_t = G_n = i) = P(G_t = i)P(G_n = i)$, is obviously not true. Thus in the second derivation, the dependence is taken into account, so that the calculation of the second term of the numerator in Equation (19.2) is modified, assuming a small probability of somatic mutation. In Equations (19.4, 19.5), i denotes one of the ten possible diploid genotypes listed before; μ is the empirical estimate of somatic mutation rate; and $P(G_t = i|G_n = i)$ is defined as the probability that the tumor sample has the same genotype as the normal sample conditioning on the genotype of the normal sample.The genotype likelihoods are taken from the MAQ algorithm [21], and the prior distribution of genotypes is specified through empirical data. That is, we have

$$P(G_t = i, G_n = i) = P(G_n = i)P(G_t = i|G_n = i), \tag{19.4}$$

$$P(G_t = i|G_n = j) = \begin{cases} \mu, & i \text{ shares one allele with } j, \\ \mu^2, & i \text{ shares no allele with } j, \\ 1 - P(G_t \neq j|G_n = j) & i = j, \end{cases}$$

$$\text{where} \quad P(G_t \neq j|G_n = j) = \sum_{k=0}^{9} I(k \neq j)P(G_t = k|G_n = j). \tag{19.5}$$

For example, the conditional probability that the tumor sample has genotype AC, given that the normal sample has genotype AA, is $P(G_t = 1|G_n = 0) = \mu$ because genotype AA (coded as 0) shares one allele with genotype AC (coded as 1). Likewise, the conditional probability that the tumor sample has genotype CC, given that the normal sample has genotype AA, is $P(G_t = 5|G_n = 0) = \mu^2$ because genotype AA (coded as 0) shares no alleles with genotype CC (coded as 5). The conditional probability that the tumor sample has genotype CC, given that the normal sample has genotype CC, is $P(G_t = 5|G_n = 5) = 1 - (3\mu + 6\mu^2)$ by Equation (19.5).

19.2.2 Statistical Modeling Based Methods

In statistical modeling approaches, the observed allele count at each genomic position is a random sample from a binomial distribution, $n \sim \text{Binomial}(d, p)$, where d is the sequencing depth at that position, and p is a function of the underlying genotypes and the error rate abstracted in both sequencing and mapping. The genotypes of tumor and normal tissues are considered as hidden variables of interest. In a Bayesian framework, the task of detecting SNAs reduces to computing $P(G_t \neq G_n|\mathbf{T}, \mathbf{N})$, the posterior probability that the tumor and normal genotype are different, given the allelic counts in tumor and normal samples. In JointSNVMix [33], a Bayesian hierarchical model is used to formulate the problem (see Equation (19.6)). The joint genotypes, (G_n, G_t), follow a multinomial distribution. There are nine possible joint genotypes, determined by two homozygous genotypes and one heterozygous genotype for both tumor and normal samples. Conditional on the joint genotypes, the allele counts in normal and tumor samples are distributed as independent binomials. Conjugate priors are used as prior distributions of the multinomial and binomial parameters π and μ:

$$\begin{aligned}
(G_n, G_t)|\pi &\sim \text{Multinomial}(1, \pi), \\
\mathbf{T} \perp \mathbf{N} &\mid (G_n, G_t), \\
t_A|d_t, \mu_t &\sim \text{Binomial}(d_t, \mu_t), \\
n_A|d_n, \mu_n &\sim \text{Binomial}(d_n, \mu_n), \\
\pi &\sim \text{Dirichlet}(\delta), \\
\mu_n|G_n &\sim \text{Beta}(\alpha_n, \beta_n|G_n), \\
\mu_t|G_t &\sim \text{Beta}(\alpha_t, \beta_t|G_t).
\end{aligned} \tag{19.6}$$

Rather than modeling the discrete "joint genotype", Strelka [35] takes the continuous "joint allelic fractions" in tumor and normal samples (see Equation (19.7)) as the observed data. In the tumor samples, the genotype at a position with somatic SNA usually deviates from the discrete genotype (AA, AB, BB) due to sample impurity, copy number variations, and existence of sub-clonal mutations, which are mutations that are only found in some of the cancer cells. For example, for an SNA in a region with copy number variation, if the ploidy (the number of sets of chromosomes in a cell) is m, the set of possible genotypes will be $\{A_k B_l, k+l=m, k \geq 0, l \geq 0\}$. In a normal human genome, the ploidy is 2. In a cancer cell undergoing whole genome duplication, the ploidy is 4. As m increases, the genotype and the allele counts will deviate more from the diploid model. In addition, for tumor samples, ploidy is not necessarily an integer, since it is confounded by contamination of normal cells and sub-clonal copy number variations. Hence, the intrinsic advantage of Strelka's approach is that it allows allelic fractions to be continuous under the diploid model, not necessarily one among $\{0, 0.5, 1\}$. However, the definition and inference of SNAs is less straightforward. The authors defined somatic state as $\{S=1\} = \{(f_t, f_n) : f_t \neq f_n\}$, and the posterior probability of a somatic event is derived in Equation (19.8).

$$\begin{aligned}
f_t &= t_B/(t_A + t_B) \\
f_n &= n_B/(n_A + n_B)
\end{aligned} \tag{19.7}$$

$$P(S=1|\mathbf{D}) = \int_{f_t, f_n} I(f_t \neq f_n) P(f_t, f_n|\mathbf{D}). \tag{19.8}$$

Thus, the problem reduces to inferring the posterior distribution of (f_t, f_n), which in turn is obtained in a standard Bayesian framework as follows:

$$P(f_t, f_n|\mathbf{D}) \propto P(\mathbf{D}|f_t, f_n) P(f_t, f_n). \tag{19.9}$$

Assuming conditional independence of tumor and normal data, given allelic fractions, the likelihood term in Equation (19.9) can be computed by incorporating any single sample SNA likelihood methods (see Equation (19.10)). In the prior term $P(f_t, f_n)$, the dependency between tumor and normal genotypes and the extensive range of tumor genotype is incorporated (see Equation (19.11)). The dependency is introduced by assuming a small probability $P(S=1)$ that allows the tumor allelic fraction to be different from normal allelic fraction. To allow non-diploid and even fractional genotype, a uniform term $U(f_t)$ is introduced, so that the distribution of tumor allelic fraction can be much dispersed:

$$P(\mathbf{D}|f_t, f_n) \sim P(\mathbf{T}|f_t)P(\mathbf{N}|f_n), \tag{19.10}$$

$$P(f_t, f_n) = \begin{cases} P(f_n)P(S=1)U(f_t), & f_t \neq f_n \\ P(f_n)P(S=0), & f_t = f_n \\ 0, & \text{otherwise.} \end{cases} \tag{19.11}$$

The prior probability $P(f_n)$ is a mixture of the expected diploid variation and a noise term,

$$P(f_n) = P_{\text{diploid}}(f_n)(1-\mu) + P_{\text{noise}}(f_n)\mu. \tag{19.12}$$

$P_{\text{diploid}}(f_n)$ is specified as follows:

$$P_{\text{diploid}}(f_n) = \begin{cases} \theta/3, & \alpha = 0.5, \\ \theta/6, & \alpha = 0, \beta = 0 \\ \theta^2/3, & \alpha = 0, \beta = 0.5 \\ 1 - 3\theta/2 + \theta^2, & \alpha = 1, \end{cases} \tag{19.13}$$

where θ is the heterozygosity term, set to 0.001. Here α is the frequency of the reference allele and β is the allele frequency ratio of the first and second most frequent alleles. The noise term is also a mixture of two components; we refer to [35] for the distribution of the two components.

Similar to Strelka, MuTect [10] also considers the possibility that tumor genotype deviates from diploid genotypes; thus the expected allelic fraction at each site is estimated from data by $\hat{f} = \frac{t_B}{d}$ instead of being pre-specified by genotype models. By modeling the sequencing reads to be generated from a binomial model, the likelihood of the tumor sample sequencing data is derived as $L(\mathbf{T}|f) \propto \prod_{i=1}^{d} P(nt_i|f)$, where nt_i is the nucleotide in the ith tumor read. The individual terms in this product are given by:

$$P(nt_i|f) = \begin{cases} fe_i/3 + (1-f)(1-e_i), & nt_i = A \\ f(1-e_i) + (1-f)e_i, & nt_i = B \\ e_i/3, & \text{otherwise.} \end{cases} \tag{19.14}$$

$P(nt_i = A|f)$ is derived from law of total probability, by conditional on whether or not the nucleotide read is an sequencing error. Under the null model, any reads with the non-reference allele is deemed as a sequencing error, corresponding to $f = 0$ in Equation (19.14). In the mutant model, the non-reference reads are a mixture of sequencing errors and SNAs, f is set to $\hat{f}$ in Equation (19.14), and e_i denotes the sequencing error. In order to detect SNAs, MuTect defined a log likelihood ratio by comparing the mutant model and the null model (Equation (19.15)). A hard-threshold is applied to the log likelihood ratio (LOD) to infer a candidate SNA.

$$\text{LOD} = \frac{L(\mathbf{T}|f = 0)}{L(\mathbf{T}|f = \hat{f})}. \tag{19.15}$$

19.2.3 Machine Learning Methods

Besides the methods mentioned above, the problem of somatic mutation detection has been approached with machine learning methods by Ding et al. [12]. In their formulation, each candidate site in the genome is assigned a label, either "somatic" or "non-somatic", by a classifier that combines 106 features derived from the mapping and genotyping of both tumor and normal reads. The features cover a broad spectrum of statistics reported by read mapping and genotyping algorithms, including allelic counts, which are the most informative source in statistical modeling approaches, and quality control metrics such as depth of coverages, allelic counts breakdown by strand, base quality, mapping quality, genotyping quality score, and others. In order to train the classifier, a set of SNAs were generated by whole exome sequencing of 48 breast cancer patients. In the initial step, variants in tumor were predicted by allelic counts and liberal thresholds. Then, the mutations identified were followed-up by re-sequencing experiments with $\sim$6000$\times$ coverage. A mutation was labeled as "somatic" if validated, and "non-somatic" if it was found to be wild-type or a germline variation. In this way, a training set of 1015 somatic mutations and 2354 non-somatic sites were compiled. Different machine learning algorithms, including Random Forests, Bayesian additive regression trees, support vector machines, and logistic regression, were employed to predict somatic mutations. In cross-validation studies, all four machine learning methods outperformed the subtraction method. However, the performance of these methods was not compared with any method based on to statistical modeling approaches.

19.3 Copy Number Aberration Detection

During carcinogenesis, there are often alterations of the dosage and/or structure of tumor suppressor genes or oncogenes in cancer cells through somatic chromosomal alterations. Identifying genomic regions with recurrent copy number alterations (gains and losses) in tumor genomes is an efficient way of tracing cancer driver genes. Ideally, such characterization should include both the precise identification of the chromosomal breakpoints of each alteration and the absolute estimation of copy numbers in each chromosomal segment. Recent advances in massively parallel sequencing provide a powerful alternative to DNA microarrays for detecting copy-number alterations. Sequencing-based approaches not only provide a comprehensive and unbiased survey of all genomic variations, but also enable the detection of both CNAs and SNAs simultaneously in one sample, which may potentially offer critical information for the understanding of cancer genome evolution.

19.3.1 GC Content and Mappability Issue

Sequencing coverage is dependent on the characteristic of the local DNA sequence. Among many factors, GC content and mappability [13] are the two main factors contributing to the inhomogeneity of the sequence depth.

It has been observed that the sequence depth has a unimodal relationship with GC-content, where regions with high or low GC-content manifest decreased sequence coverage [47]. Such bias makes the sequence depth fluctuate even when there is no change in copy number. To differentiate the true deletions/amplifications from under/over-sequenced regions, it is necessary to adjust for the baseline fluctuation in the sequencing data before applying any CNA detection method. Most published methods correct for GC-content by adjusting the sequencing depth in the binned window using the GC-content of that window [2, 47]. More specifically, a curve describing the conditional mean count per GC value is estimated, which leads to the prediction of the count for each bin based on its GC value. This strategy may be inadequate as the choice of bin sizes is often set arbitrarily to accommodate downstream analyses. Because no prior knowledge about the GC effect is utilized in these binning approaches, sometimes the key features of the GC curve have been overlooked or even completely missed in the estimation [5].

To address this limitation, Benjamini and Speed [5] recently proposed a method that produces count predictions at the base pair level for Illumina sequencing data, which allows strand-specific GC-effect correction regardless of downstream smoothing or binning. More specifically, they consider "single position models" to estimate the effect of GC on the fragment counts for individual locations and seek a parsimonious description for this family of models. In their models, the expected count of fragments starting (5' end) at x depends on the GC count in a window starting a bp from x. Each single position model can be characterized by the shift a and the length l of its "driving" window.

Let $W_{a,l}$ denote the model in which the fragment count starting at x depends on the GC between $x+a$ and $x+a+l$. Let $GC_{x+a,l}$ denote the GC count of the l bp window starting at $x+a$. The model $W_{a,l}$ has $l+1$ rate parameters, $\lambda_0,\ldots,\lambda_l$, corresponding to windows with GC count $g=0,\ldots,l$. To estimate those rate parameters, they first take a large random sample of mappable locations (millions) from the genome and remove regions with either zero or extremely high counts from the sample. Then the sample is stratified according to the GC of the reference genome: if $g = GC(x+a,l)$ then position x is assigned to stratum S_g. Using N_g to denote the number of positions assigned to S_g and F_g denote the total number of fragments (reads) starting at the x's in S_g, the ratio λ_g can be estimated by $\hat{\lambda}_g = F_g/N_g$. Each choice of GC window will correspond to a prediction for counts in genomic regions. To compare the quality of correction from different models, they propose to use the normalized 'total variation distance' (TV) between the stratified estimate ($W_{a,l}$) and a uniform rate (U, equal to the global mean rate in the sample and estimated by $\hat{\lambda} = F/n$, where F denotes the total number of reads and n denotes the total number of mappable locations.):

$$TV(W_{a,l}, U) = \frac{1}{2*\hat{\lambda}} \sum_{gc=0}^{l} \frac{N_{gc}}{n} |\hat{\lambda}_{gc} - \hat{\lambda}|. \tag{19.16}$$

This metric can be seen as the total variation distance between the empirical distribution for a single fragment under specific GC categories and a uniform distribution, which measures the proportion of fragments influenced by the stratification. Thus a model with high TV indicates that counts are strongly dependent on GC under this particular stratification. In other words, correcting for this model would best correct the GC dependence. Note that the use of TV scores enables the search of the best correction model for each dataset separately. The final prediction of mean rate λ_x for position x using model $W_{a,l}$ is $\hat{\lambda}_{GC(x+a,l)}$ if x is uniquely mappable, i.e., the average of all such numbers in x's with the same value of $GC(x+a,l)$. Therefore, the corrected counts will be the observed counts divided by λ_x. To take into account the effect of read lengths, one can further use fragment length models to fit separate single position models for fragment (reads) of different length. The above method is implemented as an R package GCcorrect and is available for download from http://www.stat.berkeley.edu/∼yuvalb. Another single position based model, BEADS [8], generates mean rates for the observed reads rather than the genomic locations. This may overlook the uncovered locations from sequencing and lead to inadequate correction in the regions of deletion.

Due to the non-linearity of the GC effect, the pair of normal and tumor samples may not have the same GC curves [5]. Thus before applying two-sample correction methods, where counts in tumor sample are corrected by counts in the normal sample, the GC effects of those samples need to be carefully studied.

Mappability is another issue that may complicate CNA detection. Next generation sequencing technology usually generates short reads less than 200bp. When aligned to the reference genome, reads that are mappable to repetitive regions may be inevitably discarded as it is difficult to determine their locations without ambiguity. Thus the sensitivity to detect the CNAs is compromised in repeated/segmental-duplicated regions [23]. This may be exacerbated when mutations or sequencing errors occur in those regions that cause reads to be mapped incorrectly. In the Pilot 2 studies (trios studies) in the 1,000 Genomes Project [4], about 20 % of the reference genome was considered inaccessible (defined as regions with many ambiguously mapped reads or unexpected high or low number of mapped reads) [1]. The detection of CNAs in those regions is still poorly studied [43].

Alkan [3] proposed a new alignment strategy, called mrFAST, which aligns sequence reads to a repeat-masked reference genome, where all loci with known highly repetitive sequences are masked before alignment, and reports all locations for multi-reads. A similar approach has been used in Sudmant [41] to identify CNAs within segmental duplications. However, these methods only work for deeply sequenced data ($>20\times$) [43]. Extending the read lengths may potentially improve the mappability in repeat regions; however even with a read length of 1kb, over 1.5 % of the human genome sequence still can not be uniquely covered [36].

19.3.2 CNA Identification by Change Point Detection Methods

Many methods for identification of CNAs with paired tumor-normal sequencing data follow the change-point detection paradigm. This is natural since change points in depth of coverage (DOC) reflect the breakpoints of CNAs. Thus those methods also called DOC methods. One drawback of DOC methods is that they can not detect copy-neutral events, such as copy-neutral Loss of Heterogeneity(LOH), where one parental copy is lost but the copy number stays unchanged. A dominant strategy to handle sequencing data is binning or imposing fixed local windows. For example, Event-Wise Testing [47] uses a fixed window to scan the GC-content-adjusted read counts; SegSeq [9] also scans the genome by a sliding fixed window and reduces counts data into the ratio of read counts in the paired tumor and normal sample; CNAseg [16] applies Hidden Markov Model segmentation using read counts in fixed windows. The strategy relies heavily on the assumption that the generative process of sequencing is uniform, where the number of reads or the read ratio in each window is assumed to follow the same parametric distribution across the genome and is proportional to the number of copies. However, this assumption hardly holds due to sequencing biases. Moreover, due to inhomogeneity of sequencing reads, fixed window size may not be the optimal for varied regions across the genome. Some methods, such as SegSeq, apply methods developed for array based data after binning, in which CNA signals are modeled as approximately normal random variables with shifts in mean. However, sequencing data are realizations of point processes and CNA signals can be represented as shift in intensity [38]. A direct modeling of this process or a nonparametric model without assuming homogeneity may be more precise and efficient. In this subsection, we describe two recent efforts to address the heterogeneity of sequencing data, namely, seqCBS [38] and BIC-seq [46].

19.3.2.1 seqCBS

SeqCBS models the sequencing process directly with non-homogeneous Poisson process (NHPP). Let $\{X_t | t \leq T\}$ and $\{Y_t | t \leq T\}$ denote the number of reads whose first bases map to the left of genomic location t for the tumor and paired normal samples, respectively. Then:

$$\{X_t\} \sim NHPP(\mu_t), \tag{19.17}$$

$$\{Y_t\} \sim NHPP(\lambda_t). \tag{19.18}$$

The NHPP model can be viewed as a Poisson process with rate parameter μ_t (or λ_t) such that the rate parameter of the process is a function of genomic location t in base pairs. The conditional probability of an event at position t being from $\{X_t\}$ given that such an event is from either $\{X_t\}$ or $\{Y_t\}$ can be expressed as:

$$p(t) = \frac{\mu_t}{\mu_t + \lambda_t} = p_k \quad \text{if} \quad t_k \leq t \leq t_{k+1}, k = 1, \ldots, K,$$

where t_k belongs to a collection of change points that lie within the observation window on the genome: $0 = t_0 < t_1 < \ldots < t_{K+1} = T$. This can be equivalently expressed as

$$\mu_t = \lambda_t f(t), \tag{19.19}$$

where $f(t) = p(t)/[1 - p(t)]$ is piecewise constant with change points $\{t_k\}$. One interpretation of the above equation is that the fluctuation of sequence depth in the tumor sample is the same as that in the paired normal sample. This assumption is reasonable when paired samples are sequenced by the same protocol and pre-processed by the same alignment procedure. Let m denote the total number of sorted sequences from both tumor and normal samples, j denote the indices of the m sequences, W_j denote the genomic location of sequence j, and Z_j be the indicator of whether the jth sequence comes from tumor sample. Since estimating the change points on genomic locations $\{W_1, \ldots, W_m\}$ is equivalent to doing so for the neighboring pair of reads, the change point model can be defined on the indices $\{1, \ldots, m\}$ of sequence reads. Thus the conditional likelihood only depends on $\{Z_j\}$:

$$p(j) = p_k \quad \text{if} \quad \tau_k \leq j < \tau_{k+1}, \tag{19.20}$$

where τ_k belongs to a collection of change points on sequence indices, i.e., $0 = \tau_0 < \tau_1 < \ldots < \tau_{K+1} = m$.

The problem of searching for change points over $\{1, \ldots, m\}$ could be solved by searching through all possible combinations of $\{\hat{\tau}_k\}$. However, this strategy cannot be scaled up for millions of sequencing reads. Instead, seqCBS adapts Circular Binary Segmentation (CBS) [28] as a greedy search method, where it finds the most significant interval over the entire chromosome, divides the chromosome into three regions, and further scans each of the intervals. For each interval (i,j), seqCBS tests whether the probability that the sequences come from tumor sample within that interval, p_{ij}, is different from the overall probability p. The authors proposed two statistics to test the null hypothesis $p_{ij} = p$. The first statistic is the difference between the number of observed and expected events under the null model given by,

$$S_{ij} = \sum_{k=1}^{m} (Z_k - \hat{p})(1_{i \le k \le j} - \frac{1}{m} \sum_{k=1}^{m} 1_{i \le k \le j}) = \sum_{i \le k \le j} Z_k - \hat{p}(j-i+1), \tag{19.21}$$

where $\hat{p} = \frac{1}{m} \sum_{k=1}^{m} Z_k$. S_{ij} needs to be standardized for comparison between different regions of different sizes. The standardized score statistic T_{ij} is approximately standard normal when $j-i$ is large. When the region has a small number of reads, this normal approximation is not accurate. A second statistic is proposed to improve the accuracy for regions with few reads. Based on the observation that $\sum_{i \le k \le j} Z_k$ is a binomial random variable, the second statistic is derived as an exact binomial generalized likelihood ratio (GLR) statistics,

$$G_{ij} = \sup_{p_0, p_{ij}} l_1(p_0, p_{ij}) - \sup_{p} l_0(p) \tag{19.22}$$

$$= \sum_{k \in [i,j]} \left\{ Z_k \log(\frac{\hat{p}_{ij}}{\hat{p}}) + (1-Z_k) \log(\frac{1-\hat{p}_{ij}}{1-\hat{p}}) \right\}$$

$$+ \sum_{k \notin [i,j]} \left\{ Z_k \log(\frac{\hat{p}_0}{\hat{p}}) + (1-Z_k) \log(\frac{1-\hat{p}_0}{1-\hat{p}}) \right\}, \tag{19.23}$$

where Equation (19.22) represents the difference in log-likelihood between the null model with probability p and the alternative model with probability p_{ij} for the interval $[i,j]$ and p_0 for outside the interval; Equation (19.23) is obtained by replacing the parameters in the binomial log likelihood with their corresponding maximum likelihood estimators (MLEs): $\hat{p} = \sum_{k=1}^{m} Z_k/m$, $\hat{p}_{ij} = \sum_{k \in [i,j]} Z_k/(j-i+1)$, $\hat{p}_0 = \sum_{k \notin [i,j]} Z_k/(m-j+i-1)$.

SeqCBS uses Modified Bayes Information Criterion (mBIC) [49] to choose the number of change points K, which is derived as a large sample approximation to the classic BIC of Schwarz [37]:

$$\text{mBIC}(K) = \log\left(\frac{\sup_{p(t),\tau} L(p(t),\tau)}{\sup_p L(p)}\right) - \frac{1}{2} \sum_{k=0}^{K} \log(\hat{\tau}_{k+1} - \hat{\tau}_k)$$

$$+ \frac{1}{2} \log(m) - K \log(m'), \tag{19.24}$$

where m' is the number of unique locations in $\{W_1, \ldots, W_m\}$. The number of change points will be selected as $\hat{K} = \arg\max_K \text{mBIC}(K)$. Searching of change points by CBS will stop when $\hat{K}$ most significant change points are collected.

19.3.2.2 BIC-seq

BIC-seq is a nonparametric model for detecting CNAs for paired tumor sequencing data. Given read i mapped to the reference genome, let Z_i be the indicator whether read i is from the tumor sample, and W_i be the genomic location. Each read can be represented by a bivariate random variable (Z,W). Thus the joint likelihood of reads $\{1,\ldots,m\}$ on a chromosome can be written as

$$L = \prod_{i=1}^{m} P(Z_i,W_i) = \prod_{i=1}^{m} P(Z_i|W_i)P(W_i) = \prod_{i=1}^{m} p_i^{Z_i}(1-p_i)^{1-Z_i} f(W_i), \tag{19.25}$$

where $p_i = P(Z_i = 1|W_i)$ is the conditional probability of a read from tumor sample given the read being mapped to W_i and $f(W_i)$ is the unknown marginal distribution of W_i. The conditional probability p_i is constant for all W_i in a region flanked by two consecutive change points. BIC-seq uses BIC to select the number of change points K. Let q_k denote the conditional probability for the region flanked by two consecutive change points τ_k and τ_{k+1}. The BIC is derived as:

$$\text{BIC} = -2\log(L) + (K+1)\lambda\log(m) \tag{19.26}$$

$$= -2\sum_{k=0}^{K+1}\left[t_k\log(\hat{q}_k) + n_k\log(1-\hat{q}_k)\right]$$

$$-2\sum_{i=0}^{m} f(W_i) + (K+1)\lambda\log(m), \tag{19.27}$$

where t_k and n_k denote the number of reads between τ_k and τ_{k+1} from tumor and normal, respectively, $\hat{q}_k = t_k/(t_k+n_k)$ is the MLE of q_k and a tuning parameter λ is introduced to give more flexibility to the method. In practice, λ can be tuned according to the sequence coverage. Note that the BIC of any two models can be compared without specifying f. The best model can be found by exhaustedly sampling and comparing models with different $\{\tau_k\}$. Due to computational complexity mentioned earlier, BIC-seq uses a heuristic greedy search procedure to find the change points. Given an initial configuration of small bins (e.g., 10 bp), BIC-seq attempts to reduce the overall BIC by merging neighboring bins. The merging is repeated until the overall BIC cannot be further reduced.

Both seqCBS and BIC-seq provide procedures to derive credible intervals for detected CNAs. Compared to seqCBS, BIC-seq is more computationally efficient since a more greedy searching strategy is used. SeqCBS is available as an R-package that can be downloaded from CRAN at http://cran.r-project.org/web/packages/seqCBS/index.html. The R-package BICseq can be obtained from http://compbio.med.harvard.edu/Supplements/PNAS11.html.

19.3.3 Copy Ratio Estimation and Accounting for Admixture Rate

After obtaining a segmentation of the chromosome, we need to estimate the copy ratio for each chromosomal segment to infer its copy number in the tumor sample. In BIC-seq, the copy ratio ρ_k in the interval of $(\tau_k, \tau_{k+1}]$ is estimated by

$$\rho_k = \frac{\hat{q}_k(1-\hat{\pi})}{\hat{\pi}(1-\hat{q}_k)}, \tag{19.28}$$

where $\hat{\pi} = \sum_k t_k / (\sum_k t_k + \sum_k n_k)$ is the proportion of reads from the tumor sample in the combined samples. However, the estimate of π is not accurate when losses or gains occur on the whole chromosome. In practice, BIC-seq first obtains an estimation from Equation (19.28), removes regions with $|\log_2(\rho_k)| < 0.2$ and re-estimates $\hat{\pi}$. The significance of detected CNA can be assessed by a normal approximation. Under the null hypothesis H_0: $q_k = \pi$,

$$\sqrt{t_k + n_k}(\hat{q}_k - \pi) \sim N(0, \pi(1-\pi)). \tag{19.29}$$

Using $\hat{\pi}$ to estimate π, the p-value is given by

$$2\Phi\left(-\sqrt{\frac{t_k+n_k}{\hat{\pi}(1-\hat{\pi})}}|\hat{q}_k - \hat{\pi}|\right). \tag{19.30}$$

In samples with high sequencing coverage ($\sim$30$\times$), we can use a normal approximation to estimate the copy ratio [7]. With a little abuse of notation, we now use j to index the genomic positions within the segment $(\tau_k, \tau_{k+1}]$. Let t_j be the read depth (RD) at the j-th position of that segment in the tumor sample, n_j be the RD at this position of that segment in the normal sample. Both t_j and n_j are Poisson distributed and approximately modeled by $t_j \sim N(\mu_t, \mu_t)$ and $n_j \sim N(\mu_n, \mu_n)$. Let the ratio at the j-th position be $r_j = t_j / n_j$. Then the Geary-Hinkley transformation

$$T_j = \frac{\mu_n r_j - \mu_t}{\sqrt{\mu_n r_j^2 + \mu_t}} \tag{19.31}$$

follows an approximately standard normal distribution. Let $\rho = \mu_t / \mu_n$ be the true copy ratio in this segment. We can estimate ρ by its MLE

$$\hat{\rho} = \arg\min_{\rho} \sum_j \left(\frac{(r_j - \rho)^2}{r_j^2 + \rho}\right). \tag{19.32}$$

However, sometimes the estimated copy ratio may not directly reflect the true copy number because of the heterogeneity of tumor samples. Compared to normal samples, tumor samples are: (i) nearly always intermixed with an unknown fraction of normal cells (admixture rate); and (ii) undergoing subclonal evolution [7] contributing to the heterogeneity of cancer cell populations. To get an accurate quantification of copy number, it is necessary to account for the admixture rate of tumor samples. Several methods have been proposed to estimate the admixture rate, using information from either CNAs or SNAs. Here we first introduce a method using information from CNAs, called SomatiCA [7], and then give an overview of alternative methods at the end of this subsection.

In tumor samples, clonal events refer to the CNAs that occur in all cancer cell population whereas subclonal events refer to CNA that only occur in a proportion of cancer cells owing to the tumor evolution. As a consequence, in the pure tumor samples, we expect copy number for clonal CNAs to be integer levels whereas copy numbers for subclonal CNAs may not be integers. Taking into account the contamination from normal samples, the copy ratios of clonal segments are centered around a certain discrete level whereas those of subclonal segments have no constraints. The basic idea in SomatiCA is that each genomic segment can be either assigned as clonal or subclonal based on its copy ratio, and the proportion of intermixed normal cells can be estimated from the shift of copy ratios of clonal SCNAs from their expectations in the pure and homogeneous tumor samples.

SomatiCA estimates the admixture rate with a Bayesian normal mixture model. With K segments called from change point methods and each segment with copy ratio ρ_k, we introduce somatic copy level for each segment as $\eta_k = 2\rho_k$ for convenience, since the expectation of which is an integer for clonal events in pure and homogeneous tumor samples. Assume there are S integer copy number levels in the tumor sample indexed by s belonging to a set L. For example, in a tumor sample with CNAs including one copy loss, one copy gain and double deletion, the number of integer copy number levels is 4 (including copy number 0, 1, 2, and 3). Each η_k is assumed to arise from one of the S integer copy number levels. Define $\{G_k : k = 1,\dots,K\}$ as indicators of copy number levels.

For each segment k, SomatiCA models G_k by

$$G_k|\Theta \sim \text{Multinomial}(\Theta), \tag{19.33}$$

where $\Theta = \{\theta_s\}$ specify the expected fraction allocation to each level. SomatiCA further puts the conjugate prior on the multinomial distribution, Dirichlet prior $\text{Dir}(1/S,\dots,1/S)$ on Θ, which means the allocation of copy levels is mainly driven by the input data. Given $G_k = s$, η_i is modeled by

$$\eta_i|(G_i = s, \nu_s) \sim N(\nu_s, \sigma^2), \tag{19.34}$$

where $\nu_s \sim N(L_s, \tau^2)$.

Under the above model, the posterior distribution of ν_s is given by:

$$\nu_s|\{\rho\},\{G\} \sim N\left(\frac{\sigma^2 L_s + \tau^2 \sum_{k:G_k=s}\rho_k}{\sigma^2 + \#\{k : G_k = s\}\tau^2}, \frac{\sigma^2\tau^2}{\sigma^2 + \#\{k : G_k = s\}\tau^2}\right), \tag{19.35}$$

where # denotes the cardinality of the set. The posterior distribution of Θ is given by:

$$\Theta|\{G\} \sim \mathrm{Dir}\Big(\frac{1}{n}+\#\{G_k=1\},\ldots,\frac{1}{n}+\#\{G_k=s\}\Big). \tag{19.36}$$

The Metropolis-Hasting algorithm is used to infer the allocation of copy levels. The segment with ambiguous assignment reflected in posterior probability will be assigned as potential subclonal events and excluded from the estimation of admixture rate. Denote the set of indices of candidate subclonal segments by E. Then the admixture rate ζ is estimated by

$$\hat{\zeta} = \arg\min_{\zeta} \sum_{k\notin E} \Big((1-\zeta)*G_k/2-\zeta+\zeta_i/2\Big)^2. \tag{19.37}$$

In practice, the number of components n can be set as the number of nearest integer levels for all η_k. Or it can be estimated from some model selection procedure. Hyperparameters σ^2 and τ^2 reflect the tolerance of the shift of the copy level of clonal events from integer levels, for example, $\sigma^2=\tau^2=0.01$ means the tolerance is about 0.1. Moreover, the minimum distance between v_s can be constrained to avoid over-fitting, for example, to restrain from reporting any case with normal contamination greater than 80 %. In addition, SomatiCA assumes that the copy ratio of 1 corresponds to the integer copy number of 2. This assumption does not hold when the paired tumor and normal sample are sequenced at very different sequencing depths.

For other methods accounting for tumor purity, ExomeCNV [34] estimates the admixture rate based on the largest Loss of Heterozygosity (LOH) region. in a genome, which likely produces a biased estimation. ASCAT [44] and ABSOLUTE [6], are two methods developed on arrays that are similar in spirit to SomatiCA, which can be also applied to sequencing data. Besides admixture rate, both methods model the global measure of tumor ploidy, and their input, copy ratio, is defined as a quantity measures the local DNA dosage conditioning on the aneuploidy of the tumor. ASCAT can be seen as directly using Equation (19.37) to find a solution to minimize the distance of nearest integer level and observed levels for all segments where subclonal heterogeneity is not considered. Moreover, it has low tolerance to normal contamination and tends to underestimates the admixture rate. ABSOLUTE uses a different formulation of a Bayesian normal mixture model, with the main difference being that it assumes a uniform distribution on subclonal events and it constrains the genomic mass allocated to each copy level. PurityEst [40] and PurBayes [19] are methods for estimating tumor purity based on SNAs, which can be applied as alternative approaches in practice. The above methods assume a single clonal cancer population and estimate tumor purity and subclonality based on an identified clonal cancer population. This assumption may be violated when there are multiple clonal cancer genomes within a sequencing profile. THetA [27] is a method developed to address this problem, which supports deconvolution of the tumor genome mixture to a normal genome and any number of cancer genomes.

19.3.4 CNA Identification by Other Methods

Other methods detect CNAs by taking advantage of sequencing design or de novo assembly, including paired-end mapping (PEM), split-read (SR) and assembly-based (AS) methods [43].

In PEM methods, reads are required to be paired with an insert size ranging from 200–300 bp to approximately 3–5 kb [46]. When the sequenced paired ends map to the reference with a distance longer than expected, it may suggest the existence of a deletion between the two ends. Similarly, if the distance is shorter than expected, it may suggest an insertion. The size of CNAs detected by PEM methods depends on the insert size. Therefore, PEM methods often identify CNAs with smaller size compared to DOC methods. SR methods also utilize paired end sequences but focus on pairs where only one read uniquely mapped to the genome. The mapped read is then used as an anchor to narrow down the search space for the unmapped read [43], the location of which may indicate the breakpoint of CNAs. AS methods rely on a genome *ab initio* assembled from the sequencing reads. The CNAs can be identified by comparing assembled tumor and normal genomes. These methods are complementary to one another and complementary to DOC methods. To discover different types of CNAs with a broad range of sizes, these methods can be jointly applied in practice. As shown in some studies [24], combination of more than one sequence signature can significantly improve the detection of CNAs. Computational methods CNVer [22], HYDRA [30] and SVDetect [48] incorporate PEM information into DOC methods. Genome STRiP [14] combines information from DOC, PEM, SR and other sequence features. We refer interested readers to reviews written by Teo [43] and Medvedev [23] for more information of these methods.

19.3.5 A Case Study

Here we use the TCGA mutation calling benchmark 4 datasets to illustrate the usage of some introduced software. This genome sequencing benchmark dataset consists of artificially mixed samples with the proportion of tumor samples (a cancer cell line, HCC1143) in a gradient from 5 % to 95 %. We focus our analysis on the sample `n20t80` (mixed with 80 % of the HCC1143 sample and 20 % of the normal sample) and its paired normal sample. The corresponding sequencing data in the BAM format `HCC1143.n20t80.bam` and `HCC1143.normal.bam` can be downloaded from https://cghub.ucsc.edu/datasets/benchmark_download.html. A BAM file is the binary version of a SAM (Sequence Alignment/Map) file, which is a tab-delimited text file that stores sequence alignment data. To manipulate alignments in the BAM format, variant data in the VCF format (Variant Call Format) and interval data in BED (Browser extensible data) format, we use utilities provided by SAMtools [20], VCFtools [11], and BEDTools [31]. In the following analysis, we assume those tools have been installed.

We demonstrate the analysis of SNAs using VarScan2. The somatic variant calling features of VarScan2 expect input in SAMtools pileup or mpileup format. Since the BAM file in this example is larger than 200GB in size, to save memory, we process the file by each chromosome separately. In the following analysis, we focus on chromosome 1.

Suppose we have reference sequences in `Homo.sapiens.assembly19-.fasta`; the following commands generate mpileup files for the tumor and the normal samples, respectively:

```
samtools mpileup -C50 \\
-f Homo.sapiens.assembly19.fasta \\
-r 1 HCC1143.n20t80.bam > tumor_chr1.mpileup
samtools mpileup -C50 \\
-f Homo.sapiens.assembly19.fasta \\
-r 1 HCC1143.normal.bam > normal_chr1.mpileup
```

where the `-C50` option reduces the effect of reads with excessive mismatches and the `-r` option specifies the chromosome.

Then we use VarScan2 to call variants and identify their somatic status (Germline /LOH /Somatic) using pileup files:

```
java -jar VarScan.jar somatic normal_ch1.mpileup \\
tumor_chr1.mpileup somatic.chr1 \\
--tumor-purity 0.8
```

where the `-tumor-purity` option sets the estimate of the percentage of cancer cells in the tumor sample and `somatic.chr1` is the base name for the output.

After running the above command, we obtain two files named `somatic.chr1.snp` and `somatic.chr1.indel` in the working directory, which store the calling results for somatic mutations and indels (short insertion and deletions), respectively. We further filter the somatic variants to remove clusters of false positives and SNAs calls near indels using VarScan2 utility `somaticFilter`:

```
java -jar VarScan.jar somaticFilter somatic.chr1.snp\\
--min-coverage 10 \\
--p-value 0.00001 \\
--indel-file somatic_chr1.indel \\
--output-file chr1.filtered
```

where the `-min-coverage` option sets the minimum read depth to support the variants and `-p-value` specifies the p-value threshold for calling variants. The minimum supporting reads for a variant, the minimum average base quality for variant-supporting reads and the minimum variant allele frequency threshold are set at the default values. The numbers of variants filtered out by each step are summarized by the following message from VarScan2:

```
Window size: 10
Window SNPs: 3
Indel margin: 3
Reading input from somatic.chr1.snp
940 cluster SNPs identified
Reading input from somatic.chr1.snp
272578 variants in input stream
11365 failed to meet coverage requirement
15416 failed to meet reads2 requirement
6372 failed to meet varfreq requirement
238624 failed to meet p-value requirement
8 in SNP clusters were removed
1 were removed near indels
792 passed filters
```

Variants that pass the filters are saved in the output file `chr1.filtered`.

We then check whether these somatic mutations affect protein coding regions by RefSeq gene annotations [29] using BEDTools. First we convert `chr1.filtered` into the BED format by:

```
grep  "Somatic" chr1.filtered | \\
awk '{print $1 "\t" $2 "\t" $2 "\t" $3 "\t" $4}' \\
> chr1.filtered.bed
```

Here we show first several lines of `chr1.filtered.bed` as an example of the BED format:

```
1 449862 449862 T C
1 990394 990394 A C
1 4800099 4800099 G A
1 5179714 5179714 C T
1 5541162 5541162 G A
```

The first three fields, chromosome, start position and end position, are required in the BED format.

Suppose *RefSeq* gene annotations are stored as the BED format in `refseq.bed`, the following command returns the gene annotation result for `chr1.filtered.bed`:

```
intersectBed -a refseq.bed \\
-b chr1.filtered.bed \\
-wa -wb > chr1.filtered.refseq.bed
```

To determine whether these identified somatic variants are deleterious, one can further consult commonly used prediction methods such as SIFT [26] and PolyPhen []. Although driver mutations might be in the coding regions of the genome, some may be in regulatory elements and other non-coding sequences. Whenever the related annotations are available in the BED format, we can use BEDTools utility `intersectBed` to annotate somatic variants as shown above.

In the second part, we demonstrate the analysis of CNAs using SomatiCA. SomatiCA is a R package that is capable of identifying, characterizing, and quantifying somatic CNAs from cancer genome sequencing. It expects both read depths and lesser allele frequencies (LAF) from SNPs for the paired tumor-normal sample.

We first call SNPs and short indels from `HCC1143.normal.bam`:

```
samtools mpileup -C50 -uf \\
-r 1 Homo.sapiens.assembly19.fasta \\
HCC1143.normal.bam | \\
bcftools view -bvcg - > normal.chr1.bcf
bcftools view normal.chr1.bcf \\
| vcfutils.pl varFilter -D300 > normal.chr1.vcf
```

where the -C50 option reduces the effect of reads with excessive mismatches and the -D option sets the maximum read depths to call a SNP.

After SNP calling, we filter out low quality variants with quality scores less than 10 by the following commands:

```
bgzip normal.chr1.vcf
tabix -p vcf normal.chr1.vcf.gz
vcf-annotate --filter Qual=10 \\
normal.chr1.vcf.gz > normal.chr1.filter.vcf
```

Then we retrieve read depths from SNPs by:

```
grep -v "INDEL" normal.chr1.filter.vcf | grep \\
-w "PASS\|#\|CHROM" | bgzip \\
-c > normal.chr1.pass.vcf.gz
vcf-query \\
-f '\%POS\t\%INFO/DP4\t\%INFO/DP\t[\%GTR\t]\n' \\
normal.chr1.pass.vcf.gz \\
> normal.chr1.pass.vcf
cut -f1-4 normal.chr1.pass.vcf | \\
awk 'NR==1{$0="POS\tTUMOR.DP4 \\
\tTUMOR.DP\tTUMOR.GT\t\n"$0}1' \\
> normal.chr1.RD.vcf
```

`normal.chr1.RD.vcf` contains genomic position, number of reads supporting A,T,C,G, total number of reads and genotype for each SNP as shown in the following:

```
POS TUMOR.DP4 TUMOR.DP TUMOR.GT
10052 10,3,2,2 18 0/1
11597 0,1,20,0 21 1/1
11637 7,1,19,0 27 0/1
11900 25,13,13,10 64 0/1
```

Applying the same procedure to sample `n20t80`, we obtain `tumor.chr1.RD.vcf`. We can combine lines of two VCF files based on genomic locations using:

```
join -j 1 <(sort normal.chr1.RD.vcf) \\
<(sort tumor.chr1.RD.vcf) | sort \\
-n > tumor.normal.chr1.vcf
```

`tumor.normal.chr1.vcf` needs to be further converted into the SomatiCA input format as the following (consulting SomatiCA manual for details):

```
chr1 16378 het 91 0.40 85 0.46
chr1 28563 hom 12 0.5 14 0.43
chr1 52238 hom 3 0 2 0
chr1 54676 het 23 0.26 17 0.41
chr1 54708 het 21 0.19 16 0.31
```

where the 7 required fields are chromosome, genomic positions, zygosity, total read counts for tumor sample, LAF for tumor sample, total read count for normal sample and germline LAF.

Suppose the required information for SNPs from all chromosomes is stored in `somatica.input.txt`, we can call CNAs and analyze the sample purity using the following commands in R.

```
> library(SomatiCA)
> y <- read.table("somatica.input.txt", as.is=T)
> colnames(y) <- c("seqnames", "start", "zygosity",
+ "tCount", "LAF", "tCountN", "germLAF")
> input <- SomatiCAFormat(y, missing = T,
+ verbose = T)
> seg <- larsCBSsegment(input, collapse.k = 0,
+ ncores = 1, verbose = T, rss=F)
> segmentwithRatio <- somaticRatio(seg$segment,
+ input, method = "mle", adjust=T)
> refined <- refineSegment(segmentwithRatio, input)
> rate <- admixtureRate(refined)$admix
```

The estimate of admixture rate for sample `n20t80` is 0.23. We can further correct the CNA calls by the estimated admixture rate using:

```
> z <- copynumberCorrected(refined, rate)
```

SomatiCA provides utilities to estimate clonality for each somatic copy number aberration.

```
> data(GCcontent)
> segmentGCcorrected <- segmentGCbiasRemoval(
+ z, input, GCcontent)
> segmentClonality <- subclonality(
+ segmentGCcorrected, rate)
```

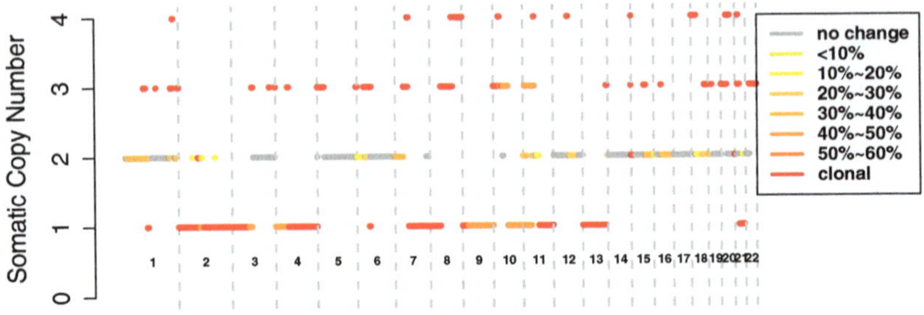

Fig. 19.1 Visualization of subclonality and somatic copy number for HCC1143 sample `n20t80`

The subclonality and somatic copy number for a sample can be visualized by `plotSubclonality` as shown in Fig. 19.1.

19.4 Discussion

In this chapter, we have surveyed some recent developments in the analysis of next generation sequencing data with paired tumor and normal samples, particularly for SNA and CNA detection. Next generation sequencing technologies provide an unprecedented opportunity to scrutinize the genome at a single base pair resolution at an economically feasible cost. Despite many methods that have been developed, there is still a great need for more efficient and accurate methods to enable the detections of SNAs and CNAs.

The detection of SNAs is an important first step in cancer genome analysis. Various SNA detection methods differ in their principles and approaches, and a comparison of their performance on sensitivity, specificity, and computational speed is greatly needed. Such results will help end users to choose appropriate computational tools for their studies. Recently, Roberts [32] compared four SNA callers, VarScan2, SomaticSniper, JointSNVMix2, and Strelka, through their applications to a whole exome sequencing dataset of a chronic myeloid leukemia patient. Without a gold standard set of somatic mutations, the comparison has been oriented to assess the consistency between methods instead of precisely addressing sensitivity and specificity. The authors concluded that output of each algorithm has "significant differences and contradictions". They suggested that the use and interpretation of the results of any individual algorithm should be made with caution. These algorithms were also evaluated by applying them to a sequence dataset of a non-cancerous sample, which was randomly split into two sets to mimic tumor/normal matched data. Thus any SNA detected in this dataset is a false positive discovery. The four algorithms detected 5, 7, 10 and 11 mutations, respectively, indicating very high specificity. In practice, the output of SNA detection algorithms is substantially filtered with various quality control metrics to remove false positive discoveries [10, 17, 32]. Although the filtering strategies differ among methods, they generally fall into the following categories: (1) base calling quality; (2) alignment quality; (3) strand bias; (4) occurrences in normal samples, such as documented in dbSNP, or platform specific control panel; and (5) depth of coverage. These filters are demonstrated to effectively reduce the number of false positive discoveries [10, 17, 32].

Currently, there is no standard protocol or quality control measures for CNV/CNA detection [43]. In the 1,000 Genomes Project, DOC, PEM and SR methods are used and each uniquely contributes to the identification of 30∼60 of the reported CNVs [3]. In addition, their performances largely depend on what alignment algorithm is used in pre-processing, especially, for repeat regions.

Here we list several statistical challenges associated with detecting CNAs using next generation sequencing data:

1. **GC-content bias correction**. Current correction methods accounts for a large portion of the variation, but additional heterogeneity such as unexplained hot spots and zero-counts needs to be incorporated. Furthermore, how the selection of the alignment algorithm affects the GC effect needs to be further investigated [5].
2. **Fast computation**. DOC methods following the change point paradigm essentially solve a combinational problem. With sequencing capacity increasing rapidly, even existing greedy search methods, such as CBS, may not be feasible. It is important to develop faster solutions to the change point detection problem.
3. **Quality/False discovery control**. New statistics are needed to control the quality of detected CNAs.
4. **Multiple sequence change point detection**. Simultaneous segmentation of multiple samples may increase the power for the detection of recurrent CNAs but raises new challenges such as modeling batch effects. Related recent developments on array data can be found in [39] and [50].
5. **Heterogeneity of tumor samples**. Accounting for normal contamination and subclonal heterogeneity is essential for characterization of CNAs. This problem is still not completely solved.

Progress in the above areas will help cancer researchers to better analyze next generation sequencing data to identify somatic mutations in cancer tissues that may lead to better understanding of cancinogenesis in the future.

Acknowledgements This work was supported in part by the National Institutes of Health grants R01-GM59507, UL1 RR024139, and P01CA154295 (L.H. and H.Z.) and a Scholarship from the Chinese Scholarship Council (M.C.).

References

[1] Abecasis, G., Altshuler, D., Auton, A., Brooks, L., Durbin, R., Gibbs, R.A., Hurles, M.E., McVean, G.A., Bentley, D., Chakravarti, A., et al.: A map of human genome variation from population-scale sequencing. Nature **467**(7319), 1061–1073 (2010)

[2] Abyzov, A., Urban, A.E., Snyder, M., Gerstein, M.: Cnvnator: an approach to discover, genotype, and characterize typical and atypical cnvs from family and population genome sequencing. Genome Res. **21**(6), 974–984 (2011)

[3] Alkan, C., Coe, B.P., Eichler, E.E.: Genome structural variation discovery and genotyping. Nat. Rev. Genet. **12**(5), 363–376 (2011)

[4] Autosomes Chromosome, X.: An integrated map of genetic variation from 1,092 human genomes. Nature **491**, 1 (2012)

[5] Benjamini, Y., Speed, T.P.: Summarizing and correcting the gccontent bias in high-throughput sequencing. Nucleic Acids Res. **40**(10), e72 (2012)

[6] Carter, S., Cibulskis, K., Helman, E., McKenna, A., Shen, H., Zack, T., Laird, P., Onofrio, R., Winckler, W., Weir, B., et al.: Absolute quantification of somatic dna alterations in human cancer. Nat. Biotechnol. **30**(5), 413–421 (2012)

[7] Chen, M., Gunel, M., Zhao, H.: Somatica: identifying, characterizing and quantifying somatic copy number aberrations from cancer genome sequencing data. PloS one 8(11), e78, 143 (2013)

[8] Cheung, M.S., Down, T.A., Latorre, I., Ahringer, J.: Systematic bias in high-throughput sequencing data and its correction by beads. Nucleic Acids Res. **39**(15), e103–e103 (2011)

[9] Chiang, D.Y., Getz, G., Jaffe, D.B., O'Kelly, M.J., Zhao, X., Carter, S.L., Russ, C., Nusbaum, C., Meyerson, M., Lander, E.S.: High-resolution mapping of copy-number alterations with massively parallel sequencing. Nat. Meth. **6**(1), 99–103 (2008)

[10] Cibulskis, K., Lawrence, M.S., Carter, S.L., Sivachenko, A., Jaffe, D., Sougnez, C., Gabriel, S., Meyerson, M., Lander, E.S., Getz, G.: Sensitive detection of somatic point mutations in impure and heterogeneous cancer samples. Nat. Biotechnol. **31**(3), 213–219 (2013)

[11] Danecek, P., Auton, A., Abecasis, G., Albers, C.A., Banks, E., DePristo, M.A., Handsaker, R.E., Lunter, G., Marth, G.T., Sherry, S.T., et al.: The variant call format and vcftools. Bioinformatics **27**(15), 2156–2158 (2011)

[12] Ding, J., Bashashati, A., Roth, A., Oloumi, A., Tse, K., Zeng, T., Haffari, G., Hirst, M., Marra, M.A., Condon, A., et al.: Feature-based classifiers for somatic mutation detection in tumour–normal paired sequencing data. Bioinformatics **28**(2), 167–175 (2012)

[13] Dohm, J.C., Lottaz, C., Borodina, T., Himmelbauer, H.: Substantial biases in ultra-short read data sets from high-throughput dna sequencing. Nucleic Acids Res. **36**(16), e105–e105 (2008)

[14] Handsaker, R.E., Korn, J.M., Nemesh, J., McCarroll, S.A.: Discovery and genotyping of genome structural polymorphism by sequencing on a population scale. Nat. Genet. **43**(3), 269–276 (2011)

[15] Hansen, N.F., Gartner, J.J., Mei, L., Samuels, Y., Mullikin, J.C.: Shimmer: detection of genetic alterations in tumors using next-generation sequence data. Bioinformatics **29**(12), 1498–1503 (2013)

[16] Ivakhno, S., Royce, T., Cox, A.J., Evers, D.J., Cheetham, R.K., Tavaré, S.: Bioinformatics **26**(24), 3051–3058 (2010)

[17] Koboldt, D.C., Zhang, Q., Larson, D.E., Shen, D., McLellan, M.D., Lin, L., Miller, C.A., Mardis, E.R., Ding, L., Wilson, R.K.: Varscan 2: somatic mutation and copy number alteration discovery in cancer by exome sequencing. Genome Res. **22**(3), 568–576 (2012)

[18] Larson, D.E., Harris, C.C., Chen, K., Koboldt, D.C., Abbott, T.E., Dooling, D.J., Ley, T.J., Mardis, E.R., Wilson, R.K., Ding, L.: Somaticsniper: identification of somatic point mutations in whole genome sequencing data. Bioinformatics **28**(3), 311–317 (2012)

[19] Larson, N.B., Fridley, B.L.: Purbayes: estimating tumor cellularity and subclonality in next-generation sequencing data. Bioinformatics **29**(14) (2013)

[20] Li, H., Handsaker, B., Wysoker, A., Fennell, T., Ruan, J., Homer, N., Marth, G., Abecasis, G., Durbin, R., et al.: The sequence alignment/map format and samtools. Bioinformatics **25**(16), 2078–2079 (2009)
[21] Li, H., Ruan, J., Durbin, R.: Mapping short dna sequencing reads and calling variants using mapping quality scores. Genome Res. **18**(11), 1851–1858 (2008)
[22] Medvedev, P., Fiume, M., Dzamba, M., Smith, T., Brudno, M.: Detecting copy number variation with mated short reads. Genome Res. **20**(11), 1613–1622 (2010)
[23] Medvedev, P., Stanciu, M., Brudno, M.: Computational methods for discovering structural variation with next-generation sequencing. Nat. Meth. **6**, S13–S20 (2009)
[24] Mills, R.E., Walter, K., Stewart, C., Handsaker, R.E., Chen, K., Alkan, C., Abyzov, A., Yoon, S.C., Ye, K., Cheetham, R.K., et al.: Mapping copy number variation by population-scale genome sequencing. Nature **470**(7332), 59–65 (2011)
[25] Muzny, D.M., Bainbridge, M.N., Chang, K., Dinh, H.H., Drummond, J.A., Fowler, G., Kovar, C.L., Lewis, L.R., Morgan, M.B., Newsham, I.F., et al.: Comprehensive molecular characterization of human colon and rectal cancer. Nature **487**, 330–337 (2012)
[26] Ng, P.C., Henikoff, S.: Sift: Predicting amino acid changes that affect protein function. Nucleic Acids Res. **31**(13), 3812–3814 (2003)
[27] Oesper, L., Mahmoody, A., Raphael, B.J.: Theta: Inferring intra-tumor heterogeneity from high-throughput dna sequencing data. Genome Biol. **14**(7), R80 (2013)
[28] Olshen, A.B., Venkatraman, E., Lucito, R., Wigler, M.: Circular binary segmentation for the analysis of array-based dna copy number data. Biostatistics **5**(4), 557–572 (2004)
[29] Pruitt, K.D., Tatusova, T., Maglott, D.R.: Ncbi reference sequences (refseq): a curated non-redundant sequence database of genomes, transcripts and proteins. Nucleic Acids Res. **35**(suppl 1), D61–D65 (2007)
[30] Quinlan, A.R., Clark, R.A., Sokolova, S., Leibowitz, M.L., Zhang, Y., Hurles, M.E., Mell, J.C., Hall, I.M.: Genome-wide mapping and assembly of structural variant breakpoints in the mouse genome. Genome Res. **20**(5), 623–635 (2010)
[31] Quinlan, A.R., Hall, I.M.: Bedtools: a flexible suite of utilities for comparing genomic features. Bioinformatics **26**(6), 841–842 (2010)
[32] Roberts, N.D., Kortschak, R.D., Parker, W.T., Schreiber, A.W., Branford, S., Scott, H.S., Glonek, G., Adelson, D.L.: A comparative analysis of algorithms for somatic snv detection in cancer. Bioinformatics **29**(18), 2223–2230 (2013)
[33] Roth, A., Ding, J., Morin, R., Crisan, A., Ha, G., Giuliany, R., Bashashati, A., Hirst, M., Turashvili, G., Oloumi, A., et al.: Jointsnvmix: a probabilistic model for accurate detection of somatic mutations in normal/tumour paired next-generation sequencing data. Bioinformatics **28**(7), 907–913 (2012)
[34] Sathirapongsasuti, J.F., Lee, H., Horst, B.A., Brunner, G., Cochran, A.J., Binder, S., Quackenbush, J., Nelson, S.F.: Exome sequencing-based copy-number variation and loss of heterozygosity detection: Exomecnv. Bioinformatics **27**(19), 2648–2654 (2011)
[35] Saunders, C.T., Wong, W.S., Swamy, S., Becq, J., Murray, L.J., Cheetham, R.K.: Strelka: accurate somatic small-variant calling from sequenced tumor–normal sample pairs. Bioinformatics **28**(14), 1811–1817 (2012)
[36] Schadt, E.E., Turner, S., Kasarskis, A.: A window into third-generation sequencing. Hum. Mol. Genet. **19**(R2), R227–R240 (2010)
[37] Schwarz, G.: Estimating the dimension of a model. Ann. Stat. **6**(2), 461–464 (1978)
[38] Shen, J.J., Zhang, N.R.: Change-point model on nonhomogeneous poisson processes with application in copy number profiling by next-generation dna sequencing. Ann. Appl. Stat. **6**(2), 476–496 (2012)
[39] Siegmund, D., Yakir, B., Zhang, N.R.: Detecting simultaneous variant intervals in aligned sequences. Ann. Appl. Stat. **5**(2A), 645–668 (2011)
[40] Su, X., Zhang, L., Zhang, J., Meric-Bernstam, F., Weinstein, J.N.: Purityest: estimating purity of human tumor samples using next-generation sequencing data. Bioinformatics **28**(17), 2265–2266 (2012)

[41] Sudmant, P.H., Kitzman, J.O., Antonacci, F., Alkan, C., Malig, M., Tsalenko, A., Sampas, N., Bruhn, L., Shendure, J., Eichler, E.E., et al.: Diversity of human copy number variation and multicopy genes. Science **330**(6004), 641–646 (2010)

[42] TCGA: Comprehensive molecular portraits of human breast tumours. Nature **490**(7418), 61–70 (2012)

[43] Teo, S.M., Pawitan, Y., Ku, C.S., Chia, K.S., Salim, A.: Statistical challenges associated with detecting copy number variations with next-generation sequencing. Bioinformatics **28**(21), 2711–2718 (2012)

[44] Van Loo, P., Nordgard, S.H., Lingjærde, O.C., Russnes, H.G., Rye, I.H., Sun, W., Weigman, V.J., Marynen, P., Zetterberg, A., Naume, B., et al.: Allele-specific copy number analysis of tumors. Proc. Natl. Acad. Sci.**107**(39), 16,910–16,915 (2010)

[45] Wei, X., Walia, V., Lin, J.C., Teer, J.K., Prickett, T.D., Gartner, J., Davis, S., Stemke-Hale, K., Davies, M.A., Gershenwald, J.E., et al.: Exome sequencing identifies grin2a as frequently mutated in melanoma. Nat. Genet. **43**(5), 442–446 (2011)

[46] Xi, R., Hadjipanayis, A.G., Luquette, L.J., Kim, T.M., Lee, E., Zhang, J., Johnson, M.D., Muzny, D.M., Wheeler, D.A., Gibbs, R.A., et al.: Copy number variation detection in whole-genome sequencing data using the bayesian information criterion. Proceed. Natl. Acad. Sci. **108**(46), E1128–E1136 (2011)

[47] Yoon, S., Xuan, Z., Makarov, V., Ye, K., Sebat, J.: Sensitive and accurate detection of copy number variants using read depth of coverage. Genome Res. **19**(9), 1586–1592 (2009)

[48] Zeitouni, B., Boeva, V., Janoueix-Lerosey, I., Loeillet, S., Legoix-Né, P., Nicolas, A., Delattre, O., Barillot, E.: Svdetect: a tool to identify genomic structural variations from paired-end and mate-pair sequencing data. Bioinformatics **26**(15), 1895–1896 (2010)

[49] Zhang, N.R., Siegmund, D.O.: A modified bayes information criterion with applications to the analysis of comparative genomic hybridization data. Biometrics **63**(1), 22–32 (2007)

[50] Zhang, N.R., Siegmund, D.O., Ji, H., Li, J.Z.: Detecting simultaneous changepoints in multiple sequences. Biometrika **97**(3), 631–645 (2010)

Chapter 20
Statistical Considerations in the Analysis of Rare Variants

Debashis Ghosh and Santhosh Girirajan

Abstract Recently, whole-genome and whole-exome sequencing has begun to demonstrate success in the identification of disease-causing genes. Many of these genes exhibit abnormal genetic behavior and low prevalence in the population; these molecules are commonly referred to as rare variants. In this chapter, we provide an overview of rare variants and their scientific relevance in medicine and public health. We then provide a review of existing methods for association, primarily focusing on the sequence kernel association test (SKAT) and related methods. These procedures are related to kernel machines, which we will also describe. Finally, we discuss the implications of rare variants in terms of multiple testing.

20.1 Introduction

Large-scale genomics has been at the forefront of science and medicine over the last decade. The advent of high-throughput technologies including single nucleotide polymorphism (SNPs) microarrays, array comparative genomic hybridization and genome sequencing have enabled rapid discovery of genetic variants varying in size and frequency [18]. Copy number variants are deletions and duplications in the genome that constitute the most genetic variation, in total base pairs, between individuals [35]. Classically, disease-association studies involved evaluation of either variants of high frequency in the population, also termed common variants, or variants of low frequency or rare variants. In this chapter, we will consider analysis of rare variants with specific focus on copy number variants. One of the key statistical challenges in the analysis of rare variants is that they have small population prevalences. If we view the rare variants as predictors that we

D. Ghosh (✉) • S. Girirajan
Penn State University, University Park, State College, PA 16801, USA
e-mail: ghoshd@psu.edu; sxg47@psu.edu

S. Datta and D. Nettleton (eds.), *Statistical Analysis of Next Generation Sequencing Data*, Frontiers in Probability and the Statistical Sciences, DOI 10.1007/978-3-319-07212-8_20,

wish to associate with a phenotype, then they in fact contain very little statistical information. To illustrate the idea, suppose we wish to regress the phenotype on a rare variant that we treat as binary, where zero indicates absence and one indicates presence. We assume that the regression model is linear. Then it can be shown analytically that the information about the regression coefficient in such a setup is maximized when half of the subjects have the rare variant and half do not. However, by definition, for rare variants, a majority of subjects will not have the rare variant, the implication being that we are in an inherently low-power situation. Thus, it is necessary to begin to think about pooling information in various ways; this will be one of the themes explicated on in the chapter.

The structure of this chapter is as follows. In Sect. 20.2, we provide some biological background to rare variants. Section 20.3 reviews association methods for the analysis of rare variants and in particular focuses on the sequence kernel association test (SKAT) [56] and its extensions. The SKAT methodology is based on the kernel machine framework originally proposed by Liu et al. [33, 34], so we also expand on this. Finally, we discuss the multiple comparisons problem and how its consideration needs to be modified for the rare variant problem in Sect. 20.4. This chapter concludes with some discussion in Sect. 20.5.

20.2 Biological Background

Association of disease genes to phenotypic traits or overt disease has been carried out with the discovery, characterization or genotyping of variants. Genetic studies have relied upon identifying causative genes by finding genetic variants, common (>1 % or >5 % in the population) or rare, and whether they are enriched in cases compared to controls. Common variants are contributed by alleles that originated during the development of humans and are therefore shared between different human populations [39]. These variants constitute most of the human genetic variation, in frequency, and are also represented as SNPs that tag specific haplotypes mapped by the HapMap project [10, 11]. While technologies and genetic methods have concentrated on implicating common or rare variants of extreme size for disease etiology, identification and characterization of variants of intermediate size and frequency remains a challenge [50].

The basis for rare variants can be best understood in a historical context. In the field of human genetics of complex traits, the dominant school of thought in the early 2000s was based on the so-called common disease-common variant (CDCV) hypothesis [45, 46]. This framework postulated that for many diseases, multiple SNPs would be needed to explain a large percentage of variation in the phenotype. Identification of SNPs in linkage disequilibrium or functional variants in the neighborhood of causative genes is the basis of genome-wide association studies (GWAS) [22]. This thought very much influenced the design of GWAS and the technology used to measure DNA variation. The dominant platform for measuring SNPs was the microarray platform, which was being used simultaneously for measuring transcript mRNA expression. The major company that developed the SNP microarray platform was Affymetrix (http://www.affymetrix.com), and the DNA variations selected to be on the chip primarily represented variants that satisfied the CDCV hypothesis, i.e., all of the variants had to be sufficiently present in the population. In particular, what tended to be excluded from the SNP microarrays were DNA variants where the less prevalent form had a population prevalence (termed minor allele frequency) that was less than 5 %.

Currently, there have been over 2000 GWAS studies that have been conducted in humans (genome.gov, 2013) with a major finding that DNA variations in the form of SNPs can only explain a limited amount of variation for several human disease associated phenotypes [37, 38]. GWAS has been only successful in studies on type-2 diabetes, age-related macular degeneration, coronary artery disease, and Crohn disease as well as for obesity and height. These studies were not successful for a majority of common complex diseases including neurodevelopmental disorders such as autism, schizophrenia, and epilepsy. This has led to consideration of reasons for the missing heritability [38]. The difficulty of achieving statistical power to identify multiple loci of small effect sizes is considered as a major factor. Other factors, not considered in traditional GWAS studies [5], such as gene-gene interactions and gene-environment interactions, have also been proposed [57].

Rare variants, on the other hand, tend to have much bigger effects than the DNA variants identified from first-generation GWAS. This alternative model is termed common disease rare variant (CDRV). From an evolutionary point of view, these variants are under strong selection and their frequency in the population is maintained by de novo mutations. In fact, new germline mutations arise constantly, based on the underlying sequencing architecture or age of the parents, at a rate of about 61 base pairs for single nucleotide mutations [4] and 16–50 kbp per diploid genome [24]. Genetic associations for copy number variants have met with higher success for variants of low frequency. These variants were classically associated with clinically recognizable syndromes such as 7q11.2 deletions in individuals with features of Williams syndrome, 22q11.2 deletions in individuals with features of DiGeorge syndrome, and 17p11.2 deletions in Smith-Magenis syndrome [19].

Developments in microarray technology and rapid incorporation of high-throughput genotyping in diagnostic laboratories have resulted in the identification of about two dozen CNVs that are strongly enriched in affected cases with neurodevelopmental impairments compared to controls. However, extensive phenotypic heterogeneity even in individuals carrying the same CNV has complicated further analysis. For example, the 16p11.2 deletion was originally associated with autism, but was later identified to be enriched in individuals with intellectual disability, epilepsy, schizophrenia, and obesity [40, 47, 55]. Comparison of CNV load (measured as the proportion of population carrying a deletion or duplication of a particular size) across cohorts of affected population suggests that the CNV load correlates with the severity of neurodevelopmental disorders [17]. Similarly, phenotypic variability and severity associated with a specific disease-associated CNV can be also explained by rare variants in the genetic background [19, 21]. These variants modulate the ultimate phenotypic expression either by additive or synergistic effect, in genetic terms, in a digenic or oligogenic manner [53] Genome sequencing has made tremendous strides in finding the missing heritability. Sequencing of the protein coding sequences in the genome for neurodevelopmental disorders has identified several rare, de novo variants that cluster in pathways related to nervous system development, maturation, and maintenance [16,41,42,48]. These studies have, however, revealed a complex genetic basis for common diseases; for example, recent estimates suggest that a minimum of 1000 genes as causal for autism. These disorders can be explained by an infinitesimal model consistent with the role of multiple rare variants in complex disease [14]. According to this model the genetic etiology can be explained by a hybrid of the two models. The challenge therefore lies in understanding how these variants work together in causing the disease rather than if they are rare or common [14].

The implications of rare variants for medicine and public health are potentially quite paradigm shifting. Both disciplines have placed a tremendous emphasis on evidence that has been gathered from consideration of population-based analyses of biomedical data. However, rare variants are predictor variables that by definition are quite individual-specific. Simply based on their prevalence, standard population-based analyses will have low power to detect them. The rare variant paradigm also is quite in tune with the notion of personalized medicine, where treatments and/or interventions would be tailored to the particular variant present in the individual. Very broadly speaking, this is consistent with the patient-centered/patient-oriented paradigm in medicine that has been developing over the last few years. Figure 20.1 describes the genetic spectrum for disease that analysts must contend with.

20.3 Kernel Machine Methodology

20.3.1 Setup and Review of Methods

We now describe tests of associations between rare variants and a phenotype. To make the ideas concrete in this chapter, we will suppose that we have available $(Y_i, \mathbf{G}_i, \mathbf{Z}_i)$, $i = 1, \ldots, n$ for n subjects, which is a random sample from $(Y, \mathbf{G}, \mathbf{Z})$. Here, Y denotes the phenotype, $\mathbf{G}$ is a $p-$dimensional vector for the genotypes for the p variants within a region, and $\mathbf{Z}$ is a $q-$dimensional vector of confounding variables to adjust for. Here and in the sequel, we will assume that each component of $\mathbf{G}$ will count the number of minor alleles. We can postulate a class of regression models for Y given $\mathbf{G}$ and $\mathbf{Z}$; a standard one would be to postulate a generalized linear model:

$$h(E[Y_i|\mathbf{G}_i, \mathbf{Z}_i]) = \alpha_0 + \alpha^T \mathbf{G}_i + \beta^T \mathbf{Z}_i, \tag{20.1}$$

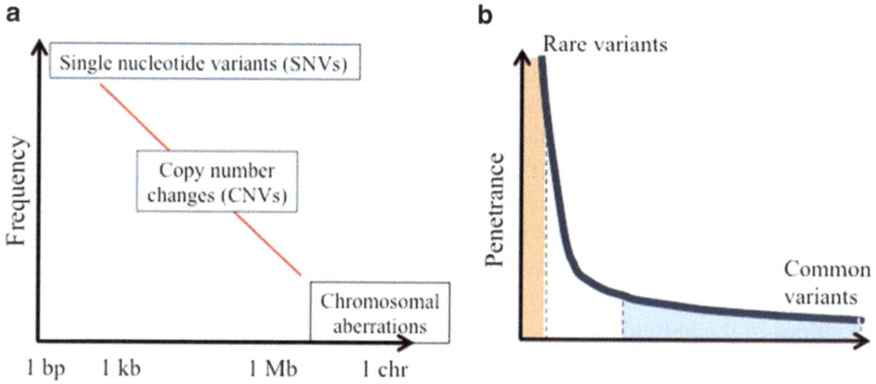

Fig. 20.1 (**a**) Size of variant. Genetic variants are ordered by size on the horizontal axis versus frequency on the vertical axis. Note that single nucleotide variants or more specifically single nucleotide polymorphisms (used for GWAS studies) are more frequent than copy number variants (i.e., deletions and duplications) in the human genome. The large chromosomal aberrations such as trisomies and monosomies are rarer and are the cause for severe developmental disabilities. (**b**) Frequency of Variants. Variants can be classified by the frequency (on the horizontal axis) and its effect, i.e. penetrance (proportion of individuals carrying a variant also manifesting a phenotype) on the vertical axis. Note that rare variants (typically <0.1 % to <5 %) are highly penetrant, associated with severe developmental disorders, while common variants have modest effect. Variants of intermediate frequency are currently missed by most studies. Current studies also suggest that multiple rare alleles interacting in common or related pathways are responsible for several human disorders

where $(\alpha_0, \alpha, \beta)$ are the regression coefficients to be estimated, and h is a link function. Note that the current model (20.1) can allow for both continuous and binary phenotypes.

While model (20.1) is quite standard in the statistical literature, new issues arise when attempting to apply it to rare variant data. First, due to the sparsity of $\mathbf{G}$ the components of α will not be estimated very well. Due to this as well as for computationally feasibility, there has been a reliance on the use of score-based tests, which will be less sensitive to this type of sparsity relative to a Wald test, for example. A second problem is one of power. Models such as (20.1) that treat the genetic effects as fixed effects will have lower power due to the number of degrees of freedom for jointly testing $\alpha = \mathbf{0}$. To circumvent this issue, two classes of approaches have been developed. The first includes methods that can broadly interpreted as collapsing methods [31, 32, 36, 44]. These tests effectively reduce $\mathbf{G}$ into a scalar quantity $\mathbf{G}^*$ and to fit model (20.1), where $\alpha^T \mathbf{G}_i$ is replaced by $\gamma \mathbf{G}_i^*$. The reduction to a one-dimensional quantity leads to a reduction in the number of parameters and a potential gain in power.

Collapsing approaches will work in situations in which the components of $\mathbf{G}$ have effects on Y that are in the same direction. However, it might be the case that this assumption is not true. The SKAT methodology of Wu et al. [56] then becomes quite useful in this regard. In particular, it generalizes (20.1):

$$h(E[Y_i|\mathbf{G}_i,\mathbf{Z}_i]) = \alpha_0 + f(\mathbf{G}_i) + \beta^T\mathbf{Z}_i, \tag{20.2}$$

where now f is a flexible non-linear function of the rare variants. This is a special case of the kernel machine framework originally proposed by Liu et al. [33, 34]. We will describe the technical details of the approach in the next section. We point out here that the rare variant effects are allowed to be much more flexible than in (20.1). Further, the test of the genetic effect in (20.2) is identical to testing for a random effect being zero for a certain linear mixed effects model. This amounts to an effective shrinking of the degrees of freedom and allows for pooling of information across the rare variants. The score test amounts to a quadratic form that takes deviations of the individual rare variant effects and squares them.

20.3.2 Kernel Machines: Technical Details

In this section, we review the technical details behind the SKAT model in the case of h in (20.2) being the identity link. This material is intended for mathematically minded readers and can be skipped upon initial reading of this chapter. Recall the model from the previous section with $\alpha_0 = 0$:

$$Y_i = \beta^T\mathbf{Z}_i + f(\mathbf{G}_i) + e_i, \tag{20.3}$$

where β is a $q \times 1$ vector of regression coefficients, $f(\mathbf{G}_i)$ is an unknown centered smooth function, and the errors e_i are assumed to be independent and follow $N(0,\sigma^2)$. Here, we are centering the response so that there is no intercept term as in (20.2). Note that when $f(\cdot) = 0$, (20.3) reduces to the standard linear regression model.

20.3.2.1 Function Space of $f(\mathbf{G})$: Specification

We assume the nonparametric function $f(\mathbf{G})$ lies in a function space $\mathscr{F}$ spanned by a set of basis functions $\{\phi_1(\mathbf{G}),\cdots,\phi_j(\mathbf{G}),\cdots,\phi_J(\mathbf{G})\}_{j=1}^J$ such that any function in the space $\mathscr{F}$ can be written as $f(\mathbf{G}) = \sum_{j=1}^J \omega_j\phi_j(\mathbf{G})$ for some constants $\{\omega_j\}_{j=1}^J$. Note that the set of basis functions can be finite ($J < \infty$) or infinite ($J = \infty$). In the machine learning literature, such basis functions are called *features*.

Specification of a function space using basis functions or features might be complicated since explicit expressions of features are required and the number of features might be high or even infinite. An alternative way to conveniently specify a function space is to use a kernel function $K(\mathbf{G},\mathbf{G}')$ instead of the basis functions. Specifically, a kernel function $K(\mathbf{G},\mathbf{G}')$ is a bounded, symmetric, positive function satisfying

$$\int K(\mathbf{G},\mathbf{G}')f(\mathbf{G})f(\mathbf{G}')d\mathbf{G}d\mathbf{G}' \geq 0, \tag{20.4}$$

for any arbitrary square integrable function $f(\mathbf{G})$ and all $\mathbf{G},\mathbf{G}' \in R^p$. The kernel function can be viewed as a measure of similarity between two values of the covariate vector $\mathbf{G}$ and $\mathbf{G}'$. Following from the Mercer Theorem (e.g., see p. 33 of [6]), any kernel function satisfying some regularity conditions implicitly specifies an unique function space spanned by a particular set of basis functions (features), and vice versa. Before formally defining such a function space, we give a few examples.

1. *The dth degree Polynomial Kernel*: $K(\mathbf{G},\mathbf{G}') = [\mathbf{G}\cdot\mathbf{G}'+1]^d$, where $\mathbf{G}\cdot\mathbf{G}' = \sum_{k=1}^{p} g_k g_k'$ denotes the dot product. Recall that g represents components of the vector $\mathbf{G}$ in (20.3). This dth degree polynomial kernel generates the function space $\mathscr{F}$ spanned by all possible dth order monomials of the components of $\mathbf{G}$. For example, if $d = 1$, the first polynomial kernel generates the linear function space with basis functions $\{z_1,\cdots,z_p\}$. If $d = 2$, the second polynomial kernel corresponds the quadratic function space with basis functions $\{z_k, z_k z_k'\}$ $(k,k' = 1,\cdots,p)$, i.e., the main effects, all two-way interactions and quadratic main effects. Note that the function space determined by the dth degree polynomial kernel is of finite dimension.
2. *The Gaussian Kernel*: $K(\mathbf{G},\mathbf{G}') = \exp\{-||\mathbf{G}-\mathbf{G}'||^2/\rho\}$, where $||\mathbf{G}-\mathbf{G}'|| = \sum_{k=1}^{p}(g_k - g_k')^2$. The Gaussian kernel generates the function space spanned by radial basis functions, whose nice properties can be found in Bühmann [3]. The function space determined by the Gaussian kernel is of infinite dimension.
3. *The identity by state kernel*: Kwee et al. [26] propose the use of the concept of identity by state to define a new kernel. The kernel is given by

$$K(\mathbf{G},\mathbf{G}') = \frac{\sum_{s=1}^{p} IBS(\mathbf{G}_s,\mathbf{G}_s')}{2p},$$

   where the IBS function denotes the number of alleles shared identically by state at position s.

The above examples suggest that the choice of a kernel function determines which function space one would like to use to approximate $f(\mathbf{G})$. The dimension of the function space defined by a kernel function $K(\cdot,\cdot)$ is determined by the dimension of the eigenfunctions of $K(\cdot,\cdot)$. The use of a kernel to specify a function space avoids specifications of complicated basis functions (features) and inner products. One will see in the next section that it has significant computational advantages in high dimensional problems. It should be noted that the term "kernel" here has a rather different meaning from that used in the kernel smoothing literature. A commonly used function space defined by a kernel is a Reproducing Kernel Hilbert Space (RKHS), which we label as $\mathscr{F}_K$. Technical details on RKHS can be found in Wahba [54] or Chapter 3 of Cristianini and Shawe-Taylor [6].

20.3.2.2 Primal and Dual Representations of $f(\mathbf{G})$

Any function $f(\mathbf{G})$ in the function space $\mathscr{F}_K$ defined by a kernel $K(\cdot,\cdot)$ can have a primal representation directly using the basis functions (features) of $\mathscr{F}_K$, and it can equivalently have a dual representation using the kernel function $K(\mathbf{G},\mathbf{G}')$ directly. Specifically, for an arbitrary function $h(\mathbf{G}) \in \mathscr{F}_K$, its primal representation takes the form

$$f(\mathbf{G}) = \sum_{j=1}^{J} \omega_j \phi_j(\mathbf{G}) = \phi(\mathbf{G})^T \omega, \tag{20.5}$$

where $\phi(\cdot) = \{\phi_1(\cdot),\cdots,\phi_J(\cdot)\}^T$ is a $J \times 1$ vector of the standardized orthogonal basis functions (features), i.e., standardized Mercer features of the function space $\mathscr{F}_K$, and $\omega \equiv (\omega_1,\cdots,\omega_J)'$ is a vector of some constants. The square norm of $f(\cdot)$ can be written as

$$\|f\|^2_{\mathscr{F}_K} = \sum_{j=1}^{J} \omega_j^2 = \omega^T \omega. \tag{20.6}$$

Alternatively, the same $f(\mathbf{G})$ can be equivalently written in a dual representation using the kernel function $K(\cdot,\cdot)$ directly as

$$f(\mathbf{G}) = \sum_{l=1}^{L} \alpha_l K(\mathbf{G}_l^*,\mathbf{G}), \tag{20.7}$$

for some integer L, some constants $\alpha_1,\ldots,\alpha_L$ and some $\{\mathbf{G}_1^*,\cdots,\mathbf{G}_L^*\} \in R^p$. For justifications of these results and more details about the RKHS, see Cristianini and Shawe-Taylor (2000[6], Chapter 3).

Estimation of β and $f(\cdot)$ proceeds by maximizing the scaled penalized likelihood function

$$-\frac{1}{2}\sum_{i=1}^{n}\{Y_i - \beta^T\mathbf{Z}_i - f(\mathbf{G}_i)\}^2 - \frac{1}{2}\lambda\|f\|^2_{\mathscr{F}_K}, \tag{20.8}$$

where λ is a tuning parameter and controls the tradeoff between goodness of fit and complexity of the model. When $\lambda = 0$, the model interpolates the data, whereas when $\lambda = \infty$, the model reduces to a simple linear model.

While the function (20.8) is hard to optimize directly, we introduce the Lagrangian multiplier (also called the dual parameter) γ to obtain

$$\mathscr{L}(\omega,\beta,e,\gamma) = -\frac{1}{2}\sum_{i=1}^{n} e_i^2 - \frac{1}{2}\lambda\omega^T\omega + \sum_{i=1}^{n}\gamma_i\{\beta^T\mathbf{Z}_i + \phi(\mathbf{G}_i)^T\omega + e_i - Y_i\}. \tag{20.9}$$

The dual problem is formulated by constructing an objective function by removing the high-dimensional primal coefficient vector ω and the constraint parameters e from $\mathscr{L}(\omega,\beta,e,\gamma)$ and writing $\mathscr{L}(\omega,\beta,e,\gamma)$ as a function of β and the dual parameter vector γ only. We will see that the resulting estimators $\hat{\beta}$ and $\hat{\gamma}$ can be expressed as a function of some kernel function $K(\cdot,\cdot)$. One can then conveniently obtain the maximizer of the original primal problem $\hat{\omega}$ and then $\hat{f}(\mathbf{G})$ at any arbitrary $\mathbf{G}$ as a function of the kernel function $K(\cdot,\cdot)$.

Specifically, the dual problem to minimizing (20.8) is

$$\min_{\beta,\gamma} \mathscr{Q}(\beta,\gamma) \tag{20.10}$$

where $\mathscr{Q}(\beta,\gamma) = \sup_{\omega,e} \mathscr{L}(\omega,\beta,e,\gamma)$. Note that (20.10) is an unconstrained optimization problem, and the number of unknown parameters depends only on β and the dual parameters γ, whose dimension is equal to the sample size n, often much smaller than J, the dimension of the primal vector ω. Therefore the dual formulation (20.10) effectively transforms the often infinite-dimensional optimization problem (20.8) into a finite-dimensional problem.

To obtain $\mathscr{Q}(\beta,\gamma)$, one differentiates $\mathscr{L}(\omega,\beta,e,\gamma)$ with respect to e and ω and sets the derivatives to zero. We have

$$\begin{aligned} \hat{e} &= \gamma \\ \hat{\omega} &= \lambda^{-1} \sum_{i=1}^{n} \gamma_i \phi(\mathbf{G}_i). \end{aligned} \tag{20.11}$$

Substituting $\hat{\omega}$ and $\hat{e}$ into $\mathscr{L}(\cdot)$, some calculations give

$$\mathscr{Q}(\beta,\gamma) = (Y - \beta^T \mathbf{Z})^T \gamma - \frac{1}{2}\gamma^T \left(I + \lambda^{-1} K\right) \gamma \tag{20.12}$$

where $Y = (Y_1,\cdots,Y_n)^T$ and $\mathbf{Z} = (\mathbf{Z}_1,\cdots,\mathbf{Z}_n)^T$, K is an $n \times n$ matrix whose (i,i')th element is $K(\mathbf{G}_i,\mathbf{G}_{i'})$, the kernel function evaluated at the pair of the design points $(\mathbf{G}_i,\mathbf{G}_{i'})$. Note that the kernel matrix K measures the similarity among the covariate values $(\mathbf{G}_1,\cdots,\mathbf{G}_n)$. One can see that even when p (the dimension of $\mathbf{G}$) or J (the dimension of the feature space) is high, the dimension of K is not affected by p and J and remains the same as the sample size n.

Differentiating $\mathscr{Q}(\beta,\gamma)$ with respect to γ and β, some calculations give

$$\hat{\beta} = \left\{\mathbf{Z}^T (I + \lambda^{-1} K)^{-1} \mathbf{Z}\right\}^{-1} \mathbf{Z}^T (I + \lambda^{-1} K)^{-1} Y \tag{20.13}$$

$$\hat{\gamma} = (I + \lambda^{-1} K)^{-1} (Y - \hat{\beta}^T \mathbf{Z}). \tag{20.14}$$

Plugging (20.14) into (20.11), we have

$$\hat{\omega}=\lambda^{-1}\{\phi(\mathbf{G}_1),\cdots,\phi(\mathbf{G}_n)\}\hat{\gamma}=\lambda^{-1}\{\phi(\mathbf{G}_1),\cdots,\phi(\mathbf{G}_n)\}(I+\lambda^{-1}K)^{-1}(Y-\hat{\beta}^T\mathbf{Z}).$$

It follows that the nonparametric function $f(\cdot)$ evaluated at the design points $(\mathbf{G}_1,\cdots,\mathbf{G}_n)^T$ is estimated as

$$\hat{f}=\lambda^{-1}K\hat{\gamma}=\lambda^{-1}K(I+\lambda^{-1}K)^{-1}(Y-\hat{\beta}^T\mathbf{Z}). \quad (20.15)$$

The estimator of the nonparametric function $f(\cdot)$ at an arbitrary $\mathbf{G}$ is

$$\hat{f}(\mathbf{G}) = \phi(\mathbf{G})^T\hat{\omega} \quad (20.16)$$

$$= \lambda^{-1}\{K(\mathbf{G},\mathbf{G}_1),\cdots,K(\mathbf{G},\mathbf{G}_n)\}(I+\lambda^{-1}K)^{-1}(Y-\hat{\beta}^T\mathbf{Z}). \quad (20.17)$$

Note that the estimators $\hat{\beta}$ and $\hat{f}(\cdot)$ in (20.13) and (20.15) are the maximizer of the original primal problem. Examination of equations (20.13) and (20.17) suggests that the estimators $\hat{\beta}$ and $\hat{f}(\cdot)$ are both conveniently evaluated using the kernel function $K(\cdot,\cdot)$ and do not require specifying the high (maybe infinite) dimensional basis functions (features) $\{\phi(\mathbf{G})\}$. This means one simply summarizes the similarity of high-dimensional covariates $(\mathbf{G}_1,\cdots,\mathbf{G}_n)$ using a kernel matrix K, then calculates $\hat{\beta}$ and $\hat{f}(\cdot)$ by inventing an $n\times n$ matrix involving the kernel matrix K, which is of the dimension of sample size and is often small in high dimensional problems, e.g., microarray problems. Using (20.14), one can easily see that $\hat{f}(\mathbf{G})$ can be rewritten as

$$\hat{f}(\mathbf{G}) = \sum_{i=1}^{n}\lambda^{-1}\hat{\gamma}_i K(\mathbf{G},\mathbf{G}_i).$$

A comparison of this equation with equation (20.7) suggests that $\hat{f}(\mathbf{G})$ takes exactly a dual representation with $L=n$, $(\mathbf{G}_1^*,\cdots,\mathbf{G}_n^*) = (\mathbf{G}_1,\cdots,\mathbf{G}_n)$ and $\alpha=\lambda^{-1}\hat{\gamma}$. Hence the estimated Lagrangian multiplier $\hat{\gamma}$ serves as the coefficients in the dual representation of $\hat{f}(\mathbf{G})$, apart from a scale factor.

In Liu et al. [33], it is shown that the estimates of f and β can be derived as estimates from a random effects model of the following form:

$$Y=\beta^T\mathbf{Z}+f+e, \quad (20.18)$$

where β is a $q\times 1$ vector of regression coefficients, f is an $n\times 1$ vector random effects following $f\sim N\{\mathbf{0},\tau K(\rho)\}$, ρ is a scale parameter, and $e\sim N(\mathbf{0},R=\sigma^2 I)$. Because of this equivalence, all regression parameters in the model can be estimated by maximum likelihood, while the variance component parameters can be estimated by restricted maximum likelihood. If we assume $f(\mathbf{G})\in\mathscr{H}_K$, one can easily see from the linear mixed model representation (20.18) of the least squares kernel machine that $H_0 : f(\mathbf{G}) = 0$ is equivalent to testing the variance component $\tau = 0$. The null hypothesis $H_0 : \tau = 0$ places τ on the boundary of the parameter space. Liu et al. [33] developed a score test for testing H_0.

20.3.3 SKAT Extensions

Since the seminal work of Wu et al. [56] on this topic, there have been several notable extensions of the SKAT methodology. One extension was by Lee et al. [28, 29], which made the observation that the collapsed approaches and SKAT could be combined into a unified framework based on a prior distribution for the linkage disequilibrium between rare variants within a genomic region of interest. An application of the SKAT statistics to meta-analysis has been developed by Lee et al. [27]. Finally, we note that Ionita-Laza et al. [23] have extended the SKAT approach to simultaneously incorporate common and rare variants.

20.3.4 SKAT Example

We now describe the application of the SKAT methodology to data from Girirajan et al. [20], in which the role of structural variants in autism was explored. The data come from the Simons Simplex Complex Foundation. For the purposes of this chapter, we will assume that the rows of the data matrix below represent statistically independent observations. A sample of the data is given below:

```
     chrom     start       end    size pheno
1     chr1   6191784   6494317 302533     0
2     chr1 108655067 108718023  62956     0
3     chr1 143636400 143700636  64236     0
4     chr1 143636400 143701095  64695     0
5     chr1 143639096 143701095  61999     0
```

In this file, start and end denote the beginning and end of the structural variant, and size denotes the length of the variant and is the difference between start and end. Finally, pheno is a coding of the phenotype as zero for control and one for case (i.e., autism). Our analyses using SKAT will use start, end and pheno.

We consider data from chromosome 1, which has been considered to be a hotspot for structural variations in autism. We have measurements from 99 cases and 76 controls. We note that the hotspots have variable length, which is why the size column shows variation. In order to implement the SKAT method, we need to convert each row of the dataset into a vector of zeroes and ones. The zero represents absence of a structural variant while one indicates its presence. We partitioned chromosome 1 into 2000 nonoverlapping windows of equal size and determined for each row of the dataset how many windows the alteration overlapped with. This is

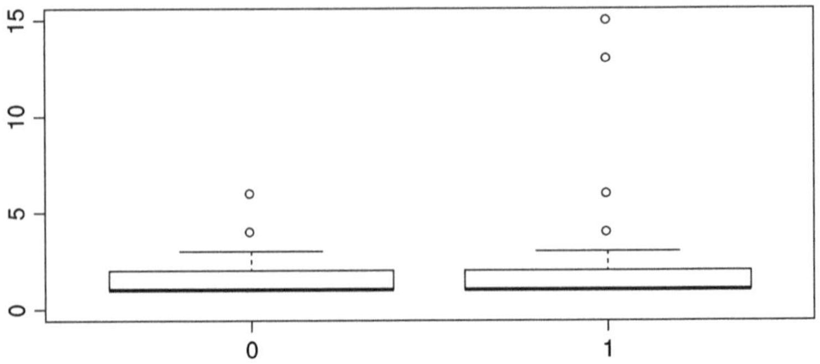

Fig. 20.2 Boxplot of distribution of copy number burden for chromosome 1 in controls (*left* boxplot) versus autism cases (*right* boxplot). The data represent the total number of structural variants from 50 windows that had at least one variant across the 175 samples

done by comparing both the `start` and `end` to the window in question. Note that this will give us a 175 by 2000 matrix with zeroes and ones. However, of the 2,000 columns, only 50 have at least one nonzero entry. This means that for each subject, we have a 50-dimensional vector of counts. It is not easy to perform descriptive statistics on this type of data. As in [20], we can define a concept of copy number burden, which means to simply add the up the counts over the 50 dimensions for each subject. A plot of the distribution of copy number burden between cases and controls is given in Fig. 20.2.

Based on Fig. 20.2, we find almost no difference between the copy number burden distribution of cases and controls, aside from two high outliers among the autism cases. However, the SKAT methodology may be able to identify differences between the controls and cases when examining the 50-dimensional count vectors that cannot be seen in the copy number burden data. To illustrate our method, we simulated a covariate $\mathbf{Z} = Z$ from a standard normal distribution and used the following R code to run SKAT.

```
# y.b = pheno variable from the dataset; Z: simulated
# normal(0,1) covariate;
# G: structural variant data, here a 50-dimensional
# vector of counts
# kernel specifies the kernel matrix needed to run
# SKAT; options include linear, IBS, quadratic
# and 2 way interaction; the first three have the
# option of being weighted by the
# inverse of the variance of the estimated
# proportion of the rare variant, as described
# in Madsen and Browning (2009)
#
# Here, we use the weighted linear kernel.
#
```

```
obj = SKAT_Null_Model(y.b~Z,out_type="D",
        kernel="linear.weighted")
skat1 = SKAT(G,obj)
```

Further details about the code can be found in the SKAT manual. We note that the default procedure of Wu et al. [56] is recommended for a sample size greater than 2000. Given that our example has a sample size of 175, SKAT performs an adjustment in terms of using higher-order approximations in order to estimate the null distribution of the test statistic. Using this adjustment, the p-value from SKAT is 5.27×10^{-5}. Thus there is strong evidence of structural variants in chromosome one being associated with autism.

20.4 Multiple Testing

Next, we discuss the impact of multiple comparisons on the analysis of rare variant data from sequencing studies. While genomics has experienced an explosion in the literature on multiple testing, there are two unique issues in the sequencing context. First, because these variations are rare by definition, the number of single variant hypothesis tests that need to be performed are actually quite small relative to numbers of tests in other problems (e.g., number of tests in common-variant GWAS studies). What is more challenging, however, is the fact that there is an inherent discreteness in the data structure. For a given rare variant, we can represent the data as in Table 20.1, where the cell entries represent the number of samples in each of the groups. We wish to test for independence of the rows and columns, and many methods exist for testing the null hypothesis of no association between presence of rare variant and group label. If the expected cell count is greater than five in all the cells, then one can safely use chi-squared statistics. However, when the cell counts are small, we then use Fisher's exact test, where the p-value is computed using a hypergeometric distribution.

While there has been a lot of work on extensions and generalizations of the FDR estimation methodology, most of the literature in this area has used the fact that under the null distribution, the p-values are uniformly distributed on (0,1) or more generally, that the test statistics have a continuous distribution. This will not apply in the case of rare variant data with respect to the presence/absence calls. The literature on multiple testing with discrete p-values is much more limited. An initial procedure was proposed by Tarone [51] which involves only considering hypotheses

Table 20.1 Rare variant presence/absence analysis

	Rare variant present	Rare variant absent	Total
$Y = 0$	a	b	a+b
$Y = 1$	c	d	c+d
	a+c	b+d	a+b+c+d

where a sufficiently small rejection probability is possible and to then perform a Bonferroni test on those selected hypotheses. This procedure has been modified to the false discovery setting in Gilbert [15], where the Bonferroni adjustment was replaced by the Benjamini-Hochberg [2] procedure. Theoretical aspects of the B-H procedure with discrete test statistics have been addressed by Ferreira [9]. An FDR-based estimation procedure in the spirit of the q-value methodology of Storey [49] was developed in Pounds and Cheng [43]. In Kulinskaya and Lewin [25], the B-H procedure was applied to so-called fuzzy p-values, whose behavior under the null hypothesis is identical to that of a Uniform(0,1) random variable so that the usual methods apply. Applications of discrete multiple testing ideas to a cancer genomics problem can be found in Ghosh [12, 13]. Some recent work of Bancroft et al. [1] uses a novel sequential permutation p-value approach to estimate FDR that would be applicable in this setting as well.

Finally, an open problem in this area is the incorporation of dependence into multiple testing procedure. While there has been a lot of recent work in the area on multiple comparisons with dependent data [8, 30], almost all of this work again assumes that the p-values are derived from continuous distribution, which is not the case here. However, the argument that rare variants operate with a network structure is less plausible than for phenomena such as gene expression, so a case could be made that dependence is not as big of an issue as in other genomic settings. Again, this topic is definitely worthy of future exploration.

20.5 Discussion

This chapter has attempted to discuss issues in the analysis of rare variant data for a statistical audience. One of the major messages from this chapter is that the phenomenon being described is one with a low probability of occurring, but given its occurrence, it can have a large effect.

One of the major challenges in this area will be development of methods that will have high power of detecting these events. A major statistical lesson that has been used here is that the score method of testing has definite merits. While classical statistical theory teaches us that the behavior of the likelihood ratio test, Wald test and score test will be identical as the sample size tends to infinity, it is also the case that we are definitely in a small-sample scenario where asymptotic theory will not hold. The score statistic provides many advantages, one of the major ones being that of avoiding having to estimate rare variant effects.

An area not discussed in this chapter is meta-analysis. This has become the *de rigueur* method for identifying candidate genes from genomewide studies. We point the reader to the recent review by Evangelou and Ioannidis [7] and note the SKAT approach to this problem that was described in Lee et al. [27].

While this area is relatively new, we should also be wise to lessons that have been learnt in many other settings. For example, it is well-known that selected variables or SNPs suffer from the so-called 'winner's curse' so that estimated effects will be biased. This will also be the case for the rare variants and is inherent to the statistical task at hand.

Finally, we believe that a tactic that will be useful in the future is what we term 'pooling information.' One of the major reasons that SKAT methods have had such a major impact in this area is that the equivalence with variance components models and the introduction of random effects models leads to the ability to pool information across estimated parameters. Statistically, this can be conceptualized using shrinkage theory, Empirical Bayes and more generally, Bayesian methods. Given the increasing availability of genomewide information from different data sources, pooling information using 'vertical integration' techniques [52] will be needed to identify and to elucidate the functionality of rare variants in the foreseeable future.

Acknowledgements The research of the authors is supported by NIH R01 CA 129102 and NSF ABI-1262538.

References

[1] Bancroft, T., Du, C., Nettleton, D.: Estimation of false discovery rate using sequential permutation p-values. Biometrics **69**, 1–7 (2013)
[2] Benjamini, Y., Hochberg, Y.: Controlling the false discovery rate: a practical and powerful approach to multiple testing. J. Roy. Stat. Soc. Ser. B **57**, 289–300 (1995)
[3] Bühmann, M.D.: Radial basis functions: theory and implementation. Cambridge University Press, Cambridge (2003)
[4] Campbell, C.D., Eichler, E.E.: Properties and rates of germline mutations in humans. Trends Genet. **29**, 575–584 (2013)
[5] Chatterjee, N., Wheeler, B., Sampson, J., Hartge, P., Chanock, S. J., Park, J. H.: Projecting the performance of risk prediction based on polygenic analyses of genome-wide association studies. Nat. Genet. **45**, 400–405 (2013)
[6] Cristianini, N., Shawe-Taylor, J.: An Introduction to Support Vector Machines and Other Kernel-based Learning Methods. Cambridge University Press, Cambridge (2000)
[7] Evangelou, E., Ioannidis, J. P.: Meta-analysis methods for genome-wide association studies and beyond. Nat. Rev. Genet. **14**, 379–389 (2013)
[8] Fan, J., Han, X., Gu, W.: Estimating false discovery proportion under arbitrary covariance dependence. J. Am. Stat. Assoc. **107**, 1019–1035 (2012)
[9] Ferreira, J. A.: The Benjamini-Hochberg method in the case of discrete test statistics. Int. J. Biostat. **3** (1), Article 11 (2007)
[10] Frazer, K.A., Ballinger, D.G., Cox, D.R., Hinds, D.A., Stuve, L.L., Gibbs, R.A., Belmont, J.W., Boudreau, A., Hardenbol, P., Leal, S.M., Pasternak, S., Wheeler, D.A., Willis, T. D., Yu, F., Yang, H., Zeng, C., Gao, Y., Hu, H., Hu, W., Li, C., Lin, W., Liu, S., Pan, H., Tang, X., Wang, J., Wang, W., Yu, J., Zhang, B., Zhang, Q., Zhao, H., Zhao, H., Zhou, J., Gabriel, S.B., Barry, R., Blumenstiel, B., Camargo, A., Defelice, M., Faggart, M., Goyette, M., Gupta, S., Moore, J., Nguyen, H., Onofrio, R.C., Parkin, M., Roy, J., Stahl, E., Winchester, E., Ziaugra, L., Altshuler, D., Shen, Y., Yao, Z., Huang, W., Chu, X., He, Y., Jin, L., Liu, Y., Shen, Y., Sun, W., Wang, H., Wang, Y., Wang, Y., Xiong, X., Xu, L., Waye, M.M., Tsui, S.K., Xue, H., Wong, J.T., Galver, L.M., Fan, J.B., Gunderson, K., Murray, S.S., Oliphant, A.R., Chee, M. S., Montpetit, A., Chagnon, F., Ferretti, V., Leboeuf, M., Olivier, J.F., Phillips, M.S., Roumy, S., Sallée, C., Verner, A., Hudson, T.J., Kwok, P.Y., Cai, D., Koboldt, D.C., Miller,

R.D., Pawlikowska, L., Taillon-Miller, P., Xiao, M., Tsui, L.C., Mak, W., Song, Y.Q., Tam, P. K., Nakamura, Y., Kawaguchi, T., Kitamoto, T., Morizono, T., Nagashima, A., Ohnishi, Y., Sekine, A., Tanaka, T., Tsunoda, T., Deloukas, P., Bird, C.P., Delgado, M., Dermitzakis, E.T., Gwilliam, R., Hunt, S., Morrison, J., Powell, D., Stranger, B.E., Whittaker, P., Bentley, D.R., Daly, M. J., de Bakker, P.I., Barrett, J., Chretien, Y.R., Maller, J., McCarroll, S., Patterson, N., Peér, I., Price, A., Purcell, S., Richter, D.J., Sabeti, P., Saxena, R., Schaffner, S.F., Sham, P.C., Varilly, P., Altshuler, D., Stein, L.D., Krishnan, L., Smith, A.V., Tello-Ruiz, M. K., Thorisson, G. A., Chakravarti, A., Chen, P. E., Cutler, D.J., Kashuk, C.S., Lin, S., Abecasis, G.R., Guan, W., Li, Y., Munro, H.M., Qin, Z. S., Thomas, D.J., McVean, G., Auton, A., Bottolo, L., Cardin, N., Eyheramendy, S., Freeman, C., Marchini, J., Myers, S., Spencer, C., Stephens, M., Donnelly, P., Cardon, L. R., Clarke, G., Evans, D.M., Morris, A.P., Weir, B.S., Tsunoda, T., Mullikin, J.C., Sherry, S.T., Feolo, M., Skol, A., Zhang, H., Zeng, C., Zhao, H., Matsuda, I., Fukushima, Y., Macer, D.R., Suda, E., Rotimi, C.N., Adebamowo, C.A., Ajayi, I., Aniagwu, T., Marshall, P.A., Nkwodimmah, C., Royal, C.D., Leppert, M.F., Dixon, M., Peiffer, A., Qiu, R., Kent, A., Kato, K., Niikawa, N., Adewole, I.F., Knoppers, B.M., Foster, M.W., Clayton, E.W., Watkin, J., Gibbs, R.A., Belmont, J.W., Muzny, D., Nazareth, L., Sodergren, E., Weinstock, G.M, Wheeler, D.A., Yakub, I., Gabriel, S.B., Onofrio, R.C, Richter, D.J., Ziaugra, L., Birren, B.W., Daly, M.J., Altshuler, D., Wilson, R.K., Fulton, L.L., Rogers, J., Burton, J., Carter, N.P., Clee, C.M., Griffiths, M., Jones, M.C., McLay, K., Plumb, R. W., Ross, M.T., Sims, S.K., Willey, D.L., Chen, Z., Han, H., Kang, L., Godbout, M., Wallenburg, J.C., L'Archeveque, P., Bellemare, G., Saeki, K., Wang, H., An, D., Fu, H., Li, Q., Wang, Z., Wang, R., Holden, A.L., Brooks, L.D., McEwen, J.E., Guyer, M.S., Wang, V.O., Peterson, J.L., Shi, M., Spiegel, J., Sung, L.M., Zacharia, L.F., Collins, F. S., Kennedy, K., Jamieson, R., Stewart, J.: A second generation human haplotype map of over 3.1 million SNPs. Nature **449**, 851–861 (2007)

[11] Gabriel, S.B., Schaffner, S.F., Nguyen, H., Moore, J. M., Roy, J., Blumenstiel, B., Higgins, J., DeFelice, M., Lochner, A., Faggart, M., Liu-Cordero, S. N., Rotimi, C., Adeyemo, A., Cooper, R., Ward, R., Lander, E.S., Daly, M.J., Altshuler, D.: The structure of haplotype blocks in the human genome. Science **296**, 2225–2229 (2002)

[12] Ghosh, D.: Discrete nonparametric algorithms for outlier detection with genomic data. J. Biopharm. Stat. **20**, 193–208 (2010)

[13] Ghosh, D: Genomic outlier detection in high-throughput data analysis. Methods Mol Biol **972**, 141–53 (2013)

[14] Gibson, G.: Rare and common variants: twenty arguments. Nat. Rev. Genet. **13**, 135–145 (2012)

[15] Gilbert, P.B.: A modified false discovery rate multiple-comparisons procedure for discrete data, applied to human immunodeficiency virus genetics. Appl. Stat. **54**, 143–158 (2005)

[16] Girard, S.L., Gauthier, J., Noreau, A., Xiong, L., Zhou, S., Jouan, L., Dionne-Laporte, A., Spiegelman, D., Henrion, E., Diallo, O., Thibodeau, P., Bachand, I., Bao, J. Y., Tong, A. H., Lin, C.H., Millet, B., Jaafari, N., Joober, R., Dion, P.A., Lok, S., Krebs, M.O., Rouleau, G.A.: Increased exonic de novo mutation rate in individuals with schizophrenia. Nat. Genet. **43**, 860–863 (2011)

[17] Girirajan, S., Brkanac, Z., Coe, B.P., Baker, C., Vives, L., Vu, T.H., Shafer, N., Bernier, R., Ferrero, G.B., Silengo, M., Warren, S.T., Moreno, C.S., Fichera, M., Romano, C., Raskind, W.H., Eichler, E.E.: Relative burden of large CNVs on a range of neurodevelopmental phenotypes. PLoS Genet **7**, e1002334 (2011)

[18] Girirajan, S., Campbell, C.D., Eichler, E.E.: Human copy number variation and complex genetic disease. Annu. Rev. Genet. **45**, 203–226 (2011)

[19] Girirajan, S., Eichler, E.E.: Phenotypic variability and genetic susceptibility to genomic disorders. Hum. Mol. Genet. **19**, R176–187 (2010)

[20] Girirajan, S., Johnson, R.L., Tassone, F., Balciuniene, J., Katiyar, N., Fox, K., Baker, C., Srikanth, A., Yeoh, K.H., Khoo, S.J., Nauth, T.B., Hansen, R., Ritchie, M., Hertz-Picciotto, I., Eichler, E.E., Pessah, I.N., Selleck, S.B.: Global increases in both common and rare copy number load associated with autism. Hum. Mol. Genet. **22**, 2870–80 (2013)

[21] Girirajan, S., Rosenfeld, J.A., Coe, B.P., Parikh, S., Friedman, N., Goldstein, A., Filipink, R. A., McConnell, J.S., Angle, B., Meschino, W.S., Nezarati, M.M., Asamoah, A., Jackson, K.E., Gowans, G.C., Martin, J.A., Carmany, E.P., Stockton, D.W., Schnur, R.E., Penney, L.S., Martin, D.M., Raskin, S., Leppig, K., Thiese, H., Smith, R., Aberg, E., Niyazov, D.M., Escobar, L.F., El-Khechen, D., Johnson, K.D., Lebel, R.R., Siefkas, K., Ball, S., Shur, N., McGuire, M., Brasington, C.K., Spence, J.E., Martin, L.S., Clericuzio, C., Ballif, B. C., Shaffer, L.G., Eichler, E.E.: Phenotypic heterogeneity of genomic disorders and rare copy-number variants. N. Engl. J. Med. **367**, 1321–31 (2012)
[22] Hirschhorn, J.N. , Daly, M.J.: Genome-wide association studies for common diseases and complex traits. Nat. Rev. Genet. **6**, 95–108 (2005)
[23] Ionita-Laza, I., Lee, S., Makarov, V., Buxbaum, J. D., Lin, X.: Sequence kernel asociation tests for the combined effect of rare and common variants. Am. J. Hum. Genet. **92**, 841–853 (2013)
[24] Itsara, A., Wu, H., Smith, J.D., Nickerson, D.A., Romieu, I., London, S.J., Eichler, E.E.: De novo rates and selection of large copy number variation. Genome Res **20**, 1469–148 (2010)
[25] Kulinskaya, E., Lewin, A: On fuzzy familywise error rate and false discovery rate procedures for discrete distributions. Biometrika **96**, 201–211 (2009)
[26] Kwee, L.C., Liu, D., Lin, X., Ghosh, D., Epstein, M. P.: A powerful and flexible multilocus association test for quantitative traits. Am. J. Hum. Genet. **82**, 386–397 (2008)
[27] Lee, S., Teslovich, T.M., Boehnke, M., Lin, X.: General framework for meta-analysis of rare variants in sequencing association studies. Am. J. Hum. Genet. **93**, 42–53 (2013)
[28] Lee, S., Emond, M.J., Bamshad, M.J., Barnes, K.C., Rieder, M.J., Nickerson, D.A., Lin, X.: Optimal unified approach for rare-variant association testing with application to small-sample case-control whole-exome sequencing studies. Am. J. Hum. Genet. **91**, 224–237 (2012)
[29] Lee, S., Wu, M.C., Lin, X.: Optimal tests for rare variant effects in sequencing association studies. Biostatistics **13**, 762–775 (2012)
[30] Leek, J.T., Storey, J.D.: A general framework for multiple testing dependence. Proc. Natl. Acad. Sci. **105**, 18718–18723 (2008)
[31] Li, B., Leal, S.M.: Methods for detecting associations with rare variants for common diseases: application to analysis of sequence data. Am. J. Hum. Genet. **83**, 311–321 (2008)
[32] Lin, D.Y., Tang, Z.Z.: A general framework for detecting disease associations with rare variants in sequencing studies. Am. J. Hum. Genet. **89**, 354–367 (2011)
[33] Liu, D., Lin, X., Ghosh, D.: Semiparametric regression of multi-dimensional genetic pathway data: least squares kernel machines and linear mixed models. Biometrics **63**, 1079–1088 (2007)
[34] Liu, D., Ghosh, D., Lin, X.: Estimation and testing for the effect of a genetic pathway on a disease outcome using logistic kernel machine regression via logistic mixed models. BMC Bioinform. **9**, 292 (2008)
[35] Lupski, J.R.: Genomic rearrangements and sporadic disease. Nat. Genet. **39**, S43–S47 (2007)
[36] Madsen, B.E., Browning, S.R.: A groupwise association test for rare mutations using a weighted sum statistic. PLoS Genet. **5**, e1000384 (2009)
[37] Maher, B.: The case of the missing heritability. Nature **456**, 18–21 (2008)
[38] Manolio, T.A., Collins, F.S., Cox, N.J., Goldstein, D.B., Hindorff, L.A., Hunter, D.J., Visscher, P.M.: Finding the missing heritability of complex diseases. Nature **461**, 747–753 (2009)
[39] McClellan, J., King, M.C.: Genetic heterogeneity in human disease. Cell **141**, 210–217 (2010)
[40] Mefford, H.C., Muhle, H., Ostertag, P., von Spiczak, S., Buysse, K., Baker, C., Franke, A., Malafosse, A., Genton, P., Thomas, P., Gurnett, C.A., Schreiber, S., Bassuk, A.G., Guipponi, M., Stephani, U., Helbig, I. and Eichler, E.E.: Genome-wide copy number variation in epilepsy: novel susceptibility loci in idiopathic generalized and focal epilepsies. PLoS Genet. **6**, e1000962 (2010)
[41] Neale, B.M., Kou, Y., Liu, L., Ma'ayan, A., Samocha, K. E., Sabo, A., Lin, C.F., Stevens, C., Wang, L. S., Makarov, V., Polak, P., Yoon, S., Maguire, J., Crawford, E.L., Campbell, N.G., Geller, E.T., Valladares, O., Schafer, C., Liu, H., Zhao, T., Cai, G., Lihm, J., Dannenfelser, R.,

Jabado, O., Peralta, Z., Nagaswamy, U., Muzny, D., Reid, J.G., Newsham, I., Wu, Y., Lewis, L., Han, Y., Voight, B.F., Lim, E., Rossin, E., Kirby, A., Flannick, J., Fromer, M., Shakir, K., Fennell, T., Garimella, K., Banks, E., Poplin, R., Gabriel, S., DePristo, M., Wimbish, J.R., Boone, B.E., Levy, S.E., Betancur, C., Sunyaev, S., Boerwinkle, E., Buxbaum, J.D., Cook, E.H. Jr, Devlin, B., Gibbs, R.A., Roeder, K., Schellenberg, G.D., Sutcliffe, J.S., Daly, M.J.: Patterns and rates of exonic de novo mutations in autism spectrum disorders. Nature **485**, 242–245 (2012)

[42] O'Roak, B.J., Vives, L., Girirajan, S., Karakoc, E., Krumm, N., Coe, B.P., Levy, R., Ko, A., Lee, C., Smith, J.D., Turner, E.H., Stanaway, I.B., Vernot, B., Malig, M., Baker, C., Reilly, B., Akey, J.M., Borenstein, E., Rieder, M.J., Nickerson, D.A., Bernier, R., Shendure, J., Eichler, E.E.: Sporadic autism exomes reveal a highly interconnected protein network of de novo mutations. Nature **485**, 246–250 (2012)

[43] Pounds, S., Cheng, C.: Robust estimation of the false discovery rate. Bioinformatics **22**, 1979–1987 (2006)

[44] Price, A.L., Kryukov, G.V., de Bakker, P.I., Purcell, S. M., Staples, J., Wei, L.J., Sunyaev, S.R.: Pooled association tests for rare variants in exon-resequencing studies. Am. J. Hum. Genet. **86**, 832–838 (2010)

[45] Pritchard, J.K., Cox, N.J.: The allelic architecture of human disease genes: common diseaseÐcommon variantÉ or not? Hum. Mol. Genet. **11**, 2417–2423 (2002)

[46] Reich, D.E., Lander, E.S.: On the allelic spectrum of human disease. Trends Genet. **17**, 502–510 (2001)

[47] Rosenfeld, J.A., Coppinger, J., Bejjani, B.A., Girirajan, S., Eichler, E.E., Shaffer, L.G., Ballif., B.C.: Speech delays and behavioral problems are the predominant features in individuals with developmental delays and 16p11.2 microdeletions and microduplications. J. Neurodevelop. Disord. **2**, 26–38 (2010)

[48] Sanders, S.J., Murtha, M.T., Gupta, A.R., Murdoch, J.D., Raubeson, M.J., Willsey, A. J., Ercan-Sencicek, A.G., DiLullo, N.M., Parikshak, N.N., Stein, J.L., Walker, M.F., Ober, G.T., Teran, N. A., Song, Y., El-Fishawy, P., Murtha, R.C., Choi, M., Overton, J.D., Bjornson, R.D., Carriero, N.J., Meyer, K. A., Bilguvar, K., Mane, S.M., Sestan, N., Lifton, R.P., Günel, M., Roeder, K., Geschwind, D.H., Devlin, B., State, M.W.: De novo mutations revealed by whole-exome sequencing are strongly associated with autism. Nature **485**, 237–241 (2013)

[49] Storey, J.D.: A direct approach to false discovery rates. J. Roy. Stat. Soc. Ser. B **64**, 479–498 (2002)

[50] Sullivan, P.F., Daly, M.J., O'Donovan, M.: Genetic architectures of psychiatric disorders: the emerging picture and its implications. Nat. Rev. Genet. **13**, 537–551 (2012)

[51] Tarone, R.E.: A modified Bonferroni method for discrete data. Biometrics **46**, 515–522 (1990)

[52] Tseng, G.C., Ghosh, D., Feingold, E.: Comprehensive literature review and statistical considerations for microarray meta-analysis. Nucleic Acids Res. **40**, 3785–99 (2012)

[53] Veltman, J.A., Brunner, H.G.: Understanding variable expressivity in microdeletion syndromes. Nat. Genet. **42**, 192–193 (2010)

[54] Wahba, G.: Spline Models for Observational Data. SIAM, Philadelphia (1990)

[55] Weiss, L.A., Shen, Y., Korn, J.M., Arking, D.E., Miller, D.T., Fossdal, R., Saemundsen, E., Stefansson, H., Ferreira, M.A., Green, T., Platt, O.S., Ruderfer, D.M., Walsh, C.A., Altshuler, D., Chakravarti, A., Tanzi, R.E., Stefansson, K., Santangelo, S.L., Gusella, J. F., Sklar, P., Wu, B.L., Daly, M. J.: Association between microdeletion and microduplication at 16p11.2 and autism. N. Engl. J. Med. **358**, 667–675 (2008)

[56] Wu, M.C., Lee, S., Cai, T., Li, Y., Boehnke, M., Lin, X.: Rare-variant association testing for sequencing data with the sequence kernel association test. Am. J. Hum. Genet. **89**, 82–93 (2011)

[57] Zuk, O., Hechter, E., Sunyaev, S.R., Lander, E.S.: The mystery of missing heritability: Genetic interactions create phantom heritability. Proc. Natl. Acad. Sci. USA **109**, 1193–1198 (2012)

Index

S. Datta and D. Nettleton (eds.), *Statistical Analysis of Next Generation Sequencing Data*, Frontiers in Probability and the Statistical Sciences, DOI 10.1007/978-3-319-07212-8,

W

Z

MIX
Papier aus verantwortungsvollen Quellen
Paper from responsible sources
FSC® C105338

If you have any concerns about our products,
you can contact us on
ProductSafety@springernature.com

In case Publisher is established outside the EU,
the EU authorized representative is:
Springer Nature Customer Service Center GmbH
Europaplatz 3, 69115 Heidelberg, Germany

Printed by Libri Plureos GmbH
in Hamburg, Germany